THE INTERNATIONAL SERIES OF MONOGRAPHS ON PHYSICS

SERIES EDITORS

J. BIRMAN — City University of New York
S. F. EDWARDS — University of Cambridge
R. FRIEND — University of Cambridge
C. H. LLEWELLYN-SMITH — University College London
M. REES — University of Cambridge
D. SHERRINGTON — University of Oxford
G. VENEZIANO — CERN, Geneva

INTERNATIONAL SERIES OF MONOGRAPHS ON PHYSICS

117. G. E. Volovik: *The universe in a helium droplet*
116. L. Pitaevskii, S. Stringari: *Bose–Einstein condensation*
115. G. Dissertori, I. G. Knowles, M. Schmelling: *Quantum chromodynamics*
114. B. DeWitt: *The global approach to quantum field theory*
113. J. Zinn-Justin: *Quantum field theory and critical phenomena, Fourth edition*
112. R. M. Mazo: *Brownian motion: fluctuations, dynamics, and applications*
111. H. Nishimori: *Statistical physics of spin glasses and information processing: an introduction*
110. N. B. Kopnin: *Theory of nonequilibrium superconductivity*
109. A. Aharoni: *Introduction to the theory of ferromagnetism, Second edition*
108. R. Dobbs: *Helium three*
107. R. Wigmans: *Calorimetry*
106. J. Kübler: *Theory of itinerant electron magnetism*
105. Y. Kuramoto, Y. Kitaoka: *Dynamics of heavy electrons*
104. D. Bardin, G. Passarino: *The standard model in the making*
103. G. C. Branco, L. Lavoura, J. P. Silva: *CP violation*
102. T. C. Choy: *Effective medium theory*
101. H. Araki: *Mathematical theory of quantum fields*
100. L. M. Pismen: *Vortices in nonlinear fields*
99. L. Mestel: *Stellar magnetism*
98. K. H. Bennemann: *Nonlinear optics in metals*
97. D. Salzmann: *Atomic physics in hot plamas*
96. M. Brambilla: *Kinetic theory of plasma waves*
95. M. Wakatani: *Stellarator and heliotron devices*
94. S. Chikazumi: *Physics of ferromagnetism*
91. R. A. Bertlmann: *Anomalies in quantum field theory*
90. P. K. Gosh: *Ion traps*
89. E. Simánek: *Inhomogeneous superconductors*
88. S. L. Adler: *Quaternionic quantum mechanics and quantum fields*
87. P. S. Joshi: *Global aspects in gravitation and cosmology*
86. E. R. Pike, S. Sarkar: *The quantum theory of radiation*
84. V. Z. Kresin, H. Morawitz, S. A. Wolf: *Mechanisms of conventional and high T_c superconductivity*
83. P. G. de Gennes, J. Prost: *The physics of liquid crystals*
82. B. H. Bransden, M. R. C. McDowell: *Charge exchange and the theory of ion–atom collision*
81. J. Jensen, A. R. Mackintosh: *Rare earth magnetism*
80. R. Gastmans, T. T. Wu: *The ubiquitous photon*
79. P. Luchini, H. Motz: *Undulators and free-electron lasers*
78. P. Weinberger: *Electron scattering theory*
76. H. Aoki, H. Kamimura: *The physics of interacting electrons in disordered systems*
75. J. D. Lawson: *The physics of charged particle beams*
73. M. Doi, S. F. Edwards: *The theory of polymer dynamics*
71. E. L. Wolf: *Principles of electron tunneling spectroscopy*
70. H. K. Henisch: *Semiconductor contacts*
69. S. Chandrasekhar: *The mathematical theory of black holes*
68. G. R. Satchler: *Direct nuclear reactions*
51. C. Møller: *The theory of relativity*
46. H. E. Stanley: *Introduction to phase transitions and critical phenomena*
32. A. Abragam: *Principles of nuclear magnetism*
27. P. A. M. Dirac: *Principles of quantum mechanics*
23. R. E. Peierls: *Quantum theory of solids*

The Universe in a Helium Droplet

GRIGORY E. VOLOVIK

Low Temperature Laboratory,
Helsinki University of Technology
and
Landau Institute for Theoretical Physics, Moscow

CLARENDON PRESS • OXFORD
2003

OXFORD
UNIVERSITY PRESS

Great Clarendon Street, Oxford OX2 6DP

Oxford University Press is a department of the University of Oxford.
It furthers the University's objective of excellence in research, scholarship,
and education by publishing worldwide in

Oxford New York

Auckland Bangkok Buenos Aires Cape Town Chennai
Dar es Salaam Delhi Hong Kong Istanbul Karachi Kolkata
Kuala Lumpur Madrid Melbourne Mexico City Mumbai Nairobi
São Paulo Shanghai Taipei Tokyo Toronto

Oxford is a registered trade mark of Oxford University Press
in the UK and in certain other countries

Published in the United States
by Oxford University Press Inc., New York

© Oxford University Press, 2003

The moral rights of the author have been asserted
Database right Oxford University Press (maker)

First published 2003

All rights reserved. No part of this publication may be reproduced,
stored in a retrieval system, or transmitted, in any form or by any means,
without the prior permission in writing of Oxford University Press,
or as expressly permitted by law, or under terms agreed with the appropriate
reprographics rights organization. Enquiries concerning reproduction
outside the scope of the above should be sent to the Rights Department,
Oxford University Press, at the address above

You must not circulate this book in any other binding or cover
and you must impose this same condition on any acquirer

British Library Cataloguing in Publication Data

Data available

Library of Congress Cataloging in Publication Data

ISBN 0 19 850782 8

10 9 8 7 6 5 4 3 2 1

Typeset by the author using LaTeX

Printed in Great Britain
on acid-free paper by
T. J. International Ltd, Padstow

FOREWORD

It is often said that the problem of the very small cosmological constant is the greatest mystery in cosmology and in particle physics, and that no one has any good ideas on how to solve it. The contents of this book make a lie of that statement. The material in this monograph builds upon a candidate solution to the problem, often dubbed 'emergence'. It is a solution so simple and direct that it can be stated here in this foreword. Visualize the vacuum of particle physics as if it were a cold quantum liquid in equilibrium. Then its pressure must vanish, unless it is a droplet – in which case there will be surface corrections scaling as an inverse power of the droplet size. But vacuum dark pressure scales with the vacuum dark energy, and thus is measured by the cosmological constant, which indeed scales as the inverse square of the 'size' of the universe. The problem is 'solved'.

But there is some bad news with the good. Photons, gravitons, and gluons must be viewed as collective excitations of the purported liquid, with dispersion laws which at high energies are not expected to be relativistic. The equivalence principle and gauge invariance are probably inexact. Many other such ramifications exist, as described in this book. And experimental constraints on such deviant behavior are extremely strong. Nevertheless, it is in my opinion not out of the question that the difficulties can eventually be overcome. If they are, it will mean that many sacrosanct beliefs held by almost all contemporary theoretical particle physicists and cosmologists will at the least be severely challenged.

This book summarizes the pioneering research of its author, Grisha Volovik, and provides a splendid guide into this mostly unexplored wilderness of emergent particle physics and cosmology. So far it is not respectable territory, so there is danger to the young researcher venturing within – working on it may be detrimental to a successful career track. But together with the danger will be high adventure and, if the ideas turn out to be correct, great rewards. I salute here those who take the chance and embark upon the adventure. At the very least they will be rewarded by acquiring a deep understanding of much of the lore of condensed matter physics. And, with some luck, they will also be rewarded by uncovering a radically different interpretation of the profound problems involving the structure of the very large and of the very small.

Stanford Linear Accelerator Center *James D. Bjorken*
August 2002

PREFACE

Topology is a powerful tool for gaining the most important information on complicated many-body systems in a very economic way. Topological classification of defects – vortices, domain walls, monopoles – allows us to elucidate which defect is stable and what will be the result of the fusion of two defects, without resort to any equations. In many cases there are no simple equations which govern such processes, while numerical simulations from the first principles – from the Theory of Everything (provided that such a theory exists) – are highly time consuming and not conclusive because of lack of generality. A number of different vortices with an intricate structure of the multi-component order parameter have been experimentally observed in the superfluid phases of ^{3}He, but the mathematics which is used to treat them is as simple as the equation 1+1=0. This equation demonstrates that the collision of two singular vortices gives rise to a continuous vortex-skyrmion, or that two soliton walls annihilate each other.

Another example is the Fermi surface: it is stable because it is a topological defect – a quantized vortex in momentum space. Again, without use of the microscopic theory, only from topology in the momentum space, one can predict all possible types of behavior of the many-body system at low energy which do not depend on details of atomic structure. The system is either fully gapped, or the Fermi surface is developed, or, what has most remarkable consequences, a singular point in the momentum space evolves – the Fermi point. If a Fermi point appears, as happens in superfluid ^{3}He-A, at low energies the system is governed by a quantum field theory describing left-handed and right-handed fermionic quasiparticles interacting with effective gauge and gravity fields. Practically all the ingredients of the Standard Model emerge, together with Lorentz invariance and other physical laws. This suggests that maybe our quantum vacuum belongs to the same universality class, and if so, the origin of the physical laws could be understood together with some puzzles such as the cosmological constant problem.

In this book we discuss the general consequences from topology on the quantum vacuum in quantum liquids and the parallels in particle physics and cosmology. This includes topological defects; emergent relativistic quantum field theory and gravity; chiral anomaly; the low-dimensional world of quasiparticles living in the core of vortices, domain walls and other 'branes'; quantum phase transitions; emergent non-trivial spacetimes; and many more.

Helsinki University of Technology   *Grigory E. Volovik*
November 2002

CONTENTS

1 Introduction: GUT and anti-GUT   1

    I   QUANTUM BOSE LIQUID

2 Gravity   11
    2.1   The Einstein theory of gravity   11
         2.1.1   Covariant conservation law   12
    2.2   Vacuum energy and cosmological term   12
         2.2.1   Vacuum energy   12
         2.2.2   Cosmological constant problem   14
         2.2.3   Vacuum-induced gravity   15
         2.2.4   Effective gravity in quantum liquids   15

3 Microscopic physics of quantum liquids   17
    3.1   Theory of Everything in quantum liquids   17
         3.1.1   Microscopic Hamiltonian   17
         3.1.2   Particles and quasiparticles   18
         3.1.3   Microscopic and effective symmetries   18
         3.1.4   Fundamental constants of Theory of Everything   20
    3.2   Weakly interacting Bose gas   21
         3.2.1   Model Hamiltonian   21
         3.2.2   Pseudorotation – Bogoliubov transformation   22
         3.2.3   Low-energy relativistic quasiparticles   23
         3.2.4   Vacuum energy of weakly interacting Bose gas   23
         3.2.5   Fundamental constants and Planck scales   25
         3.2.6   Vacuum pressure and cosmological constant   26
    3.3   From Bose gas to Bose liquid   27
         3.3.1   Gas-like vs liquid-like vacuum   27
         3.3.2   Model liquid state   28
         3.3.3   Real liquid state   29
         3.3.4   Vanishing of cosmological constant in liquid ^{4}He   29

4 Effective theory of superfluidity   32
    4.1   Superfluid vacuum and quasiparticles   32
         4.1.1   Two-fluid model as effective theory of gravity   32
         4.1.2   Galilean transformation for particles   32
         4.1.3   Superfluid-comoving frame and frame dragging   33
         4.1.4   Galilean transformation for quasiparticles   34
         4.1.5   Broken Galilean invariance. Momentum vs pseudomomentum   35
    4.2   Dynamics of superfluid vacuum   36

		4.2.1 Effective action	36
		4.2.2 Continuity and London equations	37
	4.3	Normal component – 'matter'	37
		4.3.1 Effective metric for matter field	37
		4.3.2 External and inner observers	39
		4.3.3 Is the speed of light a fundamental constant?	40
		4.3.4 'Einstein equations'	41
5	**Two-fluid hydrodynamics**		42
	5.1	Two-fluid hydrodynamics from Einstein equations	42
	5.2	Energy–momentum tensor for 'matter'	42
		5.2.1 Metric in incompressible superfluid	42
		5.2.2 Covariant and contravariant components of 4-momentum	43
		5.2.3 Energy–momentum tensor of 'matter'	44
		5.2.4 Particle current and quasiparticle momentum	44
	5.3	Local thermal equilibrium	45
		5.3.1 Distribution function	45
		5.3.2 Normal and superfluid densities	45
		5.3.3 Energy–momentum tensor	46
		5.3.4 Temperature 4-vector	47
		5.3.5 When is the local equilibrium impossible?	47
	5.4	Global thermodynamic equilibrium	48
		5.4.1 Tolman temperature	48
		5.4.2 Global equilibrium and event horizon	49
6	**Advantages and drawbacks of effective theory**		51
	6.1	Non-locality in effective theory	51
		6.1.1 Conservation and covariant conservation	51
		6.1.2 Covariance vs conservation	52
		6.1.3 Paradoxes of effective theory	52
		6.1.4 No canonical Lagrangian for classical hydrodynamics	53
		6.1.5 Novikov–Wess–Zumino action for ferromagnets	54
	6.2	Effective vs microscopic theory	56
		6.2.1 Does quantum gravity exist?	56
		6.2.2 What effective theory can and cannot do	56
	6.3	Superfluidity and universality	58
	II	QUANTUM FERMIONIC LIQUIDS	
7	**Microscopic physics**		65
	7.1	Introduction	65
	7.2	BCS theory	66
		7.2.1 Fermi gas	66
		7.2.2 Model Hamiltonian	67

		7.2.3 Bogoliubov rotation	68
		7.2.4 Stable point nodes and emergent 'relativistic' quasiparticles	69
		7.2.5 Nodal lines are not generic	70
	7.3	Vacuum energy of weakly interacting Fermi gas	70
		7.3.1 Vacuum in equilibrium	70
		7.3.2 Axial vacuum	71
		7.3.3 Fundamental constants and Planck scales	73
		7.3.4 Vanishing of vacuum energy in liquid ^{3}He	74
		7.3.5 Vacuum energy in non-equilibrium	74
		7.3.6 Vacuum energy and cosmological term in bi-metric gravity	75
	7.4	Spin-triplet superfluids	76
		7.4.1 Order parameter	76
		7.4.2 Bogoliubov–Nambu spinor	77
		7.4.3 ^{3}He-B – fully gapped system	78
		7.4.4 From Bogoliubov quasiparticle to Dirac particle. From ^{3}He-B to Dirac vacuum	79
		7.4.5 Mass generation for Standard Model fermions	80
		7.4.6 ^{3}He-A – superfluid with point nodes	81
		7.4.7 Axiplanar state – flat directions	82
		7.4.8 ^{3}He-A$_1$ – Fermi surface and Fermi points	83
		7.4.9 Planar phase – marginal Fermi points	84
		7.4.10 Polar phase – unstable nodal lines	85
8	**Universality classes of fermionic vacua**		**86**
	8.1	Fermi surface as topological object	87
		8.1.1 Fermi surface is the vortex in momentum space	87
		8.1.2 **p**-space and **r**-space topology	89
		8.1.3 Topological invariant for Fermi surface	90
		8.1.4 Landau Fermi liquid	91
		8.1.5 Collective modes of Fermi surface	91
		8.1.6 Volume of the Fermi surface as invariant of adiabatic deformations	92
		8.1.7 Non-Landau Fermi liquids	93
	8.2	Systems with Fermi points	94
		8.2.1 Chiral particles and Fermi point	94
		8.2.2 Fermi point as hedgehog in **p**-space	95
		8.2.3 Topological invariant for Fermi point	96
		8.2.4 Topological invariant in terms of Green function	97
		8.2.5 A-phase and planar state: the same spectrum but different topology	99
		8.2.6 Relativistic massless chiral fermions emerging near Fermi point	99

	8.2.7	Collective modes of fermionic vacuum – electromagnetic and gravitational fields	100
	8.2.8	Fermi points and their physics are natural	101
	8.2.9	Manifolds of zeros in higher dimensions	103

9 Effective quantum electrodynamics in ^{3}He-A — 105
9.1 Fermions — 105
9.1.1 Electric charge and chirality — 105
9.1.2 Topological invariant as generalization of chirality — 106
9.1.3 Effective metric viewed by quasiparticles — 106
9.1.4 Superfluid velocity in axial vacuum — 107
9.1.5 Spin from isospin, isospin from spin — 108
9.1.6 Gauge invariance and general covariance in fermionic sector — 109
9.2 Effective electromagnetic field — 109
9.2.1 Why does QED arise in ^{3}He-A? — 109
9.2.2 Running coupling constant — 111
9.2.3 Zero-charge effect in ^{3}He-A — 111
9.2.4 Light – orbital waves — 112
9.2.5 Does one need the symmetry breaking to obtain massless bosons? — 113
9.2.6 Are gauge equivalent states physically indistinguishable? — 114
9.3 Effective $SU(N)$ gauge fields — 114
9.3.1 Local $SU(2)$ from double degeneracy — 114
9.3.2 Role of discrete symmetries — 115
9.3.3 Mass of W-bosons, flat directions and supersymmetry — 116
9.3.4 Different metrics for different fermions. Dynamic restoration of Lorentz symmetry — 117

10 Three levels of phenomenology of superfluid ^{3}He — 118
10.1 Ginzburg–Landau level — 119
10.1.1 Ginzburg–Landau free energy — 119
10.1.2 Vacuum states — 120
10.2 London level in ^{3}He-A — 120
10.2.1 London energy — 120
10.2.2 Particle current — 121
10.2.3 Parameters of London energy — 121
10.3 Low-temperature relativistic regime — 122
10.3.1 Energy and momentum of vacuum and 'matter' — 122
10.3.2 Chemical potential for quasiparticles — 123
10.3.3 Double role of counterflow — 123
10.3.4 Fermionic charge — 124

		10.3.5 Normal component at zero temperature	125

- 10.3.5 Normal component at zero temperature — 125
- 10.3.6 Fermi surface from Fermi point — 126
- 10.4 Parameters of effective theory in London limit — 128
  - 10.4.1 Parameters of effective theory from BCS theory — 128
  - 10.4.2 Fundamental constants — 129
- 10.5 How to improve quantum liquid — 130
  - 10.5.1 Limit of inert vacuum — 130
  - 10.5.2 Effective action in inert vacuum — 131
  - 10.5.3 Einstein action in ^{3}He-A — 132
  - 10.5.4 Is G fundamental? — 132
  - 10.5.5 Violation of gauge invariance — 133
  - 10.5.6 Origin of precision of symmetries in effective theory — 134

11 Momentum space topology of 2+1 systems — 135
- 11.1 Topological invariant for 2+1 systems — 135
  - 11.1.1 Universality classes for 2+1 systems — 135
  - 11.1.2 Invariant for fully gapped systems — 135
- 11.2 2+1 systems with non-trivial **p**-space topology — 136
  - 11.2.1 **p**-space skyrmion in p-wave state — 136
  - 11.2.2 Topological invariant and broken time reversal symmetry — 137
  - 11.2.3 d-wave states — 138
- 11.3 Fermi point as diabolical point and Berry phase — 138
  - 11.3.1 Families (generations) of fermions in 2+1 systems — 138
  - 11.3.2 Diabolical points — 139
  - 11.3.3 Berry phase and magnetic monopole in **p**-space — 140
- 11.4 Quantum phase transitions — 141
  - 11.4.1 Quantum phase transition as change of momentum space topology — 141
  - 11.4.2 Dirac vacuum is marginal — 142

12 Momentum space topology protected by symmetry — 143
- 12.1 Momentum space topology of planar phase — 143
  - 12.1.1 Topology protected by discrete symmetry — 143
  - 12.1.2 Dirac mass from violation of discrete symmetry — 144
- 12.2 Quarks and leptons — 145
  - 12.2.1 Fermions in Standard Model — 146
  - 12.2.2 Unification of quarks and leptons — 147
  - 12.2.3 Spinons and holons — 149
- 12.3 Momentum space topology of Standard Model — 149
  - 12.3.1 Generating function for topological invariants constrained by symmetry — 150
  - 12.3.2 Discrete symmetry and massless fermions — 151

12.3.3 Relation to chiral anomaly	151
12.3.4 Trivial topology below electroweak transition and massive fermions	152
12.4 Reentrant violation of special relativity	153
12.4.1 Discrete symmetry in ^{3}He-A	153
12.4.2 Violation of discrete symmetry	153
12.4.3 Violation of 'Lorentz invariance' at low energy	154
12.4.4 Momentum space topology of exotic fermions	155
12.4.5 Application to RQFT	156

III TOPOLOGICAL DEFECTS

13 Topological classification of defects	**159**
13.1 Defects and homotopy groups	159
13.1.1 Vacuum manifold	160
13.1.2 Symmetry G of physical laws in ^{3}He	160
13.1.3 Symmetry breaking in ^{3}He-B	161
13.2 Analogous 'superfluid' phases in high-energy physics	162
13.2.1 Chiral superfluidity in QCD	162
13.2.2 Chiral superfluidity in QCD with three flavors	164
13.2.3 Color superfluidity in QCD	164
14 Vortices in ^{3}He-B	**165**
14.1 Topology of ^{3}He-B defects	165
14.1.1 Fundamental homotopy group for ^{3}He-B defects	165
14.1.2 Mass vortex vs axion string	165
14.1.3 Spin vortices vs pion strings	166
14.1.4 Casimir force between spin and mass vortices and composite defect	167
14.1.5 Spin vortex as string terminating soliton	168
14.1.6 Topological confinement of spin–mass vortices	170
14.2 Symmetry of defects	171
14.2.1 Topology of defects vs symmetry of defects	171
14.2.2 Symmetry of hedgehogs and monopoles	172
14.2.3 Spherically symmetric objects in superfluid ^{3}He	174
14.2.4 Enhanced superfluidity in the core of hedgehog. Generation of Dirac mass in the core	174
14.2.5 Continuous hedgehog in B-phase	176
14.2.6 Symmetry of vortices: continuous symmetry	176
14.2.7 Symmetry of vortices: discrete symmetry	177
14.3 Broken symmetry in B-phase vortex core	178
14.3.1 Most symmetric vortex and its instability	178
14.3.2 Ferromagnetic core with broken parity	179
14.3.3 Double-core vortex as Witten superconducting string	179

	14.3.4 Vorton – closed loop of the vortex with twisted core	180
15	**Symmetry breaking in ^{3}He-A and singular vortices**	**182**
15.1	^{3}He-A and analogous phases in high-energy physics	182
	15.1.1 Broken symmetry	182
	15.1.2 Connection with electroweak phase transition	182
	15.1.3 Discrete symmetry and vacuum manifold	183
15.2	Singular defects in ^{3}He-A	184
	15.2.1 Hedgehog in ^{3}He-A and 't Hooft–Polyakov magnetic monopole	184
	15.2.2 Pure mass vortices	185
	15.2.3 Disclination – antigravitating string	186
	15.2.4 Singular doubly quantized vortex vs electroweak Z-string	186
	15.2.5 Nielsen–Olesen string vs Abrikosov vortex	187
15.3	Fractional vorticity and fractional flux	189
	15.3.1 Half-quantum vortex in ^{3}He-A	189
	15.3.2 Alice string	189
	15.3.3 Fractional flux in chiral superconductor	191
	15.3.4 Half-quantum flux in d-wave superconductor	193
16	**Continuous structures**	**195**
16.1	Hierarchy of energy scales and relative homotopy group	195
	16.1.1 Soliton from half-quantum vortex	195
	16.1.2 Relative homotopy group for soliton	196
	16.1.3 How to destroy topological solitons and why singularities are not easily created in ^{3}He	198
16.2	Continuous vortices, skyrmions and merons	199
	16.2.1 Skyrmion – Anderson–Toulouse–Chechetkin vortex texture	199
	16.2.2 Continuous vortices in spinor condensates	201
	16.2.3 Continuous vortex as a pair of merons	202
	16.2.4 Semilocal string and continuous vortex in superconductors	205
	16.2.5 Topological transition between continuous vortices	205
16.3	Vortex sheet	206
	16.3.1 Kink on a soliton as a meron-like vortex	206
	16.3.2 Formation of a vortex sheet	208
	16.3.3 Vortex sheet in rotating superfluid	208
	16.3.4 Vortex sheet in superconductor	211
17	**Monopoles and boojums**	**212**
17.1	Monopoles terminating strings	212
	17.1.1 Composite defects	212

	17.1.2 Hedgehog and continuous vortices	213
	17.1.3 Dirac magnetic monopoles	214
	17.1.4 't Hooft–Polyakov monopole	214
	17.1.5 Nexus	215
	17.1.6 Nexus in chiral superconductors	217
	17.1.7 Cosmic magnetic monopole inside superconductors	217
17.2	Defects at surfaces	218
	17.2.1 Boojum and relative homotopy group	218
	17.2.2 Boojum in ^{3}He-A	220
17.3	Defects on interface between different vacua	221
	17.3.1 Classification of defects in presence of interface	221
	17.3.2 Symmetry classes of interfaces	222
	17.3.3 Vacuum manifold for interface	222
	17.3.4 Topological charges of linear defects	223
	17.3.5 Strings across AB-interface	224
	17.3.6 Boojum as nexus	226
	17.3.7 AB-interface in rotating vessel	226
	17.3.8 AB-interface and monopole 'erasure'	228
	17.3.9 Alice string at interface	229

IV ANOMALIES OF CHIRAL VACUUM

18 Anomalous non-conservation of fermionic charge — 235

18.1	Chiral anomaly	236
	18.1.1 Pumping the charge from the vacuum	236
	18.1.2 Chiral particle in magnetic field	237
	18.1.3 Adler–Bell–Jackiw equation	238
18.2	Anomalous non-conservation of baryonic charge	239
	18.2.1 Baryonic asymmetry of Universe	239
	18.2.2 Electroweak baryoproduction	240
18.3	Analog of baryogenesis in ^{3}He-A	241
	18.3.1 Momentum exchange between superfluid vacuum and quasiparticle matter	241
	18.3.2 Chiral anomaly in ^{3}He-A	243
	18.3.3 Spectral-flow force acting on a vortex-skyrmion	244
	18.3.4 Topological stability of spectral-flow force. Spectral-flow force from Novikov–Wess–Zumino action	246
	18.3.5 Dynamics of Fermi points and vortices	246
	18.3.6 Vortex texture as a mediator of momentum exchange. Magnus force	247
18.4	Experimental check of Adler–Bell–Jackiw equation in ^{3}He-A	248

19	**Anomalous currents**	251
	19.1 Helicity in parity-violating systems	251
	19.2 Chern–Simons energy term	252
	19.2.1 Chern–Simons term in Standard Model	252
	19.2.2 Chern–Simons energy in ^{3}He-A	254
	19.3 Helical instability and 'magnetogenesis' due to chiral fermions	255
	19.3.1 Relevant energy terms	255
	19.3.2 Mass of hyperphoton due to excess of chiral fermions (counterflow)	256
	19.3.3 Helical instability condition	257
	19.3.4 Mass of hyperphoton due to symmetry-violating interaction	258
	19.3.5 Experimental 'magnetogenesis' in ^{3}He-A	259
20	**Macroscopic parity-violating effects**	260
	20.1 Mixed axial–gravitational Chern–Simons term	260
	20.1.1 Parity-violating current	260
	20.1.2 Parity-violating action in terms of gravimagnetic field	261
	20.2 Orbital angular momentum in ^{3}He-A	262
	20.3 Odd current in ^{3}He-A	263
21	**Quantization of physical parameters**	266
	21.1 Spin and statistics of skyrmions in 2+1 systems	266
	21.1.1 Chern–Simons term as Hopf invariant	266
	21.1.2 Quantum statistics of skyrmions	267
	21.2 Quantized response	269
	21.2.1 Quantization of Hall conductivity	269
	21.2.2 Quantization of spin Hall conductivity	270
	21.2.3 Induced Chern–Simons action in other systems	271
	V FERMIONS ON TOPOLOGICAL OBJECTS AND BRANE WORLD	
22	**Edge states and fermion zero modes on soliton**	275
	22.1 Index theorem for fermion zero modes on soliton	276
	22.1.1 Chiral edge state – 1D Fermi surface	276
	22.1.2 Fermi points in combined $(\mathbf{p}, \mathbf{r})$ space	277
	22.1.3 Spectral asymmetry index	278
	22.1.4 Index theorem	279
	22.1.5 Spectrum of fermion zero modes	281
	22.1.6 Current inside the domain wall	282
	22.1.7 Edge states in d-wave superconductor with broken T	282
	22.2 3+1 world of fermion zero modes	283

	22.2.1 Fermi points of co-dimension 5	284
	22.2.2 Chiral 5+1 particle in magnetic field	284
	22.2.3 Higher-dimensional anomaly	285
	22.2.4 Quasiparticle world within domain wall in 4+1 film	285

23 Fermion zero modes on vortices — 288
23.1 Anomalous branch of chiral fermions — 288
  23.1.1 Minigap in energy spectrum — 288
  23.1.2 Integer vs half-odd integer angular momentum of fermion zero modes — 290
  23.1.3 Bogoliubov–Nambu Hamiltonian for fermions in the core — 291
  23.1.4 Fermi points of co-dimension 3 in vortex core — 292
  23.1.5 Andreev reflection and Fermi point in the core — 292
  23.1.6 From Fermi point to fermion zero mode — 294
23.2 Fermion zero modes in quasiclassical description — 295
  23.2.1 Hamiltonian in terms of quasiclassical trajectories — 295
  23.2.2 Quasiclassical low-energy states on anomalous branch — 296
  23.2.3 Quantum low-energy states and W-parity — 297
  23.2.4 Fermions on asymmetric vortices — 297
  23.2.5 Majorana fermion with $E=0$ on half-quantum vortex — 299
23.3 Interplay of **p**- and **r**-topologies in vortex core — 300
  23.3.1 Fermions on a vortex line in 3D systems — 300
  23.3.2 Topological equivalence of vacua with Fermi points and with vortex — 301
  23.3.3 Smooth core of ^{3}He-B vortex — 301
  23.3.4 **r**-space topology of Fermi points in the vortex core — 303

24 Vortex mass — 305
24.1 Inertia of object moving in superfluid vacuum — 305
  24.1.1 Relativistic and non-relativistic mass — 305
  24.1.2 'Relativistic' mass of the vortex — 306
24.2 Fermion zero modes and vortex mass — 307
  24.2.1 Effective theory of Kopnin mass — 307
  24.2.2 Kopnin mass of smooth vortex: chiral fermions in magnetic field — 308
24.3 Associated hydrodynamic mass of a vortex — 309
  24.3.1 Associated mass of an object — 309
  24.3.2 Associated mass of smooth-core vortex — 310

25 Spectral flow in the vortex core	312
25.1 Analog of Callan–Harvey mechanism of cancellation of anomalies	312
25.1.1 Analog of baryogenesis by cosmic strings	312
25.1.2 Level flow in the core	313
25.1.3 Momentum transfer by level flow	313
25.2 Restricted spectral flow in the vortex core	314
25.2.1 Condition for free spectral flow	314
25.2.2 Kinetic equation for fermion zero modes	315
25.2.3 Solution of Boltzmann equation	316
25.2.4 Measurement of Callan–Harvey effect in ^{3}He-B	317
VI NUCLEATION OF QUASIPARTICLES AND TOPOLOGICAL DEFECTS	
26 Landau critical velocity	321
26.1 Landau critical velocity for quasiparticles	321
26.1.1 Landau criterion	321
26.1.2 Supercritical superflow in ^{3}He-A	322
26.1.3 Landau velocity as quantum phase transition	322
26.1.4 Landau velocity, ergoregion and horizon	324
26.1.5 Landau velocity, ergoregion and horizon in case of superluminal dispersion	324
26.1.6 Landau velocity, ergoregion and horizon in case of subluminal dispersion	325
26.2 Analog of pair production in strong fields	325
26.2.1 Pair production in strong fields	326
26.2.2 Experimental pair production	326
26.3 Vortex formation	328
26.3.1 Landau criterion for vortices	328
26.3.2 Thermal activation. Sphaleron	330
26.3.3 Hydrodynamic instability as mechanism of vortex formation	331
26.4 Nucleation by macroscopic quantum tunneling	332
26.4.1 Instanton in collective coordinate description	332
26.4.2 Action for vortices and quantization of particle number in quantum vacuum	333
26.4.3 Volume law for vortex instanton	335
27 Vortex formation by Kelvin–Helmholtz instability	339
27.1 Kelvin–Helmholtz instability in classical and quantum liquids	339
27.1.1 Classical Kelvin–Helmholtz instability	339
27.1.2 Kelvin–Helmholtz instabilities in superfluids at low T	340
27.1.3 Ergoregion instability and Landau criterion	344

27.1.4 Crossover from ergoregion instability to Kelvin–Helmholtz instability	345
27.2 Interface instability in two-fluid hydrodynamics	346
27.2.1 Thermodynamic instability	346
27.2.2 Non-linear stage of instability	347

28 Vortex formation in ionizing radiation — 351

28.1 Vortices and phase transitions	351
28.1.1 Vortices in equilibrium phase transitions	351
28.1.2 Vortices in non-equilibrium phase transitions	354
28.1.3 Vortex formation by neutron radiation	356
28.1.4 Baked Alaska vs KZ scenario	357
28.2 Vortex formation at normal–superfluid interface	358
28.2.1 Propagating front of second-order transition	358
28.2.2 Instability region in rapidly moving interface	359
28.2.3 Vortex formation behind the propagating front	360
28.2.4 Instability of normal–superfluid interface	362
28.2.5 Interplay of KH and KZ mechanisms	363
28.2.6 KH instability as generic mechanism of vortex nucleation	364

VII VACUUM ENERGY AND VACUUM IN NON-TRIVIAL GRAVITATIONAL BACKGROUND

29 Casimir effect and vacuum energy — 369

29.1 Analog of standard Casimir effect in condensed matter	369
29.2 Interface between two different vacua	372
29.2.1 Interface between vacua with different broken symmetry	372
29.2.2 Why the cosmological phase transition does not perturb the zero value of the the cosmological constant	374
29.2.3 Interface as perfectly reflecting mirror	375
29.2.4 Interplay between vacuum pressure and pressure of matter	375
29.2.5 Interface between vacua with different speeds of light	376
29.3 Force on moving interface	377
29.3.1 Andreev reflection at the interface	377
29.3.2 Force acting on moving mirror from thermal relativistic fermions	378
29.3.3 Force acting on moving AB-interface	379
29.4 Vacuum energy and cosmological constant	380
29.4.1 Why is the cosmological constant so small?	380
29.4.2 Why is the cosmological constant of order of the present mass of the Universe?	382

	29.4.3 Vacuum energy from Casimir effect	383
	29.4.4 Vacuum energy induced by texture	384
	29.4.5 Vacuum energy due to Riemann curvature and Einstein Universe	386
	29.4.6 Why is the Universe flat?	387
	29.4.7 What is the energy of false vacuum?	388
	29.4.8 Discussion: why is vacuum not gravitating?	389
29.5	Mesoscopic Casimir force	390
	29.5.1 Vacuum energy from 'Theory of Everything'	391
	29.5.2 Leakage of vacuum through the wall	391
	29.5.3 Mesoscopic Casimir force in 1D Fermi gas	393
	29.5.4 Mesoscopic Casimir forces in a general condensed matter system	394
	29.5.5 Discussion	394

30 Topological defects as source of non-trivial metric — 397
30.1 Surface of infinite red shift — 397
- 30.1.1 Walls with degenerate metric — 397
- 30.1.2 Vierbein wall in ^{3}He-A film — 398
- 30.1.3 Surface of infinite red shift — 399
- 30.1.4 Fermions across static vierbein wall — 401
- 30.1.5 Communication across the wall via non-linear superluminal dispersion — 403

30.2 Conical space and antigravitating string — 404

31 Vacuum under rotation and spinning strings — 406
31.1 Sagnac effect using superfluids — 406
- 31.1.1 Sagnac effect — 406
- 31.1.2 Superfluid gyroscope under rotation. Macroscopic coherent Sagnac effect — 407

31.2 Vortex, spinning string and Lense–Thirring effect — 408
- 31.2.1 Vortex as vierbein defect — 408
- 31.2.2 Lense–Thirring effect — 409
- 31.2.3 Spinning string — 410
- 31.2.4 Asymmetry in propagation of light — 411
- 31.2.5 Vortex as gravimagnetic flux tube — 412

31.3 Gravitational Aharonov–Bohm effect and Iordanskii force on a vortex — 412
- 31.3.1 Symmetric scattering from the vortex — 413
- 31.3.2 Asymmetric scattering from the vortex — 414
- 31.3.3 Classical derivation of asymmetric cross-section — 414
- 31.3.4 Iordanskii force on spinning string — 415

31.4 Quantum friction in rotating vacuum — 416
- 31.4.1 Zel'dovich–Starobinsky effect — 416

	31.4.2 Effective metric for quasiparticles under rotation	417
	31.4.3 Ergoregion in rotating superfluids	418
	31.4.4 Radiation to the ergoregion as a source of rotational quantum friction	420
	31.4.5 Emission of rotons	421
	31.4.6 Discussion	422

32 Analogs of event horizon — 424
32.1 Event horizons in vierbein wall and Hawking radiation — 424
- 32.1.1 From infinite red shift to horizons — 424
- 32.1.2 Vacuum in the presence of horizon — 427
- 32.1.3 Dissipation due to horizon — 430
- 32.1.4 Horizons in a tube and extremal black hole — 431

32.2 Painlevé–Gullstrand metric in superfluids — 434
- 32.2.1 Radial flow with event horizon — 434
- 32.2.2 Ingoing particles and initial vacuum — 436
- 32.2.3 Outgoing particle and gravitational red shift — 438
- 32.2.4 Horizon as the window to Planckian physics — 440
- 32.2.5 Hawking radiation — 440
- 32.2.6 Preferred reference frames: frame for Planckian physics and absolute spacetime — 441
- 32.2.7 Schwarzschild metric in effective gravity — 443
- 32.2.8 Discrete symmetries of black hole — 445

32.3 Horizon and singularity on AB-brane — 446
- 32.3.1 Effective metric for modes living on the AB-brane — 447
- 32.3.2 Horizon and singularity — 449
- 32.3.3 Brane instability beyond the horizon — 450

32.4 From 'acoustic' black hole to 'real' black hole — 452
- 32.4.1 Black-hole instability beyond the horizon — 452
- 32.4.2 Modified Dirac equation for fermions — 454
- 32.4.3 Fermi surface for Standard Model fermions inside horizon — 456
- 32.4.4 Thermodynamics of 'black-hole matter' — 458
- 32.4.5 Gravitational bag — 459

33 Conclusion — 461

References — 469

Index — 497

1
INTRODUCTION: GUT AND ANTI-GUT

There are fundamental relations between three vast areas of physics: particle physics, cosmology and condensed matter. These relations constitute a successful example of the unity of physics. The fundamental links between cosmology and particle physics, in other words, between macro- and micro-worlds, have been well established. There is a unified system of laws governing all scales from subatomic particles to the cosmos and this principle is widely exploited in the description of the physics of the early Universe (baryogenesis, cosmological nucleosynthesis, etc.). The connection of these two fields with the third ingredient of modern physics – condensed matter – is the main goal of the book.

This connection allows us to simulate the least understood features of high-energy physics and cosmology: the properties of the quantum vacuum (also called ether, spacetime foam, quantum foam, Planck medium, etc.). In particular, the vacuum energy estimated using the methods of particle physics is in disagreement with modern cosmological experiments. This is the famous cosmological constant problem. A major advantage of condensed matter is that it is described by a quantum field theory in which the properties of the quantum vacuum are completely known from first principles: they can be computed (at least numerically) and they can be measured experimentally in a variety of quantum condensed matter systems, such as quantum liquids, superconductors, superfluids, ferromagnets, etc.

The analogy between the quantum vacuum in particle physics and in condensed matter could give an insight into trans-Planckian physics and thus help in solving the cosmological constant problem and other outstanding problems in high-energy physics and cosmology, such as the origin of matter–antimatter asymmetry, the formation of the cosmological magnetic field, the problem of the flatness of present Universe, the formation of large-scale structure, the physics of the event horizon in black holes, etc.

The traditional Grand Unification view of the nature of physical laws is that the low-energy symmetry of our world is the remnant of a larger symmetry, which exists at high energy, and is broken when the energy is reduced. According to this philosophy the higher the energy the higher is the symmetry. At very high energy there is the Grand Unification Theory (GUT), which unifies strong, weak and hypercharge interactions into one big group such as the $SO(10)$ group or its G(224) subgroup of the Pati–Salam model (Sec. 12.2.2). At about 10^{15} GeV this big symmetry is spontaneously broken (probably with intermediate stages) into the symmetry group of the Standard Model, which contains three subgroups corresponding to separate symmetry for each of three interactions:

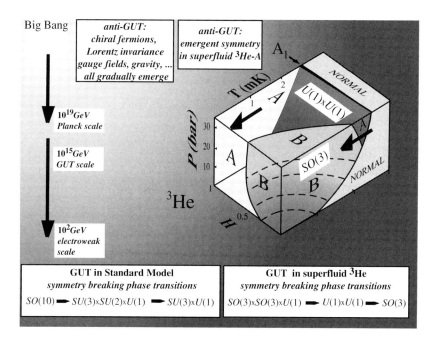

FIG. 1.1. Grand Unification and anti-Grand-Unification schemes in the Standard Model and in superfluid ^{3}He.

$U(1) \times SU(2) \times SU(3)$. Below about 200 GeV the electroweak symmetry $U(1) \times SU(2)$ is violated, and only the group of electromagnetic and strong interactions, $U(1) \times SU(3)$, survives (see Fig. 1.1 and Sec. 12.2).

The less traditional view is quite the opposite: it is argued that starting from some energy scale (probably the Planck energy scale) one finds that the higher the energy the poorer are the symmetries of the physical laws, and finally even the Lorentz invariance and gauge invariance will be smoothly violated (Froggatt and Nielsen 1991; Chadha and Nielsen 1983). From this point of view, the relativistic quantum field theory (RQFT) is an effective theory (Polyakov 1987; Weinberg 1999; Jegerlehner 1998). It is an emergent phenomenon arising as a fixed point in the low-energy corner of the physical vacuum whose nature is inaccessible from the effective theory. In the vicinity of the fixed point the system acquires new symmetries which it did not have at higher energy. It is quite possible that even such symmetries as Lorentz symmetry and gauge invariance are not fundamental, but gradually appear when the fixed point is approached. From this viewpoint it is also possible that Grand Unification schemes make no sense if the unification occurs at energies where the effective theories are no longer valid.

Both scenarios occur in condensed matter systems. In particular, superfluid ^{3}He-A provides an instructive example. At high temperature the ^{3}He gas and at lower temperature the normal ^{3}He liquid have all the symmetries that ordinary condensed matter can have: translational invariance, global $U(1)$ group and two global $SO(3)$ symmetries of spin and orbital rotations. When the temperature decreases further the liquid ^{3}He reaches the superfluid transition temperature T_c about 1 mK (Fig. 1.1), below which it spontaneously loses each of its symmetries except for the translational one – it is still liquid. This breaking of symmetry at low temperature, and thus at low energy, reproduces that in particle physics. Though the 'Grand Unification' group in ^{3}He, $U(1) \times SO(3) \times SO(3)$, is not as big as in particle physics, the symmetry breaking is nevertheless the most important element of the ^{3}He physics at low temperature.

However, this is not the whole story. When the temperature is reduced further, the opposite 'anti-Grand-Unification' (anti-GUT) scheme starts to work: in the limit $T \to 0$ the superfluid ^{3}He-A gradually acquires from nothing almost all the symmetries which we know today in high-energy physics: (an analog of) Lorentz invariance, local gauge invariance, elements of general covariance, etc. It appears that such an enhancement of symmetry in the limit of low energy happens because ^{3}He-A belongs to a special universality class of Fermi systems. For condensed matter in this class, the chiral (left-handed and right-handed) fermions and gauge bosons arise as fermionic quasiparticles and bosonic collective modes together with the corresponding symmetries. Even the left-handedness and right-handedness are the emergent low-energy properties of quasiparticles.

The quasiparticles and collective bosons perceive the homogeneous ground state of condensed matter as an empty space – a vacuum – since they do not scatter on atoms comprising this vacuum state: quasiparticles move in a quantum liquid or in a crystal without friction just as particles move in empty space. The inhomogeneous deformations of this analog of the quantum vacuum is seen by the quasiparticles as the metric field of space in which they live. It is an analog of the gravitational field. This conceptual similarity between condensed matter and the quantum vacuum gives some hint on the origin of symmetries and also allows us to simulate many phenomena in high-energy physics and cosmology related to the quantum vacuum using quantum liquids, Bose–Einstein condensates, superconductors and other materials.

We shall exploit here both levels of analogies: GUT and anti-GUT. Each of the two levels has its own energy and temperature range in ^{3}He. According to Fig. 1.1, in ^{3}He the GUT scheme works at a higher energy than the anti-GUT one. However, it is possible that at very low temperatures, after the anti-GUT scheme gives rise to new symmetries, these symmetries will be spontaneously broken again according to the GUT scheme.

The GUT scheme shows an important property of the asymmetry of the quantum vacuum arising due to the phenomenon of spontaneous symmetry breaking. Just as the symmetry is often broken in condensed matter systems when the temperature is reduced, it is believed that the Universe, when cooling down, would have undergone a series of symmetry-breaking phase transitions. One of the most important consequences of such symmetry breaking is the existence of topological defects in both systems. Cosmic strings, monopoles, domain walls and solitons, etc., have their counterparts in condensed matter: namely, quantized vortices, hedgehogs, domain walls and solitons, etc., which we shall discuss in detail in Part III.

The topological defects formed at early-Universe phase transitions may in turn have cosmological implications. Reliable observational input in cosmology to test these ideas is scarce and the ability to perform controlled experiments, of course, absent. However, such transitions exhibit many generic features which are also found in symmetry-breaking transitions in condensed matter systems at low temperatures and thus can be tested in that context (see Part VI).

However, mostly we shall be concerned with the anti-GUT phenomenon of gradually emerging symmetries. On a microscopic level, the analogs of the quantum vacuum – quantum condensed matter systems, such as ^{3}He and ^{4}He quantum liquids, superconductors and magnets – consist of strongly correlated and/or strongly interacting quantum elements (atoms, electrons, spins, etc.). Even in its ground state, such a system is usually rather complicated: its many-body wave function requires extensive analytic studies and numerical simulations. However, it appears that such calculations are not necessary if one wishes to study low-energy phenomena in these systems. When the energy scale is reduced, one can no longer resolve the motion of isolated elements. The smaller the energy the better is the system described in terms of its zero modes – the states whose energy is close to zero. All the particles of the Standard Model have energies which are extremely small compared to the 'Planck' energy scale, that is why one may guess that all of them originate from the fermionic or bosonic zero modes of the quantum vacuum. If so, our goal must be to describe and classify the possible zero modes of quantum vacua.

These zero modes are are represented by three major components: (i) the bosonic collective modes of the quantum vacuum; (ii) the dilute gas of the particle-like excitations – quasiparticles – which play the role of elementary particles; (iii) topological defects have their own bosonic and fermionic zero modes. The dynamics of the zero modes is described within what we now call 'the effective theory'.

In superfluid ^{4}He, for example, this effective theory incorporates the collective motion of the ground state – the superfluid quantum vacuum; the dynamics of quasiparticles in the background of the moving vacuum – phonons, which form the normal component of the liquid; and the dynamics of the topological defects – quantized vortices interacting with the other two subsystems. It appears that in the low-energy limit the dynamics of phonons is the same as the dynamics of relativistic particles in the metric field $g_{\mu\nu}$. This effective acoustic metric is dynamical and is determined by the superfluid motion of the vacuum of superfluid ^{4}He. For quasiparticles, the essentially Galilean space and time of the world of the laboratory are combined into an entangled spacetime continuum with properties determined by what they think of as gravity. In this sense, the two-fluid hydrodynamics describing the interdependent dynamics of superfluid quantum vacuum and quasiparticles represents the metric theory of gravity and its interaction with matter (Part I). This metric theory is a caricature of the Einstein gravity, since the metric field does not obey the Einstein equations. But this caricature is very useful for investigating many problems of RQFT in curved space

(Part VII), including the cosmological constant problem and the behavior of the quantum matter field in the presence of the event horizon. For example, for the horizon problem it is only important that the effective metric $g_{\mu\nu}$ exhibits the event horizon for (quasi)particles, and it does not matter from which equations, Einstein or Landau–Khalatnikov, such a metric has been obtained.

An effective theory in condensed matter does not depend on details of microscopic (atomic) structure of the substance. The type of effective theory is determined by the symmetry and topology of zero modes of the quantum vacuume, and the role of the underlying microscopic physics is only to choose among different universality classes of quantum vacua on the basis of the minimum energy consideration. Once the universality class is determined, the low-energy properties of the condensed matter system are completely described by the effective theory, and the information on the underlying microscopic physics is lost.

In Part II we consider universality classes in the fermionic vacua. It is assumed that the quantum vacuum of the Standard Model is also a fermionic system, while the bosonic modes are the secondary quantities which are the collective modes of this vacuum. The universality classes are determined by the momentum space topology of fermion zero modes, in other words by the topological defects in momentum space – the points, lines or surfaces in **p**-space at which the energy of quasiparticles becomes zero. There are two topologically different classes of vacua with gapless fermions. Zeros in the energy spectrum form in 3-dimensional **p**-space either the Fermi surface or the Fermi points. The vacua with a Fermi surface are more abundant in condensed matter; the Fermi points exist in superfluid ^{3}He-A, in the planar and axiplanar phases of spin-triplet superfluids and in the Standard Model. The effective theory in the vacua with Fermi points is remarkable. At low energy, the quasiparticles in the vicinity of the Fermi point – the fermion zero modes – represent chiral fermions; the collective bosonic modes represent gauge and gravitational fields acting on the chiral quasiparticles. This emerges together with the 'laws of physics', such as Lorentz invariance and gauge invariance, and together with such notions as chirality, spin, isotopic spin, etc.

All this reproduces the main features of the Standard Model and general relativity, and supports the anti-GUT viewpoint that the Standard Model of the electroweak and strong interactions, and general relativity, are effective theories describing the low-energy phenomena emerging from the fermion zero modes of the quantum vacuum. The nature of this Planck medium – the quantum vacuum – and its physical structure on a 'microscopic' trans-Planckian scale remain unknown, but from topological properties of elementary particles of the Standard Model one might suspect that the quantum vacuum belongs to the same universality class as ^{3}He-A. More exactly, to reproduce all the bosons and fermions of the Standard Model, one needs several Fermi points in momentum space, related by some discrete symmetries. In this respect the Standard Model is closer to the so-called planar phase of spin-triplet p-wave superfluidity, which will be also discussed.

The conceptual similarity between the systems with Fermi points allows us to use ^{3}He-A as a laboratory for the simulation and investigation of those phenomena related to the chiral quantum vacuum of the Standard Model, such as axial (or chiral) anomalies, electroweak baryogenesis, electroweak genesis of primordial magnetic fields, helical instability of the vacuum, fermionic charge of the bosonic fields and topological objects, quantization of physical parameters, etc. (Part IV). As a rule these phenomena are determined by the combined topology: namely, topology in the combined **r**- and **p**-spaces. The exotic terms in the action describing these effects represent the product of the topological invariants in **r**- and **p**-spaces.

The combined topology in the extended (**r**, **p**)-space is also responsible for the fermion zero modes present within different topological defects (Part V). Because of the similarity in topology, the world of quasiparticles in the vortex core or in the core of the domain wall in condensed matter, has a similar emergent physics as the brane world – the 3+1 world of particles and fields in the core of the topological membranes (branes) in multi-dimensional spacetime – the modern development of the Kaluza–Klein theory of extra dimensions.

We hope that condensed matter can show us possible routes from our present low-energy corner of the effective theory to the 'microscopic' physics at Planckian and trans-Planckian energies. The relativistic quantum field theory of the Standard Model, which we have now, is incomplete due to ultraviolet divergences at small length scales. The ultraviolet divergences in the quantum theory of gravity are even more crucial, and after 70 years of research quantum gravity is still far from realization in spite of numerous magnificent achievements (Rovelli 2000). This represents a strong indication that gravity, both classical and quantum, and the Standard Model are not fundamental: both are effective field theories which are not applicable at small length scales where the 'microscopic' physics of a vacuum becomes important, and, according to the anti-GUT scenario, some or all of the known symmetries in nature are violated. The analogy between the quantum vacuum and condensed matter could give an insight into this trans-Planckian physics since it provides examples of the physically imposed deviations from Lorentz and other invariances at higher energy. This is important in many different areas of high-energy physics and cosmology, including possible CPT violation and black holes, where the infinite red shift at the horizon opens a route to trans-Planckian physics.

As we already mentioned, condensed matter teaches us (see e.g. Anderson 1984; Laughlin and Pines 2000) that the low-energy properties of different vacua (magnets, superfluids, crystals, liquid crystals, superconductors, etc.) are robust, i.e. they do not depend much on the details of microscopic (atomic) structure of these substances. The principal role is played by the symmetry and topology of condensed matter: they determine the soft (low-energy) hydrodynamic variables, the effective Lagrangian describing the low-energy dynamics, topological defects and quantization of physical parameters. The microscopic details provide us only with the 'fundamental constants', which enter the effective phenomenological Lagrangian, such as the speed of 'light' (say, the speed of sound), superfluid density, modulus of elasticity, magnetic susceptibility, etc. Apart from these 'fundamental constants', which can be rescaled, the systems behave similarly in the infrared limit if they belong to the same universality and symmetry classes, irrespective of their microscopic origin.

The detailed information on the system is lost in such an acoustic or hydrodynamic limit. From the properties of the low-energy collective modes of the system – acoustic waves in the case of crystals – one cannot reconstruct the atomic structure of the crystal since all the crystals have similar acoustic waves described by the same equations of the same effective theory; in crystals it is the classical theory of elasticity. The classical fields of collective modes can be quantized to obtain quanta of acoustic waves – phonons. This quantum field remains the effective field which is applicable only in the long-wavelength limit, and does not give detailed information on the real quantum structure of the underlying crystal (except for its symmetry class). In other words, one cannot construct the full quantum theory of real crystals using the quantum theory of elasticity. Such a theory would always contain divergences on an atomic scale, which cannot be regularized in a unique way.

It is quite probable that in the same way the quantization of classical gravity, which is one of the infrared collective modes of the quantum vacuum, will not add much to our understanding of the 'microscopic' structure of the vacuum (Hu 1996; Padmanabhan 1999; Laughlin and Pines 2000). Indeed, according to this anti-GUT analogy, properties of our world, such as gravitation, gauge fields, elementary chiral fermions, etc., all arise in the low-energy corner as low-energy soft modes of the underlying 'Planck condensed matter'. At high energy (on the Planck scale) these modes merge with the continuum of all the high-energy degrees of freedom of the 'Planck condensed matter' and thus can no longer be separated from each other. Since gravity is not fundamental, but appears as an effective field in the infrared limit, the only output of its quantization would be the quanta of the low-energy gravitational waves – gravitons. The deeper quantization of gravity makes no sense in this philosophy.

The main advantage of the condensed matter analogy is that in principle we know the condensed matter structure at any relevant scale, including the interatomic distance, which plays the part of one of the Planck length scales in the hierarchy of scales. Thus condensed matter can show the limitation of the effective theories: what quantities can be calculated within the effective field theory using, say, a renormalization group approach, and what quantities depend essentially on the details of trans-Planckian physics. For such purposes the real strongly interacting systems such as ^{3}He and ^{4}He liquids are too complicated for calculations. It is more instructive to choose the simplest microscopic models, which in the low-energy corner leads to the vacuum of the proper universality class. Such a choice of model does not change the 'relativistic' physics arising at low energy, but the model allows all the route from sub-Planckian to trans-Planckian physics to be visualized. We shall use the well known models of weakly interacting Bose (Part I) and Fermi (Part II) gases, such as the BCS model with the proper interaction providing in the low-energy limit the universality class with Fermi points.

Using these models, we can see what results for the vacuum energy when it is calculated within the effective theory and in an exact microscopic theory. The difference between the two approaches just reflects the huge discrepancy between the estimation of the cosmological constant in RQFT and its observational limit which is smaller by about 120 orders. The exact microscopic theory shows that the equilibrium vacuum 'is not gravitating', which immediately follows from the stability of the vacuum state of the weakly interacting quantum gas. Moreover, the same stability condition applied to the quantum liquids shows that the energy of their ground state is exactly zero irrespective of the microscopic structure of the liquid. This implies that either the energy of our quantum vacuum is zero (in a full equilibrium), or the vacuum energy is non-gravitating (in a full equilibrium). This paradigm of the non-gravitating equilibrium vacuum, which is easily derived in condensed matter when we know the microscopic 'trans-Planckian' physics, can be considered as one of the postulates of the effective phenomenological theory of general relativity. This principle cannot be derived within the effective theory. It can follow only from the still unknown fundamental level.

On the other hand, the quantum liquid examples demonstrate that this principle does not depend on the details of the fundamental physics. This is actually our goal: to extract from the well-determined and complete microscopic many-body systems the general principles which do not depend on details of the system. Then we can hope that these principles will be applicable to such a microscopic system as the quantum vacuum.

By investigating quantum liquids we can get a hint on what kind of condensed matter the 'Planck matter' is. The quantum Fermi liquids belonging to the universality class of Fermi points nicely reproduce the main features of the Standard Model. But the effective gravity still remains a caricature of the Einstein theory, though it is a much better caricature than the effective gravity in the Bose liquids. From the model it follows that to bring the effective gravity closer to Einstein theory, i.e. to reproduce the Einstein–Hilbert action for the $g_{\mu\nu}$ field, one must suppress all the effects of the broken symmetry: the 'Planck matter' must be in the non-superfluid disordered phase without Goldstone bosons, but with Fermi points in momentum space.

The condensed matter analogy is in some respect similar to string theory, where the gauge invariance and general covariance are not imposed, and fermions, gravitons and gauge quanta are the emergent low-energy properties of an underlying physical object – the fundamental string. At the moment string theory is viewed as the most successful attempt to quantize gravity so far. However, as distinct from string theory which requires higher dimensions, the 'relativistic' Weyl fermions and gauge bosons arise in the underlying non-relativistic condensed matter in an ordinary 3+1 spacetime, provided that the condensed matter belongs to the proper universality class.

Part I

Quantum Bose liquid

2
GRAVITY

Since we are interested in the analog of effective gravity arising in the low-energy corner of the quantum Bose liquids, let us first recall some information concerning the original Einstein theory of the gravitational field.

2.1 The Einstein theory of gravity

The Einstein theory of gravitation consists of two main elements:

(1) Gravity is related to a curvature of spacetime in which particles move along geodesic curves in the absence of non-gravitational forces. The geometry of spacetime is described by the metric $g_{\mu\nu}$ which is the dynamical field of gravity. The action S_M for matter in the presence of a gravitational field is obtained from the special relativity action for the matter fields by replacing everywhere the flat Minkowski metric by the dynamical metric $g_{\mu\nu}$ and the partial derivative by the g-covariant derivative. This follows from the principle that the equations of motion do not depend on the choice of the coordinate system (the so-called general covariance). This also means that motion in the non-inertial frame can be described in the same manner as motion in some gravitational field – this is the equivalence principle. Another consequence of the equivalence principle is that spacetime geometry is the same for all particles: gravity is universal. Since the action S_M for matter fields depends on $g_{\mu\nu}$, it simultaneously describes the coupling between gravity and the matter fields (all other fields except gravity).

(2) The dynamics of the gravitational field itself is determined by adding to S_M the action functional S_G for $g_{\mu\nu}$, describing the propagation and self-interaction of the gravitational field:

$$S = S_\mathrm{G} + S_\mathrm{M} \; . \tag{2.1}$$

The general covariance requires that S_G is a functional of the curvature. In the original Einstein theory only the first-order curvature term is retained:

$$S_\mathrm{G} = -\frac{1}{16\pi G} \int d^4x \sqrt{-g}\mathcal{R} \; , \tag{2.2}$$

where G is the Newton gravitational constant and $\mathcal{R}$ is the Ricci scalar curvature.

In the following chapters we shall exploit the analogy with the effective Lorentz invariance and effective gravity arising in quantum liquids. In these systems the 'speed of light' c – the maximum attainable speed of the low-energy quasiparticles – is not a fundamental constant, but a material parameter. In order to describe gravity in a manner also applicable to quantum liquids, the action should not contain explicitly the material parameters. That is why eqn (2.2) is written in such a way that it does not contain the speed of light c: it is absorbed by the Newton constant and by the metric. It is assumed that in flat spacetime, the Minkowski metric has the form $g_{\mu\nu} = \mathrm{diag}(-1, c^{-2}, c^{-2}, c^{-2})$. This form can be extended for anisotropic superfluids, where the 'speed of light' depends on the direction of propagation, and the metric of the effective 'Minkowski' spacetime is $g_{\mu\nu} = \mathrm{diag}(-1, c_x^{-2}, c_y^{-2}, c_z^{-2})$. The dimension of the Newton constant is now $[G]^{-1} = [E]^2/[\hbar]$, where $[E]$ means the dimension of energy. Further, if the Planck constant $\hbar$ is not written explicitly, we assume that $\hbar = 1$, and correspondingly $[G]^{-1} = [E]^2$.

The Einstein–Hilbert action is thus

$$S_{\mathrm{Einstein}} = -\frac{1}{16\pi G} \int d^4 x \sqrt{-g} R + S_{\mathrm{M}} \ . \qquad (2.3)$$

Variation of this action over the metric field $g^{\mu\nu}$ gives the Einstein equations:

$$\frac{\delta S_{\mathrm{Einstein}}}{\delta g^{\mu\nu}} = \frac{1}{2}\sqrt{-g}\left[-\frac{1}{8\pi G}\left(R_{\mu\nu} - \frac{1}{2}R g_{\mu\nu}\right) + T_{\mu\nu}^{\mathrm{M}}\right] = 0 \ , \qquad (2.4)$$

where $T_{\mu\nu}^{\mathrm{M}}$ is the energy–momentum tensor of the matter fields.

2.1.1 Covariant conservation law

Using Bianchi identities for the curvature tensor one obtains from eqn (2.4) the 'covariant' conservation law for matter:

$$T_{\nu;\mu}^{\mu\mathrm{M}} = 0 \ \text{ or } \ \partial_\mu \left(T_\nu^{\mu\mathrm{M}}\sqrt{-g}\right) = \frac{1}{2}\sqrt{-g} T^{\alpha\beta\mathrm{M}} \partial_\nu g_{\alpha\beta} \ . \qquad (2.5)$$

Equation (2.5) is the check for the self-consistency of the Einstein equations, since the 'covariant' conservation law for matter can be obtained directly from the field equations for matter without invoking the dynamics of the gravitational field. The only input is that the field equations for matter obey the general covariance. We shall see later that in the effective gravity arising in the low-energy corner of quantum liquids, eqn (2.5) does hold for the 'matter fields' (i.e. for quasiparticles), though the effective 'gravity field' $g_{\mu\nu}$ does not obey the Einstein equations.

2.2 Vacuum energy and cosmological term

2.2.1 Vacuum energy

In particle physics, field quantization allows a zero-point energy, the constant vacuum energy when all fields are in their ground states. In the absence of gravity the constant energy can be ignored, since only the difference between the energies of the field in the excited and ground states is meaningful. If the vacuum is distorted, as for example in the Casimir effect (Casimir 1948), one can measure the difference in energies between distorted and original vacua. But the absolute value of the vacuum energy is unmeasurable.

On the contrary, in the Einstein theory of gravity the gravitational field reacts to the total value of the energy–momentum tensor of the matter fields, and thus the absolute value of the vacuum energy becomes meaningful. If the energy–momentum tensor of the vacuum is non-zero it must be added to the Einstein equations. The corresponding contribution to the action is given by the so-called cosmological term, which was introduced by Einstein (1917):

$$S_\Lambda = -\rho_\Lambda \int d^4x \sqrt{-g} \,, \quad T^\Lambda_{\mu\nu} = \frac{2}{\sqrt{-g}} \frac{\delta S_\Lambda}{\delta g^{\mu\nu}} = \rho_\Lambda g_{\mu\nu} \,. \tag{2.6}$$

The energy–momentum tensor of the vacuum shows that the quantity $\rho_\Lambda \sqrt{-g}$ is the vacuum energy density, while the partial pressure of the vacuum is $-\rho_\Lambda \sqrt{-g}$. Thus the equation of state of the vacuum is

$$P_\Lambda = -\rho_\Lambda \,. \tag{2.7}$$

The cosmological term modifies the Einstein equations (2.4) adding the energy–momentum tensor of the vacuum to the right-hand side:

$$\frac{1}{8\pi G}\left(R_{\mu\nu} - \frac{1}{2}R g_{\mu\nu}\right) = T^M_{\mu\nu} + T^\Lambda_{\mu\nu} \,. \tag{2.8}$$

At the moment there is no law of nature which forbids the cosmological constant, and what is not forbidden is compulsory. Moreover, the idea that the vacuum is full of various fluctuating fields (such as electromagnetic and Dirac fields) and condensates (such as Higgs fields) is widely accepted. Using the Standard Model or its extension one can easily estimate the vacuum energy of fluctuating fields: the positive contribution comes from the zero-point energy $(1/2)E(p)$ of quantum fluctuations of bosonic fields and the negative one from the occupied negative energy levels in the Dirac sea. Since the largest contribution comes from the fluctuations with ultrarelativistic momenta $p \gg mc$, where the masses m of particles can be neglected, the energy spectrum of particles can be considered as massless, $E = cp$. Then the energy density of the vacuum is simply

$$\rho_\Lambda \sqrt{-g} = \frac{1}{V}\left(\nu_{\text{bosons}} \sum_p \frac{1}{2} cp - \nu_{\text{fermions}} \sum_p cp\right) \,. \tag{2.9}$$

Here V is the volume of the system; ν_{bosons} is the number of bosonic species and ν_{fermions} is the number of fermionic species in the vacuum. The vacuum energy is divergent and the natural ultraviolet cut-off is provided by the gravity itself. The cut-off Planck energy is determined by the Newton constant:

$$E_{\text{Planck}} = \left(\frac{\hbar}{G}\right)^{1/2} \,. \tag{2.10}$$

It is on the order of 10^{19} GeV. If there is no supersymmetry – the symmetry between the fermions and bosons – the Planck energy-scale cut-off provides the following estimate for the vacuum energy density:

$$\rho_\Lambda \sqrt{-g} \sim \pm \frac{1}{c^3 \hbar^3} E_{\text{Planck}}^4 = \pm \frac{1}{\hbar^3} \sqrt{-g}\, E_{\text{Planck}}^4\ . \tag{2.11}$$

The sign of the vacuum energy is determined by the fermionic and bosonic content of the quantum field theory.

The vacuum energy was calculated for flat spacetime with the Minkowski metric in the form $g_{\mu\nu} = \text{diag}(-1, c^{-2}, c^{-2}, c^{-2})$, so that $\sqrt{-g} = c^{-3}$. The right-hand side of eqn (2.11) does not depend on the 'material parameter' c, and thus (as we see later) is also applicable to quantum liquids.

2.2.2 *Cosmological constant problem*

The 'cosmological constant problem' refers to a huge disparity between the naively expected value in eqn (2.11) and the range of actual values. The experimental observations show that the energy density in our Universe is close to the critical density ρ_c, corresponding to a flat Universe. The energy content is composed of baryonic matter (very few percent), the so-called dark matter which does not emit or absorb light (about 30%) and vacuum energy (about 70%). Such a distribution of energy leads to an accelerated expansion of the Universe which is revealed by recent supernovae Ia observations (Perlmutter *et al.* 1999; Riess *et al.* 2000). The observed value of the vacuum energy $\rho_\Lambda \sim 0.7 \rho_c$ is thus on the order of $10^{-123} E_{\text{Planck}}^4$ in contradiction with the theoretical estimate in eqn (2.11). This is probably the largest discrepancy between theory and experiment in physics.

This huge discrepancy can be cured by supersymmetry, the symmetry between fermions and bosons. If there is a supersymmetry, the positive contribution of bosons and negative contribution of fermions exactly cancel each other. However, it is well known that there is no supersymmetry in our low-energy world. This means that there must by an energy scale E_{SuSy} below which the supersymmetry is violated and thus there is no balance between bosons and fermions in the vacuum energy . This scale is providing the cut-off in eqn (2.9), and the theoretical estimate of the cosmological constant becomes $\rho_\Lambda(\text{theor}) \sim E_{\text{SuSy}}^4$. Though E_{SuSy} can be much smaller than the Planck scale, the many orders of magnitude disagreement between theory and reality still persists, and we must accept the experimental fact that the vacuum energy in eqn (2.9) is not gravitating. This is the most severe problem in the marriage of gravity and quantum theory (Weinberg 1989), because it is in apparent contradiction with the general principle of equivalence, according to which the inertial and gravitating masses coincide.

One possible way to solve this contradiction is to accept that the theoretical criteria for setting the absolute zero point of energy are unclear within the effective theory, i.e. the vacuum energy is by no means the energy of vacuum fluctuations of effective fields in eqn (2.9). Its estimation requires physics beyond the Planck scale and thus beyond general relativity. To clarify this issue we can consider such quantum systems where the elements of the gravitation are at least partially reproduced, but where the structure of the quantum vacuum beyond the 'Planck scale' is known. Quantum liquids are the right systems, and later we shall show how 'trans-Planckian physics' solves the cosmological constant problems at least in quantum liquids.

2.2.3 Vacuum-induced gravity

Why does the Planck energy in eqn (2.10) serve as the natural cut-off in quantum field theory? This is based on the important observation made by Sakharov (1967a) that the second element of the Einstein theory (the action for the gravity field) can follow from the first element (the action of the matter field in the presence of gravity). Sakharov showed that vacuum fluctuations of the matter fields in the presence of the classical background metric field $g_{\mu\nu}$ make the metric field dynamical inducing the curvature term in the action for $g_{\mu\nu}$.

According to Sakharov's calculations, the magnitude of the induced Newton constant is determined by the value of the ultraviolet cut-off: $G \sim \hbar/E_{\text{cutoff}}^2$. Actually $G^{-1} \sim NE_{\text{cutoff}}^2/\hbar$, where N is the number of (fermionic) matter fields contributing to the gravitational constant. Thus in Sakharov's theory of gravity induced by quantum fluctuations, the causal connection between gravity and the cut-off is reversed: the physical high-energy cut-off determines the gravitational constant. Gravity is so small compared to the other forces just because the cut-off energy is so big. Electromagnetic, weak and strong forces also depend on the ultraviolet cut-off E_{cutoff}. But the corresponding coupling 'constants' have only a mild logarithmic dependence on E_{cutoff}, as compared to the $1/E_{\text{cutoff}}^2$ dependence of G.

2.2.4 Effective gravity in quantum liquids

Sakharov's theory does not explain the first element of the Einstein theory: it shows how the Einstein–Hilbert action for the metric field $g_{\mu\nu}$ arises from the vacuum fluctuations, but it does not indicate how the metric field itself appears. The latter can be given only by the fundamental theory of the quantum vacuum, such as string theory where gravity appears as a low-energy mode. Quantum liquids also provide examples of how the metric field naturally emerges as the low-energy collective mode of the quantum vacuum. The action for this mode is provided by the dynamics of quantum vacuum in accordance with Sakharov's theory, and even the curvature term can be reproduced in some condensed matter systems in some limiting cases. From this point of view gravity is not the fundamental force, but is determined by the properties of the quantum vacuum: gravity is one of the collective modes of the quantum vacuum.

The metric field naturally arises in many condensed matter systems. An example is the motion of acoustic phonons in a distorted crystal lattice, or in the background flow field of superfluid condensate. This motion is described by the effective acoustic metric (see below in Chapter 4). For this 'relativistic matter field' (acoustic phonons with dispersion relation $E = cp$, where c is the speed of sound, simulate relativistic particles) the analog of the equivalence principle is fulfilled. As a result the covariant conservation law in eqn (2.5) does hold for the acoustic mode, if $g_{\mu\nu}$ is replaced by the acoustic metric (Sec. 6.1.2).

The second element of the Einstein gravity is not easily reproduced in condensed matter. In general, the dynamics of the effective metric $g_{\mu\nu}$ does not obey the equivalence principle. In existing quantum liquids, the action for $g_{\mu\nu}$ is induced not only by quantum fluctuations of the 'relativistic matter fields', but also by the high-energy degrees of freedom of the quantum vacuum, which are 'non-relativistic'. The latter are typically dominating, as a result the effective action for $g_{\mu\nu}$ is non-covariant. Of course, one can find some very special cases where the Einstein action for the effective metric is dominating, but this is not the rule in condensed matter.

Nevertheless, despite the incomplete analogy with the Einstein theory, the effective gravity in quantum liquids can be useful for investigating many problems that are not very sensitive to whether or not the gravitational field satisfies the Einstein equations. In particular, these are the problems related to the behavior of the quantum vacuum in the presence of a non-trivial metric field (event horizon, ergoregion, spinning string, etc.); and also the cosmological constant problem.

3
MICROSCOPIC PHYSICS OF QUANTUM LIQUIDS

There are two ways to study quantum liquids:

(1) The fully microscopic treatment. This can be realized completely either by numerical simulations of the many-body problem, or for some special ranges of the material parameters analytically, for example, in the limit of weak interaction between the particles.

(2) A phenomenological approach in terms of effective theories. The hierarchy of the effective theories corresponds to the low-frequency, long-wavelength dynamics of quantum liquids in different ranges of frequency. Examples of effective theories: Landau theory of the Fermi liquid; Landau–Khalatnikov two-fluid hydrodynamics of superfluid ^{4}He; the theory of elasticity in solids; the Landau–Lifshitz theory of ferro- and antiferromagnetism; the London theory of superconductivity; the Leggett theory of spin dynamics in superfluid phases of ^{3}He; effective quantum electrodynamics arising in superfluid ^{3}He-A; etc. The last example indicates that the existing Standard Model of electroweak and strong interactions, and the Einstein gravity too, are the phenomenological effective theories of high-energy physics which describe its low-energy edge, while the microscopic theory of the quantum vacuum is absent.

3.1 Theory of Everything in quantum liquids

3.1.1 *Microscopic Hamiltonian*

The microscopic Theory of Everything for quantum liquids and solids – 'a set of equations capable of describing all phenomena that have been observed' (Laughlin and Pines 2000) in these quantum systems – is extremely simple. On the 'fundamental' level appropriate for quantum liquids and solids, i.e. for all practical purposes, the ^{4}He or ^{3}He atoms of these quantum systems can be considered as structureless: the ^{4}He atoms are the structureless bosons and the ^{3}He atoms are the structureless fermions with spin 1/2. The simplest Theory of Everything for a collection of a macroscopic number N of interacting ^{4}He or ^{3}He atoms is contained in the non-relativistic many-body Hamiltonian

$$\mathcal{H} = -\frac{\hbar^2}{2m} \sum_{i=1}^{N} \frac{\partial^2}{\partial \mathbf{r}_i^2} + \sum_{i=1}^{N} \sum_{j=i+1}^{N} U(\mathbf{r}_i - \mathbf{r}_j) , \qquad (3.1)$$

acting on the many-body wave function $\Psi(\mathbf{r}_1, \mathbf{r}_2, \ldots, \mathbf{r}_i, \ldots, \mathbf{r}_j, \ldots)$. Here m is the bare mass of the atom; $U(\mathbf{r}_i - \mathbf{r}_j)$ is the pair interaction of the bare atoms i and j.

The many-body physics can be described in the second quantized form, where the Schrödinger many-body Hamiltonian (3.1) becomes the Hamiltonian of the quantum field theory (Abrikosov *et al.* 1965):

$$\mathcal{H} - \mu \mathcal{N} = \int d\mathbf{x} \psi^\dagger(\mathbf{x}) \left[-\frac{\nabla^2}{2m} - \mu \right] \psi(\mathbf{x}) + \frac{1}{2} \int d\mathbf{x} d\mathbf{y} U(\mathbf{x}-\mathbf{y}) \psi^\dagger(\mathbf{x}) \psi^\dagger(\mathbf{y}) \psi(\mathbf{y}) \psi(\mathbf{x}). \tag{3.2}$$

In ^{4}He, the bosonic quantum fields $\psi^\dagger(\mathbf{x})$ and $\psi(\mathbf{x})$ are the creation and annihilation operators of the ^{4}He atoms. In ^{3}He, $\psi^\dagger(\mathbf{x})$ and $\psi(\mathbf{x})$ are the corresponding fermionic quantum fields and the spin indices must be added. Here $\mathcal{N} = \int d\mathbf{x}\, \psi^\dagger(\mathbf{x})\psi(\mathbf{x})$ is the operator of the particle number (number of atoms); μ is the chemical potential – the Lagrange multiplier introduced to take into account the conservation of the number of atoms. Note that the introduction of creation and annihilation operators for helium atoms does not imply that we really can create atoms from the vacuum: this is certainly highly prohibited since the relevant energies in the liquid are of order 10 K, which is many orders of magnitude smaller than the GeV energy required to create the atom-antiatom pair from the vacuum.

3.1.2 *Particles and quasiparticles*

In quantum liquids, the analog of the quantum vacuum – the ground state of the quantum liquid – has a well-defined number of atoms. Existence of bare particles (atoms) comprising the quantum vacuum of quantum liquids represents the main difference from the relativistic quantum field theory (RQFT). In RQFT, particles and antiparticles which can be created from the quantum vacuum are similar to quasiparticles in quantum liquids. What is the analog of the bare particles which comprise the quantum vacuum of RQFT is not clear today. At the moment we simply do not know the structure of the vacuum, and whether it is possible to describe it in terms of some discrete elements – bare particles – whose number is conserved.

In the limit when the number N of bare particles in the vacuum is large, one might expect that the difference between two quantum field theories, with and without conservation of particle number, disappears completely. However, this is not so. We shall see that the mere fact that there is a conservation law for the number of particles comprising the vacuum leads to a definite conclusion on the value of the relevant vacuum energy : it is exactly zero in equilibrium. Also, as we shall see below in Chapter 29, the discreteness of the quantum vacuum can be revealed in the mesoscopic Casimir effect.

3.1.3 *Microscopic and effective symmetries*

The Theory of Everything (3.2) has a very restricted number of symmetries: (i) The Hamiltonian is invariant under translations and $SO(3)$ rotations in 3D space. (ii) There is a global $U(1)_N$ group originating from the conservation of the number N of atoms: $\mathcal{H}$ is invariant under global gauge rotation $\psi(\mathbf{x}) \to e^{i\alpha}\psi(\mathbf{x})$ with constant α. The particle number operator serves as the generator of the gauge rotations: $e^{-i\alpha\mathcal{N}}\psi e^{i\alpha\mathcal{N}} = \psi e^{i\alpha}$. (iii) In ^{3}He, the spin–orbit coupling is relatively weak. If it is ignored, then $\mathcal{H}$ is also invariant under separate rotations of spins, $SO(3)_\mathbf{S}$ (later we shall see that the symmetry violating spin–orbit interaction plays an important role in the physics of fermionic and bosonic zero modes in all of the superfluid phases of ^{3}He). At low temperature the phase transition to the superfluid or to the quantum crystal state occurs where some of these symmetries are broken spontaneously.

In the ^{3}He-A state all of the symmetries of the Hamiltonian, except for the translational one, are broken. However, when the temperature and energy decrease further the symmetry becomes gradually enhanced in agreement with the anti-Grand-Unification scenario (Froggatt and Nielsen 1991; Chadha and Nielsen 1983). At low energy the quantum liquid or solid is well described in terms of a dilute system of quasiparticles. These are bosons (phonons) in ^{4}He and fermions and bosons in ^{3}He, which move in the background of the effective gauge and/or gravity fields simulated by the dynamics of the collective modes. In particular, as we shall see below, phonons propagating in the inhomogeneous liquid are described by the effective Lagrangian for the scalar field α in the presence of the effective gravitational field:

$$L_{\text{effective}} = \sqrt{-g}g^{\mu\nu}\partial_\mu\alpha\partial_\nu\alpha . \qquad (3.3)$$

Here $g^{\mu\nu}$ is the effective acoustic metric provided by the inhomogeneity of the liquid and by its flow (Unruh 1981, 1995; Stone 2000b).

These quasiparticles serve as the elementary particles of the low-energy effective quantum field theory. They represent the analog of matter. The type of the effective quantum field theory – the theory of interacting fermionic and bosonic quantum fields – depends on the universality class emerging in the low-energy limit. In normal Fermi liquids, the effective quantum field theory describing dynamics of fermion zero modes in the vicinity of the Fermi surface which interact with the collective bosonic fields is the Landau theory of Fermi liquid. In superfluid ^{3}He-A, which belongs to different universality class, the effective quantum field theory contains chiral 'relativistic' fermions, while the collective bosonic modes interact with these 'elementary particles' as gauge fields and gravity. All these fields emerge together with the Lorentz and gauge invariances and with elements of the general covariance from the fermionic Theory of Everything in eqn (3.2). The vacuum of the Standard Model belong to the same universality class, and the RQFT of the Standard Model is the corresponding effective theory.

The emergent phenomena do not depend much on the details of the Theory of Everything (Laughlin and Pines 2000) – in our case on the details of the pair potential $U(\mathbf{x}-\mathbf{y})$. Of course, the latter determines the universality class in which the system finds itself at low energy. But once the universality class is established, the physics remains robust to deformations of the pair potential. The details of $U(\mathbf{x}-\mathbf{y})$ influence only the 'fundamental' parameters of the effective theory ('speed of light', 'Planck' energy cut-off, etc.) but not the general structure of the theory. Within the effective theory the 'fundamental' parameters are considered as phenomenological.

3.1.4 Fundamental constants of Theory of Everything

The original number of fundamental parameters of the microscopic Theory of Everything is big: these are all the relevant Fourier components of the pair potential $U(r = |\mathbf{x} - \mathbf{y}|)$. However, one can properly approximate the shape of the potential. Typically the Lennard-Jones potential $U(r) = \epsilon_0 \left((r_0/r)^{12} - (r_0/r)^6 \right)$ is used, which simulates the hard-core repulsion of two atoms at small distances and their van der Waals attraction at large distances. This $U(r)$ contains only two parameters, the characteristic depth ϵ_0 of the potential well and the length r_0 which characterizes both the hard core of the atom and the dimension of the potential well.

Thus the microscopic Theory of Everything in a quantum liquid can be expressed in terms of four parameters: $\hbar$, ϵ_0, r_0 and the mass of the atom m. These 'fundamental constants' of the Theory of Everything determine the 'fundamental constants' of the descending effective theory at lower energy. On the other hand, we know that at least two of them, ϵ_0 and r_0, can be derived from the more fundamental Theory of Everything – atomic physics – whose 'fundamental constants' are $\hbar$, electric charge e and the mass of the electron m_e. In turn, e and m_e are determined by the higher-energy Theory of Everything – the Standard Model etc. Such a hierarchy of 'fundamental constants' indicates that the ultimate set of fundamental constants probably does not exist at all.

The Theory of Everything for liquid ^{3}He or ^{4}He does not contain a small parameter: the dimensionless quantity, which can be constructed from the four constants $r_0\sqrt{m\epsilon_0}/\hbar$, appears to be of order unity for ^{3}He and ^{4}He atoms. As a result the quantum liquids ^{3}He and ^{4}He are strongly correlated and strongly interacting systems. The distance between atoms in equilibrium liquids is determined by the competition of the attraction and the repulsing zero-point oscillations of atoms, and is thus also of order r_0. Zero-point oscillations of atoms prevent solidification: in equilibrium both systems are liquids. Solidification occurs when rather mild external pressure is applied.

Since there is no small parameter, it is a rather difficult task to derive the effective theory from first principles, though it is possible if one has enough computer time and memory. That is why it is instructive to consider the microscopic theory for some special model potentials $U(r)$ which contain a small parameter, but leads to the same universality class of effective theories in the low-energy limit. This allows us to solve the problem completely or perturbatively. In the case of the Bose–liquids the proper model is the Bogoliubov model (1947) of weakly interacting Bose gas; for the superfluid phases of ^{3}He it is the Bardeen–Cooper–Schrieffer (BCS) model.

Such models are very useful, since they simultaneously cover the low-energy edge of the effective theory and the Theory of Everything, i.e. high-energy 'trans-Planckian' physics. In particular, this allows us to check the validity of different regularization schemes elaborated within the effective theory.

3.2 Weakly interacting Bose gas

3.2.1 *Model Hamiltonian*

In the Bogoliubov theory of the weakly interacting Bose gas the pair potential in eqn (3.2) is weak. As a result, in the vacuum state most particles are in the Bose–Einstein condensate, i.e. in the state with momentum $\mathbf{p} = 0$. The vacuum with Bose condensate is characterized by the scalar order paramater – the non-zero vacuum expectation value (vev) of the particle annihilation operator at $\mathbf{p} = 0$:

$$\langle a_{\mathbf{p}=0} \rangle = \sqrt{N_0} e^{i\Phi}, \quad \langle a^\dagger_{\mathbf{p}=0} \rangle = \sqrt{N_0} e^{-i\Phi}. \tag{3.4}$$

Here $N_0 < N$ is the particle number in the Bose condensate, and Φ is the phase of the condensate. The vacuum state is not invariant under $U(1)_N$ global gauge rotations, and thus the vacuum states are degenerate: vacua with different Φ are distinguishable but have the same energy. Further we choose a particular vacuum state with $\Phi = 0$. Since the number of particles in the condensate is large, one can treat operators $a_{\mathbf{p}=0}$ and $a^\dagger_{\mathbf{p}=0}$ as classical fields, merely replacing them by their vev in the Hamiltonian.

If there is no interaction between the particles (an ideal Bose gas), the vacuum is completely represented by the Bose condensate particles, $N_0 = N$. The interaction pushes some fraction of particles from the $\mathbf{p} = 0$ state. If the interaction is small, the fraction of the non-condensate particles in the vacuum is also small, and they have small momenta $\mathbf{p}$. As a result, only the zero Fourier component of the pair potential is relevant, and the potential can be approximated by a δ-function, $U(\mathbf{r}) = U\delta(\mathbf{r})$. The Theory of Everything in eqn (3.2) then acquires the following form:

$$\mathcal{H} - \mu \mathcal{N} = -\mu N_0 + \frac{N_0^2 U}{2V} \tag{3.5}$$

$$+ \sum_{\mathbf{p} \neq 0} \left(\frac{p^2}{2m} - \mu \right) a^\dagger_{\mathbf{p}} a_{\mathbf{p}} \tag{3.6}$$

$$+ \frac{N_0 U}{2V} \sum_{\mathbf{p} \neq 0} \left(2a^\dagger_{\mathbf{p}} a_{\mathbf{p}} + 2a^\dagger_{-\mathbf{p}} a_{-\mathbf{p}} + a_{\mathbf{p}} a_{-\mathbf{p}} + a^\dagger_{\mathbf{p}} a^\dagger_{-\mathbf{p}} \right). \tag{3.7}$$

We ignore quantum fluctuations of the operator a_0 considering it as a c-number: $N_0 = a^\dagger_0 a_0 = a_0 a^\dagger_0 = a^\dagger_0 a^\dagger_0 = a_0 a_0$. Note that the last two terms in eqn (3.7) do not conserve particle number: this is the manifestation of the broken $U(1)_N$ symmetry in the vacuum.

Minimization of the main part of the energy in eqn (3.5) over N_0 gives $UN_0/V = \mu$ and one obtains

$$\mathcal{H} - \mu \mathcal{N} = -\frac{\mu^2}{2U} V + \sum_{\mathbf{p} \neq 0} \mathcal{H}_{\mathbf{p}}, \tag{3.8}$$

$$\mathcal{H}_{\mathbf{p}} = \frac{1}{2} \left(\frac{p^2}{2m} + \mu \right) \left(a^\dagger_{\mathbf{p}} a_{\mathbf{p}} + a^\dagger_{-\mathbf{p}} a_{-\mathbf{p}} \right) + \frac{\mu}{2} \left(a_{\mathbf{p}} a_{-\mathbf{p}} + a^\dagger_{\mathbf{p}} a^\dagger_{-\mathbf{p}} \right). \tag{3.9}$$

3.2.2 Pseudorotation – Bogoliubov transformation

At each $\mathbf{p}$ the Hamiltonian $\mathcal{H}_\mathbf{p}$ can be diagonalized using the following consideration. Three operators

$$\mathcal{L}_3 = \frac{1}{2}(a_\mathbf{p}^\dagger a_\mathbf{p} + a_{-\mathbf{p}}^\dagger a_{-\mathbf{p}} + 1) \;,\; \mathcal{L}_1 + i\mathcal{L}_2 = a_\mathbf{p}^\dagger a_{-\mathbf{p}}^\dagger \;,\; \mathcal{L}_1 - i\mathcal{L}_2 = a_\mathbf{p} a_{-\mathbf{p}} \quad (3.10)$$

form the group of pseudorotations, $SU(1,1)$ (the group which conserves the form $x_1^2 + x_2^2 - x_3^2$), with the commutation relations:

$$[\mathcal{L}_3, \mathcal{L}_1] = i\mathcal{L}_2 \;,\; [\mathcal{L}_2, \mathcal{L}_3] = i\mathcal{L}_1 \;,\; [\mathcal{L}_1, \mathcal{L}_2] = -i\mathcal{L}_3 \;. \quad (3.11)$$

The Hamiltonian in eqn (3.9) can be written in terms of these operators in the following general form:

$$\mathcal{H}_\mathbf{p} = g^i(\mathbf{p})\mathcal{L}_i + g^0(\mathbf{p}) \;. \quad (3.12)$$

In more complicated cases, when the spin or other internal degrees of freedom are important, the corresponding Hamiltonian $\mathcal{H}_\mathbf{p}$ is expressed in terms of generators of a higher symmetry group. For example, the $SO(5)$ group naturally arises for triplet Cooper pairing (see e.g. Murakami *et al.* (1999)).

In the particular case under discussion and for non-zero phase Φ of the condensate, one has

$$g^1 = \mu\cos(2\Phi) \;,\; g^2 = \mu\sin(2\Phi) \;,\; g^3 = \frac{p^2}{2m} + \mu \;,\; g^0 = -\frac{1}{2}\left(\frac{p^2}{2m} + \mu\right) \;. \quad (3.13)$$

The diagonalization of this Hamiltonian is achieved first by phase rotation by angle 2Φ around axis x_3 to obtain $g^1 = \mu$ and $g^2 = 0$, and then by the Lorentz transformation – pseudorotation around axis x_2:

$$\mathcal{L}_3 = \tilde{\mathcal{L}}_3\mathrm{ch}\chi + \tilde{\mathcal{L}}_1\mathrm{sh}\chi \;,\; \mathcal{L}_1 = \tilde{\mathcal{L}}_1\mathrm{ch}\chi + \tilde{\mathcal{L}}_3\mathrm{sh}\chi \;,\; \mathrm{th}\chi = \frac{\mu}{\frac{p^2}{2m} + \mu} \;. \quad (3.14)$$

This corresponds to Bogoliubov transformation and gives the diagonal Hamiltonian as a set of uncoupled harmonic oscillators:

$$\mathcal{H}_\mathbf{p} = \tilde{\mathcal{L}}_3\sqrt{\left(\frac{p^2}{2m} + \mu\right)^2 - \mu^2} - \frac{1}{2}\left(\frac{p^2}{2m} + \mu\right) \quad (3.15)$$

$$= \frac{1}{2}E(\mathbf{p})\left(\tilde{a}_\mathbf{p}^\dagger \tilde{a}_\mathbf{p} + \tilde{a}_{-\mathbf{p}}^\dagger \tilde{a}_{-\mathbf{p}}\right) + \frac{1}{2}\left(E(\mathbf{p}) - \left(\frac{p^2}{2m} + \mu\right)\right) \;, \quad (3.16)$$

$$E(\mathbf{p}) = \sqrt{\left(\frac{p^2}{2m} + \mu\right)^2 - \mu^2} = \sqrt{p^2c^2 + \frac{p^4}{4m^2}} \;,\; c^2 = \frac{\mu}{m} \;. \quad (3.17)$$

Here $\tilde{a}_\mathbf{p}$ and $E(\mathbf{p})$ are the annihilation operator of quasiparticles and the quasiparticle energy spectrum respectively.

The above procedure makes sense if the interaction between the original atoms is repulsive, i.e. $U > 0$ and thus $\mu > 0$, otherwise $c^2 < 0$ and the gas will collapse to form a condensed state such as a liquid, or explode.

3.2.3 Low-energy relativistic quasiparticles

The total Hamiltonian now represents the ground state – the vacuum determined as $\tilde{a}_\mathbf{p}|\text{vac}\rangle = 0$ – and the system of quasiparticles above the ground state

$$\mathcal{H} - \mu\mathcal{N} = \langle\mathcal{H} - \mu\mathcal{N}\rangle_\text{vac} + \sum_\mathbf{p} E(\mathbf{p})\tilde{a}^\dagger_\mathbf{p}\tilde{a}_\mathbf{p} \,. \tag{3.18}$$

The Hamiltonian (3.18) has important properties. First, it conserves the number of quasiparticles (this conservation law is, however, approximate: the higher-order terms do not conserve quasiparticle number).

Second, though the vacuum state contains many particles (atoms), a quasiparticle cannot be scattered on them. Quasiparticles do not feel the homogeneous vacuum, it is an empty space for them; in other words, it is impossible to scatter a quasiparticle on quantum fluctuations. A quasiparticle can be scattered by inhomogeneity of the vacuum, such as a flow field of the vacuum, or by other quasiparticles. However, the lower the energy or the temperature T of the system, the more dilute is the gas of thermal quasiparticles and thus the weaker is the average interaction between them.

This description in terms of the vacuum state and dilute system of quasiparticles is generic for quantum liquids and is valid even if the interaction of the initial bare particles is strong. The phenomenological effective theory in terms of vacuum state and quasiparticles was developed by Landau for both Bose and Fermi liquids. Quasiparticles (not the bare particles) play the role of elementary particles in such effective quantum field theories.

In a weakly interacting Bose gas in eqn (3.17), the spectrum of quasiparticles at low energy (i.e. at $p \ll mc$) becomes that of the massless relativistic particle, $E = cp$. The maximum attainable speed here coincides with the speed of sound $c^2 = N(d\mu/dN)/m$. This can be obtained from the leading term in energy: one has $N = -d(E-\mu N)/d\mu = \mu V/U$ and $c^2 = N(d\mu/dN)/m = \mu/m$. These quasiparticles are thus phonons, quanta of sound waves.

The same quasiparticle spectrum occurs in the real superfluid liquid ^{4}He, where the interaction between the bare particles is strong. This shows that the low-energy properties of the system do not depend much on the microscopic (trans-Planckian) physics. The latter determines only the 'fundamental constant' of the effective theory – the speed of sound c. One can say that weakly interacting Bose gas and strongly interacting superfluid liquid ^{4}He belong to the same universality class, and thus have the same low-energy properties. One cannot distinguish between the two systems if one investigates only the low-energy effects well below the corresponding 'Planck energy scale', since they are described by the same effective theory.

3.2.4 Vacuum energy of weakly interacting Bose gas

In systems with conservation of the particle number two kinds of vacuum energy can be determined: the vacuum energy $\langle\mathcal{H}-\mu\mathcal{N}\rangle_\text{vac}$ – the thermodynamic potential expressed in terms of the chemical potential μ; and the vacuum energy

$\langle \mathcal{H} \rangle_{\text{vac}}$ – the thermodynamic potential expressed in terms of particle number N. For the weakly interacting Bose gas these vacuum energies come from eqns (3.8), (3.15) and (3.16):

$$\langle \mathcal{H} - \mu \mathcal{N} \rangle_{\text{vac}} = -\frac{\mu^2}{2U}V + \frac{1}{2}\sum_{\mathbf{p}}\left(E(\mathbf{p}) - \frac{p^2}{2m} - mc^2 + \frac{m^3 c^4}{p^2}\right), \quad (3.19)$$

and

$$\langle \mathcal{H} \rangle_{\text{vac}} = E_{\text{vac}}(N) = \frac{1}{2}Nmc^2 + \frac{1}{2}\sum_{\mathbf{p}}\left(E(\mathbf{p}) - \frac{p^2}{2m} - mc^2 + \frac{m^3 c^4}{p^2}\right). \quad (3.20)$$

The last term in both equations m^3c^4/p^2 is added to take into account the perturbative correction to the matrix element U; as a result the sum becomes finite.

Let us compare these equations to the naive estimation of the vacuum energy (2.9) in RQFT. First of all, the effective theory is unable to resolve between two kinds of vacuum energy. Second, none of these two vacuum energies is determined by the zero-point motion of the 'relativistic' phonon field. Of course, eqn (3.19) and eqn (3.20) contain the term $\frac{1}{2}\sum_{\mathbf{p}} E(\mathbf{p})$, which at low energy is $\frac{1}{2}\sum_{\mathbf{p}} cp$, and this can be considered as zero-point energy of relativistic field. But it represents only a part of the vacuum energy, and its separate treatment is meaningless.

This divergent 'zero-point energy' is balanced by three counterterms in eqn (3.20) coming from the microscopic physics, that is why they explicitly contain the microscopic parameter – the mass m of the atom. These counterterms cannot be properly constructed within the effective theory, which is not aware of the existence of the microscopic parameter. Actually the theory of weakly interacting Bose gas can be considered as one of the regularization schemes, which naturally arise in the microscopic physics. After such 'regularization', the contribution of the 'zero-point energy' in eqn (3.20) becomes finite

$$\frac{1}{2}\sum_{\mathbf{p}\ \text{reg}} E(\mathbf{p}) = \frac{1}{2}\sum_{\mathbf{p}} E(\mathbf{p}) - \frac{1}{2}\sum_{\mathbf{p}}\left(\frac{p^2}{2m} + mc^2 - \frac{m^3c^4}{p^2}\right) = \frac{8}{15\pi^2}Nmc^2\frac{m^3c^3}{n\hbar^3},$$
(3.21)

where $n = N/V$ is the particle density in the vacuum. Thus the total vacuum energy in terms of N is

$$E_{\text{vac}}(N) \equiv \int d^3r\, \epsilon(n), \quad (3.22)$$

$$\epsilon(n) = \frac{1}{2}mc^2\left(n + \frac{16}{15\pi^2}\frac{m^3c^3}{\hbar^3}\right) = \frac{1}{2}Un^2 + \frac{8}{15\pi^2\hbar^3}m^{3/2}U^{5/2}n^{5/2}. \quad (3.23)$$

Here, we introduce the vacuum energy density $\epsilon(n)$ as a function of particle density n (note that the 'speed of light' $c(n)$ also depends on n). This energy was first calculated by Lee and Yang (1957).

In this model, the 'zero-point energy', even after 'regularization' (the second terms in eqn (3.23)), is much smaller than the leading contribution to the vacuum energy, which comes from the interaction (the first term in eqn (3.23)).

3.2.5 Fundamental constants and Planck scales

In Sec. 3.1.4 we found that the reasonable Theory of Everything for liquid ^{4}He is described by four 'fundamental' constants. The Theory of Everything for weakly interacting Bose gas also contains four 'fundamental' constants. They can be chosen as $\hbar$, m, c and n. In this theory there is a small parameter, which regulates the perturbation theory in the above procedure of derivation of the low-energy properties of the system. This is $mca_0/\hbar \ll 1$, where a_0 is the interatomic distance related to the particle density n: $a_0 \sim n^{-1/3}$.

In this choice of units, however, the 'speed of light' c is actually computed in terms of more fundamental units, $\hbar$, m, U and n, which characterize the microscopic physics: $c = \sqrt{Un/m}$. That is why according to the Weinberg (1983) criterion it is not fundamental (see also the discussion by Duff et al. (2002) and Okun (2002) and the review paper on fundamental constants by Uzan (2002)). However, the 'speed of light' c becomes the fundamental constant of the effective theory arising in the low-energy corner. A small speed of sound reflects the smallness of the pair interaction U. In the microscopic 'fundamental' units the small parameter is $a_0\sqrt{mU/a_0^3}/\hbar \ll 1$.

Even among these microscopic 'fundamental' units, some constants are more 'fundamental' than others. For instance, in the set $\hbar$, m, U and n, the most fundamental is $\hbar$, since quantum mechanics is fundamental in the high-energy atomic world and does not change in the low-energy world. The least fundamental is the particle density n, which is a dynamical variable rather than a constant. Both c and n are local quantities rather than global ones.

The 'fundamentalness' of constants depends on the energy scale. Using four 'fundamental' constants one can construct two energy parameters, which play the role of the Planck energy scales:

$$E_{\text{Planck 1}} = mc^2 \quad , \quad E_{\text{Planck 2}} = \frac{\hbar c}{a_0} , \qquad (3.24)$$

with $E_{\text{Planck 1}} \ll E_{\text{Planck 2}}$ in the Bose gas.

Below the first Planck scale $E \ll E_{\text{Planck 1}}$, the energy spectrum of quasiparticles in eqn (3.17) is linear, $E = cp$, and the effective 'relativistic' quantum field theory arises in the low-energy corner with c being the fundamental constant. This Planck scale not only marks the border where the 'Lorentz' symmetry is violated, but also provides the natural cut-off for the 'zero-point energy' in eqn (3.21), which can be written as

$$\frac{1}{2} \sum_{\mathbf{p} \text{ reg}} E(\mathbf{p}) = \frac{8}{15\pi^2 \hbar^3} \sqrt{-g} E_{\text{Planck 1}}^4 . \qquad (3.25)$$

Here $g = -1/c^6$ is the determinant of the acoustic 'Minkowski' metric $g_{\mu\nu} = \text{diag}(-1, c^{-2}, c^{-2}, c^{-2})$. This contribution to the vacuum energy has the same structure as the cosmological term in eqn (2.11), with the factor $8/15\pi^2$ provided by the microscopic theory. The Planck mass corresponding to the first Planck scale $E_{\text{Planck 1}}$ is the mass m of Bose particles that comprise the vacuum.

The second Planck scale $E_{\text{Planck 2}}$ marks the border where the discreteness of the vacuum becomes important: the microscopic parameter which enters this scale is the number density of particles and thus the distance between the particles in the vacuum. This scale corresponds to the Debye temperature in solids. In the model of weakly interacting particles one has $E_{\text{Planck 1}} \ll E_{\text{Planck 2}}$; this shows that the distance between the particles in the vacuum is so small that quantum effects are stronger than interactions. This is the limit of strong correlations and weak interactions. Because of that, the leading term in the vacuum energy density is

$$\frac{1}{2}Un^2 = \frac{1}{2\hbar^3}\sqrt{-g}E_{\text{Planck 2}}^3 E_{\text{Planck 1}} \,. \tag{3.26}$$

It is much larger than the 'conventional' cosmological term in eqn (3.25). This example clearly shows that the naive estimate of the vacuum energy in eqn (2.9), as the zero-point energy of relativistic bosonic fields or as the energy of the Dirac sea in the case of fermionic fields, can be wrong.

3.2.6 Vacuum pressure and cosmological constant

The pressure in the vacuum state is the variation of the vacuum energy over the vacuum volume at a given number of particles:

$$P = -\frac{d\langle \mathcal{H}\rangle_{\text{vac}}}{dV} = -\frac{d(V\epsilon(N/V))}{dV} = -\epsilon(n) + n\frac{d\epsilon}{dn} = -\frac{1}{V}\langle \mathcal{H} - \mu\mathcal{N}\rangle_{\text{vac}} \,. \tag{3.27}$$

In the last equation we used the fact that the quantity $d\epsilon/dn$ is the chemical potential μ of the system, which can be obtained from the equilibrium condition for the vacuum state at $T = 0$. The equilibrium state of the liquid at $T = 0$ (equilibrium vacuum) corresponds to the minimum of the energy functional $\int d^3 r \epsilon(n)$ as a function of n at a given number $N = \int d^3 r n$ of bare atoms. This corresponds to the extremum of the following thermodynamic potential

$$\int d^3 r \tilde{\epsilon}(n) \equiv \int d^3 r (\epsilon(n) - \mu n) = \langle \mathcal{H} - \mu\mathcal{N}\rangle_{\text{vac}} \,, \tag{3.28}$$

where the chemical potential μ plays the role of the Lagrange multiplier. In equilibrium one has

$$\frac{d\tilde{\epsilon}}{dn} = 0 \quad \text{or} \quad \frac{d\epsilon}{dn} = \mu \,. \tag{3.29}$$

It is important that it is the thermodynamic potential $\tilde{\epsilon}$ which provides the action for effective fields arising in the quantum liquids at low energy including the effective gravity. That is why $\tilde{\epsilon}$ is responsible also for the 'cosmological constant'. According to eqn (3.27) the pressure of the vacuum is minus $\tilde{\epsilon}$:

$$P_{\text{vac}} = -\tilde{\epsilon}_{\text{vac eq}} \,. \tag{3.30}$$

The thermodynamic relation between the energy $\tilde{\epsilon}$ and pressure in the ground state of the quantum liquid is the same as the equation of state for the vacuum in RQFT obtained from the Einstein cosmological term in eqn (2.7). Such an equation of state is actually valid for any homogeneous vacuum at $T=0$.

The pressure in the vacuum state of the weakly interacting Bose gas is given by

$$P = \frac{U}{2}n^2 - \frac{8m^{3/2}U^{5/2}}{15\pi^2\hbar^3}n^{5/2} = \frac{1}{2\hbar^3}\sqrt{-g}\left(E^3_{\text{Planck 2}}E_{\text{Planck 1}} - \frac{16}{15\pi^2}E^4_{\text{Planck 1}}\right). \tag{3.31}$$

Two terms in eqn (3.31) represent two partial contributions to the vacuum pressure. The 'zero-point energy' of the phonon field, the second term in eqn (3.31), gives the negative vacuum pressure. This is just what is expected from the bosonic vacuum according to eqn (2.9), which follows from the effective theory and our intuition. However, the magnitude of this negative pressure is smaller than the positive pressure coming from the microscopic 'trans-Planckian' degrees of freedom (the first term in eqn (3.31)). The weakly interacting Bose gas can exist only under external positive pressure.

Equation (3.31) demonstrates the counterintuitive property of the quantum vacuum: using the effective theory one cannot even predict the sign of the vacuum pressure.

3.3 From Bose gas to Bose liquid

3.3.1 *Gas-like vs liquid-like vacuum*

In a real liquid, such as liquid ^{4}He, the interaction between the bare particles (atoms) is not small. It is a strongly correlated and strongly interacting system, where the two Planck scales are of the same order, $mc^2 \sim \hbar c/a_0$. The interaction energy and the energy of the zero-point motion of atoms (which is only partly represented by the zero-point motion of the phonon field) are of the same order in equilibrium. Each of them depends on the particle density n, and it appears that one can find a value of n at which the two contributions to the vacuum pressure compensate each other. This means that the system can be in equilibrium at zero external pressure, $P = 0$, i.e. it can exist as a completely autonomous isolated system without any interaction with the environment: a droplet of quantum liquid in empty space. The ^{4}He atoms in such a droplet do not fly apart as happens for gases, but are held together forming the state with a finite mean particle density n. This density n is fixed by the attractive interatomic interaction and repulsive zero-point oscillations of atoms.

To ensure that atoms do not evaporate from the droplet to the surrounding empty space, the chemical potential, counted from the energy of an isolated ^{4}He atom, must be negative. This is not the case for the weakly interacting Bose gas, where μ is positive, but it is the case in real ^{4}He where $\mu \sim -7.2$ K [510, 114] (Woo 1976; Dobbs 2000).

Thus there are two principal factors distinguishing liquid ^{4}He (or liquid ^{3}He) from the weakly interacting gases. Liquid ^{4}He and ^{3}He can be in equilibrium at zero pressure, and their chemical potentials at $T = 0$ are negative.

3.3.2 Model liquid state

Using our intuitive understanding of the liquid state, let us try to construct a simple model of the quantum liquid. The model energy density describing a stable isolated liquid at $T = 0$ must satisfy the following conditions: (i) it must be attractive (negative) at small n and repulsive (positive) at large n to provide equilibrium density of liquid; (ii) the chemical potential must be negative to prevent evaporation; (iii) the liquid must be locally stable, i.e. either the eigenfrequencies of collective modes must be real, or their imaginary part must be negative. All these conditions can be satisfied if we modify eqn (3.23) for a weakly interacting gas in the following way. Let us change the sign of the first term describing interaction and leave the second term coming from vacuum fluctuations intact, assuming that it is valid even at high density of particles. Due to the attractive interaction at low density the Bose gas collapses forming the liquid state. Of course, this is a rather artificial construction, but it qualitatively describes the liquid state. So we have the following model:

$$\epsilon(n) = -\frac{1}{2}\alpha n^2 + \frac{2}{5}\beta n^{5/2} , \tag{3.32}$$

where $\alpha > 0$ and $\beta > 0$ are fitting parameters. In principle, one can use the exponents of n as other fitting parameters.

Now we can use the equilibrium condition to obtain the equilibrium particle density $n_0(\mu)$ as a function of the chemical potential:

$$\frac{d\epsilon}{dn} = \mu \quad \to \quad -\alpha n_0 + \beta n_0^{3/2} = \mu . \tag{3.33}$$

From the equation of state one finds the pressure as a function of the equilibrium density:

$$P(n_0) = -\tilde{\epsilon}(n_0) = \mu n_0(\mu) - \epsilon(n_0(\mu)) = -\frac{1}{2}\alpha n_0^2 + \frac{3}{5}\beta n_0^{5/2} . \tag{3.34}$$

The two contributions to the vacuum pressure cancel each other for the following value of the particle density:

$$n_0(P = 0) = \left(\frac{5\alpha}{6\beta}\right)^2 . \tag{3.35}$$

A droplet of the liquid with such a density will be in complete equilibrium in empty space if $\mu < 0$ and $c^2 > 0$. These conditions are satisfied since the chemical potential and speed of sound are

$$\mu(P = 0) = -\frac{1}{6}n_0\alpha , \tag{3.36}$$

$$mc^2 = \left(\frac{dP}{dn_0}\right)_{P=0} = \left(n\frac{d^2\epsilon}{dn^2}\right)_{P=0} = \frac{7}{8}n_0\alpha = 5.25\,|\mu|\,. \qquad (3.37)$$

Equation $\tilde{\epsilon}(n_0) = -P = 0$ means that the vacuum energy density which is relevant for effective theories is zero in the equilibrium state at $T = 0$. This corresponds to zero cosmological constant in effective gravity (see Sec. 3.3.4 below).

3.3.3 Real liquid state

The parameters of liquid ^{4}He at $P = 0$ have been calculated using the exact microscopic Theory of Everything (3.2) with the realistic pair potential (Woo 1976; Dobbs 2000). The many-body ground state wave function of ^{4}He atoms has been constructed at $P = 0$ (and thus with zero vacuum energy $\tilde{\epsilon} = \epsilon - \mu n = 0$). It gave the following values for the equilibrium particle density, chemical potential and speed of sound:

$$n_0(P=0) \sim 2 \cdot 10^{22}\,\mathrm{cm}^{-3}\,, \quad \mu(P=0) = \frac{\epsilon(P=0)}{n_0(P=0)} \sim -7\,\mathrm{K}\,,$$
$$c \sim 2.5 \cdot 10^4\,\mathrm{cm\,s}^{-1}\,, \qquad (3.38)$$

in good agreement with experimental values.

The values of the two Planck energy scales in eqn (3.24) are thus

$$E_{\mathrm{Planck}\,1} = mc^2 \sim 30\,\mathrm{K}\,, \quad E_{\mathrm{Planck}\,2} = \frac{\hbar c}{a_0} \sim 5\,\mathrm{K}\,. \qquad (3.39)$$

It is interesting that the ratio $mc^2/|\mu| \sim 4$ is close to the value 5.25 of our oversimplified model in eqn (3.37).

Let us compare the partial pressures in the vacuum of the Bose gas (3.31) and in that of the liquid state (3.34). The quantum zero-point energy produces a negative vacuum pressure in the Bose gas, but a positive vacuum pressure in the liquid. The contributions to the vacuum pressure from the interaction of bare particles also have opposite signs: interaction leads to a positive vacuum pressure in the Bose gas and to a negative vacuum pressure in the liquid.

The reason is that in the liquid there is a self-sustained equilibrium state which is obtained from the competition of two effects: attractive interaction of bare atoms (corresponding to the negative vacuum pressure in eqn (3.34)) and their zero-point motion which leads to repulsion (corresponding to the positive vacuum pressure in eqn (3.34)). Because of the balance of the two effects in equilibrium, the two Planck scales in eqn (3.39) are of the same order of magnitude.

This example of the quantum vacuum in a liquid shows that even the sign of the energy of the zero-point motion in the vacuum contradicts the naive treatment of the vacuum energy in effective theory, eqn (2.9).

3.3.4 Vanishing of cosmological constant in liquid ^{4}He

In the equilibrium vacuum state of the liquid, which occurs when there is no interaction with the environment (e.g. for a free droplet of liquid ^{4}He), the pressure

$P = 0$. Thus the relevant vacuum energy density $\tilde{\epsilon} \equiv V^{-1} \langle \mathcal{H} - \mu \mathcal{N} \rangle_{\text{vac}} = -P$ in eqn (3.30) is also zero: $\tilde{\epsilon} = 0$. This can now be compared to the cosmological term in RQFT, eqn (3.30). Vanishing of both the energy density and the pressure of the vacuum, $P_\Lambda = -\rho_\Lambda = 0$, means that, if the effective gravity arises in the liquid, the cosmological constant would be identically zero without any fine tuning. The only condition for this vanishing is that the liquid must be in complete equilibrium at $T = 0$ and isolated from the environment.

Note that no supersymmetry is needed for exact cancellation. The symmetry between the fermions and bosons is simply impossible in ^{4}He, since there are no fermionic fields in this Bose liquid.

This scenario of vanishing vacuum energy survives even if the vacuum undergoes a phase transition. According to conventional wisdom, the phase transition, say to the broken-symmetry vacuum state, is accompanied by a change of the vacuum energy, which must decrease in a phase transition. This is what usually follows from the Ginzburg–Landau description of phase transitions. However, if the liquid is isolated from the environment, its chemical potential μ will be automatically adjusted to preserve the zero external pressure and thus the zero energy $\tilde{\epsilon}$ of the vacuum. Thus the relevant vacuum energy is zero above the transition and (after some transient period) below the transition, meaning that $T = 0$ phase transitions do not disturb the zero value of the cosmological constant. We shall see this in the example of phase transitions between two superfluid states at $T = 0$ in Sec. 29.2.

The vacuum energy vanishes in liquid-like vacua only. For gas-like states, the chemical potential is positive, $\mu > 0$, and thus these states cannot exist without an external pressure. That is why one might expect that the solution of the cosmological constant problem can be provided by the mere assumption that the vacuum of RQFT is liquid-like rather than gas-like. However, as we shall see later, the gas-like states suggest their own solution of the cosmological constant problem. One finds that, though the vacuum energy is not zero in the gas-like vacuum, in the effective gravity arising there the vacuum energy is not gravitating if the vacuum is in equilibrium. It follows from the local stability of the vacuum state (see Sec. 7.3.6).

Thus in both cases it is the principle of vacuum stability which leads to the (almost) complete cancellation of the cosmological constant in condensed matter. It is possible that the same arguement of vacuum stability can be applied to the 'cosmological fluid' – the quantum vacuum of the Standard Model. If so, then this generates the principle of non-gravitating vacuum, irrespective of what are the internal variables of this fluid. According to Brout (2001) the role of the variable n in a quantum liquid can be played by the inflaton field – the filed which causes inflation.

Another important lesson from the quantum liquid is that the naive estimate of the vacuum energy density from the zero-point fluctuations of the effective bosonic or fermionic modes in eqn (2.9) never gives the correct magnitude or even sign. The effective theory suggesting such an estimate is valid only in the sub-Planckian region of energies, in other words for fermionic and bosonic zero modes, and knows nothing about the microscopic trans-Planckian degrees of freedom. In quantum liquids, we found that it is meaningless to consider separately such a zero-point energy of effective fields. This contribution to the vacuum energy is ambiguous, being dependent on the regularization scheme. However, irrespective of how the contribution of low-energy degrees of freedom to the cosmological constant is estimated, because of the vacuum stability it will be exactly cancelled by the high-energy degrees without any fine tuning. Moreover, the microscopic many-body wave function used for calculations of the physical parameters of liquid ^{4}He (Woo 1976; Dobbs 2000) contains, in principle, all the information on the quantum vacuum of the system. It automatically includes the quantum fluctuations of the low-energy phononic degrees of freedom, which are usually considered within the effective theory in eqn (2.9). That is why separate consideration of the zero-point energy of the effective fields can lead at best to the double counting of the contribution of fermion zero modes to the vacuum energy.

4
EFFECTIVE THEORY OF SUPERFLUIDITY

4.1 Superfluid vacuum and quasiparticles

4.1.1 Two-fluid model as effective theory of gravity

Here we discuss how the effective theory incorporates the low-energy dynamics of the superfluid vacuum and the dynamics of the system of quasiparticles in Bose liquids. The effective theory – two-fluid hydrodynamics – was developed by Landau and Khalatnikov (see the book by Khalatnikov (1965)). According to the general ideas of Landau a weakly excited state of the quantum system can be considered as a small number of elementary excitations. Applying this to the quantum liquid ^{4}He, the dense system of strongly interacting ^{4}He atoms can be represented in the low-energy corner by a dilute system of weakly interacting quasiparticles (phonons and rotons). In addition, the state without excitation – the ground state, or vacuum – has its own degrees of freedom: it can experience the coherent collective motion. The superfluid vacuum can move without friction with a superfluid velocity generated by the gradient of the condensate phase $\mathbf{v}_s = (\hbar/m)\nabla\Phi$. The two-fluid hydrodynamics incorporates the dynamics of the superfluid vacuum, the dynamics of quasiparticles which form the so-called normal component of the liquid, and their interaction.

We can compare this to the Einstein theory of gravity, which incorporates the dynamics of both the gravitational and matter fields, and their interaction. We find that in superfluids, the inhomogeneity of the flow of the superfluid vacuum serves as the effective metric field acting on quasiparticles representing the matter field. Thus the two-fluid hydrodynamics serves as an example of the effective field theory which incorporates both the gravitational field (collective motion of the superfluid background) and the matter (quasiparticle excitations). Though this effective gravity is different from the Einstein gravity (the equations for the effective metric $g_{\mu\nu}$ are different), such effective gravity is useful for the simulation of many phenomena related to quantum field theory in curved space.

4.1.2 Galilean transformation for particles

We have already seen that it is necessary to distinguish between the bare particles and quasiparticles. The particles are the elementary objects of the system on a microscopic 'trans-Planckian' level, these are the atoms of the underlying liquid (^{3}He or ^{4}He atoms). The many-body system of the interacting atoms forms the quantum vacuum – the ground state. The non-dissipative collective motion of the superfluid vacuum with zero entropy is determined by the conservation laws experienced by the atoms and by their quantum coherence in the superfluid state. The quasiparticles are the particle-like excitations above this vacuum state, and serve as elementary particles in the effective theory. The bosonic excitations in superfluid ^{4}He and fermionic and bosonic excitations in superfluid ^{3}He represent matter in our analogy. In superfluids they form the viscous normal component responsible for the thermal and kinetic low-energy properties of superfluids.

The liquids considered here are non-relativistic: under laboratory conditions their velocity is much less than the speed of light. That is why they obey the Galilean transformation law with great precision. Under the Galilean transformation to a coordinate system moving with a velocity $\mathbf{u}$, the superfluid velocity – the velocity of the quantum vacuum – transforms as $\mathbf{v}_s \to \mathbf{v}_s + \mathbf{u}$.

The transformational properties of bare particles comprising the vacuum and quasiparticles (matter) are essentially different. Let us start with bare particles, say ^{4}He atoms. If $\mathbf{p}$ and $E(\mathbf{p})$ are the momentum and energy of the bare particle (atom with mass m) measured in the system moving with velocity $\mathbf{u}$, then from the Galilean invariance it follows that its momentum and energy measured by the observer at rest are correspondingly $\mathbf{p} + m\mathbf{u}$ and $E(\mathbf{p} + m\mathbf{u}) = E(\mathbf{p}) + \mathbf{p} \cdot \mathbf{u} + (1/2)m\mathbf{u}^2$. This transformation law contains the mass m of the bare atom.

However, where the quasiparticles are concerned, one cannot expect that such a characteristic of the microscopic world as the bare mass m enters the transformation law for quasiparticles. Quasiparticles in effective low-energy theory have no information on the trans-Planckian world of atoms comprising the vacuum state. All the information on the quantum vacuum which a low-energy quasiparticle has, is encoded in the effective metric $g_{\mu\nu}$. Since the mass m must drop out, one may expect that the transformation law for quasiparticles is modified putting $m = 0$. Thus the momentum of the quasiparticle must be invariant under the Galilean transformation $\mathbf{p} \to \mathbf{p}$, while the quasiparticle energy must simply be Doppler shifted: $E(\mathbf{p}) \to E(\mathbf{p}) + \mathbf{p} \cdot \mathbf{u}$. Further, using the simplest system – the system of non-interacting atoms – we shall check that this is the correct law.

4.1.3 Superfluid-comoving frame and frame dragging

Such a transformation law allows us to write the energy of a quasiparticle when the superfluid vacuum is moving.

Let us first note that at any point $\mathbf{r}$ of the moving superfluid vacuum (or in any other liquid) one can find the preferred reference frame. It is the local frame in which the superfluid vacuum (liquid) is at rest, i.e. where $\mathbf{v}_s = 0$. In its motion, the superfluid vacuum drags this frame with it. We call this local frame the superfluid-comoving frame. This frame is important, because it determines the invariant characteristics of the liquid: the physical properties of the superfluid vacuum do not depend on its local velocity if they are measured in this local frame. This concerns the quasiparticle energy spectrum too: in the superfluid-comoving frame it does not depend on $\mathbf{v}_s$.

Let $\mathbf{p}$ and $E(\mathbf{p})$ be the $\mathbf{v}_s$-independent quasiparticle momentum and energy measured in the frame comoving with the superfluid. If the superfluid vacuum itself moves with velocity $\mathbf{v}_s$ with respect to some frame of the environment (say, the laboratory frame) then according to the modified Galilean transformation its momentum and energy in this frame will be

$$\tilde{\mathbf{p}} = \mathbf{p} \, , \, \tilde{E}(\mathbf{p}) = E(\mathbf{p}) + \mathbf{p} \cdot \mathbf{v}_\text{s} \, . \tag{4.1}$$

4.1.4 Galilean transformation for quasiparticles

The difference in the transformation properties of bare particles and quasiparticles comes from their different status. While the momentum and energy of bare particles are determined in 'empty' spacetime, the momentum and energy of quasiparticles are counted from that of the quantum vacuum. This difference is essential even in case when the quasiparticles have the same energy spectrum as bare particles. The latter happens for example in the weakly interacting Bose gas: according to eqn (3.17) $E(\mathbf{p})$ approaches the spectrum of bare particles, $E(\mathbf{p}) \to p^2/2m$, in the limit of large momentum $p \gg mc$. At first glance the difference between particles and quasiparticles completely disappears in this limit. Why then are their transformation properties so different?

To explain the difference we consider an even more confusing case – an ideal Bose gas of non-interacting bare particles, i.e. $c = 0$. Now quasiparticles have exactly the same spectrum as particles: $E(\mathbf{p}) = p^2/2m$. Let us start with the ground state of the ideal Bose gas in the superfluid-comoving reference frame. In this frame all N particles comprising the Bose condensate have zero momentum and zero energy. If the Bose condensate moves with velocity $\mathbf{v}_\text{s}$ with respect to the laboratory frame, then its momentum and energy measured in that frame are correspondingly

$$\langle \mathcal{P} \rangle_\text{vac} = Nm\mathbf{v}_\text{s} \, , \tag{4.2}$$

$$\langle \mathcal{H} \rangle_\text{vac} = N \frac{m\mathbf{v}_\text{s}^2}{2} \, . \tag{4.3}$$

Now let us consider the excited state, which is the vacuum state + one quasiparticle with momentum $\mathbf{p}$. In the superfluid-comoving reference frame it is the state in which $N - 1$ particles have zero momenta, while one particle has the momentum $\mathbf{p}$. The momentum and energy of such a state are correspondingly $\langle \mathcal{P} \rangle_\text{vac+1qp} = \mathbf{p}$ and $\langle \mathcal{H} \rangle_\text{vac+1qp} = E(\mathbf{p}) = p^2/2m$. In the laboratory frame the momentum and energy of such a state are obtained by the Galilean transformation

$$\langle \mathcal{P} \rangle_\text{vac+1qp} = (N-1)m\mathbf{v}_\text{s} + (\mathbf{p} + m\mathbf{v}_\text{s}) = \langle \mathcal{P} \rangle_\text{vac} + \mathbf{p} \, , \tag{4.4}$$

$$\langle \mathcal{H} \rangle_\text{vac+1qp} = (N-1)\frac{m\mathbf{v}_\text{s}^2}{2} + \frac{(\mathbf{p} + m\mathbf{v}_\text{s})^2}{2m} = \langle \mathcal{H} \rangle_\text{vac} + E(\mathbf{p}) + \mathbf{p} \cdot \mathbf{v}_\text{s} \, . \tag{4.5}$$

The right-hand sides of eqns (4.4) and (4.5) reproduce eqn (4.1) for a quasiparticle spectrum in a moving vacuum. They show that, since the energy and the momentum of quasiparticles are counted from that of the quantum vacuum, the transformation properties of quasiparticles are different from the Galilean transformation law. The part of the Galilean transformation which contains the mass of the atom is absorbed by the quantum vacuum.

The right-hand sides of eqns (4.4) and (4.5) have universal character and remain valid for the interacting system as well (provided, of course, that the quasiparticle energy spectrum $E(\mathbf{p})$ in the superfluid-comoving frame remains well determined, i.e. provided that quasiparticles still exist).

4.1.5 Broken Galilean invariance. Momentum vs pseudomomentum

The modified Galilean transformation law for quasiparticles is a consequence of the fact that the mere presence of the underlying liquid (the superfluid vacuum) breaks the Galilean invariance for quasiparticles. The Galilean invariance is a true symmetry for the total system only, i.e. for the quantum vacuum + quasiparticles. But it is not a true symmetry for 'matter' only, i.e. for the subsystem of quasiparticles. Thus in superfluid ^{4}He two symmetries are broken: the superfluid vacuum breaks the global $U(1)$ symmetry and the Galilean invariance.

On the other hand, the homogeneous quantum vacuum does not violate the translational invariance. As a result there are two different types of translational invariance as discussed by Stone (2000b, 2002): (i) The energy of a box which contains the whole system – the superfluid vacuum and quasiparticles – does not depend on the position of the box in the empty space. This is the original symmery of the empty space. (ii) One can shift the quasiparticle system with respect to the quantum vacuum, and if the box is big enough the energy of the whole system also does not change.

In case (i) we used the fact that empty space is translationally invariant. This is the translational invariance for particles comprising the vacuum, i.e. the symmetry of the Theory of Everything in eqn (3.1) with respect to the shift of all particles, $\mathbf{r}_i \to \mathbf{r}_i + \mathbf{a}$. The corresponding conserved quantity is the momentum. Thus the bare particles in empty space are characterized by the momentum.

The operation (ii) is a symmetry operation provided the superfluid vacuum is homogeneous, and thus infinite. In this case the quasiparticles do not see the superfluid vacuum: for them the homogeneous vacuum is an empty space. As distinct from (i) the translational symmetry (ii) is approximate, since it does not follow from the Theory of Everything in eqn (3.1). It becomes exact only in the limit of infinite volume; in the case of the finite volume, by moving a quasiparticle we will finally drag the whole box. The symmetry of quasiparticles under translations with respect to the effectively empty space gives rise to another conserved quantity called the pseudomomentum (Stone 2000b, 2002). The pseudomomentum characterizes excitations of the superfluid vacuum – quasiparticles. The difference between momentum and pseudomomentum is the reason for the different Galilean–transformation properties of particles and quasiparticles.

The Galilean invariance is a symmetry of the underlying microscopic physics of atoms in empty space. It is broken and fails to work for quasiparticles. Instead, it produces the transformation law in eqn (4.1), in which the microscopic quantity – the mass m of bare particles – drops out. This is an example of how the memory of the microscopic physics is erased in the low-energy corner. Furthermore, when the low-energy corner is approached and the effective field theory emerges, these modified transformations gradually become part of the more general coordinate transformations appropriate for the Einstein theory of gravity.

The difference between momentum in empty space and pseudomomentum in the quantum vacuum poses the question. What do we measure when we measure the momentum of the elementary particle such as electron: momentum or pseudomomentum?

4.2 Dynamics of superfluid vacuum

4.2.1 Effective action

In the simplest superfluid, such as bosonic superfluid ^{4}He, the coherent motion of the superfluid vacuum is characterized by two soft collective (hydrodynamic) variables: the mean particle number density $n(\mathbf{r}, t)$ of atoms comprising the liquid, and the condensate phase Φ of atoms in the superfluid vacuum. The superfluid velocity $\mathbf{v}_s(\mathbf{r}, t)$ – the velocity of the coherent motion of superfluid vacuum – is determined by the gradient of the phase: $\mathbf{v}_s = (\hbar/m)\nabla\Phi$, where m is the bare mass of the particle – the mass of the ^{4}He atom. The simple hand-waving argument why it is so comes from the observation that if the whole system moves with constant velocity $\mathbf{v}_s$, the phase Φ of each atom is shifted by $\mathbf{p}\cdot\mathbf{r}/\hbar$, where the momentum $\mathbf{p} = m\mathbf{v}_s$.

The flow of such superfluid vacuum is curl-free: $\nabla \times \mathbf{v}_s = 0$. This is not the rule however. There can be quantized vortices – topologically singular lines at which the phase Φ is not determined. Also, we shall see in Sec. 9.1.4 that in some other superfluids, e.g. in ^{3}He-A, the macroscopic coherence is more complicated: it is not determined by the phase alone. As a result the flow of superfluid vacuum is not potential: it can have a continuous vorticity, $\nabla \times \mathbf{v}_s \neq 0$.

The particle number density $n(\mathbf{r}, t)$ and the condensate phase $\Phi(\mathbf{r}, t)$ are canonically conjugated variables (like angular momentum and angle). This observation dictates the effective action for the vacuum motion at $T = 0$:

$$S_G = \int d^4x \left(\hbar n \dot{\Phi} + \frac{1}{2} m n \mathbf{v}_s^2 + \epsilon(n) \right) \ , \ \mathbf{v}_s = \frac{\hbar}{m} \nabla \Phi \ . \tag{4.6}$$

The last two terms represent the energy density of the liquid: the kinetic energy of superflow and the vacuum energy density as a function of particle density, $\epsilon(n)$. The factor in front of the kinetic energy of the flow of the vacuum is dictated by the Galilean invariance.

Since the variables $\mathbf{v}_s(\mathbf{r}, t)$ and $n(\mathbf{r}, t)$ provide the effective metric for quasiparticle motion, the above effective action is the superfluid analog of the Einstein action for the metric in eqn (2.2). The effective action in eqn (4.6) does not depend much on the microscopic physics. The contribution of the Theory of Everything is very mild: it only determines the function $\epsilon(n)$ (after extensive numerical simulation). The structure of the effective theory is robust to details of the interaction between the original particles, unless the interaction is so strong that the system undergoes a quantum phase transition to another universality class: it can become a crystal, a quantum liquid without superfluidity, a superfluid crystal, etc.

4.2.2 Continuity and London equations

The variation of the effective action over Φ and n gives the closed system of the phenomenological equations which govern the dynamics of the quantum vacuum at $T = 0$. These are the continuity equation, which manifests the conservation law for the particle number in the liquid:

$$\frac{\partial n}{\partial t} + \nabla \cdot (n\mathbf{v}_s) = 0 , \qquad (4.7)$$

where $\mathbf{J} = n\mathbf{v}_s$ is the particle current; and the London equation:

$$\hbar\dot{\Phi} + \frac{1}{2}m\mathbf{v}_s^2 + \frac{\partial \epsilon}{\partial n} = 0 . \qquad (4.8)$$

Taking the derivative of both sides of eqn (4.8) one obtains the more familiar Euler equation for an ideal irrotational liquid:

$$\frac{\partial \mathbf{v}_s}{\partial t} = -\nabla \left(\frac{1}{2}\mathbf{v}_s^2 + \frac{1}{m}\frac{\partial \epsilon}{\partial n} \right) . \qquad (4.9)$$

From eqn (4.8) it follows that the value of the effective action (4.6) calculated at its extremum is

$$S_G(\text{extremum}) = \int d^4x \left(-n\frac{\partial \epsilon}{\partial n} + \epsilon(n) \right) = \int d^4x \, \tilde{\epsilon}(n) = -\int d^4x \, P , \quad (4.10)$$

where P is the pressure in the liquid. The action in equilibrium coincides with the thermodynamic potential of a grand canonical ensemble, which again shows that $\tilde{\epsilon} = \epsilon(n) - \mu n$ is the relevant vacuum energy density responsible for the cosmological constant in quantum liquids.

If the liquid is in complete equilibrium at $T = 0$ in the absence of any contact with the environment, i.e. at zero external pressure, and if the liquid is homogeneous, one finds that the value of the effective action is exactly zero. This supports our conclusion that the cosmological constant in the equilibrium quantum vacuum in the absence of inhomogeneity and matter fields is exactly zero without any fine tuning.

4.3 Normal component – 'matter'

4.3.1 Effective metric for matter field

The structure of the quasiparticle spectrum in superfluid ^{4}He becomes more and more universal the lower the energy. In the low-energy limit one obtains the linear spectrum, $E(\mathbf{p}, n) \to c(n)p$, which characterizes the phonon modes – quanta of sound waves. Their spectrum depends only on the 'fundamental constant', the speed of 'light' $c(n)$ obeying $c^2(n) = (n/m)(d^2\epsilon/dn^2)$. All other information on the microscopic atomic nature of the liquid is lost. Since phonons have long wave-length and low frequency, their dynamics is within the responsibility of the effective theory. The effective theory is unable to describe the high-energy part of the spectrum – the rotons – which can be determined in a fully microscopic theory only.

The action for sound wave perturbations of the vacuum, propagating above the smoothly varying superfluid background, can be obtained from the action (4.6) by decomposition of the vacuum variables, the condensate phase Φ and the particle density n, into the smooth and fluctuating parts: $\Phi \to \Phi + (m/\hbar)\tilde{\Phi}$, $n \to n + \tilde{n}$, $\mathbf{v}_s \to \mathbf{v}_s + \nabla\tilde{\Phi}$ (Unruh 1981, 1995; Visser 1998; Stone 2000a). Substituting this into the action (4.6) one finds that the linear terms in perturbations, $\tilde{\Phi}$ and $\tilde{n}$, are canceled due to the equations of motion for the vacuum fields, and one has

$$S = S_G + S_M \,, \qquad (4.11)$$

where the action for the 'matter' field is

$$S_M = \frac{m}{2} \int d^4 x \, n \left((\nabla\tilde{\Phi})^2 - \frac{\left(\dot{\tilde{\Phi}} + (\mathbf{v}_s \cdot \nabla)\tilde{\Phi}\right)^2}{c^2} \right) \qquad (4.12)$$

$$+ \int d^4 x \, \frac{mc^2}{2n} \left(\tilde{n} - \frac{nm}{\hbar c^2}\left(\dot{\tilde{\Phi}} + \mathbf{v}_s \cdot \nabla\tilde{\Phi}\right) \right)^2 \,. \qquad (4.13)$$

Variation of this action over $\tilde{n}$ eliminates the second term, eqn (4.13). The remaining action in eqn (4.12) can be written in terms of the effective metric induced by the smooth velocity and density fields of the vacuum:

$$S_M = \frac{1}{2} \int d^4 x \sqrt{-g} g^{\mu\nu} \partial_\mu \tilde{\Phi} \partial_\nu \tilde{\Phi} \,. \qquad (4.14)$$

The effective Riemann metric experienced by the sound wave, the so-called acoustic metric, is simulated by the smooth parts describing the vacuum fields (here and below $\mathbf{v}_s$, n and c mean the smooth parts of the velocity, density and 'speed of light', respectively). The contravariant components of the acoustic metrics are

$$g^{00} = -\frac{1}{mnc} \,, \quad g^{0i} = -\frac{1}{mnc} v_s^i \,, \quad g^{ij} = \frac{1}{mnc}(c^2 \delta^{ij} - v_s^i v_s^j) \,. \qquad (4.15)$$

The non-diagonal elements of this acoustic metric are provided by the superfluid velocity.

The inverse (covariant) metric is

$$g_{00} = -\frac{nm}{c}(c^2 - \mathbf{v}_s^2) \,, \quad g_{ij} = \frac{mn}{c}\delta_{ij} \,, \quad g_{0i} = -g_{ij} v_s^j \,, \quad \sqrt{-g} = \frac{m^2 n^2}{c} \,. \qquad (4.16)$$

It gives rise to the effective spacetime experienced by the sound waves. This provides a typical example of how an enhanced symmetry and an effective Lorentzian metric appear in condensed matter in the low-energy corner. In this particular example the matter is represented by the fluctuations of the 'gravity field' itself, i.e. phonons play the role of gravitons, and the matter consists of gravitons only.

In the more complicated quantum liquids, such as ^{3}He-A, the relevant matter fields are fermions and gauge bosons.

The second quantization of the Lagrangian for the matter field gives the energy spectrum of sound wave quanta – phonons – determined by the effective metric:

$$g^{\mu\nu} p_\mu p_\nu = 0 , \quad (4.17)$$

or

$$-(\tilde{E} - \mathbf{p} \cdot \mathbf{v}_s)^2 + c^2 p^2 = 0 . \quad (4.18)$$

This gives the following quasiparticle energy in the laboratory frame:

$$\tilde{E}(\mathbf{p}, \mathbf{r}) = E(\mathbf{p}, n(\mathbf{r})) + \mathbf{p} \cdot \mathbf{v}_s(\mathbf{r}) . \quad (4.19)$$

This is in accordance with eqn (4.1), which states that due to the modified Galilean invariance, the quasiparticle energy in the laboratory frame can be obtained by the Doppler shift from the invariant quasiparticle energy $E(\mathbf{p}, n(\mathbf{r}))$ in the superfluid-comoving frame, i.e. in the frame locally comoving with the superfluid vacuum.

4.3.2 External and inner observers

The action (4.14) for the matter field obeys Lorentz invariance and actually general covariance: invariance under general coordinate transformations. Of course, the 'gravitational' part (4.6) of the action, which governs the dynamics of the $g^{\mu\nu}$-field, lacks general covariance. But if the gravitational field is established (it does not matter from which equations), the motion of a quasiparticle in the effective metric field is determined by the same relativistic equations as the motion of a particle in a real gravitational field. That is why many effects experienced by a particle (and by low-energy degrees of quantum vacuum) in the non-trivial metric field can be reproduced using the inhomogeneous quantum liquid.

Since there are particles and quasiparticles, there are two observers, 'external' and 'inner', who belong to different worlds. An 'external' observer is made of particles and belongs to the microscopic world. This is, for example, the experimentalist who lives in the 'trans-Planckian' Galilean world of atoms. The world is Galilean because typically all the relevant velocities related to the quantum liquids are non-relativistic and thus the quantum liquid itself obeys Galilean physics. The atoms of the liquid and an external observer live in absolute Galilean space.

An 'inner' observer is made of low-energy quasiparticles and lives in the effective 'relativistic' world. For the inner observer the liquid in its ground state is an empty space. This observer views the smooth inhomogeneity of the underlying liquid as the effective spacetime in which free quasiparticles move along geodesics. This spacetime, determined by the acoustic metric, does not reflect the real absolute space and absolute time of the world of atoms. In measurements here, the inner observer uses clocks and rods also made of the 'relativistic' low-energy quasiparticles. This observer can synchronize such clocks using the sound signals, and define distances by an acoustic radar procedure (Liberati et al. 2002). The clocks and rods are 'flexible', being determined by the local acoustic metric, as distinct from 'rigid' clocks and rods used by an external observer.

Let us suppose that both observers are at the same point **r** of space and both are at rest in the laboratory frame. Do they see the world in the same way? Let two events at this point **r** occur at moments t_1 and t_2. What is the time interval between the events from the point of view of the two observers? For the external observer the time between two events is simply $\Delta t = t_2 - t_1$. For the inner observer this is not so simple: the observer's watches are sensitive to the flow velocity of the superfluid vacuum. The time between two events (the proper time) is $\Delta \tau = \Delta t \sqrt{-g_{00}} \propto \Delta t \sqrt{1 - v_s^2/c^2}$. The closer the velocity to the speed of sound/light, the slower are the clocks. When the velocity exceeds c, the inner observer cannot be at rest in the laboratory frame, and will be dragged by the vacuum flow.

4.3.3 *Is the speed of light a fundamental constant?*

When an external observer measures the propagation of 'light' (sound, or other massless low-energy quasiparticles), he or she finds that the speed of light is coordinate-dependent. Moreover, it is anisotropic: for instance, it depends on the direction of propagation with respect to the flow of the superfluid vacuum.

On the contrary, the inner observer always finds that the 'speed of light' (the maximum attainable speed for low-energy quasiparticles) is an invariant quantity. This observer does not know that this invariance is the result of the flexibility of the clocks and rods made of quasiparticles: the physical Lorentz–Fitzgerald contraction of length of such a rod and the physical Lorentz slowing down of such a clock (the time dilation) conspire to produce an effective special relativity emerging in the low-energy corner. These physical effects experienced by low-energy instruments do not allow the inner observer to measure the 'ether drift', i.e. the motion of the superfluid vacuum: the Michelson–Morley-type measurements of the speed of massless quasiparticles in moving 'ether' would give a negative result. The low-energy rods and clocks also follow the anisotropy of the vacuum and thus cannot record this anisotropy. As a result, all the inner observers would agree that the speed of light is the fundamental constant. Living in the low-energy corner, they are unable to believe that in the broader world the external observer finds that, say, in ^{3}He-A the 'speed of light' varies from about 3 cm s^{-1} to 100 m s^{-1} depending on the direction of propagation.

The invariance of the speed of sound in inhomogeneous, anisotropic and moving liquid as measured by a local inner observer is very similar to the invariance of the speed of light in special and general relativity. In the same manner, the invariance of the speed holds only if the measurement is purely local. If the measurement is extended to distances at which the gradients of c and $\mathbf{v}_s$ (the gravitational field) become important, the measured speed of light differs from its local value. It is called the 'coordinate speed of light' in general relativity.

4.3.4 'Einstein equations'

The action in eqn (4.11) with the action for the matter field in eqn (4.14) is the analog of the Einstein action for the gravity field $g^{\mu\nu}$ expressed in terms of n and Φ, and for the matter field $\tilde{\Phi}$. Variation of this 'Einstein action' over vacuum variables n and Φ gives the analog of the Einstein equations:

$$0 = \frac{\delta S}{\delta \Phi} = \frac{\delta S_{\rm G}}{\delta \Phi} + \frac{\delta S_{\rm M}}{\delta g^{\mu\nu}} \frac{\delta g^{\mu\nu}}{\delta \Phi} \rightarrow$$

$$-m\left(\dot{n} + \nabla(n\mathbf{v}_{\rm s})\right) - \frac{1}{2}\nabla\left(\frac{\partial g^{\mu\nu}}{\partial \mathbf{v}_{\rm s}} T^{\rm M}_{\mu\nu} \sqrt{-g}\right) = 0 , \qquad (4.20)$$

$$0 = \frac{\delta S}{\delta n} = \frac{\delta S_{\rm G}}{\delta n} + \frac{\delta S_{\rm M}}{\delta g^{\mu\nu}} \frac{\delta g^{\mu\nu}}{\delta n} \rightarrow$$

$$m\dot{\Phi} + \frac{1}{2}m\mathbf{v}_{\rm s}^2 + \frac{d\epsilon}{dn} + \frac{1}{2}\frac{\partial g^{\mu\nu}}{\partial n} T^{\rm M}_{\mu\nu} \sqrt{-g} = 0 . \qquad (4.21)$$

Here the energy–momentum tensor of matter is

$$T^{\rm M}_{\mu\nu} = \frac{2}{\sqrt{-g}} \frac{\delta S_{\rm matter}}{\delta g^{\mu\nu}} = \partial_\mu \tilde{\Phi} \partial_\nu \tilde{\Phi} - \frac{1}{2} g_{\mu\nu} g^{\alpha\beta} \partial_\alpha \tilde{\Phi} \partial_\beta \tilde{\Phi} . \qquad (4.22)$$

This energy-momentum tensor obeys the covariant conservation law $T^{\mu\rm M}_{\nu;\mu} = 0$. As distinct from the general relativity the covariant conservation does not follow from the 'Einstein equations' (4.20) and (4.21) because of the lack of covariance. It follows from the covariant equation for the matter field obtained by variation of the action over the matter variable $\tilde{\Phi}$:

$$\frac{\delta S}{\delta \tilde{\Phi}} = 0 \rightarrow \partial_\nu \left(\sqrt{-g} g^{\mu\nu} \partial_\mu \tilde{\Phi}\right) = 0 . \qquad (4.23)$$

For perturbations propagating in the background of a conventional fluid the covariant conservation law has been derived by Stone (2000b).

5
TWO-FLUID HYDRODYNAMICS

5.1 Two-fluid hydrodynamics from Einstein equations

At non-zero, but small temperature the 'matter' consists of quanta of the $\tilde{\Phi}$-field – phonons (or gravitons) – which form a dilute gas. In superfluids, this gas of quasiparticles represents the so-called normal component of the liquid. This component bears all the entropy of the liquid. In a local equilibrium, the normal component is characterized by temperature T and its velocity $\mathbf{v}_\mathrm{n}$. The two-fluid hydrodynamics is the system of equations describing the motion of the superfluid vacuum and normal component in terms of four collective variables n, $\mathbf{v}_\mathrm{s}$ (or Φ), T and $\mathbf{v}_\mathrm{n}$.

It is not an easy problem to write the action for matter in terms of T and $\mathbf{v}_\mathrm{n}$, simply because the action for collective variables is not necessarily well-defined. That is why, in the same way as in general relativity, we shall use the 'Einstein equations' (4.20) and (4.21) directly without invoking the 'Einstein action'. Since these equations do not automatically produce the covariant conservation law (2.5), they must be supplemented by the equation $T^{\mu\mathrm{M}}_{\nu;\mu} = 0$, which is actually the equation for the matter field expressed in terms of the collective variables T and $\mathbf{v}_\mathrm{n}$.

5.2 Energy–momentum tensor for 'matter'

5.2.1 *Metric in incompressible superfluid*

To complete the derivation of the two-fluid hydrodynamics in the 'relativistic' regime, we must express the tensor $T^{\mu\mathrm{M}}_{\nu;\mu}$ in terms of collective variables. For simplicity we consider an incompressible liquid, where the particle density n is constant, and the gravity is simulated by the superfluid velocity only. Since we ignore the spacetime dependence of the density n and thus of the speed of sound c, the constant factor mnc can be removed from the effective acoustic metric in eqns (4.15–4.16). Keeping in mind that later we shall deal with the anisotropic superfluid ^{3}He-A, we generalize the effective metric to incorporate the anisotropy:

$$g^{00} = -1 \, , \ g^{0i} = -v_\mathrm{s}^i \, , \ g^{ij} = g^{ij}_\mathrm{SCF} - v_\mathrm{s}^i v_\mathrm{s}^j \, , \tag{5.1}$$

$$g_{00} = -1 + g_{ij} v_\mathrm{s}^i v_\mathrm{s}^j \, , \ g_{ij} g^{jk}_\mathrm{SCF} = \delta^k_i \, , \ g_{0i} = -g_{ij} v_\mathrm{s}^j \, ,$$

$$\sqrt{-g} = \left(\det g^{ij}_\mathrm{SCF} \right)^{-1/2} . \tag{5.2}$$

Here g_{SCF}^{ik} is the effective (acoustic) metric in the superfluid-comoving frame (SCF). It determines the general form of the energy spectrum of quasiparticles in the superfluid-comoving frame:

$$E^2(\mathbf{p}) = p_i p_k g_{SCF}^{ik} . \qquad (5.3)$$

An inverse matrix $(g_{SCF}^{ik})^{-1}$ gives the effective covariant spatial metric g_{ik} in the laboratory frame.

In the isotropic superfluid ^{4}He one has

$$g_{SCF}^{ik} = c^2 \delta^{ik} , \quad g_{ik} = c^{-2} \delta_{ik} , \quad \sqrt{-g} = \frac{1}{c^3} , \qquad (5.4)$$

while in the anisotropic ^{3}He-A one has

$$g_{SCF}^{ik} = c_\parallel^2 \hat{l}^i \hat{l}^k + c_\perp^2 \left(\delta^{ik} - \hat{l}^i \hat{l}^k \right) , \quad g_{ik} = \frac{1}{c_\parallel^2} \hat{l}^i \hat{l}^k + \frac{1}{c_\perp^2} \left(\delta^{ik} - \hat{l}^i \hat{l}^k \right) , \quad \sqrt{-g} = \frac{1}{c_\parallel c_\perp^2} , \qquad (5.5)$$

where $\hat{\mathbf{l}}$ is the direction of the axis of uniaxial anisotropy.

In anisotropic superfluids the 'speed of light' – the maximum attainable velocity of quasiparticles – depends on the direction of propagation. In ^{3}He-A it varies between the speed $c_\perp \sim 3$ cm s^{-1} for quasiparticles propagating in a direction transverse to $\hat{\mathbf{l}}$ and the speed $c_\parallel \sim 60$ m s^{-1} along $\hat{\mathbf{l}}$.

5.2.2 Covariant and contravariant components of 4-momentum

Let us write down covariant and contravariant components of the 4-momentum of quasiparticles – the massless 'relativistic' particles, whose spectrum in the superfluid-comoving frame is given by eqn (5.3):

$$p_\mu = (p_0, p_i) , p_0 = -\tilde{E}(\mathbf{p}) = -E(\mathbf{p}) - \mathbf{p} \cdot \mathbf{v}_s , \qquad (5.6)$$

$$p^0 = g^{0\mu} p_\mu = E(\mathbf{p}) , \quad p^i = g^{i\mu} p_\mu . \qquad (5.7)$$

We also introduce the 4-vector of the group velocity of quasiparticles, $v_G^\mu = p^\mu/E$, and $v_{G\mu} = p_\mu/E$:

$$v_G^\mu = (1, v_G^i) , \quad v_G^i = \frac{\partial \tilde{E}}{\partial p_i} = \frac{\partial E}{\partial p_i} + v_s^i , \quad v_{G\mu} = g_{\mu\nu} v_G^\nu , \qquad (5.8)$$

$$v_{Gi} = g_{ij} \frac{\partial E}{\partial p_j} , v_{G0} = -1 - v_{Gi} v_s^i , \quad v_G^\mu v_{G\mu} = -1 + g_{ij} \frac{\partial E}{\partial p_i} \frac{\partial E}{\partial p_j} = 0 . \qquad (5.9)$$

The group velocity is a null vector as well as a 4-momentum: $v_G^\mu v_{G\mu} = p_\mu p^\mu = 0$.

The physical meaning of the covariant and contravariant vectors is fairly evident: $p^0 = E(\mathbf{p})$ and $-p_0 = E(\mathbf{p}) + \mathbf{p} \cdot \mathbf{v}_s$ represent the invariant quasiparticle energy in the superfluid-comoving frame and the Doppler-shifted quasiparticle energy in the laboratory frame, respectively. The energy E is an invariant quantity, since, being determined in the superfluid-comoving frame, it is velocity independent. That is why it plays the role of invariant mass under the Galilean transformation from the superfluid-comoving to laboratory frame.

Similarly v_G^i is the group velocity of quasiparticles in the laboratory frame, while v_{Gi} represents the phase velocity in the superfluid-comoving frame. Covariant momentum p_i is the canonical momentum of a quasiparticle, and its quantum mechanical operator expression is $\mathcal{P}_i = -i\hbar\partial_i$. The contravariant momentum $p^i = E v_G^i$ is the energy (mass) E times the group velocity in the laboratory frame.

5.2.3 Energy–momentum tensor of 'matter'

Introducing the distribution function $f(\mathbf{p})$ of quasiparticles, one can represent the energy–momentum tensor of 'matter' in a general relativistic form (Fischer and Volovik 2001; Stone 2000b, 2002)

$$\sqrt{-g}T^\mu{}_\nu = \int \frac{d^3p}{(2\pi\hbar)^3} f v_G^\mu p_\nu \ . \tag{5.10}$$

Here we omit the index M in the energy–momentum tensor everywhere except for the momentum of quasiparticles. Let us write down some components of the energy–momentum tensor which have a definite physical meaning:

$$\sqrt{-g}T^{00} = \int fE \qquad \text{energy density in superfluid-comoving frame}, \tag{5.11}$$

$$-\sqrt{-g}T^0{}_0 = \int f\tilde{E} \qquad \text{energy density in laboratory frame}, \tag{5.12}$$

$$-\sqrt{-g}T^i{}_0 = \int f\tilde{E}v_G^i \qquad \text{energy flux in laboratory frame}, \tag{5.13}$$

$$\sqrt{-g}T^0{}_i = \int f p_i = P_i^{\mathrm{M}} \qquad \text{momentum density in either frame}, \tag{5.14}$$

$$\sqrt{-g}T^k{}_i = \int f p_i v_G^k \qquad \text{momentum flux in laboratory frame}. \tag{5.15}$$

Here $\int$ means $\int d^3p/(2\pi\hbar)^3$.

5.2.4 Particle current and quasiparticle momentum

Let us insert the energy–momentum tensor in the first of two Einstein equations, eqn (4.20). Using

$$\frac{1}{2}\frac{\partial g^{\mu\nu}}{\partial \mathbf{v}_s^k}T_{\mu\nu} = -\left(T_{0k} + v_s^k T_{ik}\right) = \left(g^{00}T_{0k} + g^{0i}T_{ik}\right) = T^0{}_k = \frac{1}{\sqrt{-g}}P_k^{\mathrm{M}} \ , \tag{5.16}$$

one finds that quasiparticles modify the conservation law for the particle number – the continuity equation (4.7):

$$m\frac{\partial n}{\partial t} + \nabla\cdot\mathbf{P} = 0 \ , \quad \mathbf{P} = mn\mathbf{v}_s + \mathbf{P}^{\mathrm{M}} \ , \quad \mathbf{P}^{\mathrm{M}}(\mathbf{r}) = \int \frac{d^3p}{(2\pi\hbar)^3} f(\mathbf{p},\mathbf{r})\mathbf{p} \ . \tag{5.17}$$

Thus in the effective theory there are two contributions to the particle current. The term $n\mathbf{v}_s$ is the particle current transferred coherently by the collective motion of the superfluid vacuum with the superfluid velocity $\mathbf{v}_s$. In equilibrium at $T=0$ this is the only particle current. In the microscopic Galilean world, the momentum $\mathbf{P}$ of the liquid coincides with the mass current, which in the monoatomic liquid is the particle current multiplied by the mass m of the atom: $\mathbf{P}=m\mathbf{J}$. That is why, if quasiparticles are excited above the ground state, their momenta $\mathbf{p}$ contribute to the mass current and thus to the particle current, giving rise to the second term in eqn (5.17).

Since the momentum of quasiparticles is invariant under the Galilean transformation, the total density of the particle current,

$$\mathbf{J} = \frac{\mathbf{P}}{m} = n\mathbf{v}_s + \frac{\mathbf{P}^M}{m}, \quad (5.18)$$

transforms in a proper way: $\mathbf{J} \to \mathbf{J} + n\mathbf{u}$.

5.3 Local thermal equilibrium

5.3.1 *Distribution function*

In a local thermal equilibrium the distribution of quasiparticles is characterized by the collective variables – the local temperature T and the local velocity of the quasiparticle gas $\mathbf{v}_n$, which is called the normal component velocity:

$$f_{\mathcal{T}}(\mathbf{p},\mathbf{r}) = \left(\exp \frac{\tilde{E}(\mathbf{p},\mathbf{r}) - \mathbf{p} \cdot \mathbf{v}_n(\mathbf{r})}{T(\mathbf{r})} \pm 1 \right)^{-1}, \quad (5.19)$$

where the + sign is for the fermionic quasiparticles in Fermi superfluids and the minus sign is for the bosonic quasiparticles in Bose superfluids; index $\mathcal{T}$ means thermal equilibrium.

After the energy–momentum tensor of 'matter' is expressed through T and $\mathbf{v}_n$, one finally obtains the closed system of six equations of the two-fluid hydrodynamics for six hydrodynamic variables: n, Φ, T, v_n^i. These are two 'Einstein equations' (4.20) and (4.21), and four equations of the covariant conservation law (2.5).

Since $\tilde{E}(\mathbf{p}) - \mathbf{p} \cdot \mathbf{v}_n = E(\mathbf{p}) - \mathbf{p} \cdot (\mathbf{v}_n - \mathbf{v}_s)$, the distribution of quasiparticles in local equilibrium is determined by the Galilean invariant quantity $\mathbf{v}_n - \mathbf{v}_s \equiv \mathbf{w}$, which is the normal component velocity measured in the frame comoving with the superfluid vacuum (superfluid-comoving frame), called the counterflow velocity.

5.3.2 *Normal and superfluid densities*

In the limit where the counterflow velocity $\mathbf{w} = \mathbf{v}_n - \mathbf{v}_s$ is small, the quasiparticle ('matter') contribution to the liquid momentum and thus to the particle current is proportional to the counterflow velocity:

$$P_i^M = m n_{nik}(v_{nk} - v_{sk}), \quad n_{nik} = -\sum_{\mathbf{p}} \frac{p_i p_k}{m} \frac{\partial f_{\mathcal{T}}}{\partial E}, \quad (5.20)$$

where the tensor n_{nik} is the so-called normal density (or density of the normal component). In this linear in velocities regime, the total current in eqn (5.18) can be represented as the sum of the currents carried by the normal and superfluid velocities

$$P_i = mJ_i = mn_{sik}v_{sk} + mn_{nik}v_{nk} , \quad (5.21)$$

where tensor $n_{sik} = n\delta_{ik} - n_{nik}$ is the so-called superfluid density.

In the isotropic superfluids ^{4}He and ^{3}He-B, where the quasiparticle spectrum in the superfluid-comoving frame in eqn (4.19) is isotropic, $E(\mathbf{p}) = E(p)$, the normal density is an isotropic tensor, $n_{nik} = n_n\delta_{ik}$. In superfluid ^{3}He-A the normal density is a uniaxial tensor which reflects a uniaxial anisotropy of the quasiparticle spectrum (5.5). At $T = 0$ the quasiparticles are frozen out and one has $n_{nik}(T=0) = 0$ and $n_{sik}(T=0) = n\delta_{ik}$ in all monoatomic superfluids. Further we shall see that this is valid only in the linear regime.

5.3.3 Energy–momentum tensor

Substituting the distribution function (5.19) into the energy–momentum tensor of the quasiparticle system (matter), one obtains that it is determined by the generic thermodynamic potential density (the pressure) defined in the superfluid-comoving frame

$$\Omega = \mp T \sum_s \int \frac{d^3p}{(2\pi\hbar)^3} \ln(1 \mp f) , \quad (5.22)$$

with the upper sign for fermions and the lower sign for bosons. For bosonic quasiparticles one has

$$\Omega = \frac{\pi^2}{90\hbar^3} T_{\text{eff}}^4 \sqrt{-g} , \quad T_{\text{eff}} = \frac{T}{\sqrt{1-w^2}} , \quad (5.23)$$

where the renormalized effective temperature T_{eff} absorbs all the dependence on the effective metric and on two velocities (normal and superfluid) of the liquid. Here

$$w^2 = g_{ik}w^iw^k , \quad (5.24)$$

where $\mathbf{w} = \mathbf{v}_n - \mathbf{v}_s$ is the counterflow velocity. If the normal component (matter) is made of phonons with the isotropic energy spectrum $E = cp$, one has $w^2 = \mathbf{w}^2/c^2$.

In the laboratory frame the energy–momentum tensor of the 'matter' in local equilibrium has a form which is standard for a gas of massless relativistic particles. It is expressed through the 4-velocity of the 'matter'

$$T^{\mu\nu} = (\varepsilon + \Omega)u^\mu u^\nu + \Omega g^{\mu\nu} , \quad \varepsilon = -\Omega + T\frac{\partial\Omega}{\partial T} = 3\Omega , \quad T^\mu{}_\mu = 0 . \quad (5.25)$$

The 4-velocity of the 'matter', u^α or $u_\alpha = g_{\alpha\beta}u^\beta$, satisfies the normalization equation $u_\alpha u^\alpha = -1$. It is expressed in terms of the superfluid and normal component velocities as

$$u^0 = \frac{1}{\sqrt{1-w^2}}, \quad u^i = \frac{v_n^i}{\sqrt{1-w^2}}, \quad u_i = \frac{g_{ik}w^k}{\sqrt{1-w^2}}, \quad u_0 = -\frac{1 + g_{ik}w^iv_s^k}{\sqrt{1-w^2}} . \quad (5.26)$$

As before for the group velocity of quasiparticles, the contravariant and covariant components, u^i and u_i, are related to the velocity of the normal component of the liquid in the laboratory and superfluid-comoving frames respectively.

5.3.4 Temperature 4-vector

The distribution of quasiparticles in local equilibrium in eqn (5.19) can be expressed via the temperature 4-vector β^μ and thus via the effective temperature T_{eff} introduced in eqn (5.23):

$$f_T = \frac{1}{1 \pm \exp[-\beta^\mu p_\mu]}, \quad \beta^\mu = \frac{u^\mu}{T_{\text{eff}}} = \left(\frac{1}{T}, \frac{\mathbf{v}_n}{T}\right), \quad g_{\mu\nu}\beta^\mu \beta^\nu = -T_{\text{eff}}^{-2}. \tag{5.27}$$

5.3.5 When is the local equilibrium impossible?

According to eqn (5.23) the local equilibrium in superfluids exists only if $w < 1$. At $w > 1$ there is no local stability of the liquid. For the isotropic superfluids, this means that the local equilibrium exists when $|\mathbf{w}| < c$, i.e. the counterflow velocity – the relative velocity between the normal and superfluid components – should not exceed the 'maximum attainable speed' c for quasiparticles.

However, each of two velocities separately, $\mathbf{v}_s$ and $\mathbf{v}_n$, can exceed c. The quantity c is the maximum attainable velocity for quasiparticles in the vacuum but not for the motion of the vacuum itself. This is why the event horizon can be constructed in superfluids: if $\mathbf{v}_s$ exceeds c in some region of liquid, quasiparticles cannot escape from this region.

Later we shall see that for Fermi superfluids the local thermal equilibrium is in principle possible even for $w > 1$, if one takes into account the corrections to the 'relativistic' energy spectrum of quasiparticles at high energy. This will be discussed later in Sec. 32.4 when the phenomena related to event horizons will be considered.

Extending the discussion in Sec. 4.3.3 on the possible fundamentality of the speed of light, we note that the 'speed of light' does not enter explicitly the criterion $w = 1$ for the violation of the local equilibrium condition. This is because none of eqns (5.23–5.26) contains c: it is absorbed by the metric field. Moreover, we know that in anisotropic superfluids there is no unique 'speed of light': as follows from eqn (5.5) the 'speed of light' depends on the direction of propagation. Since the 'speed of light' is not a fundamental quantity, but is determined by the material parameters of the liquid, it cannot enter explicitly into any physical result or equation written in covariant form.

As follows from Sec. 2.1, general relativity satisfies this requirement: all the equations of general relativity can be written in such a way that they do not contain c explicitly. Written in such a form, these equations can be applied to the effective low-energy theory arising in anisotropic ^{3}He-A, where there is no unique speed of light. This concerns the equations of quantum electrodynamics too; we shall discuss this later in Chapter 9 where the effective electrodynamics emerging in ^{3}He-A is considered.

5.4 Global thermodynamic equilibrium

The global thermal equilibrium is the thermodynamic state in which no dissipation occurs. The global equilibrium requires the following conditions: (i) There is a reference frame of environment in which the system does not depend on time, i.e. superfluid velocity field, textures, boundaries, etc., are stationary. This is typically the laboratory frame. But if the walls of container are very far the role of the environment can be played by texture (the texture-comoving frame). (ii) In the environment frame the velocity of the normal component must be zero, $\mathbf{v}_n = 0$. (iii) The temperature T is constant everywhere throughout the system.

For a relativistic system, the true equilibrium with vanishing entropy production is established if the 4-temperature β^μ is a so-called time-like Killing vector. A Killing vector is a 4-vector along which the metric does not change. For instance, if the metric is time independent, then the 4-vector $K^\mu = (1, 0, 0, 0)$ is a Killing vector, since $K^\alpha \partial_\alpha g_{\mu\nu} = 0$. In general, a Killing vector must satisfy the following equations:

$$K^\alpha \partial_\alpha g_{\mu\nu} + (g_{\mu\alpha} \partial_\nu + g_{\nu\alpha} \partial_\mu) K^\alpha = 0 \quad \text{or} \quad K_{\mu;\nu} + K_{\nu;\mu} = 0 \ . \tag{5.28}$$

A time-like Killing vector satisfies in addition the condition $g_{\mu\nu} \beta^\mu \beta^\nu < 0$, so that according to eqn (5.27) the effective temperature T_{eff} is well-defined.

Let us apply these equations to β^μ in the environment frame, where the metric is stationary. The $\mu = \nu = 0$ component of this equation gives $\beta_{0;0} = \beta^i \partial_i g_{00} = 0$. To satisfy this condition in the general case, where g_{00} depends on the space coordinates, one must require that $\beta^i = 0$, i.e. $\mathbf{v}_n = 0$ in the environment frame. Similarly the other components of eqn (5.28) are satisfied when $1/T = \beta^0 =$ constant. Thus the global equilibrium conditions for superfluids can be obtained from the requirement that β^μ is a time-like Killing vector determined using the 'acoustic' metric $g_{\mu\nu}$.

5.4.1 Tolman temperature

From the equilibrium conditions $T =$ constant and $\mathbf{v}_n = 0$ it follows that in global equilibrium, the effective temperature in eqn (5.23) is space dependent according to

$$T_{\text{eff}}(\mathbf{r}) = \frac{T}{\sqrt{1 - v_s^2(\mathbf{r})}} = \frac{T}{\sqrt{-g_{00}(\mathbf{r})}} \ . \tag{5.29}$$

Here $v_s^2 = g_{ik} v_s^i v_s^k$ in general and $v_s^2 = \mathbf{v}_s^2/c^2$ for the isotropic superfluid vacuum.

According to eqn (5.27) the effective temperature T_{eff} corresponds to the 'covariant relativistic' temperature in general relativity. It is an apparent temperature as measured by a local inner observer, who 'lives' in a superfluid vacuum and uses sound for communication as we use light signals. Equation (5.29) is exactly Tolman's (1934) law in general relativity, which shows how the locally measured temperature (T_{eff}) changes in the gravity field in a global equilibrium. The role of the constant Tolman temperature is played by the temperature T of the liquid measured by an external observer living in the Galilean world of the laboratory. This is real thermodynamic temperature since it is constant throughout the entire liquid.

Note that Ω is the pressure created by quasiparticles ('matter'). In superfluids this pressure is supplemented by the pressure of the superfluid component – the vacuum pressure discussed in Sec. 3.2.6, so that the total pressure in equilibrium is

$$P = P_{\rm vac} + P_{\rm matter} = P_{\rm vac} + \frac{\pi^2}{90\hbar^3} T_{\rm eff}^4 \sqrt{-g} \ . \tag{5.30}$$

For the liquid in the absence of an interaction with the environment, the total pressure of the liquid is zero in equilibrium, which means that the vacuum pressure compensates the pressure of matter. In the non-equilibrium situation this compensation is not complete, but the two pressures are of the same order of magnitude. Perhaps this can provide the natural solution of the second cosmological constant problem: why the vacuum energy is on the order of the energy density of matter. A detailed discussion of the cosmological constant problems will be presented in Sec. 29.4.

Finally let us mention that if the quantum liquid has several soft modes a, each described by its own effective Lorentzian metric $g_{\mu\nu}^{(a)}$, the total 'matter' pressure of the liquid is

$$P_{\rm matter} = \frac{\pi^2 T^4}{90\hbar^3} \sum_a \gamma_a \frac{\sqrt{-g^{(a)}}}{\left(g_{00}^{(a)}\right)^2} \ , \tag{5.31}$$

where γ_a are dimensionless quantities depending on spin and the statistics of massless modes: $\gamma = 1$ for a massless scalar field; $\gamma = 2$ for the analog of electromagnetic waves emerging in ^{3}He-A; and $\gamma = (7/4)N_F$ for N_F chiral fermionic quasiparticles also emerging in ^{3}He-A. In the absence of counterflow the sum becomes $\sum_a \gamma_a/c_a^3$, where c_a are speeds of corresponding 'relativistic' quasiparticles.

5.4.2 Global equilibrium and event horizon

According to Tolman's law in eqn (5.29) the global thermal equilibrium is possible if $v_s < 1$, i.e. the superfluid velocity in the environment frame does not exceed the speed of 'light'. Within the discussed relativistic domain, the global thermal equilibrium is not possible in the presence of the ergosurface where g_{00} crosses zero. In the ergoregion beyond the ergosurface $g_{00} > 0$ (i.e. $v_s > 1$) and thus β^μ becomes space-like, $g_{\mu\nu}\beta^\mu\beta^\nu > 0$, so that the effective relativistic temperature $T_{\rm eff}$ in eqn (5.27) is not determined. When the ergosurface, where $v_s(\mathbf{r}) = 1$, is approached the effective relativistic temperature increases leading to the increasing energy of thermally excited quasiparticles.

At some moment the linear 'relativistic' approximation becomes invalid, and the non-linear non-relativistic corrections to the energy spectrum become important. This shows that the ergosurface or horizon is the place where Planckian physics is invoked. In principle, Planckian physics can modify the relativistic criterion for the global thermodynamic stability in such a way that it can be satisfied even beyond the horizon. This can occur in the case of the superluminal dispersion of the quasiparticle energy spectrum, i.e. if the high-energy quasiparticles are propagating with velocity higher than the maximum attainable velocity c of the relativistic low-energy quasiparticles. Though in the low-energy relativistic world β^μ becomes space-like beyond the horizon, it remains time-like on the fundamental level. We shall discuss this in Chapter 32.

In conclusion of this section, the normal part of superfluid ^{4}He fully reproduces the dynamics of relativistic matter in the presence of a gravity field. Though the corresponding 'Einstein equations' for 'gravity' itself are not covariant, by using the proper superflow fields we can simulate many phenomena related to the classical and quantum behavior of matter in curved spacetime, including black-hole physics.

6
ADVANTAGES AND DRAWBACKS OF EFFECTIVE THEORY

6.1 Non-locality in effective theory

6.1.1 Conservation and covariant conservation

As is known from general relativity, the equation $T^\mu{}_{\nu;\mu} = 0$ or

$$\partial_\mu \left(T^\mu{}_\nu \sqrt{-g}\right) = \frac{\sqrt{-g}}{2} T^{\alpha\beta} \partial_\nu g_{\alpha\beta} \tag{6.1}$$

does not represent any conservation in a strict sense, since the covariant derivative is not a total derivative (Landau and Lifshitz 1975). In superfluid ^{4}He it acquires the form

$$\partial_\mu \left(T^\mu{}_\nu \sqrt{-g}\right) = \int \frac{d^3p}{(2\pi\hbar)^3} f \partial_\nu \tilde{E} = P_i^M \partial_\nu v_s^i + \int \frac{d^3p}{(2\pi\hbar)^3} f |\mathbf{p}| \partial_\nu c . \tag{6.2}$$

This does not mean that energy and momentum are not conserved in superfluids. One can check that the momentum and the energy of the whole system (superfluid vacuum + quasiparticles)

$$\int d^3r \left(m n \mathbf{v}_s + \int \frac{d^3p}{(2\pi\hbar)^3} \mathbf{p} f \right) , \quad \int d^3r \left(\frac{m}{2} n v_s^2 + \epsilon(n) + \int \frac{d^3p}{(2\pi\hbar)^3} \tilde{E} f \right) , \tag{6.3}$$

are conserved. For example, for the density of the total momentum of the liquid one has the conservation law

$$\partial_t(P_i) + \nabla_k \Pi_{ik} = 0 , \quad P_i = m n v_{si} + P_i^M , \tag{6.4}$$

with the following stress tensor:

$$\Pi_{ik} = P_i v_{sk} + v_{si} P_k^M + \delta_{ik} \left(n \left(\frac{\partial \epsilon}{\partial n} + \int \frac{d^3p}{(2\pi\hbar)^3} f \frac{\partial E}{\partial n} \right) - \epsilon \right) + \int \frac{d^3p}{(2\pi\hbar)^3} p_k f \frac{\partial E}{\partial p_i} . \tag{6.5}$$

Equation (6.4) together with the corresponding equation for the density of the total energy can be written in the form

$$\partial_\mu \left(T^\mu{}_\nu(\text{vacuum}) + \sqrt{-g} T^\mu{}_\nu(\text{matter})\right) = 0 . \tag{6.6}$$

This is the true conservation law for the energy and momentum, while the covariant conservation law (6.1) or (6.2) simply demostrates that the energy and momentum are not conserved for the quasiparticle subsystem alone: there is an energy and momentum exchange between the vacuum and 'matter'. The right-hand sides of eqn (6.2) or (6.1) represent the 'gravitational' force acting on the 'matter' from the inhomogeneity of the superfluid vacuum, which simulates the gravity field.

6.1.2 *Covariance vs conservation*

In the true conservation law eqn (6.6) the energy–momentum tensor for the vacuum field (gravity) is evidently non-covariant. This can be seen, for example, from eqn (6.3) for energy and momentum. Of course, this happens because the dynamics of the superfluid background is not covariant. However, even for the fully covariant dynamics of gravity in Einstein theory the corresponding quantity – the energy–momentum tensor for the gravitational field – cannot be presented in the covariant form. This is the famous problem of the energy–momentum tensor in general relativity. One must sacrifice either covariance of the theory, or the true conservation law. In general relativity usually the covariance is sacrificed and one introduces the non-covariant energy–momentum pseudotensor for the gravitational field (Landau and Lifshitz 1975).

From the condensed matter point of view, the inconsistency between the covariance and the conservation law for the energy and momentum, is an aspect of the much larger problem of the non-locality of effective theories. Inconsistency between the effective and exact symmetries is one particular example of non-locality. In general relativity the symmetry under translations, which is responsible for the conservation laws for energy and momentum, is inconsistent with the symmetry under general coordinate transformations. From this point of view, this is a clear indication that the Einstein gravity is really an effective theory, with exact translational invariance at the fundamental level and emerging general covariance in the low-energy limit. In the effective theories of condensed matter such paradoxes come from the fact that the description of the many-body system in terms of a few collective fields is always approximate. In many cases the fully local effective theory cannot be constructed, since there is still some exchange with the microscopic degrees of freedom, which is not covered by local theory.

Let us consider some simple examples of condensed matter, where the paradoxes related to the non-locality of the effective theory emerge.

6.1.3 *Paradoxes of effective theory*

There are many examples of apparent inconsistencies in the effective theories of condensed matter: in various condensed matter systems the low-energy dynamics cannot be described by a well-defined local action expressed in terms of the collective variables; the momentum density determined as a variation of the hydrodynamic energy over $\mathbf{v}_s$ does not coincide with the canonical momentum in most condensed matter systems; in the case of an axial (chiral) anomaly, which is also reproduced in condensed matter (Chapter 18), the classical conservation of the baryonic charge is incompatible with quantum mechanics; etc. All such paradoxes are naturally built into the effective theory; they necessarily arise when the fully microscopic description is reduced by coarse graining to a restricted number of collective degrees of freedom.

The paradoxes of the effective theory disappear completely at the fundamental atomic level, sometimes together with the effective symmetries of the low-energy physics. In a fully microscopic description where all the degrees of freedom are taken into account, the dynamics of atoms is fully determined either by the well-defined microscopic Lagrangian which respects all the symmetries of atomic physics, or by the canonical Hamiltonian formalism for pairs of canonically conjugated variables, the coordinates and momenta of atoms. This microscopic 'Theory of Everything' does not contain the above paradoxes. But the other side of the coin is that the 'Theory of Everything' fails to describe the low-energy physics just because of the enormous number of degrees of freedom. In such cases the low-energy physics cannot be derived from first principles without extensive numerical simulations, while the effective theory operating with the restricted number of soft variables can incorporate the most important phenomena of the low-energy physics, which sometimes are too exotic (the quantum Hall effect (QHE) is an example) to be predicted by 'The Theory of Everything' (Laughlin and Pines 2000).

Thus we must choose between the uncomfortable life of microscopic physics without paradoxes, and the comfortable life of the effective theory with its unavoidable paradoxes. Probably this refers to quantum mechanics too.

Let us discuss two examples of the effective theory: the Euler equations for a perfect liquid and the dynamics of ferromagnets.

6.1.4 *No canonical Lagrangian for classical hydrodynamics*

Let us consider the hydrodynamics of a normal liquid. We suppose that there is no superfluid transition up to a very low temperature, so that the liquid is fully normal, $n_n = n$. Then the hydrodynamics of the liquid is described by two variables: the mass density $\rho = mn$ and the velocity $\mathbf{v}_n$, which is now the velocity of the whole liquid (we denote it by $\mathbf{v}$ as in conventional hydrodynamics). As distinct from the superfluid velocity $\mathbf{v}_s$ in superfluid ^{4}He, the velocity $\mathbf{v}$ is not curl-free: $\nabla \times \mathbf{v} \neq 0$. The direct consequence of that is that even the non-dissipative hydrodynamic equations, the Euler and continuity equations, cannot be derived from a local action expressed through the hydrodynamic variables n and $\mathbf{v}$ only. Such action does not exist: the original Lagrange principle is violated in the effective theory after coarse graining.

The only completely local theory of hydrodynamics is presented by the Hamiltonian formalism. In this approach, the hydrodynamic equations are obtained from the Hamiltonian using the Poisson brackets:

$$\partial_t \rho = \{\mathcal{H}, \rho\} \ , \ \partial_t \mathbf{v} = \{\mathcal{H}, \mathbf{v}\} \ . \tag{6.7}$$

The Poisson brackets between the hydrodynamic variables are universal, are determined by the symmetry of the system and do not depend on the Hamiltonian (see the review paper by Dzyaloshinskii and Volovick 1980). In the case of the hydrodynamics of a normal liquid these are (see the book by Khalatnikov (1965) and Novikov 1982)

$$\{\rho(\mathbf{r}_1), \rho(\mathbf{r}_2)\} = 0 , \qquad (6.8)$$

$$\{\mathbf{v}(\mathbf{r}_1), \rho(\mathbf{r}_2)\} = \nabla \delta(\mathbf{r}_1 - \mathbf{r}_2) , \qquad (6.9)$$

$$\{v_i(\mathbf{r}_1), v_j(\mathbf{r}_2)\} = -\frac{1}{\rho} e_{ijk} (\nabla \times \mathbf{v})_k \delta(\mathbf{r}_1 - \mathbf{r}_2) . \qquad (6.10)$$

The Hamiltonian is simply the energy of the liquid expressed in terms of hydrodynamic variables (compare with eqn (4.6)):

$$\mathcal{H} = \int d^3 x \left(\frac{1}{2} \rho \mathbf{v}^2 + \epsilon(\rho) \right) . \qquad (6.11)$$

Then the Hamilton equations (6.7) become continuity and Euler equations:

$$\frac{\partial \rho}{\partial t} + \nabla \cdot (\rho \mathbf{v}) = 0 , \quad \frac{\partial \mathbf{v}}{\partial t} + (\mathbf{v} \cdot \nabla) \mathbf{v} + \nabla \frac{\partial \epsilon}{\partial \rho} = 0 . \qquad (6.12)$$

The hydrodynamic variables do not form pairs of canonically conjugated variables, and thus there is no well-defined Lagrangian which can be expressed in terms of well-defined variables. The action can be introduced, say, in terms of the non-local Clebsch variables which are not applicable for description of the general class of the flow field – the flow with non-zero fluid helicity $\int d^3 x \mathbf{v} \cdot (\nabla \times \mathbf{v})$. The absence of the local action for the soft collective variables in many condensed matter systems (Novikov 1982; Dzyaloshinskii and Volovick 1980) is one of the consequences of the reduction of the degrees of freedom in effective field theory, as compared to a fully microscopic description where the Lagrangian exists at the fundamental level. When the high-energy microscopic degrees are integrated out, the non-locality of the remaining coarse-grained action is a typical phenomenon, which shows up in many faces.

6.1.5 Novikov–Wess–Zumino action for ferromagnets

In ferromagnets, the magnetization vector $\mathbf{M}$ has three components. The odd number of variables cannot produce pairs of canonically conjugated variables, and as a result the action for $\mathbf{M}$ cannot be written as an integral over spacetime $(\mathbf{r}, t)$ of any local integrand. But as in Sec. 6.1.4, instead of the Lagrangian one can use the Hamiltonian description introducing the Poisson brackets,

$$\{M_i(\mathbf{r}_1), M_j(\mathbf{r}_2)\} = -e_{ijk} M_k(\mathbf{r}_1) \delta(\mathbf{r}_1 - \mathbf{r}_2) . \qquad (6.13)$$

$M^2 = \mathbf{M} \cdot \mathbf{M}$ commutes with all three variables, thus the magnitude of magnetization is a constant of motion (if the dissipation is neglected). In slow hydrodynamic motion M equals its equilibrium value, so that the true slow (hydrodynamic) variable is the unit vector $\hat{\mathbf{m}} = \mathbf{M}/M$. Since the 2D manifold of a unit vector $\hat{\mathbf{m}}$ – the sphere S^2 – is compact, this variable also cannot produce a well-defined canonical pair. Of course, one can introduce spherical coordinates (θ, ϕ) of unit vector $\hat{\mathbf{m}}$, and find that $\cos\theta$ and ϕ do form a canonical pair. But these variables are not well defined: the azimuthal angle ϕ is ill defined at the poles of the unit sphere of the $\hat{\mathbf{m}}$-vector, i.e. at $\theta = 0$ and $\theta = \pi$.

There is another way to treat the problem: one introduces the non-local and actually multi-valued action in terms of well-defined variables (Novikov 1982). Such an action is given by the Novikov–Wess–Zumino term, which contains an extra coordinate τ. For ferromagnets this term is (Volovik 1986b, 1987)

$$S_{\text{NWZ}} = \int d^D x \, dt \, d\tau \, M \, \hat{\mathbf{m}} \cdot (\partial_t \hat{\mathbf{m}} \times \partial_\tau \hat{\mathbf{m}}) \,, \tag{6.14}$$

which must be added to the free energy F of the ferromagnet. The integral here is over the $D+1+1$ disk $(\mathbf{r}, t, \tau)$, whose boundary is the physical $D+1$ spacetime $(\mathbf{r}, t)$. Though the action is written in a fictitious $D+1+1$ space, its variation is a total derivative and thus depends on the physical field $\mathbf{M}(\mathbf{r}, t)$ defined in physical spacetime:

$$\delta S_{\text{NWZ}} = \int d^D x \, dt \, M \, \hat{\mathbf{m}} \cdot (\partial_t \hat{\mathbf{m}} \times \delta \mathbf{m}) \,. \tag{6.15}$$

As a result the variation of the Novikov–Wess–Zumino term together with the free energy F gives rise to the Landau–Lifshitz equation describing the dynamics of the magnetization:

$$M \, \hat{\mathbf{m}} \times \partial_t \hat{\mathbf{m}} = \frac{\delta F}{\delta \hat{\mathbf{m}}} - \hat{\mathbf{m}} \left(\hat{\mathbf{m}} \cdot \frac{\delta F}{\delta \hat{\mathbf{m}}} \right) \,. \tag{6.16}$$

The same equation is obtained from the Poisson brackets (6.13) if F is considered as the Hamiltonian.

Since a well-defined action is absent, the energy–momentum tensor is also ill defined in ferromagnets. This is the result of momentum exchange with the microscopic degrees of freedom (see Volovik 1987). As distinct from the conventional dissipation of the collective motion to the microscopic degrees of freedom, which leads to the production of entropy, this exchange can be reversible. Later in Sec. 18.3.1 we shall discuss a similar reversible momentum exchange between the moving texture (collective motion) and the system of quasiparticles. This exchange, which is described by the same equations as the phenomenon of axial anomaly and also by the Wess–Zumino term in action, leads to a reversible non-dissipative force acting on the moving texture (the spectral-flow or Kopnin force).

According to the condensed matter analogy, the presence of a non-local Novikov–Wess–Zumino term in RQFT would indicate that such theory is effective. Probably the same happens in gravity: the absence of the covariant energy–momentum tensor simply reflects the existence of underlying 'microscopic' degrees of freedom, which are responsible for non-locality of the energy and momentum of the 'collective' gravitational field.

6.2 Effective vs microscopic theory

6.2.1 *Does quantum gravity exist?*

Gravity is the low-frequency (and actually the classical) output of all the quantum degrees of freedom of the 'Planck condensed matter'. The condensed matter analogy supports the extreme point of view expressed by Hu (1996) that one should not quantize gravity again. One can quantize gravitons but one should not use the low-energy quantization for the construction of Feynman loop diagrams containing integration over high momenta. In particular, the effective field theory (RQFT) is not appropriate for the calculation of the vacuum energy and thus of the cosmological constant.

General relativity in the quantum vacuum as well as quantum hydrodynamics in quantum liquids are not renormalizable theories. In both of them the effective theory can be used only at a tree level. The use of the effective theory at a loop level is (with rare exceptions when one can isolate the infrared contribution) forbidden, since it gives rise to catastrophic ultraviolet divergence whose treatment is well beyond the effective theory. The effective theory is the product of the more fundamental microscopic (or 'trans-Planckian') physics. It is important that it is already the final product which (if the infrared edge is not problematic) does not require further renormalization within the effective theory.

If gravity emerges in the low-energy corner as a low-energy soft mode (zero mode) of the underlying quantum Planck matter, then it would indicate that quantum gravity simply does not exist. If there are low-energy modes which can be identified with gravity, it does not mean that these modes will survive at high energy. Most probably they will merge with the continuum of all other high-energy degrees of freedom of the Planck condensed matter (corresponding to the motion of separate atoms of the liquid in the case of ^{4}He and ^{3}He) and thus can no longer be identified as gravitational modes. What is allowed in effective theory is to quantize the low-energy modes to produce phonons from sound waves and gravitons from gravitational waves. The deeper quantum theory of gravity makes no sense in this philosophy. Our knowledge of the physics of phonons/gravitons does not allow us to make predictions on the microscopic (atomic/Planck) structure of the bosonic or fermionic vacuum.

6.2.2 *What effective theory can and cannot do*

The vacuum energy density in the effective theory (2.11) is of fourth order in the cut-off energy. Such a huge dependence on the cut-off indicates that the vacuum energy is not within the responsibility of the effective theory, and microscopic physics is required. We have already seen that, regardless of the real 'microscopic' structure of the vacuum, the mere existence of the 'microscopic' physics ensures that the energy of the equilibrium vacuum is not gravitating. If we nevertheless want to use the effective theory, we must take it for granted that the diverging energy of quantum fluctuations of the effective fields and thus the cosmological term must be regularized to zero in a full and fully homogeneous equilibrium.

This certainly does not exclude the Casimir effect, which appears if the vacuum is not homogeneous. The long-wavelength perturbations of the vacuum can be described in terms of the change in the zero-point oscillations of the collective modes, since they do not disturb the high-energy degrees (see Chapter 29). The smooth deviations from the homogeneous equilibrium vacuum, due to, say, boundary conditions, are within the responsibility of the low-energy domain. These deviations can be successfully described by the effective field theory, and thus their energy can gravitate.

The Einstein action in Sakharov's (1967a) theory, obtained by integration over the fermionic and/or bosonic vacuum fields in the gravitational background, is quadratically divergent:

$$\mathcal{L}_{\text{Einstein}} = -\frac{1}{16\pi G}\sqrt{-g}\mathcal{R} \ , \quad G^{-1} \sim E_{\text{Planck}}^2 \ . \tag{6.17}$$

Such dependence on the cut-off also indicates that the Einstein action is not within the responsibility of the effective theory used in derivation. Within the effective theory we cannot even resolve between two possible formulation of the Einstein action: one is eqn (6.17), while the other one is

$$\mathcal{L}_{\text{Einstein}} = -\frac{1}{16\pi}\sqrt{-g}Rg^{\mu\nu}\Theta_\mu\Theta_\nu \ , \tag{6.18}$$

where Θ_μ is the cut-off 4-momentum. Each of the two expressions has its plus and minus points. The familiar eqn (6.17) is covariant but does not but obey the global scale invariance – the invariance under multiplication of $g_{\mu\nu}$ by a constant factor. The latter symmetry is present in eqn (6.18), but the general covariance is lost because of the cut-off which introduces a preferred reference frame. This is a typical contradiction between the symmetries in effective theories, which we discussed in Sec. 6.1.

The effective action for the gravity field must also contain the higher-order derivative terms, which are quadratic in the Riemann tensor,

$$\sqrt{-g}(q_1 R_{\mu\nu\alpha\beta}R^{\mu\nu\alpha\beta} + q_2 R_{\mu\nu}R^{\mu\nu} + q_3 R^2) \ln\left(\frac{g^{\mu\nu}\Theta_\mu\Theta_\nu}{R}\right) \ . \tag{6.19}$$

The parameters q_i depend on the matter content of the effective field theory. If the 'matter' consists of scalar fields, phonons or spin waves, the integration over these collective modes gives $q_1 = -q_2 = (2/5)q_3 = 1/(180 \cdot 32\pi^2)$ (see e.g. Frolov and Fursaev 1998) These terms are non-analytic: they depend logarithmically on both the ultraviolet and infrared cut-off. As a result their calculation in the framework of the infrared effective theory is justified. This is the reason why they obey (with logarithmic accuracy) all the symmetries of the effective theory including the general covariance and the invariance under rescaling of the metric. That is why these terms are the most appropriate for the self-consistent effective theory of gravity. However, they are small compared to the regular terms in Einstein action.

This is the general rule that the logarithmically diverging terms in the action play a special role, since they can always be obtained within the effective theory. As we shall see below, the logarithmic terms arise in the effective action for the effective gauge fields, which appear in superfluid ^{3}He-A in the low-energy corner (Sec. 9.2.2). With the logarithmic accuracy they dominate over the non-renormalizable terms. These logarithmic terms in superfluid ^{3}He-A were obtained first in microscopic calculations; however, it appeared that their physics can be completely determined by the low-energy tail and thus they can be calculated within the effective theory. This is well known in particle physics as dimensionless running coupling constants exhibiting either the zero-charge effect or asymptotic freedom.

The non-analytic terms (6.19) coming from infrared physics must exist even in superfluid ^{4}He. But being of higher order in gradient, they are always small compared to the leading hydrodynamic terms coming from Planckian physics, and thus they almost always play no role in the dynamics of superfluids.

6.3 Superfluidity and universality

The concept of superfluidity was introduced by Kapitza (1938). Below a critical temperature, which was known as the λ-point, $T_\lambda \sim 2.2$ K, from the peculiar shape of its specific heat anomaly, liquid ^{4}He transforms to a new state, He-II (Kamerlingh Onnes 1911b). According to Kapitza, this new phase of liquid ^{4}He which he called superfluid was similar to the superconducting state of metals, which had also been discovered by Kamerlingh Onnes (1911a). analogous phenomena: the vanishing ohmic resistance in the motion of conduction electrons in superconductors and the frictionless motion of atoms in superfluid He-II.

It is now believed that most systems displaying free motion of particles well below some characteristic degeneracy temperature transform to the superfluid or superconducting state. However, the concept of inviscid motion was so alien to us accustomed to the world at 300 K that it took 30 years for the low-temperature physicists of the first half of the past century to accept the new concept. The paradox created by the complete absence of viscosity in some experiments and the conflicting evidence from seemingly normal dissipative flow in other experiments was not a simple problem to resolve.

An important step in understanding the nature of the superfluid state was taken by F. London (1938) who associated it with the properties of an ideal Bose gas close to its ground state, following the theory which had been worked out by Einstein (1924, 1925). Einstein had found that in a gas of non-interacting bosons an unusual kind of condensation occurs below some characteristic critical temperature: the atoms condense in configurational space to their common ground state, i.e. a macroscopic fraction of the atoms occupies the same quantum state, which is the state of minimum energy. It is interesting that in the spirit of that time London considered He-II not as a liquid but as 'liquid crystal', a crystal with such a strong zero-point motion caused by quantum uncertainty that atoms are no more fixed at the lattice sites but nearly freely move in the periodic potential produced by other atoms. Now we know that periodicity does not happen in He-II, but such 'liquid-crystal' behavior occurs in solid ^{4}He, which in modern terminology is known as a quantum crystal. However, superfluidity in quantum crystals has not yet been observed.

Although the bridge from a simple non-interacting Bose gas to an interacting system like liquid He-II was not developed at that time, nevertheless this association provided the cornerstone on which Tisza (1938, 1940) built his famous picture of two-fluid motion. According to the two-fluid hypothesis, He-II consists of intermixed normal and superfluid components. Kapitza's experiment on the superflow in a narrow channel demonstrated that the normal component was locked by its viscous interactions to the bounding surfaces, while the superfluid component moved freely and was responsible for the observed absence of viscosity. Tisza interpreted the superfluid component as consisting of atoms condensed in the ground state. The normal component in turn corresponds to particles in excited states and is assumed to behave like a rarefied gas of normal viscous He-I above the λ-point. Qualitatively this picture provides a good working understanding of the properties of a superfluid. Tisza himself demonstrated this by supplying an explanation for the fountain effect and by predicting the existence of thermal waves, known today as second sound by the name coined by Landau a few years later. Second-sound motion was ultimately demonstrated by Kapitza's student V. P. Peshkov (1944).

The rigorous physical basis for the two-fluid concept was developed by Landau (1941) who derived the complete and self-consistent set of equations of two-fluid hydrodynamics – the effective theory of bosonic zero modes which we now use. Landau did not use the intriguing connection with the Bose–Einstein condensate, since He-II in reality is a complicated system of strongly interacting particles, and based his argument instead on the character of the low-energy spectrum of the excitations of the system, quasiparticles. According to Landau, He-II corresponds to a flowing vacuum state, similar to a cosmic ether, in which quasiparticles – bosomic zero modes of quantum vacuum – move. The quanta of these collective modes of the vacuum form a rarefied gas which is responsible for the thermal and viscous effects ascribed to the presence of the normal component. The motion of the quasiparticle gas produces the normal–superfluid counterflow, the relative motion of the normal component (velocity $\mathbf{v}_n$) with respect to the vacuum (velocity $\mathbf{v}_s$).

The microscopic analysis by Bogoliubov in his model of a weakly interacting Bose gas, which was discussed in Sec. 3.2, fully supported Landau's idea. According to Bogoliubov, in the interacting system the number of particles in the condensate N_0, i.e. those with momenta $\mathbf{p} = 0$, is less than the total number of particles N even at $T = 0$. Nevertheless, at $T = 0$ the normal component is absent: the vacuum – the state without quasiparticles – is made of all N particles of the liquid. The superfluid vacuum is thus more complicated than its predecessor – the Bose–Einstein condensate. It is the coherent ground state of N atoms which includes the atoms with $\mathbf{p} = 0$ as well as those with $\mathbf{p} \neq 0$. The number density N_0/V of atoms with $\mathbf{p} = 0$ does not enter the two-fluid hydrodynamics; instead the total number density $n = N/V$ is the variable in

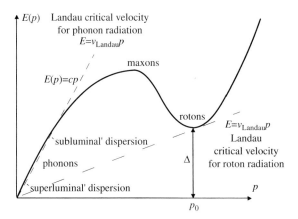

FIG. 6.1. Schematic illustration of phonon–roton spectrum.

this effective theory. In modern language, in the interacting system the number N_0 of atoms in the Bose–Einstein condensate is effectively renormalized to the total number N of particles in the coherent many-body state of the superfluid vacuum, thus supporting the Landau picture.

Landau's prediction for the spectrum of quasiparticles as corresponding to the sound wave spectrum was also confirmed in the model. In modern language, Landau established the universality class of Bose liquids whose behavior does not depend on the details of the interaction.

Two important hypotheses were introduced by Landau to justify the superfluidity of the quantum vacuum:

(1) The non-viscous flow of the vacuum should be potential (irrotational). Landau (1941) suggested that in liquid ^{4}He the rotational degrees of freedom have higher energy levels than the levels of potential motion, and these rotational levels are separated from the vacuum state by the energy gap which was estimated as $\sim \hbar^2/ma_0^2$, where a_0 is the interatomic distance. The corresponding quanta of rotational motion were initially named rotons. Later it was found that the quasiparticle energy spectrum, which starts at small p as a phonon spectrum, has a local minimum at finite p (Fig. 6.1). The name roton was assigned to a quasiparticle close to the local minimum, but this quasiparticle has nothing to do with rotational degrees of freedom. Later, after the works of Onsager (1949) and Feynman (1955), it was understood that rotational degrees of freedom were related to quantized vortices. The lowest energy level related to rotational motion is provided by a vortex ring of minimum possible size whose energy $\sim \hbar^2/ma_0^2$ is in agreement with Landau's suggestion.

(2) The excitation spectrum $E(p)$ of the quasiparticles, rotons and quanta of sound – phonons – places an upper limit on the smooth irrotational flow of the vacuum. The so-called Landau critical velocity is $v_{\text{Landau}} = (E(p)/p)_{min}$ (see Chapter 26). If the relative velocity $|\mathbf{v}_s - \mathbf{v}_n|$ of the motion of the vacuum with respect to the excitations or boundaries is less than v_{Landau}, then superflow is

dissipationless because the vacuum cannot transfer momentum p to the excitations. If $|\mathbf{v}_s - \mathbf{v}_n| > v_{\text{Landau}}$, new excitations can be created from the vacuum, such that the momentum of superfluid motion is dissipated in the momenta p of the newly created excitations, and friction arises.

At low temperature, the effective theory of two-fluid hydrodynamics rests on a general principle, namely that the properties of an interacting system are determined in the low-energy limit by the spectrum of low-energy excitations – zero modes. Later Landau (1956) applied the same principle also to liquid ^{3}He, which at the temperatures where it can be described in terms of quasiparticles is still in a normal (non-superfluid) state. ^{3}He atoms obey Fermi–Dirac statistics and have a single-particle-like spectrum of essentially different character from the phonons and rotons in ^{4}He-II. The result is known as the Landau theory of Fermi liquids, which is one of the most successful cornerstones of condensed matter theory. Thus a new universality class of quantum vacua and its effective theory have been established. Below, in Sec. 8.1, we shall see that Fermi liquids belong to the most powerful universality class of quantum vacua, were the fermion zero modes form the Fermi surface being protected by the topology in the momentum space.

Alongside Tisza's idea (1940) of the Bose–Einstein condensate and Landau's idea (1941) of superfluid vacuum, the physical origin of two-fluid motion was suggested by Kapitza (1941). Later it was recognized that both Tisza and Landau had added important pieces to the description of superfluidity in He-II, while Kapitza's suggestion was long considered as an oddity. However, the discovery of anisotropic superfluidity in liquid ^{3}He-A by Osheroff et al. (1972) has completely absolved Kapitza's hypothesis: it now appears to be quite to the point, while the Tisza and Landau theories were to be modified.

Kapitza (1941) postulated on the basis of his ingenious new experiments that only a thin surface layer of liquid He-II coating the solid walls was superfluid while bulk He-II behaved like a normal fluid. However, although the superfluid ^{4}He film is very mobile and responsible for many striking phenomena, superfluidity is clearly not only restricted to the surface film in He-II, as Kapitza himself later recognized. Nevertheless, this idea is close to the actual situation in superfluid ^{3}He-A: experiments demonstrate (see the book by Vollhardt and Wölfle (1990)): that bulk superflow of ^{3}He-A is unstable, while near the surface of the channel the superflow is stabilized due to boundary conditions on the order parameter. Moreover, the origin of superfluidity in ^{3}He is different from that in He-II, but actually is the same as superconductivity in metals: it results from the formation of coherent Cooper pairs instead of the London–Tisza scenario of the Bose–Einstein condensation of bosons to a common ground state. Also it turns out, as we shall see, that the superflow of ^{3}He-A is not irrotational, i.e. it is not restricted to the simple potential flow of isotropic superfluids (for which $\vec{\nabla} \times \mathbf{v}_s = 0$, as is the case for ^{4}He-II). In other words, as distinct from superfluid ^{4}He, in ^{3}He-A there is no energy gap between rotational and irrotational motions. This is the reason why the bulk ^{3}He-A behaves as normal fluid (see Fig.

17.9). Also the Landau critical velocity $v_{\rm Landau}$ is vanishingly small in ^{3}He-A because of the gapless fermionic energy spectrum (see Sec. 26.1.1). Thus none of the conditions representing the signatures of superfluidity in Landau's sense are strictly valid in ^{3}He-A. This superfluid opened another important universality class of Fermi systems with gapless excitations protected by the topology in momentum space (Sec. 8.2).

Part II

Quantum fermionic liquids

7
MICROSCOPIC PHYSICS

7.1 Introduction

Now we proceed to the Fermi systems, where the low-energy effective theory involves both bosonic and fermionic fields. First we consider the simplest models where different types of fermionic quasiparticle spectra arise. Then we show that the types of fermionic spectra considered are generic. They are determined by universality classes according to the momentum space topology of the fermionic systems.

Above the transition temperature T_c to the superconducting or superfluid state, the overwhelming majority of systems consisting of fermionic particles (electrons in metals, neutrons in neutron stars, ^{3}He atoms in ^{3}He liquid, etc.) form a so-called Fermi liquid (Fig. 7.1). As was pointed out by Landau, the Fermi liquids share the properties of their simplest representative – weakly interacting Fermi gas: the low-energy physics of the interacting particles in a Fermi liquid is equivalent to the physics of a gas of quasiparticles moving in collective Bose fields produced by all other particles. We shall see below in Chapter 8 that this reduction from Fermi liquid to Fermi gas is possible because they belong to the same universality class determined by momentum space topology. The topology is robust to details and, in particular, to the strength of the interaction between the particles, which distinguishes Fermi liquids from the Fermi gases. We shall see that it is the momentum space topology which distributes Fermi systems into different universality classes. Fermi liquids belong to one of these universality classes, which is of most importance in condensed matter. This class is characterized by the so-called Fermi surface – the topologically stable surface in momentum space (Sec. 8.1).

The class of fermionic vacua with a Fermi surface is most powerful, since it is described by the lowest-order homotopy group π_1, called the fundamental group (Sec. 8.1). The overwhelming majority of mobile fermionic systems have a Fermi surface. The other important class of fermionic vacua (which contains the vacuum of the Standard Model and ^{3}He-A) is described by the higher (and thus weaker) group – the third hompotopy group π_3 (Sec. 8.2.3). If the vacuum of the Standard Model is sufficiently perturbed, e.g. by the appearance of an event horizon, which introduces Planckian physics near and beyond the horizon, the Fermi surface almost necessarily appears (see Sec. 32.4).

In this Chapter we discuss the BCS theory of the weakly interacting Fermi gas in the same manner as the weakly interacting Bose gas in Sec. 3.2, and demonstrate how different universality classes arise in the low-energy corner.

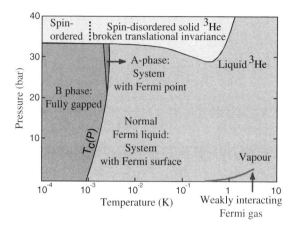

FIG. 7.1. ^{3}He phase diagram demonstrating all three major universality classes of those fermionic vacua which are translationally invariant. Normal liquid ^{3}He represents the class where the quasiparticle spectrum is gapless on the Fermi surface. ^{3}He-B belongs to the class with trivial topology in momentum space: the quasiparticle spectrum is fully gapped. ^{3}He-A has topologically stable Fermi points in momentum space, where fermionic quasiparticles have zero energy.

7.2 BCS theory

7.2.1 *Fermi gas*

Let us start with an ideal (non-interacting) non-relativistic Fermi gas of, say, electrons in metals or ^{3}He atoms. In this simplest case of an isotropic system, the energy spectrum of particles is

$$E(p) = \frac{p^2}{2m} - \mu \,, \tag{7.1}$$

where μ, as before, is the chemical potential for particles, and it is assumed that $\mu > 0$. For $\mu > 0$ there is a Fermi surface which bounds the volume in momentum space where the energy is negative, $E(p) < 0$, and where the particle states are all occupied at $T = 0$. For an isotropic system the Fermi surface is a sphere of radius $p_F = \sqrt{2m\mu}$. At low T all the low-energy properties of the Fermi gas come from the particles in the vicinity of the Fermi surface, where the energy spectrum is linearized: $E(p) \approx v_F(p - p_F)$, where $v_F = \partial_p E|_{p=p_F}$ is the Fermi velocity. Fermi particles have spin 1/2 and thus there are actually two Fermi surfaces, one for particles with spin up and another for particles with spin down.

Later in Sec. 8.1 we shall see that the Fermi surface is a topologically stable singularity of the Green function in momentum space. Due to its topological stability, the Fermi surface survives in non-ideal Fermi gases and even in Fermi liquids, where the interaction between particles is strong, unless a total non-perturbative reconstruction of quantum vacuum occurs and the system transfers to another universality class.

In particular, such reconstruction takes place when Cooper pairing occurs between the particles, and the superfluid or superconducting state arises. Later we shall be mostly interested in such superfluid phases, where the Cooper pairing occurs between the particles within the same Fermi surface, i.e. which have the same spin projection. For such cases of so-called equal spin pairing (ESP) we can forget about the spin degrees of freedom and consider only single spin projection, say, spin up. In other words, our fermions are 'spinless' but obey Fermi–Dirac statistics. Thus in our original 'Theory of Everything' there is no spin–statistic theorem at the 'fundamental' level; this theorem will appear in the low-energy corner together with Lorentz invariance and the corresponding relativistic spin.

7.2.2 Model Hamiltonian

To make life easier we assume a special type of the BSC model with the interaction leading to pairing in a p-wave state consistent with the equal spin pairing:

$$\mathcal{H} - \mu \mathcal{N} = \sum_{\mathbf{p}} \left(\frac{p^2}{2m} - \mu \right) a^\dagger_{\mathbf{p}} a_{\mathbf{p}} - \frac{\lambda}{V} \left(\sum_{\mathbf{p}'} \mathbf{p}' a^\dagger_{-\mathbf{p}'} a^\dagger_{\mathbf{p}'} \right) \cdot \left(\sum_{\mathbf{p}} \mathbf{p} a_{\mathbf{p}} a_{-\mathbf{p}} \right) . \quad (7.2)$$

Here λ is the small parameter of interaction. In the superfluid state the quantity $\sum_{\mathbf{p}} \mathbf{p} a_{\mathbf{p}} a_{-\mathbf{p}}$ acquires a non-zero vacuum expectation value (vev), $\left\langle \sum_{\mathbf{p}} \mathbf{p} a_{\mathbf{p}} a_{-\mathbf{p}} \right\rangle_{\text{vac}}$. This vev is not identically zero because both $\mathbf{p}$ and $a_{\mathbf{p}} a_{-\mathbf{p}}$ are odd under parity transformation: $\mathrm{P}\mathbf{p} = -\mathbf{p}$ and $\mathrm{P}(a_{\mathbf{p}} a_{-\mathbf{p}}) = a_{-\mathbf{p}} a_{\mathbf{p}} = -a_{\mathbf{p}} a_{-\mathbf{p}}$. The minus sign in the last equation is due to the Fermi statistics. The odd factor $\mathbf{p}$ in vev allows the equal spin pairing (ESP) pairing within the same Fermi surface.

This vev is similar to the vev of the operator of annihilation of the Bose particle with zero momentum in eqn (3.4). In our case the two Fermi particles involved also form a Bose particle – the Cooper pair – with the total momentum also equal to zero. However, in the case of p-wave pairing the order parameter is not a complex scalar as in eqn (3.4) but a complex vector:

$$\frac{2\lambda}{V} \left\langle \sum_{\mathbf{p}} \mathbf{p} a_{\mathbf{p}} a_{-\mathbf{p}} \right\rangle_{\text{vac}} = \mathbf{e}_1 + i \mathbf{e}_2 . \quad (7.3)$$

Here $\mathbf{e}_1$ and $\mathbf{e}_2$ are real vectors.

Equilibrium values of these vectors are determined by minimization of the vacuum energy. However, symmetry considerations are enough to find the possible structure of the order parameter: there are two structures that are always the extrema of the energy functional. In the first case the two vectors, $\mathbf{e}_1$ and $\mathbf{e}_2$, are perpendicular to each other and have the same magnitude:

$$\mathbf{e}_1 \cdot \mathbf{e}_2 = 0 , \quad |\mathbf{e}_1| = |\mathbf{e}_2| . \quad (7.4)$$

Such a structure of the orbital part of the order parameter occurs in superfluid ^{3}He-A (Sec. 7.4.6), ^{3}He-A$_1$ (Sec. 7.4.8) and in the planar phase (Sec. 7.4.9). In all these cases $|\mathbf{e}_1| = |\mathbf{e}_2| = \Delta_0/p_F$, where Δ_0 is the amplitude of the gap. The three phases differ by their spin structure, which we do not discuss here in the model of spinless fermions.

In the second case the two vectors are parallel to each other, $\mathbf{e}_1 \parallel \mathbf{e}_2$, and can differ only by phase factor. Thus we have

$$\mathbf{e}_1 + i\mathbf{e}_2 = \mathbf{e}\, e^{i\Phi}\,, \tag{7.5}$$

where $\mathbf{e}$ is a real vector along the common direction. The orbital part of the order parameter in the polar phase of the spin-triplet superfluid in Sec. 7.4.10 has the same structure.

For the moment we shall not specify the two vectors, and consider the general case of arbitrary $\mathbf{e}_1$ and $\mathbf{e}_2$. In the limit of weak interaction $\lambda \to 0$, the quantum fluctuations of the order parameter – deviations of the operator $\sum_{\mathbf{p}} \mathbf{p} a_{\mathbf{p}} a_{-\mathbf{p}}$ from its vev – are small. Neglecting the term quadratic in deviations from vev, one obtains the following Hamiltonian:

$$\mathcal{H} - \mu \mathcal{N} = \frac{V}{4\lambda}\left(\mathbf{e}_1^2 + \mathbf{e}_2^2\right) + \sum_{\mathbf{p}} \mathcal{H}_{\mathbf{p}}\,, \tag{7.6}$$

$$\mathcal{H}_{\mathbf{p}} = \left(\frac{p^2}{2m} - \mu\right)\frac{a^\dagger_{\mathbf{p}} a_{\mathbf{p}} + a^\dagger_{-\mathbf{p}} a_{-\mathbf{p}}}{2} + \mathbf{p}\cdot\frac{\mathbf{e}_1 - i\mathbf{e}_2}{2} a_{-\mathbf{p}} a_{\mathbf{p}} + \mathbf{p}\cdot\frac{\mathbf{e}_1 + i\mathbf{e}_2}{2} a^\dagger_{\mathbf{p}} a^\dagger_{-\mathbf{p}}\,. \tag{7.7}$$

7.2.3 Bogoliubov rotation

For each $\mathbf{p}$ the Hamiltonian $\mathcal{H}_{\mathbf{p}}$ can be diagonalized using the operators $\mathcal{L}_i$

$$\mathcal{L}_3 = \frac{1}{2}(a^\dagger_{\mathbf{p}} a_{\mathbf{p}} + a^\dagger_{-\mathbf{p}} a_{-\mathbf{p}} - 1)\,,\quad \mathcal{L}_1 - i\mathcal{L}_2 = a^\dagger_{\mathbf{p}} a^\dagger_{-\mathbf{p}}\,,\quad \mathcal{L}_1 + i\mathcal{L}_2 = a_{\mathbf{p}} a_{-\mathbf{p}}\,. \tag{7.8}$$

As distinct from the Bose case in eqn (3.11) these operators are equivalent to the generators of the group $SO(3)$ of conventional rotations with commutation relations appropriate for the angular momentum: $[\mathcal{L}_i, \mathcal{L}_j] = i\epsilon_{ijk}\mathcal{L}_k$. In terms of $\mathcal{L}_i$ one has

$$\mathcal{H}_{\mathbf{p}} = g^i(\mathbf{p})\mathcal{L}_i + g^0(\mathbf{p})\,, \tag{7.9}$$

with

$$g^1 = \mathbf{p}\cdot\mathbf{e}_1\,,\quad g^2 = \mathbf{p}\cdot\mathbf{e}_2\,,\quad g^3 = \frac{p^2}{2m} - \mu\,,\quad g^0 = \frac{1}{2}\left(\frac{p^2}{2m} - \mu\right)\,. \tag{7.10}$$

The Hamiltonian is diagonalized by two $SO(3)$ rotations. First we rotate by the angle $\tan^{-1}(\mathbf{p}\cdot\mathbf{e}_2/\mathbf{p}\cdot\mathbf{e}_1)$ around the x_3 axis. This is the $U(1)$ global gauge rotation which transforms $g^1 \to \sqrt{(\mathbf{p}\cdot\mathbf{e}_2)^2 + (\mathbf{p}\cdot\mathbf{e}_1)^2}$, $g^2 \to 0$. Then we rotate by the angle $\tan^{-1}(\sqrt{(\mathbf{p}\cdot\mathbf{e}_2)^2 + (\mathbf{p}\cdot\mathbf{e}_1)^2}/(\frac{p^2}{2m} - \mu))$ around the x_2 axis; this is the Bogoliubov transformation. As a result one obtains the diagonal Hamiltonian

$$\mathcal{H}_\mathbf{p} = E(\mathbf{p})\tilde{\mathcal{L}}_3 + \frac{1}{2}\left(\frac{p^2}{2m} - \mu\right)$$

$$= \frac{1}{2}E(\mathbf{p})\left(\tilde{a}^\dagger_\mathbf{p}\tilde{a}_\mathbf{p} + \tilde{a}^\dagger_{-\mathbf{p}}\tilde{a}_{-\mathbf{p}}\right) + \frac{1}{2}\left(\frac{p^2}{2m} - \mu\right) - \frac{1}{2}E(\mathbf{p}) . \quad (7.11)$$

Here $\tilde{a}_\mathbf{p}$ is the annihilation operator of fermionic quasiparticles (the so-called Bogoliubov quasiparticles), whose energy spectrum $E(\mathbf{p})$ is

$$E(\mathbf{p}) = \sqrt{\left(\frac{p^2}{2m} - \mu\right)^2 + (\mathbf{p}\cdot\mathbf{e}_1)^2 + (\mathbf{p}\cdot\mathbf{e}_2)^2} . \quad (7.12)$$

The total Hamiltonian represents the energy of the vacuum (the state without quasiparticles) and that of fermionic quasiparticles in the background of the vacuum:

$$\mathcal{H} - \mu\mathcal{N} = \langle\mathcal{H} - \mu\mathcal{N}\rangle_{\text{vac}} + \sum_\mathbf{p} E(\mathbf{p})\tilde{a}^\dagger_\mathbf{p}\tilde{a}_\mathbf{p} , \quad (7.13)$$

where the vacuum energy is

$$\langle\mathcal{H} - \mu\mathcal{N}\rangle_{\text{vac}} = \frac{V}{4\lambda}\left(\mathbf{e}_1^2 + \mathbf{e}_2^2\right) - \frac{1}{2}\sum_\mathbf{p} E(\mathbf{p}) + \frac{1}{2}\sum_\mathbf{p}\left(\frac{p^2}{2m} - \mu\right) . \quad (7.14)$$

7.2.4 Stable point nodes and emergent 'relativistic' quasiparticles

The equilibrium state of the vacuum will be later determined by minimization of the vacuum energy eqn (7.14) over vectors $\mathbf{e}_1$ and $\mathbf{e}_2$, but it is instructive to consider first the general vacuum state with arbitrary vectors $\mathbf{e}_1$ and $\mathbf{e}_2$. In general, if we disregard for a moment the very exceptional, degenerate case of the order parameter when these vectors are exactly parallel to each other, the quasiparticle energy spectrum in eqn (7.12) has two points $\mathbf{p}^{(a)}$ ($a = 1, 2$) in momentum space where the energy is zero:

$$\mathbf{p}^{(a)} = q_a p_F \hat{\mathbf{l}} , \quad \hat{\mathbf{l}} = \frac{\mathbf{e}_1 \times \mathbf{e}_2}{|\mathbf{e}_1 \times \mathbf{e}_2|} , \quad q_a = \pm 1 . \quad (7.15)$$

Here, as before $p_F = \sqrt{2m\mu}$, is the Fermi momentum of an ideal Fermi gas. Close to each of these two points, which we refer to as Fermi points in analogy with the Fermi surface, the energy spectrum becomes

$$E_a^2(\mathbf{p}) \approx (\mathbf{e}_1 \cdot (\mathbf{p} - q_a\mathbf{A}))^2 + (\mathbf{e}_2 \cdot (\mathbf{p} - q_a\mathbf{A}))^2 + (\mathbf{e}_3 \cdot (\mathbf{p} - q_a\mathbf{A}))^2 , \quad (7.16)$$

$$\mathbf{e}_3 = \frac{p_F}{m}\hat{\mathbf{l}} , \quad \mathbf{A} = p_F\hat{\mathbf{l}} . \quad (7.17)$$

This corresponds to the energy spectrum of two massless 'relativistic' particles with electric charges $q_a = \pm 1$ in the background of an electromagnetic field with the vector potential $\mathbf{A} = p_F \hat{\mathbf{l}}$, in the space with the anisotropic contravariant metric tensor $g^{\mu\nu}$:

$$g^{ik} = e_1^i e_1^k + e_2^i e_2^k + e_3^i e_3^k \;,\; g^{00} = -1 \;,\; g^{0i} = 0 \;. \qquad (7.18)$$

Thus for the typical vacuum state the quasiparticle energy spectrum becomes relativistic and massless in the low-energy corner. This is an emergent property which follows solely from the existence of the Fermi points. In turn, according to eqn (7.15), the existence of the Fermi points is robust to any deformation of the order parameter vectors $\mathbf{e}_1$ and $\mathbf{e}_2$ of the vacuum, excluding the exceptional case when $\mathbf{e}_1 \parallel \mathbf{e}_2$ and the metric $g^{\mu\nu}$ becomes degenerate. This illustrates the topological stability of Fermi points, which we shall discuss later. Because of this stability, emergent relativity is really a generic phenomenon in the universality class of Fermi systems to which this model belongs – systems with topologically stable Fermi points in momentum space (Sec. 8.2).

7.2.5 Nodal lines are not generic

In the exceptional case of $\mathbf{e}_1 \parallel \mathbf{e}_2 \equiv \mathbf{e}$ (this corresponds to the so-called polar state discussed in Sec. 7.4.9) the energy spectrum is zero on the line ($p = p_F$, $\mathbf{p} \perp \mathbf{e}$) in 3D momentum space. The fact that the line of zeros occurs only as an exception to the rule shows that nodal lines are topologically unstable. Under general deformations of the order parameter the nodal lines disappear. In our simplified model where the spin degrees of freedom are suppressed, the line of zeros disappears leaving behind the pairs of Fermi points. But in a more general case there is an alternative destiny for a nodal line: zeros can disappear completely so that the system becomes fully gapped.

7.3 Vacuum energy of weakly interacting Fermi gas

7.3.1 Vacuum in equilibrium

An equilibrium value of the order parameter is obtained by minimization of the vacuum energy in eqn (7.14) over $\mathbf{e}_1$ and $\mathbf{e}_2$

$$\frac{V}{\lambda} \mathbf{e}_{1,2} - \sum_{\mathbf{p}} \frac{\partial E(\mathbf{p})}{\partial \mathbf{e}_{1,2}} = 0 \;. \qquad (7.19)$$

Excluding the volume V of the system one obtains two non-linear equations for vectors $\mathbf{e}_1$ and $\mathbf{e}_2$:

$$\mathbf{e}_{1,2} = \lambda \int \frac{d^3 p}{(2\pi\hbar)^3} \frac{\mathbf{p}(\mathbf{p} \cdot \mathbf{e}_{1,2})}{E(\mathbf{p})} \;. \qquad (7.20)$$

Substituting eqn (7.20) into eqn (7.14) one obtains that the minimum of the vacuum energy (i.e. the energy of the equilibrium vacuum) has the form:

$$\langle \mathcal{H} - \mu \mathcal{N} \rangle_{\text{eq vac}} = \frac{1}{2} \sum_{\mathbf{p}} \left[-E(\mathbf{p}) + \left(\frac{p^2}{2m} - \mu \right) + \frac{(\mathbf{p} \cdot \mathbf{e}_1)^2 + (\mathbf{p} \cdot \mathbf{e}_2)^2}{2E(\mathbf{p})} \right] \;. \qquad (7.21)$$

Compare this with eqn (3.19) for the vacuum energy of the weakly interacting Bose gas and with eqn (2.9) for vacuum energy in RQFT. The first term $-\frac{1}{2}\sum_{\mathbf{p}} E(\mathbf{p})$ can be recognized as the energy of the Dirac vacuum for fermionic quasiparticles. It has the sign opposite to that of the corresponding term in the vacuum energy of Bose gas, which contains zero-point energy of bosonic fields. This Dirac-vacuum term is natural from the point of view of the effective theory, where the quasiparticles (not the bare particles) are the physical objects.

The somewhat unusual factor 1/2 is the result of the specific properties of Bogoliubov quasiparticles. The Bogoliubov quasiparticle represents the hybrid of the bare particle and bare hole; as a result the quasiparticle equals its antiquasiparticle as in the case of Majorana fermions. That is why the naive summation of the negative energies in the Dirac vacuum leads to the double counting, which is just compensated by the factor 1/2 in eqn (7.21). We shall consider this in more detail in the next chapter.

As in the case of the Bose gas, the other two terms in eqn (7.21) represent the counterterms that make the total vacuum energy convergent. They are naturally provided by the microscopic physics of bare particles. In other words, the trans-Planckian physics determines the regularization scheme, which depends on the details of the trans-Planckian physics. Within the effective theory there are many regularization schemes, which may correspond to different trans-Planckian physics. There is no principle which allows us to choose between these schemes unless we know the microscopic theory.

The counterterms in eqn (7.21) can be regrouped to reflect different Planck energy scales. The natural 'regularization' is achieved by consideration of the difference between the energy of weakly interacting gas and that of an ideal Fermi gas with $\lambda = 0$. This effectively removes the largest contribution to the vacuum energy. The energy of the ideal Fermi gas is

$$\langle \mathcal{H} - \mu \mathcal{N} \rangle_{\text{vac } \lambda=0} = \frac{1}{2} \sum_{\mathbf{p}} \left[-\left| \frac{p^2}{2m} - \mu \right| + \left(\frac{p^2}{2m} - \mu \right) \right] = -V \frac{p_F^5}{30m\pi^2\hbar^3} \,. \quad (7.22)$$

The rest part of the vacuum energy is considerably smaller:

$$\langle \mathcal{H} - \mu \mathcal{N} \rangle_{\text{eq vac}} - \langle \mathcal{H} - \mu \mathcal{N} \rangle_{\text{vac } \lambda=0}$$

$$= -\frac{1}{4} \sum_{\mathbf{p}} \frac{\left[E(\mathbf{p}) - \left| \frac{p^2}{2m} - \mu \right| \right]^2}{E(\mathbf{p})} = -V \frac{p_F^3 m}{24\pi^2 \hbar^3}(\mathbf{e}_1^2 + \mathbf{e}_2^2) \,, \quad (7.23)$$

since the order parameter contains the small coupling constant λ.

7.3.2 Axial vacuum

It is easy to check that the order parameter structure, which realizes the vacuum state with the lowest energy, has the form of eqn (7.4):

$$\frac{2\lambda}{V} \left\langle \sum_{\mathbf{p}} p a_{\mathbf{p}} a_{-\mathbf{p}} \right\rangle_{\text{eq vac}} = (\mathbf{e}_1 + i\mathbf{e}_2)_{\text{eq}} = c_\perp (\hat{\mathbf{m}} + i\hat{\mathbf{n}}) \,, \quad (7.24)$$

$$\hat{\mathbf{m}}^2 = \hat{\mathbf{n}}^2 = 1 \ , \ \hat{\mathbf{m}} \cdot \hat{\mathbf{n}} = 0. \tag{7.25}$$

The condition on the order parameter in equilibrium in eqn (7.25) still allows some freedom: the $SO(3)$ rotations of the pair of unit vectors $\hat{\mathbf{m}}$ and $\hat{\mathbf{n}}$ do not change the vacuum energy. Thus the vacuum is degenerate. We choose for simplicity one particular vacuum state with $\hat{\mathbf{m}} = \hat{\mathbf{x}}$ and $\hat{\mathbf{n}} = \hat{\mathbf{y}}$, so that the energy spectrum of quasiparticles becomes

$$E(\mathbf{p}) = \sqrt{\left(\frac{p^2}{2m} - \mu\right)^2 + c_\perp^2 (\mathbf{p} \times \hat{\mathbf{l}})^2} = \sqrt{\left(\frac{p^2}{2m} - \mu\right)^2 + c_\perp^2 (p_x^2 + p_y^2)} \ . \tag{7.26}$$

It is axisymmetric and has uniaxial anisotropy along the axis $\hat{\mathbf{l}} = \hat{\mathbf{m}} \times \hat{\mathbf{n}}$, which according to eqn (7.15) shows the direction to the Fermi points in momentum space. This demonstrates that the vacuum itself is axisymmetric and has uniaxial anisotropy along $\hat{\mathbf{l}}$. We shall see later that $\hat{\mathbf{l}}$ also marks the direction of the angular momentum of Cooper pairs, and thus is an axial vector. Such a vacuum state is called axial. The corresponding superconducting states in condensed matter are called chiral superconductors.

The parameter $c_\perp$ is the maximum attainable speed of low-energy quasiparticles propagating transverse to $\hat{\mathbf{l}}$, since in the low-energy limit (i.e. close to point nodes) the energy spectrum becomes

$$E_a^2(\mathbf{p}) \approx c_\parallel^2 (p_z - q_a p_F)^2 + c_\perp^2 (p_x^2 + p_y^2) \ , \ c_\parallel = \frac{p_F}{m} \equiv v_F = \sqrt{\frac{2\mu}{m}} \ . \tag{7.27}$$

The Fermi velocity v_F represents the maximum attainable speed of low-energy quasiparticles propagating along $\hat{\mathbf{l}}$.

The difference in vacuum energies between the weakly interacting and non-interacting Fermi gas in eqn (7.23) is also completely determined by $c_\perp$:

$$\langle \mathcal{H} - \mu \mathcal{N} \rangle_{\text{eq vac}} - \langle \mathcal{H} - \mu \mathcal{N} \rangle_{\text{vac } \lambda=0} = -V \frac{p_F^3 m}{12\pi^2 \hbar^3} c_\perp^2 \ . \tag{7.28}$$

The value of the parameter $c_\perp$ is determined from eqn (7.20):

$$1 = \lambda \int \frac{d^3 p}{(2\pi\hbar)^3} \frac{p_x^2 + p_y^2}{2E(\mathbf{p})} \ . \tag{7.29}$$

The integral on the rhs is logarithmically divergent and one must introduce the physical ultraviolet cut-off at which our model becomes inapplicable. This cut-off influences only the value of $c_\perp = (E_{\text{cutoff}}/p_F) e^{-mn/\lambda}$. Once the value of $c_\perp$ is established, we can forget its origin (except that it is small compared to v_F) and consider it as a phemomenological parameter of the effective theory. For superfluid ^{3}He-A the ratio $c_\perp^2/v_F^2$ is about 10^{-5}.

Since the order parameter magnitude $c_\perp$ is relatively small, its influence on the relation between the particle number density n and the Fermi momentum

p_F is also small. That is why for the system with single spin projection one has $n \approx p_F^3/6\pi^2\hbar^3$. Actually $|n - p_F^3/6\pi^2\hbar^3|/n \sim c_\perp^2/c_\parallel^2 \ll 1$. As a result eqn (7.28) becomes

$$\langle \mathcal{H} - \mu \mathcal{N} \rangle_{\text{eq vac}} - \langle \mathcal{H} - \mu \mathcal{N} \rangle_{\text{vac }\lambda=0} \approx -\frac{1}{2}Nmc_\perp^2 , \qquad (7.30)$$

where $N = nV$.

7.3.3 Fundamental constants and Planck scales

As in the case of the weakly interacting Bose gas, the Theory of Everything for weakly interacting Fermi gas contains four 'fundamental' constants. They can be chosen as $\hbar$, m, $c_\perp$ and n (or p_F, which is related to n). The small factor, which determines the relative smallness of the superfluid energy with respect to the energy of an ideal Fermi gas in eqn (7.22), is again $mc_\perp a_0/\hbar \ll 1$, where $a_0 \sim \hbar/p_F \sim n^{-1/3}$ is the interparticle distance in the Fermi gas vacuum state.

Because of the anisotropy of the energy spectrum there are three important energy scales – the 'Planck' scales:

$$E_{\text{Planck 1}} = mc_\perp^2 , \quad E_{\text{Planck 2}} = c_\perp p_F \sim \frac{\hbar c_\perp}{a_0} , \quad E_{\text{Planck 3}} = mc_\parallel^2 = \frac{p_F^2}{m} , \qquad (7.31)$$

with $E_{\text{Planck 1}} \ll E_{\text{Planck 2}} \ll E_{\text{Planck 3}}$.

Below the first Planck scale $E \ll E_{\text{Planck 1}} = mc_\perp^2$, the energy spectrum of quasiparticles has the relativistic form in eqn (7.16) (or in eqn (7.27)), and the effective 'relativistic' quantum field theory arises in the low-energy corner with $c_\perp$ and $c_\parallel$ being the fundamental constants. This Planck scale marks the border where 'Lorentz' symmetry is violated.

The second Planck scale $E_{\text{Planck 2}}$ is responsible for superfluidity. Below this scale one can distinguish the superfluid state of the Fermi gas from its normal state. The temperature of the superfluid phase transition $T_c \sim E_{\text{Planck 2}}$. It provides the natural cut-off for the contribution of the Dirac vacuum to the vacuum energy, and thus determines the part of the vacuum energy in eqn (7.28), which is the difference between the energies of normal and superfluid states:

$$\langle \mathcal{H} - \mu \mathcal{N} \rangle_{\text{eq vac}} - \langle \mathcal{H} - \mu \mathcal{N} \rangle_{\text{vac }\lambda=0} = -\frac{1}{12\pi^2}V\sqrt{-g}E_{\text{Planck 2}}^4 . \qquad (7.32)$$

Here $g = -c_\parallel^{-2}c_\perp^{-4}$ is the determinant of the effective metric for quasiparticles $g_{\mu\nu} = \text{diag}(-1, c_\perp^{-2}, c_\perp^{-2}, c_\parallel^{-2})$.

This can be compared to the analog contribution to the vacuum energy in the Bose gas in eqn (3.25). Both contributions are proportional to the square root of the determinant of the effective Lorentzian metric, which arises in the low-energy corner, and thus they represent the Einstein cosmological term. Both contributions agree with the point of view of an inner observer, who knows only the low-energy excitations and believes that the vacuum energy comes from the Dirac vacuum of fermions and from the zero-point energy of bosons. They have correspondingly negative and positive signs as in eqn (2.11) for the cosmological

7.3.4 Vanishing of vacuum energy in liquid ^{3}He

In the Fermi system under consideration, both contributions to the vacuum energy density $\tilde{\epsilon}$ are negative:

$$\tilde{\epsilon} = -\frac{1}{12\pi^2 \hbar^3}\sqrt{-g}E^4_{\text{Planck 2}} - \frac{1}{30\pi^2 \hbar^3}\sqrt{-g}E^3_{\text{Planck 3}}E_{\text{Planck 2}} . \tag{7.33}$$

Thus the weakly interacting Fermi gas can exist only under external positive pressure $P = -\tilde{\epsilon}$. However, when the interaction between the bare particles (atoms) increases, the quantum effects become less important.

In liquids the quantum effects and interaction are of the same order. When those high-energy degrees of freedom are considered that mostly contribute to the construction of the vacuum state, one finds that the symmetry breaking (superfluidity) and even the difference in quantum statistics of ^{3}He and of ^{4}He atoms have a minor effect (see Fig. 3-6 in the book by Anderson 1984). For the main contribution to the vacuum energy the difference between the strongly interacting Bose system of ^{4}He atoms and the strongly interacting Fermi system of ^{3}He atoms is not very big. Both systems represent quantum liquids which can exist without an external pressure. The chemical potential of liquid ^{3}He, if counted from the energy of an isolated ^{3}He atom, is also negative ($\mu \sim -2.47$ K (Woo 1976; Dobbs 2000), i.e. the superfluid ^{3}He is a liquid-like (not a gas-like) substance. Thus in both systems the equilibrium value of the vacuum energy is exactly zero, $\langle \mathcal{H} - \mu \mathcal{N} \rangle_{\text{eq vac}} = \tilde{\epsilon}V = 0$, if there are no external forces acting on the liquid. On the other hand, the inner observer believes that the vacuum energy essentially depends on the fermionic and bosonic content of the effective theory.

7.3.5 Vacuum energy in non-equilibrium

We know that, if the system is liquid and thus can exist in the absence of an environment, its vacuum energy density $\tilde{\epsilon}_{vac}$ is zero in complete equilibrium. What is the value of the vacuum energy, and thus of the 'cosmological' term, if the order parameter (and thus the vacuum) is out of equilibrium? To understand this we first note that the BCS theory is applicable to the real liquid too. Though the particle–particle interaction in the liquid is strong, the effective interaction which leads to the Cooper pairing and thus to superfluidity can be relatively weak. This is just what occurs in liquid ^{3}He, which is manifested by a relatively small value of the order parameter: $c_\perp^2/v_F^2 \sim 10^{-5}$. Thus we can apply the BCS scheme discussed above to the liquids.

We must take a step back and consider the superfluid part of the vacuum energy, eqn (7.14), before the complete minimization over the order parameter.

For simplicity we fix the vector structure of the order parameter as in equilibrium, and vary only its amplitude $c_\perp$. Then, taking into account that in the isolated liquid the vacuum energy density $\tilde\epsilon$ must be zero when $c_\perp$ equals its equilibrium value $c_{\perp 0}$, one arrives at a rather simple expression for the vacuum energy density in terms of $c_\perp$:

$$\tilde\epsilon_{\rm vac}(c_\perp) = m^* n \left[c_\perp^2 \ln \frac{c_\perp^2}{c_{\perp 0}^2} + c_{\perp 0}^2 - c_\perp^2 \right]. \tag{7.34}$$

Equation (7.34) contains four 'fundamental' parameters of different degrees of fundamentality: $\hbar$, which is really fundamental; p_F and v_F, which are fundamental at the Fermi liquid level (here $m^* = p_F/v_F$ and $n = p_F^3/6\pi^2\hbar^2$); and the equilibrium value of the transverse 'speed of light' $c_{\perp 0}$, which is fundamental for the effective RQFT at low energy. Let us stress again that this equation is applicable to real liquids; the only requirement is that $p_F c_\perp$ is small compared to the energy scales relevant for the liquid state, in particular $p_F c_\perp \ll p_F v_F$.

7.3.6 Vacuum energy and cosmological term in bi-metric gravity

Equation (7.34) can be rewritten in terms of the effective metric in equilibrium $g_{\mu\nu(0)} = \mathrm{diag}(-1, c_{\perp 0}^{-2}, c_{\perp 0}^{-2}, c_\parallel^{-2})$, and the effective dynamical metric $g_{\mu\nu} = \mathrm{diag}(-1, c_\perp^{-2}, c_\perp^{-2}, c_\parallel^{-2})$:

$$\tilde\epsilon_{\rm vac}(g) = \rho_\Lambda \sqrt{-g_0} \left[\frac{\sqrt{-g_0}}{\sqrt{-g}} \ln \frac{\sqrt{-g_0}}{\sqrt{-g}} - \frac{\sqrt{-g_0}}{\sqrt{-g}} + 1 \right] , \quad \rho_\Lambda = \frac{(p_F c_{\perp 0})^4}{6\pi^2 \hbar^3}. \tag{7.35}$$

Close to equilibrium the vacuum energy density is quadratic in deviation from equilibrium:

$$\tilde\epsilon_{\rm vac}(g) \approx \frac{\rho_\Lambda}{2\sqrt{-g_0}} \left(\sqrt{-g} - \sqrt{-g_0} \right)^2. \tag{7.36}$$

Let us play with the result (7.35) obtained for a quantum liquid, considering it as a guess for the vacuum energy in the effective theory of gravity. Then the energy–momentum tensor of the vacuum – the cosmological term which enters the Einstein equation (2.8) – is obtained by variation of eqn (7.35) over $g^{\mu\nu}$:

$$T_{\mu\nu}^\Lambda = \rho_\Lambda g_{\mu\nu} \frac{g_0}{g} \ln \frac{\sqrt{-g_0}}{\sqrt{-g}}. \tag{7.37}$$

Thus in a quantum liquid the vacuum energy (7.35) and the cosmological term (7.37) have a different dependence on the metric field. But both of them are zero in equilibrium, where $g = g_0$. Moreover, the reasons for the vanishing of the cosmological term and for the vanishing of the vacuum energy are also different.

The zero value of the cosmological term in eqn (7.37) in equilibrium, which indicates that in eqiuilibrium the vacuum is not gravitating, is a consequence of the local stability of the vacuum. The latter implies that the correction to the vacuum energy due to non-equilibrium is the positively definite quadratic form of the deviations from equilibrium, such as eqn (7.36). This condition of local stability of the vacuum is valid for any system, i.e. not only for liquids but for weakly interacting gases too. The cosmological term is the variation of the vacuum energy over the metric, that is why it is linear in the deviation from equilibrium, and must be zero in equilibrium.

As for the vacuum energy itself it is zero if, in addition, the equilibrium vacuum is 'isolated from the environment'. This is valid for the quantum liquid but not for a gas-like substance such as the weakly interacting Fermi gas.

In both equations, (7.35) and (7.37), the overall factor – the 'bare' cosmological constant ρ_Λ – has the naturally expected Planck scale value of order E_{Planck}^4. Thus, in spite of a huge value for the bare cosmological constant, one obtains that the equilibrium quantum vacuum is not gravitating. The cosmological term and in the case of the liquid even the vacuum energy itself are zero in equilibrium without any fine tuning.

The price for this solution of the cosmological constant problem is that we now have the analog of the bi-metric theory of gravity: the metric $g_{\mu\nu(0)}$ characterizes the equilibrium vacuum, while the metric $g_{\mu\nu}$ is responsible for the dynamical gravitational field acting on quasiparticles. Actually the metric $g_{\mu\nu(0)}$ is also dynamical though at a deeper fundamental level: it depends on the characteristics of the underlying liquid, such as the density n of atoms.

7.4 Spin-triplet superfluids

7.4.1 Order parameter

In Sec. 7.2 we considered the equal spin pairing (ESP) of particles, i.e. the pairing of atoms/electrons having the same spin projection. We discussed only single spin projection, and thus the particles were effectively spinless. Now we turn to the spin structure of the Cooper pairs in ^{3}He liquids. In a particular case of ESP, when particles forming the Cooper pair have the same spin projection, the total spin projection of the Cooper pair is either $+1$ or -1. These are particular states arising in spin-triplet pairing, the pairing into the $S = 1$ state. The general form of the order parameter for triplet pairing is the direct modification of eqn (7.3) when the spin indices are taken into account:

$$\frac{2\lambda}{V}\left\langle\sum_{\mathbf{p}}\mathbf{p}a_{\mathbf{p}\alpha}a_{-\mathbf{p}\beta}\right\rangle_{\text{vac}} = \mathbf{e}_\mu(\sigma^{(\mu)}\mathbf{g})_{\alpha\beta}\ ,\ \mathbf{g} = i\sigma^{(2)}. \tag{7.38}$$

Here $\alpha = (\uparrow,\downarrow)$ and $\beta = (\uparrow,\downarrow)$ denote the spin projections of a particle; and $\sigma^{(\mu)}$ with $\mu = 1, 2, 3$ are the 2×2 Pauli matrices.

Instead of the complex vector in the spinless case, the order parameter now is triplicated and becomes a 3×3 complex matrix $\mathbf{e}_\mu \equiv e_{\mu i}$, with $\mu = 1, 2, 3$ and $i = 1, 2, 3$. This order parameter $e_{\mu i}$ belongs to the vector representation $L = 1$ of the orbital rotation group $SO(3)_{\mathbf{L}}$ (whence the name p-wave pairing), and to the vector representation $S = 1$ of the spin rotation group $SO(3)_{\mathbf{S}}$ (whence the name spin-triplet pairing). In other words, $e_{\mu i}$ transforms as a vector under $SO(3)_{\mathbf{S}}$ spin rotations (the first index) and as a vector under $SO(3)_{\mathbf{L}}$ orbital rotations (the second index). Also, the order parameter is not invariant under $U(1)_N$

symmetry operations, which is responsible for the conservation of the particle number N (^{3}He atoms or electrons): under global $U(1)_N$ gauge transformation ($a_\mathbf{p} \to e^{i\alpha} a_\mathbf{p}$) the order parameter transforms as

$$e_{\mu i} \to e^{2i\alpha} e_{\mu i} \,. \tag{7.39}$$

The factor 2 is because the order parameter $e_{\mu i}$ is the vev of two annihilation operators $a_\mathbf{p}$.

Thus all the symmetry groups of the liquid in the normal state

$$G = SO(3)_\mathbf{L} \times SO(3)_\mathbf{S} \times U(1)_N \tag{7.40}$$

are completely or partially broken in spin-triplet p-wave superfluids and superconductors.

7.4.2 Bogoliubov–Nambu spinor

The modification of the spinless BCS Hamiltonian (7.6) to the spin-triplet case is straightforward. We mention only that there is another useful way to treat the BCS Hamiltonian. Since in this Hamiltonian the states with $N+1$ particles (quasiparticle) and with $N-1$ particles (quasihole) are hybridized by the order parameter, it is instructive to double the number of degrees of freedom adding the antiparticle for each particle as an independent field. This is accomplished by constructing the Bogoliubov–Nambu field operator χ, which is a spinor in the new particle–hole space (Bogoliubov–Nambu space) as well as the spinor in conventional spin space:

$$\chi_\mathbf{p} = \begin{pmatrix} a_\mathbf{p} \\ i\sigma^{(2)} a^\dagger_{-\mathbf{p}} \end{pmatrix} \,. \tag{7.41}$$

Under a $U(1)_N$ symmetry operation this spinor transforms as

$$\chi \to e^{i\check{\tau}^3 \alpha} \chi \,, \tag{7.42}$$

where $\check{\tau}^b$ (with $b = 1, 2, 3$) are Pauli matrices in the Bogoliubov–Nambu space. $\check{\tau}^3$ is the particle number operator N with eigenvalues $+1$ for the particle component of the Bogoliubov quasiparticle and -1 for its hole component.

Two components of the Bogoliubov–Nambu spinor $\chi_\mathbf{p}$ are not independent: they form a so-called Majorana spinor. However, in what follows we ignore the connection and consider the two components of the spinor as independent. When calculating different quantities of the liquids using the Bogoliubov–Nambu fermions we must divide the final result by two in order to compensate the double counting. Then for each $\mathbf{p}$ the Bogoliubov–Nambu Hamiltonian $\mathcal{H}_\mathbf{p}$ in eqn (7.6) has the form

$$\mathcal{H}_\mathbf{p} = \check{\tau}^3 M(\mathbf{p}) + \check{\tau}^1 \sigma_\mu p_i \operatorname{Re} e_{\mu i} - \check{\tau}^2 \sigma_\mu p_i \operatorname{Im} e_{\mu i} \,, \tag{7.43}$$

where

$$M(\mathbf{p}) = \frac{p^2}{2m} - \mu \approx v_F(p - p_F) \,. \tag{7.44}$$

Since the order parameter is typically much smaller than p_F^2/m, in all the effects related to superfluidity the bare energy of particles $M(\mathbf{p})$ is concentrated close to p_F. (Note that here the chemical potential μ is counted from the bottom of the band and thus is positive; in liquids it is negative when counted from the energy of the isolated atoms.)

It is instructive to consider the following six vacuum states of p-wave superfluidity: ^{3}He-B, ^{3}He-A, ^{3}He-A$_1$, axiplanar, planar and polar states. Some of the states realize the true vacuum if the parameters of the Theory of Everything in eqn (3.2) are favourable for that. The other states correspond either to local minima (i.e. to a false vacuum) or to saddle points of the energy functional. The first three of them are realized in nature as superfluid phases of ^{3}He.

7.4.3 ^{3}He-B – fully gapped system

^{3}He-B is the only isotropic phase possible for spin-triplet p-wave superfluidity. In its simplest form the matrix order parameter is

$$e^{(0)}_{\mu i} = \Delta_0 \delta_{\mu i} \,. \tag{7.45}$$

This state has the quantum number $J = 0$, where $\mathbf{J} = \mathbf{L} + \mathbf{S}$ is the total angular momentum of the Cooper pair. $J = 0$ means that the quantum vacuum of the B-phase is isotropic under simultaneous rotations in spin and coordinate space, and thus ensures the isotropy of the liquid. The phase transition to ^{3}He-B corresponds to the symmetry-breaking scheme (see Sec. 13.1 for definition of the group G of the symmetry of physical laws and the group H of the symmetry of the degenerate quantum vacuum)

$$G = U(1)_N \times SO(3)_\mathbf{L} \times SO(3)_\mathbf{S} \to H_B = SO(3)_{\mathbf{S+L}} \tag{7.46}$$

All the degenerate quantum vacuum states are obtained from the simplest state (7.45) by symmetry transformations of the group G:

$$e_{\mu i} = R^S_{\mu\nu} R^L_{ik} e^{2i\alpha} e^{(0)}_{\nu k} = \Delta_0 R_{\mu i} e^{i\Phi} \,, \tag{7.47}$$

where $R_{\mu i}$ is the real orthogonal matrix (rotation matrix) resulting from the combined action of spin and orbital rotations: $R_{\mu i} = R^S_{\mu\nu} R^L_{ik} \delta_{\nu k}$; Φ is the phase of the condensate.

The simplest state in eqn (7.45) has the following Bogoliubov–Nambu Hamiltonian:

$$\mathcal{H}_B = \begin{pmatrix} M(\mathbf{p}) & c(\boldsymbol{\sigma} \cdot \mathbf{p}) \\ c(\boldsymbol{\sigma} \cdot \mathbf{p}) & -M(\mathbf{p}) \end{pmatrix} = M(\mathbf{p}) \check{\tau}^3 + c(\boldsymbol{\sigma} \cdot \mathbf{p}) \check{\tau}^1 \,, \; c = \frac{\Delta_0}{p_F} \,, \tag{7.48}$$

with $M(\mathbf{p})$ given in eqn (7.44). The square of the energy of fermionic quasiparticles in this pair-correlated state is

$$E_B^2(\mathbf{p}) = \mathcal{H}^2 = M^2(\mathbf{p}) + c^2 p^2 \,. \tag{7.49}$$

This equation is invariant under the group G and thus is valid for any degenerate state in eqn (7.47).

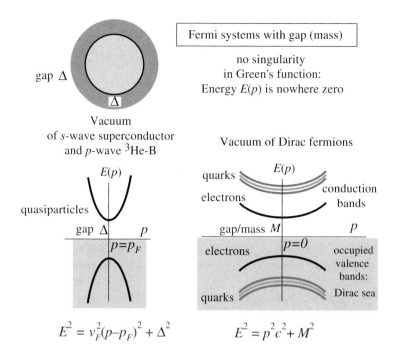

FIG. 7.2. Fermi systems with an energy gap, or mass. *Top left*: The gap appearing on the Fermi surface in conventional superconductors and in ^{3}He-B. *Bottom left*: The quasiparticle spectrum in conventional superconductors and in ^{3}He-B. *Bottom right*: The spectrum of Dirac particles and of quasiparticles in semiconductors.

7.4.4 *From Bogoliubov quasiparticle to Dirac particle. From ^{3}He-B to Dirac vacuum*

If M were independent of $\mathbf{p}$, eqns (7.48) and (7.49) would represent the Hamiltonian and the energy spectrum of Dirac particles with mass M. However, in our case $M(\mathbf{p})$ depends on $\mathbf{p}$ in an essential way. In the weak-coupling approximation (which corresponds to the BCS theory) the magnitude Δ_0 of the order parameter is small, so that the minimum of the energy is located close to $p = p_F$. Near the minimum of the energy, one has

$$E_{\mathrm{B}}^2(\mathbf{p}) \approx v_F^2(p - p_F)^2 + \Delta_0^2 \ , \qquad (7.50)$$

showing that Δ_0 plays the role of the gap in the quasiparticle energy spectrum.

One can continuously deform the energy spectrum of the ^{3}He-B Bogoliubov quasiparticle in eqn (7.49) to the spectrum of Dirac particles. This can be done, for example, by increasing c. In the limit of large c one obtains the Dirac Hamiltonian with the mass determined by the chemical potential:

$$mc^2 \gg |\mu| \; : \; M(\mathbf{p}) \approx -\mu \, , \; \mathcal{H} \to \mathcal{H}_{\text{Dirac}} = cp_i\alpha^i - \mu\beta \, , \; \mathcal{H}^2_{\text{Dirac}} = E^2 = \mu^2 + c^2p^2 \, ,$$
(7.51)

where α^i and β are Dirac matrices. This shows that ^{3}He-B and the Dirac vacuum belong to the same universality class. This is the class with trivial topology in momentum space: there are no points, lines or surfaces in momentum space where the energy vanishes (Fig. 7.2).

Actually, the absence of zeros in the energy spectrum does not necessarily imply trivial topology in momentum space. We shall see on examples of 2+1 systems (films) in Chapter 11 that $\mathbf{p}$-space topology can be non-trivial even for the fully gapped spectrum, so that the non-zero topological charge leads to quantization of physical parameters. Such a situation is fairly typical for even space dimensions (2+1 and 4+1), but can in principle occur in 3+1 dimensions too.

^{3}He-B is topologically equivalent to the Dirac vacuum and contains interacting fermionic and bosonic (propagating oscillations of the order parameter $e_{\mu i}$) quantum fields, whose dynamics is determined by the quantum field theory. However, ^{3}He-B does not provide the *relativistic* quantum field theory needed for the simulation of the quantum vacuum: there is no Lorentz invariance and some components of the order parameter only remotely resemble gravitons. Nevertheless, the analogy with the Dirac vacuum can be useful, and ^{3}He-B can serve as a model system for simulations of different phenomena in particle physics and cosmology. In particular, the symmetry-breaking pattern in superfluid ^{3}He-B was used for the analysis of color superconductivity in quark matter (Alford *et al.* 1998; Wilczek 1998). Nucleation of quantized vortices observed in non-equilibrium phase transitions by Ruutu *et al.* (1996a, 1998) and nucleation of other topological defects – spin–mass vortices – observed by Eltsov *et al.* (2000) served as experimental simulations in ^{3}He-B of the Kibble (1976) mechanism, describing the formation of cosmic strings during a symmetry-breaking phase transition in the expanding Universe. Vortices in ^{3}He-B were also used by Bevan *et al.* (1997b) for the experimental simulation of baryon production by cosmic strings mediated by spectral flow (see Chapter 23).

7.4.5 *Mass generation for Standard Model fermions*

Finally, let us discuss the connection with the electroweak phase transition, where the originally gapless fermions (see Sec. 8.2.1 below) also acquire mass. In eqn (7.48) for ^{3}He-B, particles with the energy spectrum $M(\mathbf{p})$ are hybridized with holes whose energy spectrum is $-M(\mathbf{p})$. The order parameter $e^{(0)}_{\mu i}$ provides the off-diagonal matrix element which mixes particles and holes, and thus does not conserve particle number. The resulting object is neither a particle nor a hole, but their combination – the Bogoliubov quasiparticle.

In the electroweak transition, the right-handed particles with the spectrum $c(\sigma \cdot \mathbf{p})$ and the left-handed particles with the spectrum $-c(\sigma \cdot \mathbf{p})$ are hybridized. For example, below the transition, the common Hamiltonian for left and right electrons becomes

$$\mathcal{H}_e = \begin{pmatrix} c(\sigma \cdot \mathbf{p}) & M \\ M^* & -c(\sigma \cdot \mathbf{p}) \end{pmatrix} , \quad E^2(\mathbf{p}) = \mathcal{H}_e^2 = c^2 p^2 + |M|^2 . \quad (7.52)$$

The off-diagonal matrix element M, mixing left and right electrons, is provided by the electroweak order parameter – the Higgs field in eqn (15.8): $M \propto \phi_2$. This gives the mass $M_e = |M|$ to the resulting electron, the quasiparticle, which is neither a left nor a right particle, but a combination of them – a Dirac particle. Thus, ignoring the difference in the relevant symmetry groups, the physics of mass generation for the Standard Model fermions is essentially the same as in superconductors and Fermi superfluids (Nambu and Jona-Lasinio 1961; Anderson 1963). For the further consideration of universality classes of fermionic vacua it is important to note that in the Standard Model, the primary excitations are not Dirac particles, but massless (gapless) fermions.

7.4.6 ^{3}He-A – superfluid with point nodes

The phase transition to ^{3}He-A corresponds to the following symmetry-breaking scheme (at the moment we consider continuous symmetries only and do not take into account discrete symmetries):

$$G = U(1)_N \times SO(3)_{\mathbf{L}} \times SO(3)_{\mathbf{S}} \to H_A = U(1)_{L_z - N/2} \times U(1)_{S_z} . \quad (7.53)$$

The order parameter with the residual symmetry H_A is

$$e^{(0)}_{\mu i} = \Delta_0 \hat{z}_\mu (\hat{x}_i + i\hat{y}_i) . \quad (7.54)$$

It is symmetric under spin rotations about the spin axis $\hat{z}_\mu$, with S_z being the generator of rotations. Orbital rotation by an angle β about the orbital axis $\hat{z}_i$ transforms $(\hat{x}_i + i\hat{y}_i)$ to $e^{-i\beta}(\hat{x}_i + i\hat{y}_i)$. This transformation can be compensated by a phase rotation (from the group $U(1)_N$) with an angle $\alpha = \beta/2$. Thus the order parameter is symmetric under combined orbital and phase rotations with the generator $L_z - N/2$.

The general form of the order parameter is obtained from eqn (7.54) by the action of the group G. It is the product of the spin part described by the real unit vector $\hat{\mathbf{d}}$, obtained from $\hat{z}_\mu$ by spin rotations $SO(3)_{\mathbf{S}}$, and the orbital part described by the complex vector $\hat{\mathbf{m}} + i\hat{\mathbf{n}}$ in eqn (7.25), obtained from $\hat{x}_i + i\hat{y}_i$ by orbital rotations $SO(3)_{\mathbf{L}}$:

$$e_{\mu i} = R^S_{\mu\nu} R^L_{ik} e^{2i\alpha} e^{(0)}_{\nu k} = \Delta_0 \hat{d}_\mu (\hat{m}_i + i\hat{n}_i) , \quad \hat{\mathbf{m}} \cdot \hat{\mathbf{m}} = \hat{\mathbf{n}} \cdot \hat{\mathbf{n}} = 1 , \quad \hat{\mathbf{m}} \cdot \hat{\mathbf{n}} = 0 . \quad (7.55)$$

Let us also mention the very important discrete Z_2 symmetry P of the A-phase vacuum. It is the symmetry of the vacuum under the following combined π rotations in spin and orbital spaces. For the simplest order parameter in eqn (7.54) this symmetry is

$$P = U^S_{2x} U^L_{2z} . \quad (7.56)$$

Here U^S_{2x} is the spin rotation by π around the axis $\hat{x}_\mu$ (or in general it is the rotation by π about any axis perpendicular to $\hat{\mathbf{d}}$). U^L_{2z} is the rotation by π about axis $\hat{z}_i$ (or in general about the $\hat{\bf l}$-vector); its action on the order parameter is equivalent to the phase rotation by $\alpha = \pi/2$. The combined action of these two groups is the symmetry operation, since each of the two symmetry operations changes the sign of the order parameter. The latter property is the reason why we call it the parity transformation. In many physical cases the combined symmetry does play the role of space inversion. It will be shown later that P is this discrete symmetry which gives rise to the Alice string (the half-quantum vortex, see Secs 15.3.1 and 15.3.2).

The Bogoliubov–Nambu Hamiltonian for the fermionic quasiparticles in ^{3}He-A has the form

$$\mathcal{H}_A = \begin{pmatrix} M({\bf p}) & \Delta({\bf p}) \\ \Delta^\dagger({\bf p}) & -M({\bf p}) \end{pmatrix} = M({\bf p})\check\tau^3 + c_\perp (\sigma\cdot\hat{\bf d})(\check\tau^1 \hat{\bf m}\cdot{\bf p} - \check\tau^2 \hat{\bf n}\cdot{\bf p})\ ,\ c_\perp = \frac{\Delta_0}{p_F}\ , \quad (7.57)$$

and the quasiparticle energy spectrum is

$$E_A^2({\bf p}) = \mathcal{H}^2 = M^2({\bf p}) + c_\perp^2 ({\bf p}\times\hat{\bf l})^2\ ,\quad \hat{\bf l} = \hat{\bf m}\times\hat{\bf n}\ . \quad (7.58)$$

The unit vector $\hat{\bf l}$ shows the anisotropy axis of the quasiparticle spectrum, and also determines the direction of the orbital momentum of Cooper pairs. This energy spectrum has two zeros – Fermi points – at ${\bf p}^{(a)} = q_a p_F \hat{\bf l}$ with $q_a = \pm 1$ (Fig. 7.3). As we discussed in Sec. 7.2.4, eqn (7.16), the relativistic spectrum emerges in the vicinity of each node.

Equation (7.57) shows that the spin projection of quasiparticles on the axis $\hat{\bf d}$ is a good quantum number. Thus the Hamiltonians for quasiparticles with spin projection $S_z = +(1/2)\sigma\cdot\hat{\bf d}$ and with spin projection $S_z = -(1/2)\sigma\cdot\hat{\bf d}$ are independent. This demonstrates that ^{3}He-A represents one of several possible ESP states, where the pairing occurs independently for each spin projection. The two Hamiltonians, $\mathcal{H}_\pm = M({\bf p})\check\tau^3 \pm c_\perp(\check\tau^1\hat{\bf m}\cdot{\bf p} - \check\tau^2\hat{\bf n}\cdot{\bf p})$, differ only by the sign in front of $\hat{\bf m}$ and $\hat{\bf n}$, i.e. they have the same direction of $\hat{\bf l}$. Thus in this particular ESP state the Cooper pairs for both spin populations have the same axial orbital structure and the same direction of the orbital momentum $\hat{\bf l}$. The vacuum of ^{3}He-A is axial.

7.4.7 Axiplanar state – flat directions

The more general ESP state, in which the direction of the orbital momentum is different for two spin projections, is represented by the axiplanar state (Fig. 7.3). The order parameter in the axiplanar state has the form

$$e_{\mu i} = \frac{1}{2}\Delta_\uparrow \left(\hat{\bf d}' + i\hat{\bf d}''\right)_\mu (\hat{\bf m}_\uparrow + i\hat{\bf n}_\uparrow)_i + \frac{1}{2}\Delta_\downarrow \left(\hat{\bf d}' - i\hat{\bf d}''\right)_\mu (\hat{\bf m}_\downarrow + i\hat{\bf n}_\downarrow)_i\ . \quad (7.59)$$

Here $\hat{\bf d}'\cdot\hat{\bf d}' = \hat{\bf d}''\cdot\hat{\bf d}'' = 1$, $\hat{\bf d}'\cdot\hat{\bf d}'' = 0$; $\Delta_\uparrow$ is the gap amplitude of the pairing of the spin-up fermions with respect to the axis $\hat{\bf s} = \hat{\bf d}'\times\hat{\bf d}''$, while $\Delta_\downarrow$ is the gap amplitude for the spin-down population; the directions of the orbital momenta of Cooper pairs are correspondingly $\hat{\bf l}_\uparrow = \hat{\bf m}_\uparrow\times\hat{\bf n}_\uparrow$ and $\hat{\bf l}_\downarrow = \hat{\bf m}_\downarrow\times\hat{\bf n}_\downarrow$. In zero

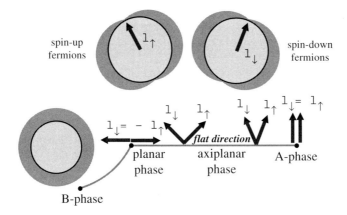

FIG. 7.3. Positions of the Fermi points in the A (^{3}He-A), axiplanar and planar phases. In the model BCS Hamiltonian used, the pairing occurs independently for spin-up and spin-down fermions leading to hidden symmetry: the energy of the state (solid line) does not depend on the mutual directions of the orbital vectors $\hat{\mathbf{l}}_\uparrow$ and $\hat{\mathbf{l}}_\downarrow$, which show the directions to the point nodes in the spectrum of spin-up and spin-down quasiparticles, respectively. This is the reason for the so-called flat direction in configurational space: the energy is the same all the way from the A-phase to the planar phase. The fully gapped B-phase (^{3}He-B) has the lowest energy among the considered p-wave spin-triplet states. The energy monotonously decreases on the way from the planar to the B-phase. In real liquid ^{3}He the hidden symmetry is approximate, so that both the B-phase and the A-phase correspond to local minima of energy. The model BCS Hamiltonian with the hidden symmetry and flat directions is believed to work best at low pressure.

external magnetic field, spin-up and spin-down pairs have the same amplitudes $\Delta_\uparrow = \Delta_\downarrow \equiv \Delta_0$

In ^{3}He-A spin-up and spin-down pairs have the same directions of orbital vectors $\hat{\mathbf{l}}_\uparrow = \hat{\mathbf{l}}_\downarrow \equiv \hat{\mathbf{l}}$. But they can have different phases: for example, $\hat{\mathbf{m}}_\uparrow + i\hat{\mathbf{n}}_\uparrow = (\hat{\mathbf{x}} + i\hat{\mathbf{y}})^{i\Phi_\uparrow}$ and $\hat{\mathbf{m}}_\downarrow + i\hat{\mathbf{n}}_\downarrow = (\hat{\mathbf{x}} + i\hat{\mathbf{y}})^{i\Phi_\downarrow}$. In this case one obtains eqn (7.55) with $\hat{\mathbf{d}} = \hat{\mathbf{d}}' \cos \frac{\Phi_\uparrow - \Phi_\downarrow}{2} - \hat{\mathbf{d}}'' \sin \frac{\Phi_\uparrow - \Phi_\downarrow}{2}$ and with $\hat{\mathbf{m}} + i\hat{\mathbf{n}} = (\hat{\mathbf{x}} + i\hat{\mathbf{y}})^{i(\Phi_\uparrow + \Phi_\downarrow)/2}$. This demonstrates that the phase difference between the two populations is equivalent to rotation of the spin part $\hat{\mathbf{d}}$ of the order parameter.

7.4.8 ^{3}He-A_1 – Fermi surface and Fermi points

In the ^{3}He-A_1 phase only fermions of one spin population are paired. The order parameter is given by eqn (7.59) with, say, $\Delta_\downarrow = 0$. Such a state occurs in the presence of an external magnetic field immediately below the phase transition from the normal state. This is an example of the coexistence of topologically different zeros in the energy spectrum: the spin-down quasiparticles have a Fermi surface, while the spin-up quasiparticles have two Fermi points. The residual symmetry of this state is

$$H_{A_1} = U(1)_{L_z - N/2} \times U(1)_{S_z - N/2} . \tag{7.60}$$

7.4.9 Planar phase – marginal Fermi points

In the so-called planar phase two spin populations haves opposite directions of the orbital angular momentum, $\hat{\mathbf{l}}_\uparrow = -\hat{\mathbf{l}}_\downarrow$, as a result the time reversal symmetry is not broken and the vacuum of the planar phase is not axial. The order parameter has the form

$$e_{\mu i} = \Delta_0 \left(\hat{d}'_\mu \hat{m}_i + \hat{d}''_\mu \hat{n}_i \right) e^{i\Phi} . \tag{7.61}$$

If $\Phi = 0$ this can be obtained from eqn (7.59) with $\Delta_\uparrow = \Delta_\downarrow = \Delta_0$, $\hat{\mathbf{m}}_\uparrow = \hat{\mathbf{m}}_\downarrow \equiv \hat{\mathbf{m}}$, $\hat{\mathbf{n}}_\uparrow = -\hat{\mathbf{n}}_\downarrow \equiv \hat{\mathbf{n}}$. The Bogoliubov–Nambu Hamiltonian for the planar state is

$$\mathcal{H}_{\text{planar}} = M(\mathbf{p})\check{\tau}^3 + c_\perp \check{\tau}^1 ((\hat{\mathbf{m}} \cdot \mathbf{p})(\sigma \cdot \hat{\mathbf{d}}') + (\hat{\mathbf{n}} \cdot \mathbf{p})(\sigma \cdot \hat{\mathbf{d}}'')) . \tag{7.62}$$

The square of the energy spectrum

$$E^2_{\text{planar}}(\mathbf{p}) = \mathcal{H}^2 = M^2(\mathbf{p}) + c_\perp^2 (\mathbf{p} \times \hat{\mathbf{l}})^2 \ , \ \hat{\mathbf{l}} = \hat{\mathbf{m}} \times \hat{\mathbf{n}} , \tag{7.63}$$

coincides with that of the A-phase in eqn (7.58). The energy spectrum also has two zeros at $\mathbf{p}^{(a)} = \pm p_F \hat{\mathbf{l}}$.

At first glance the planar state has the same fermionic spectrum with the same point gap nodes as ^{3}He-A. At least this is what follows from comparison of eqn (7.58) and eqn (7.63). However, the 'square roots' of E^2 for the planar phase (7.62) and for ^{3}He-A (7.57) are different. We shall see in Sec. 12.1.1 that the topology of the point nodes is different in the two systems: the topological charges of the gap nodes of two spin populations are added up in ^{3}He-A, but compensate each other in the planar phase. This means that in contrast to ^{3}He-A, the planar state is marginal, being topologically unstable toward the closing of the nodes.

The planar phase is energetically unstable toward the fully gapped B-phase (Fig. 7.3), say, along the following path: $e_{\mu i}(M) = \Delta_0 (\hat{x}_\mu \hat{x}_i + \hat{y}_\mu \hat{y}_i) + M\hat{z}_\mu \hat{z}_i$, where the parameter M changes from $M = 0$ in the planar state to $M = \Delta_0$ in the B-phase. At any non-zero M the quasiparticle spectrum $E^2(\mathbf{p}) = M^2(\mathbf{p}) + c_\perp^2 (p_x^2 + p_y^2 + M^2 p_z^2/p_F^2)$ acquires the finite gap M. It is an analog of the Dirac mass, since for small M the low-energy spectrum has the form

$$E_a^2(\mathbf{p}) \approx v_F^2 (p_z - q_a p_F)^2 + c_\perp^2 (p_x^2 + p_y^2) + M^2 . \tag{7.64}$$

In complete analogy with the Standard Model fermions, the left-handed and right-handed quasiparticles of the planar state are hybridized to form the Dirac particle with the mass M.

The residual symmetry of the planar state includes the continuous and discrete symmetries

$$H_{\text{planar}} = U(1)_{L_z+S_z} \times P ,\qquad(7.65)$$

where the element of the discrete symmetry P is

$$P = U_{2z}^S e^{(\pi/2)N} .\qquad(7.66)$$

It contains the spin rotation by angle π about the $\hat{z}_\mu$ axis combined with $U(1)_N$ phase rotation by angle $\alpha = \pi/2$; each of them changes the sign of the order parameter and thus plays the part of space inversion. The discrete symmetry P fixes $M = 0$ and thus protects the gapless (massless) Dirac quasiparticles.

The vacuum of the planar phase has similar properties to that of the Standard Model of electroweak interactions. In both systems the Fermi points are marginal being described by similar momentum space topological invariants protected by symmetry. When the symmetry is broken, fermion zero modes disappear and elementary particles acquire mass. We discuss this later in Sec. 12.3.2.

7.4.10 Polar phase – unstable nodal lines

In the so-called polar phase the residual symmetry is

$$H_{\text{polar}} = U(1)_{L_z} \times U(1)_{S_z} \times P_1 \times P_2 ,\qquad(7.67)$$

where the discrete symmetries are $P_1 = U_{2x}^S U_{2x}^L$ and $P_2 = U_{2x}^S e^{(\pi/2)N}$. The general form of the order parameter is

$$e_{\mu i} = \Delta_0 \hat{d}_\mu \hat{e}_i\, e^{i\Phi} ,\qquad(7.68)$$

with real unit vectors $\hat{e}$ and $\hat{d}$. From the square of the energy spectrum,

$$E^2_{\text{polar}}(\mathbf{p}) = M^2(\mathbf{p}) + c_\perp^2 (\mathbf{p} \cdot \hat{e})^2 ,\qquad(7.69)$$

one finds that the energy of quasiparticles is zero on a line in $\mathbf{p}$-space. Here it is the circumference in the equatorial plane: $p = p_F$, $\mathbf{p} \cdot \hat{e} = 0$. We already know (see Sec. 7.2.5) that the polar state is marginal, since the lines of zeros are not protected by topology. These fermion zero modes are, however, protected by the symmetry H_{polar} of the polar state in eqn (7.67). When the symmetry is violated, nodal lines disappear.

8
UNIVERSALITY CLASSES OF FERMIONIC VACUA

Now we proceed to effective theories of quantum fermionic liquids. In the low-energy limit the type of the effective theory depends on the structure of the quasiparticle spectrum, which in turn is determined by the universality class of the Fermi system.

In the previous chapter several different types of fermionic quasiparticle spectra have been discussed: (i) ^{3}He-A$_1$ has a Fermi surface for one of the two spin populations – the 2D surface in 3D **p**-space where the energy is zero. (ii) In ^{3}He-A and in the axiplanar states the spectrum has stable point nodes. The planar state also has point nodes in the quasiparticle spectrum, but these point zeros disappear at arbitrarily small perturbations violating the symmetry of the planar state. (iii) ^{3}He-B as well as conventional s-wave superconductor has a fully gapped spectrum. (iv) Finally the polar phase has lines of zeros, which can be destroyed by small perturbations of the order parameter.

Why do some types of zeros seem to be stable, while others are destroyed by perturbations? The question is very similar to the problem of stability of extended structures, such as vortices and domain walls: Why do some defects appear to be stable, while others can be continuously unwound by deformations? The answer is given by topology which studies the properties robust to continuous deformations. For extended objects it is the topology operating in real **r**-space; in the case of the energy spectrum it is the **p**-space topology.

The **p**-space topology distinguishes three main generic classes of the stable fermionic spectrum in the quantum vacuum of a 3+1 fermionic system: (i) vacua with Fermi surfaces; (ii) vacua with Fermi points; and (iii) vacua with a fully gapped fermionic spectrum. Systems with the Fermi lines (nodal lines) in the spectrum are topologically unstable and by small perturbations can be transformed to one of the three classes; that is why they do not enter this classification scheme.

The same topological classification is applicable to the fermionic vacua in high-energy physics. The vacuum of the Weyl fermions in the Standard Model, with the excitation spectrum $E^2(\mathbf{p}) = c^2|\mathbf{p}|^2$, belongs to the class (ii); as we shall see below, this class is very special: within this class RQFT with chiral fermions emerges in the low-energy corner even in the originally non-relativistic fermionic system. The vacuum of Dirac fermions, with the excitation spectrum $E^2(\mathbf{p}) \to M^2 + c^2|\mathbf{p}|^2$, belongs to the class (iii) together with conventional superconductors and ^{3}He-B; this similarity allowed the methods developed for superconductors to be applied to RQFT (Nambu and Jona-Lasinio 1961). And finally, in strong fields the vacuum can acquire a Fermi surface (an example of the Fermi surface arising beyond the event horizon will be discussed in Sec. 32.4).

^{3}He liquids present examples of all three classes of homogeneous fermionic vacua (Fig. 7.1). The normal ^{3}He liquid at $T > T_c$ and also the superfluid ^{3}He phases in the 'high-energy' limit, i.e. at energy $E \gg \Delta_0$, are representative of the class (i). Below the superfluid transition temperature T_c one has either an anisotropic superfluid ^{3}He-A, which belongs to the class (ii), where RQFT with chiral fermions gradually arises at low temperature, or an isotropic superfluid ^{3}He-B of the class (iii).

A universality class unites systems of different origin and with different interactions. Within each of these universality classes one can find the system of non-interacting fermions: namely, Fermi gas of non-interacting non-relativistic fermions in class (i); non-interacting Weyl fermions in class (ii); and non-interacting Dirac fermions in class (iii).

We shall not consider here the class (iii) with trivial momentum space topology. Since there are no fermion zero modes in the vacuum of this class, the generic excitations of such a vacuum must have a mass of order of the Planck energy scale. Considering the other two classes of quantum vacua, which contain fermion zero modes, we start in each case with the non-interacting systems, and show how the fermion zero modes are protected by topology when the interaction is turned on.

The object whose topology is relevant must be related to the energy spectrum of the propagating particle or quasiparticle. It turns out to be the propagator – the Green function, and more precisely its Fourier components. For the non-interacting or simplified systems the Green function is expressed in terms of the Hamiltonian in **p**-space. That is why in these cases the universality classes can be expressed in terms of the topologically different classes of the Hamiltonians in **p**-space. But in general it is the topology of the propagator of the fermionic field which distinguishes classes of the fermionic vacua.

Let us start with the universality class (i).

8.1 Fermi surface as topological object

8.1.1 *Fermi surface is the vortex in momentum space*

A Fermi surface naturally arises in Fermi gases, where it marks the boundary in **p**-space between the occupied states ($n_{\bf p} = 1$) and empty states ($n_{\bf p} = 0$). It is clear that in the ideal Fermi gas the Fermi surface is a stable object: if the energy of particles is deformed the boundary between the occupied and empty states does not disappear. Small deformations lead only to the change of shape of the Fermi surface. Thus the Fermi surface is locally stable resembling the stability of the domain wall in Ising ferromagnets, which separates domains with spin up and spin down. The role of the Ising variable is played by $I \equiv n_{\bf p} - 1/2 = \pm 1/2$.

If the interaction between particles is introduced, the distribution function $n_{\bf p}$ of particles in the ground state of the system is no longer exactly 1 or 0. Nevertheless, it appears that the Fermi surface survives as the singularity in $n_{\bf p}$. Such stability of the Fermi surface comes from a topological property of

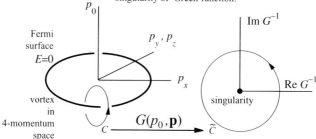

FIG. 8.1. Fermi surface as a topological object in momentum space. *Top*: In a Fermi gas the Fermi surface bounds the solid Fermi sphere of the occupied negative energy states. *Bottom*: The Fermi surface survives even if an interaction between the particles is introduced. The reason is that the Fermi surface is a topologically stable object: it is a vortex in the 4-momentum space $(p_0, \mathbf{p})$.

the Feynman quantum mechanical propagator for fermionic particles – the one-particle Green function

$$\mathcal{G} = (z - \mathcal{H})^{-1} \ . \tag{8.1}$$

Let us write the propagator for a given momentum $\mathbf{p}$ and for an imaginary frequency, $z = ip_0$. The imaginary frequency is introduced to avoid the conventional singularity of the propagator 'on the mass shell', i.e. at $z = E(p)$. For non-interacting particles the propagator has the form

$$G = \frac{1}{ip_0 - v_F(p - p_F)} \ . \tag{8.2}$$

Obviously there is still a singularity: on the 2D hypersurface $(p_0 = 0, p = p_F)$ in the 4-momentum space $(p_0, \mathbf{p})$ the propagator is not well defined. This singularity is stable, i.e. it cannot be eliminated by small perturbations. The reason is that the phase Φ of the Green function $G = |G|e^{i\Phi}$ changes by 2π around the path C embracing the element of the 2D hypersurface in the 4-momentum space.

This can be easily visualized if we skip one spatial dimension; then the Fermi surface is the closed line in the 2D space (p_x, p_y). The singularities of the propagator are lying on a closed line in the 3-momentum space (p_0, p_x, p_y) at the bottom of Fig. 8.1. The phase Φ of the propagator changes by 2π around any path C embracing any element of this vortex loop in the 3-momentum space.

The phase winding number $N_1 = 1$ cannot change continuously; that is why it is robust toward any perturbation. The singularity of the Green function on the 2D surface in the frequency–momentum space and thus the fermion zero modes near the Fermi surface are preserved, even when interactions between particles are introduced.

The properties of the systems which are robust under deformations are usually described by topology. All the configurations (in momentum, coordinate or mixed momentum–coordinate spaces and spacetimes) can be distributed into classes. The configurations within a given topological class can be continuously deformed into each other, while the configurations of different classes cannot. These classes typically form a group and can be described by the group elements. In most of the cases the group (the homotopy group) is Abelian and the classes can be characterized by integer numbers, called topological charges. We shall discuss this in more detail in Part III which is devoted to topological defects.

In the simplest case of the complex scalar Green function for particles with a single spin projection, the topological chargewhich determines the stability of fermion zero mode is the winding number N_1 of the phase field $\Phi(p_0, \mathbf{p})$ in the 4-momentum space. The phase $\Phi(p_0, \mathbf{p})$ of the Green function realizes the mapping of the closed contours C in the 4-momentum space $(p_0, \mathbf{p})$ to the closed contours $\tilde{C}$ in the space of the phase Φ – the circumference S^1. The topological charge N_1 distinguishes the classes of homotopically equivalent contours $\tilde{C}$ on S^1. The group, whose elements are classes of contours, is called the fundamental homotopy group and is denoted as π_1. The homotopy group π_1 of the space S^1 is thus $\pi_1(S^1) = Z$ – it is the group of integers N_1.

Exactly the same fundamental group $\pi_1(S^1) = Z$ leads to the stability of quantized vortices in simple superfluids and superconductors, described by the complex order parameter $\Psi = |\Psi| e^{i\Phi}$. The phase Φ of the order parameter changes by $2\pi n_1$ around the path embracing the vortex line in 3D $\mathbf{r}$-space (or embracing the vortex sheet in 3+1 spacetime $(\mathbf{r}, t)$). The only difference is that, in the case of vortices, the phase is determined in $\mathbf{r}$-space or in $(\mathbf{r}, t)$ spacetime, instead of the 4-momentum space. Thus in systems with a Fermi surface the manifold of singularities of the Green function in $(p_0, \mathbf{p})$ space is topologically equivalent to a quantized vortex in 3+1 spacetime $(\mathbf{r}, t)$.

8.1.2 $\mathbf{p}$-space and $\mathbf{r}$-space topology

As a rule, in strongly correlated fermionic systems, there are no small parameters which allow us to treat these system perturbatively. Also there are not many models which can be solved exactly. Hence the qualitative description based on the universal features coming from the symmetry and topology of the ground state is instructive. In particular, it allows us to construct the effective low-energy theory of fermion zero modes in strongly correlated fermionic systems of a given universality class, which incorporates all the important features of this class. All the information on the symmetry and topology of the fermion zero modes is contained in the low-energy asymptote of the Green function of the fermionic fields, which is characterized by topological quantum numbers (see the book by Thouless (1998) for a review on the role of the topological quantum numbers in physics).

The Fermi surface and quantized vortex in Sec. 8.1.1 are two examples of the simplest topological classes of configurations. There are a lot of other configurations which are described by more complicated π_1 groups and also by higher homotopy groups (the second homotopy group π_2, the third homotopy group π_3, etc.) and by relative homotopy groups. However, the main scheme of the distributions of the configurations into topologically different classes is preserved. We shall not discuss the technical calculations of the corresponding groups, which can be found in the books and review papers by Mermin (1979), Michel (1980), Kléman (1983) and Mineev (1998) on the application of topological methods to the classification of defects in condensed matter systems.

These two examples also show that one can consider on the same grounds the topology of configurations in coordinate space and in momentum space. There can also be an interconnection between spacetime topology and the topology in the 4-momentum space (see e.g. Volovik and Mineev 1982). If the vacuum is inhomogeneous in spacetime, the propagator in the semiclassical approximation depends both on 4-momentum $(p_0, \mathbf{p})$ and on spacetime coordinates, i.e. $G(p_0, \mathbf{p}, t, \mathbf{r})$. The topology in the (4+4)-dimensional phase space describes: the momentum space topology of fermion zero modes in homogeneous vacua, which we discuss here; topological defects of the order parameter in spacetime (vortices, strings, monopoles, domain walls, solitons, etc. in Part III); the topology of the energy spectrum of fermion zero modes within the topological defects and edge states (Part V); the quantization of physical parameters (see Chapter 21) and the fermionic charges of the topologically non-trivial extended objects. Using the topological properties of the spectrum of fermion zero modes of the quantum vacuum or inside the topological object we are able to construct the effective theory which describes the low-frequency dynamics of the system, and using it to investigate many phenomena including the axial anomaly discussed in Part IV, vortex dynamics discussed in Part V, etc. The combined $(p_0, \mathbf{p}, t, \mathbf{r})$ space can be extended even further to include the space of internal or external parameters which characterize, say, the ground state of the system (particle density, pressure, magnetic field, etc.). When such a parameter changes it can cross the point of the quantum phase transition at which the topology of the quantum vacuum and its fermion zero modes changes abruptly (Sec. 8.2.8 and Chapter 21).

8.1.3 Topological invariant for Fermi surface

Let us return to the topology of the Fermi surface. In a more general situation the Green function is a matrix with spin indices. In addition, it has the band indices in the case of electrons in the periodic potential of crystals, and indices related to internal symmetry and to different families of quarks and leptons in the Standard Model. In such a case the phase of the Green function becomes meaningless; however, the topological property of the Green function remains robust. But now the integer momentum space topological invariant, which is responsible for the stability of the Fermi surface, is written in the general matrix form

$$N_1 = \mathbf{tr} \oint_C \frac{dl}{2\pi i} \mathcal{G}(p_0, p) \partial_l \mathcal{G}^{-1}(p_0, p) \,. \tag{8.3}$$

Here the integral is again taken over an arbitrary contour C in the 4-momentum space $(\mathbf{p}, p_0)$, which encloses the Fermi hypersurface (Fig. 8.1 *bottom*), and $\mathbf{tr}$ is the trace over the spin, band or other indices.

8.1.4 Landau Fermi liquid

The topological class of systems with a Fermi surface is rather broad. In particular it contains conventional Landau Fermi liquids, in which the propagator preserves the pole. Close to the pole the propagator is

$$G = \frac{Z}{i p_0 - v_F (p - p_F)} \,. \tag{8.4}$$

As distinct from the Fermi gas in eqn (8.2) the Fermi velocity v_F no longer equals p_F/m, but is a separate 'fundamental constant' of the Fermi liquid. It determines the effective mass of the quasiparticle, $m^* = p_F/v_F$. Also the residue is different: one now has $Z \neq 1$, but this does not influence the low-energy properties of the liquid. One can see that that neither the change of v_F, nor the change of Z, changes the topological invariant for the propagator, eqn (8.3), which remains $N_1 = 1$. This is essential for the Landau theory of an interacting Fermi liquid; it confirms the conjecture that there is one-to-one correspondence between the low-energy quasiparticles in Fermi liquids and particles in a Fermi gas.

Thus (if there are no infrared peculiarities, which occur in low-dimensional systems, see below) in (isotropic) Fermi liquids the spectrum of fermionic quasiparticles approaches at low energy the universal behavior

$$E(\mathbf{p}) \to v_F (|\mathbf{p}| - p_F) \,, \tag{8.5}$$

with two 'fundamental constants', the Fermi velocity v_F and Fermi momentum p_F. The values of these parameters are governed by the microscopic physics, but in the effective theory of Fermi liquids they are the fundamental constants. The energy of the fermionic quasiparticle in eqn (8.5) is zero on a 2D manifold $|\mathbf{p}| = p_F$ in 3D momentum space – the Fermi surface.

8.1.5 Collective modes of Fermi surface

The topological stability of the Fermi surface determines also the possible bosonic collective modes of a Landau Fermi liquid. Smooth perturbations of the vacuum cannot change the topology of the fermionic spectrum, but they produces effective fields acting on a given particle due to the other moving particles. This effective field cannot destroy the Fermi surface, due to its topological stability, but it can locally shift the position of the Fermi surface. Therefore a collective motion of the vacuum is seen by an individual quasiparticle as dynamical modes

Zero sound – propagating oscillations of shape of Fermi surface

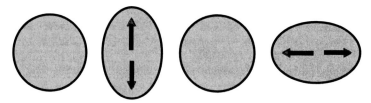

FIG. 8.2. Bosonic collective modes in the fermionic vacuum of the Fermi surface universality class. Collective motion of particles comprising the vacuum is seen by an individual quasiparticle as dynamical modes of the Fermi surface. Here the propagating elliptical deformations of the Fermi surface are shown.

of the Fermi surface (Fig. 8.2). These bosonic modes are known as the different harmonics of zero sound (see the book by Khalatnikov 1965).

Dynamics of fermion zero modes interacting with the collective bosonic fields represents the quantum field theory. The effective quantum field theory in the vacuum of the Fermi surface universality class is the Landau theory of Fermi liquid.

8.1.6 *Volume of the Fermi surface as invariant of adiabatic deformations*

Topological stability means that any continuous change of the system will leave the system within the same universality class. Such a continuous perturbation can include the adiabatic, i.e. slow in time, change of the interaction strength between the particles, or adiabatic deformation of the Fermi surface, etc. Under adiabatic perturbations, no spectral flow of the quasiparticle energy levels occurs across the Fermi surface (if the deformation is slow enough, of course). The state without excitations transforms to another state, in which excitations are also absent, i.e. one vacuum transforms into another vacuum. The absence of the spectral flow leads in particular to Luttinger's (1960) theorem stating that the volume of the Fermi surface is an adiabatic invariant, i.e. the volume is invariant under adiabatic deformations of the Fermi surface, if the total number of particles is kept constant.

For the isotropic Fermi liquid, where the Fermi surface is spherical, the Luttinger theorem means that the volume of the Fermi surface is invariant under adiabatic switching on the interaction between the particles. In other words, in the Fermi liquid the relation between the particle density and the Fermi momentum remains the same as in an ideal non-interacting Fermi gas:

$$n = \frac{p_F^3}{3\pi^2 \hbar^3} \ . \tag{8.6}$$

Here we take into acoount two spin projections.

Actually the theorem must be applied to the total system, say, to the whole isolated droplet of the ^{3}He liquid. The corresponding Fermi surface is the 5D manifold in the 6D $(\mathbf{p},\mathbf{r})$ space, where the quasiparticle energy becomes zero: $E(\mathbf{p},\mathbf{r}) = 0$. One can state that there is a relation between the total number of particles and the volume $V_{\text{phase space}}$ of the 6D phase space inside the 5D hypersurface of zeros:

$$N = 2\frac{V_{\text{phase space}}}{(2\pi\hbar)^3} \ . \tag{8.7}$$

A topological approach to Luttinger's theorem has also been discussed by Oshikawa (2000). The processes related to the spectral flow of quasiparticle energy levels will be considered in Chapters 18 and 23 in connection with the phenomenon of axial anomaly. Sometimes it happens that the spectral flow can occur on the boundary of the system even during the slow switching of the interaction, and this can violate the Luttinger theorem even in the limit of an infinite system.

8.1.7 Non-Landau Fermi liquids

The Fermi hypersurface described by the topological invariant N_1 exists for any spatial dimension. In 2+1 dimensions the Fermi hypersurface is a line in 2D momentum space, which corresponds to the vortex loop in the 3D frequency-momentum space in Fig. 8.1. In 1+1 dimensions the Fermi surface is a point vortex in momentum space.

In low-dimensional systems the Green function can deviate from its canonical Landau form in eqn (8.4). For example, in 1+1 dimensions it loses its pole due to infrared divergences. Nevertheless, the Fermi surface and fermion zero modes are still there (Volovik 1991; Blagoev and Bedell 1997). Though the Landau Fermi liquid transforms to another state with different infrared properties, this occurs within the same topological class with given N_1. An example is provided by the so-called Luttinger liquid. Close to the Fermi surface the Green function for the Luttinger liquid can be approximated by (see Wen 1990b; Volovik 1991; Schulz et al. 2000)

$$G(z,p) \sim (ip_0 - v_1\tilde{p})^{\frac{g-1}{2}}(ip_0 + v_1\tilde{p})^{\frac{g}{2}}(ip_0 - v_2\tilde{p})^{\frac{g-1}{2}}(ip_0 + v_2\tilde{p})^{\frac{g}{2}} \ , \tag{8.8}$$

where v_1 and v_2 correspond to Fermi velocities of spinons and holons and $\tilde{p} = p - p_F$. The above equation is not exact but illustrates the robustness of the momentum space topology. Even if $g \neq 0$ and $v_1 \neq v_2$, the singularity of the Green function occurs at the point $(p_0 = 0, \tilde{p} = 0)$. One can easily check that the momentum space topological invariant in eqn (8.3) remains the same, $N_1 = 1$, as for the conventional Landau Fermi liquid. Thus the Fermi surface (as the geometrical object in the 4-momentum space, where singularities of the Green function are lying) persists even if the Landau state is violated.

The difference from the Landau Fermi liquid occurs only at real frequency z: the quasiparticle pole is absent and one has branch-cut singularities instead of a mass shell, so that the quasiparticles are not well-defined. The population of the particles has no jump on the Fermi surface, but has a power-law singularity in the derivative (Blagoev and Bedell 1997).

Another example of a non-Landau Fermi liquid is the Fermi liquid with exponential behavior of the residue found by Yakovenko (1993). It also has a Fermi surface with the same topological invariant, but the singularity at the Fermi surface is exponentially weak.

A factorization of quasiparticles in terms of spinons and holons will be discussed in Sec. 12.2.3 for elementary particles in the Standard Model. Such a factorization of the Green function (8.8) in terms of the fractional factors has also a counterpart in $\mathbf{r}$-space, thus providing the analogy between composite fermions and composite defects. We shall see that fractional topological defects in $\mathbf{r}$-space such as Alice string – the half-quantum vortex – in Secs 15.3.1 and 15.3.2. are the consequence of the factorization of the order parameter into two parts each with the fractional winding number, see e.g. eqn (15.18).

8.2 Systems with Fermi points

8.2.1 *Chiral particles and Fermi point*

In ^{3}He-A the energy spectrum of fermionic quasiparticles has point nodes. Close to each of the nodes the spectrum has a relativistic form given in eqn (7.16). The Bogoliubov–Nambu Hamiltonian for ^{3}He-A in eqn (7.57) written for one spin projection (along $\hat{\mathbf{d}}$) and close to the nodes at $\mathbf{p}^{(a)} = q_a p_F \hat{\mathbf{z}}$ (with $q_a = \pm 1$) has the form

$$\mathcal{H}_A = q_a \check{\tau}^3 c_\| \tilde{p}_z + \check{\tau}^1 c_\perp \tilde{p}_x - \check{\tau}^2 c_\perp \tilde{p}_y \ , \ \tilde{\mathbf{p}} = \mathbf{p} - q_a p_F \hat{\mathbf{z}} \ , \quad (8.9)$$

where $|\tilde{\mathbf{p}}| \ll p_F$.

In particle physics eqn (8.9) represents the Weyl Hamiltonian which describes the Standard Model fermions, leptons and quarks, above the electroweak transition where they are massless and chiral. The Weyl Hamiltonian for the free massless spin-1/2 particle is a 2×2 matrix

$$\mathcal{H} = \pm c \sigma^i p_i \ , \quad (8.10)$$

where c is the speed of light, and σ^i are the Pauli matrices. The plus and minus signs refer to right-handed and left-handed particles, with their spin oriented along or opposite to the particle momentum $\mathbf{p}$, respectively (Fig. 8.3).

The anisotropy of 'speeds of light' in eqn (8.9) can be removed by rescaling the coordinate along the $\hat{\mathbf{l}}$ direction: $z = (c_\|/c_\perp)\tilde{z}$. We can also perform rotation in momentum space to equalize the signs in front of the Pauli matrices $\check{\tau}^b$ and obtain $\mathcal{H}_A = -q_a c_\perp \check{\tau}^i \tilde{p}_i$. Then the only difference remaining between eqn (8.10) and eqn (8.9) is that the Hamiltonian for Bogoliubov quasiparticles in ^{3}He-A is expressed in terms of the Pauli matrices $\check{\tau}^b$ in the Bogoliubov–Nambu particle–hole space, instead of matrices σ^i in the spin space in the Hamiltonian for quarks and leptons. This simply means that Bogoliubov–Nambu isospin plays the same role for quasiparticles as the conventional spin for matter. Later we shall see the inverse relation: the conventional spin of Bogoliubov quasiparticles which comes from the spin of the ^{3}He atom plays the same role as the weak isospin for Standard Model fermions.

SYSTEMS WITH FERMI POINTS

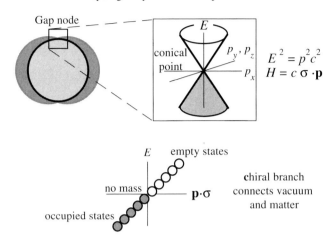

FIG. 8.3. *Top*: Gap node in superfluid ^{3}He-A is the conical point in the 4-momentum space. Quasiparticles in the vicinity of the nodes in ^{3}He-A and elementary particles in the Standard Model above the electroweak transition are chiral fermions. Particles occupying the negative energy states of the vacuum can leak through the conical point to the positive energy world of matter. *Bottom*: The spectrum of the right-handed chiral particle as a function of momentum along the spin. For particles with positive energy the spin σ is oriented along the momentum **p**. The negative energy states are occupied.

The point $\mathbf{p} = 0$ in eqn (8.10) is the exceptional point in **p**-space, since the direction of the spin of the particle is not determined at this point. At this point also the energy $E = cp$ is zero. Thus distinct from the case of the Fermi surface, where the energy of the quasiparticle is zero at the surface in 3D **p**-space, the energy of a chiral particle is zero at isolated points – the Fermi points. The fermion zero modes in the vacua of this universality class are chiral quasiparticles.

Let us show that such singular points in momentum space also have topological signature, which is, however, different from that of the Fermi surface.

8.2.2 *Fermi point as hedgehog in* **p**-*space*

Let us again start with the non-interacting particles, which can be described in terms of a one-particle Hamiltonian, and consider the simplest equation (8.10). Let us consider the behavior of the particle spin $\mathbf{s}(\mathbf{p})$ as a function of the particle momentum **p** in the 3D space $\mathbf{p} = (p_x, p_y, p_z)$ (Fig. 8.4). For the right-handed particle, whose spin is parallel to the momentum, one has $\mathbf{s}(\mathbf{p}) = \mathbf{p}/2p$, while for left-handed ones $\mathbf{s}(\mathbf{p}) = -\mathbf{p}/2p$. In both cases the spin distribution in momentum space looks like a hedgehog, whose spines are represented by spins. Spines point

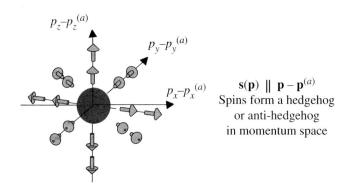

FIG. 8.4. Illustration of the meaning of the topological invariant for the simplest case: the Fermi point as a hedgehog in 3D momentum space. For each momentum $\mathbf{p}$ we draw the direction of the quasiparticle spin, or its equivalent in ^{3}He-A – the Bogoliubov spin. The topological invariant for the hedgehog is the mapping $S^2 \to S^2$ with integer winding number N_3 which is $N_3 = +1$ for the drawn case of a right-handed particle. The topological invariant N_3 is robust to any deformation of the spin field $\sigma(\mathbf{p})$: one cannot comb the hedgehog smooth.

outward for the right-handed particle producing the mapping of the sphere S^2 in 3D $\mathbf{p}$-space onto the sphere S^2 of the spins with the winding number $N_3 = +1$. For the left-handed particle the spines of the hedgehog look inward and one has the mapping with $N_3 = -1$. In 3D space the hedgehogs are topologically stable, which means that the deformation of the Hamiltonian cannot change the winding number and thus cannot destroy the singularity (the fermion zero mode) of the Hamiltonian.

8.2.3 Topological invariant for Fermi point

Let us consider the general 2×2 Hamiltonian

$$\mathcal{H} = \check{\tau}^b g_b(\mathbf{p}) , \quad b = (1, 2, 3) , \tag{8.11}$$

where $g_b(\mathbf{p})$ are three arbitrary functions of $\mathbf{p}$. An example of such a Hamiltonian is provided by the p-wave pairing of 'spinless' fermions in eqn (7.10)

$$g_1 = \mathbf{p} \cdot \mathbf{e}_1 , \ g_2 = -\mathbf{p} \cdot \mathbf{e}_2 , \ g_3 = \frac{p^2}{2m} - \mu , \ \mathbf{e}_1 \times \mathbf{e}_2 = \hat{\mathbf{l}} \, |\mathbf{e}_1 \times \mathbf{e}_2| \neq 0 . \tag{8.12}$$

The singular points in momentum space are the Fermi points $\mathbf{p}^{(a)}$ where $g_b(\mathbf{p}^{(a)}) = 0$. In case of eqn (8.12) these are $\mathbf{p}^{(a)} = q_a p_F \hat{\mathbf{l}} = \pm p_F \hat{\mathbf{l}}$. The topological invariant which describes these points is the winding number of the mapping of the sphere σ_2 around the singular point in $\mathbf{p}$-space to the 2-sphere of the unit vector $\hat{\mathbf{g}} = \mathbf{g}/|\mathbf{g}|$:

$$N_3 = \frac{1}{8\pi} e_{ijk} \int_{\sigma_2} dS^k \, \hat{\mathbf{g}} \cdot \left(\frac{\partial \hat{\mathbf{g}}}{\partial p_i} \times \frac{\partial \hat{\mathbf{g}}}{\partial p_j} \right) . \tag{8.13}$$

One can check that the topological charge of the hedgehog at $\mathbf{p} = 0$ in the Hamiltonian (8.10) describing the right-handed particle (sign +) is $N_3 = +1$, the same as of the Fermi point with $q = -1$ (i.e. at $\mathbf{p} = -p_F \hat{\mathbf{l}}$) in the Hamiltonian for quasiparticles in a 'spinless' p-wave superfluid.

Correspondingly, $N_3 = -1$ for the left-handed fermions (eqn (8.10) with sign $-$) and for quasiparticles in 'spinless' p-wave superfluids in the vicinity of the Fermi point at $\mathbf{p} = p_F \hat{\mathbf{l}}$.

8.2.4 Topological invariant in terms of Green function

In the general case the system is not described by a simple 2×2 matrix, since there can be other degrees of freedom. For example, the Hamiltonian eqn (7.57) for quasiparticles in ^{3}He-A is a 4×4 matrix. Also, if the interaction between the fermions is included, the system cannot be described by a single-particle Hamiltonian. However, even in this complicated case the topological invariant can be written analytically if we proceed from $\mathbf{p}$-space to the 4-momentum space $(p_0, \mathbf{p})$ by introducing the Green function. Since the invariant should not depend on the deformation, we can consider it on an example of the Green function for non-interacting fermions. Let us again introduce the Green function on the imaginary frequency axis, $z = ip_0$ (Fig. 8.5):

$$\mathcal{G} = (ip_0 - \check{\tau}^b g_b(\mathbf{p}))^{-1} . \tag{8.14}$$

One can see that this propagator has a singularity at the points in the 4-momentum space: $(p_0 = 0, \mathbf{p} = \mathbf{p}^{(a)})$. The generalization of the $\mathbf{p}$-space invariant N_3 in eqn (8.13) to the 4-momentum space gives

$$N_3 = \frac{1}{24\pi^2} e_{\mu\nu\lambda\gamma} \, \mathrm{tr} \int_{\sigma_3} dS^\gamma \, \mathcal{G} \partial_{p_\mu} \mathcal{G}^{-1} \mathcal{G} \partial_{p_\nu} \mathcal{G}^{-1} \mathcal{G} \partial_{p_\lambda} \mathcal{G}^{-1} . \tag{8.15}$$

Now the integral is over the 3D surface σ_3 embracing the singular point $(p_0 = 0, \mathbf{p} = \mathbf{p}^{(a)})$ – the Fermi point. It is easy to check that substitution of eqn (8.14) into eqn (8.15) gives the $\mathbf{p}$-space invariant in eqn (8.13). However, in contrast to eqn (8.13) the equation (8.15) is applicable for the interacting systems where the single-quasiparticle Hamiltonian is not determined: the only requirement for the Green function matrix $\mathcal{G}(p_0, \mathbf{p})$ is that it is continuous and differentiable outside the singular point.

One can check that under continuous variation of the matrix function the integrand changes by a total derivative. That is why the integral over the closed 3-surface does not change, i.e. N_3 is invariant under continuous deformations of the Green function, and also it is independent of the choice of closed 3-surface σ_3 around the singularity.

The possible values of the invariant can be found from the following consideration. If one considers the matrix Green function for the massless chiral

UNIVERSALITY CLASSES OF FERMIONIC VACUA

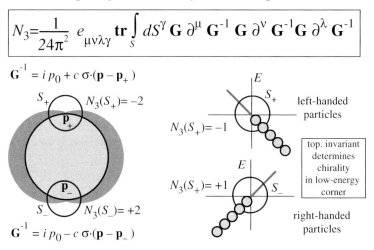

FIG. 8.5. The Green function for fermions in ^{3}He-A and in the Standard Model have point singularities in the 4-momentum space, which are described by the integer-valued topological invariant N_3. The Fermi points in ^{3}He-A at $\mathbf{p}^{(a)} = \pm p_F \hat{\mathbf{l}}$ have $N_3 = \mp 2$. The Fermi point at $\mathbf{p} = 0$ for the chiral relativistic particle in eqn (8.10) has $N_3 = C_a$, where $C_a = \pm 1$ is the chirality. The chirality, however, appears only in the low-energy corner together with the Lorentz invariance. Thus the topological index N_3 is the generalization of the chirality to the Lorentz non-invariant case.

fermions in eqn (8.10), one finds that it is proportional to a unitary 2×2 matrix: $\mathcal{G}^{-1} = ip_0 - c\sigma^i p_i = i\sqrt{p_0^2 + c^2 p^2}\, \mathcal{U}$ with $\mathcal{U}\mathcal{U}^+ = 1$. Let us discuss the more complicated case, when this unitary matrix is an arbitrary function of the 4-momentum: $\mathcal{U} = u_0(p_\mu) + i\sigma^i u_i(p_\mu)$, where $u_0^2 + \mathbf{u}^2 = 1$. In this particular case the invariant N_3 becomes

$$N_3 = \frac{1}{(4\pi)^2} e_{\mu\nu\lambda\gamma} \operatorname{tr} \int_{\sigma_3} dS^\gamma\, \mathcal{U}\partial_{p_\mu}\mathcal{U}^+ \mathcal{U}\partial_{p_\nu}\mathcal{U}^+ \mathcal{U}\partial_{p_\lambda}\mathcal{U}^+ \tag{8.16}$$

$$= \frac{1}{(4\pi)^2} e_{\mu\nu\lambda\gamma}\, e^{\alpha\beta\epsilon\kappa} \int_{\sigma_3} dS^\gamma\, u_\alpha \partial_{p_\mu} u_\beta \partial_{p_\nu} u_\epsilon \partial_{p_\lambda} u_\kappa\, . \tag{8.17}$$

It describes the mapping of the S^3 sphere σ_3 surrounding the singular point in 4D p_μ-space onto the S^3 sphere $u_0^2 + \mathbf{u}^2 = 1$. The classes of such mappings form the non-trivial third homotopy group $\pi_3(S^3) = Z$, where Z is the group of integer numbers N_3. The same integer values N_3 are preserved for any continuous deformations of the Green function matrix, if it is well determined outside the singularity where $\det \mathcal{G}^{-1} \neq 0$. The integral-valued index N_3 thus represents topologically different classes of Fermi points – the singular points in 4-momentum space.

8.2.5 A-phase and planar state: the same spectrum but different topology

For quasiparticles in ^{3}He-A, where the Hamiltonian in eqn (7.57) is a 4×4 matrix, one obtains that the topological charges of the Fermi points at $\mathbf{p} = -p_F\hat{\mathbf{l}}$ and $\mathbf{p} = +p_F\hat{\mathbf{l}}$ are $N_3 = +2$ and $N_3 = -2$ correspondingly. On the other hand, in the planar phase in eqn (7.62) one obtains $N_3 = 0$ for both Fermi points. Thus, though quasiparticles in the A-phase and planar state have the same energy spectrum $E(\mathbf{p}) = \sqrt{M^2(\mathbf{p}) + c_\perp^2 (\mathbf{p} \times \hat{\mathbf{l}})^2}$, the topology of the Green function is essentially different. Because of the zero value of the topological invariant the Fermi points of the planar state is marginal and can be destroyed by small perturbations, while the perturbations of the ^{3}He-A can only split the Fermi point, say, with the topological charge $N_3 = +2$ into two Fermi points with $N_3 = +1$. This is what actually occurs when the A-phase transforms to the axiplanar phase in Fig. 7.3.

8.2.6 Relativistic massless chiral fermions emerging near Fermi point

Systems with elementary Fermi points, i.e. Fermi points with topological charge $N_3 = +1$ or $N_3 = -1$, have remarkable properties. In such a system, even if it is non-relativistic, the Lorentz invariance always emerges at low energy. Since the topological invariant is non-zero, there must be a singularity in the Green function or in the Hamiltonian, which means that the quasiparticle spectrum remain gapless: fermions are massless even in the presence of interaction. For the cases $N_3 = \pm 1$, the function $\mathbf{g}(\mathbf{p})$ in eqn (8.11) is linear in deviations from the Fermi point, $p_i - p_i^{(a)}$. Expanding this function $\mathbf{g}(\mathbf{p})$ one obtains the general form in the vicinity of each Fermi point: $g_b(\mathbf{p}) \approx e_b^i(p_i - p_i^{(a)})$. After the shift of the momentum and the diagonalization of the 3×3 matrix e_b^i, one again obtains $\mathcal{H} = \pm c\sigma^i p_i$, where the sign is determined by the sign of the determinant of the matrix e_b^i. Thus the Hamiltonian for relativistic chiral particles is the emergent property of the Fermi point universality class.

If the interactions are introduced, and the one-particle Hamiltonian is no longer valid, the same can be obtained using the Green function: in the vicinity of the Fermi point $p_\mu^{(a)}$ in the 4-momentum space one can expand the propagator in terms of the deviations from this Fermi point, $p_\mu - p_\mu^{(a)}$. If the Fermi point has the lowest non-zero value of the topological charge, i.e. $N_3 = \pm 1$, then close to the Fermi point the linear deviations are dominating. As a result the general form of the inverse propagator is expressed in terms of the tetrad field e_b^μ (vierbein):

$$\mathcal{G}^{-1} = \check{\tau}^b e_b^\mu (p_\mu - p_\mu^{(a)}) \ , \quad b = (0, 1, 2, 3) \ . \tag{8.18}$$

Here we have returned from the imaginary frequency axis to the real energy, so that $z = E = -p_0$ instead of $z = ip_0$; and $\check{\tau}^b = (1, \check{\tau}^1, \check{\tau}^2, \check{\tau}^3)$. The quasiparticle spectrum $E(\mathbf{p})$ is given by the poles of the propagator, and thus by the following equation:

$$g^{\mu\nu}(p_\mu - p_\mu^{(a)})(p_\nu - p_\nu^{(a)}) = 0 \ , \quad g^{\mu\nu} = \eta^{bc} e_b^\mu e_c^\nu \ , \tag{8.19}$$

where $\eta^{bc} = \text{diag}(-1, 1, 1, 1)$.

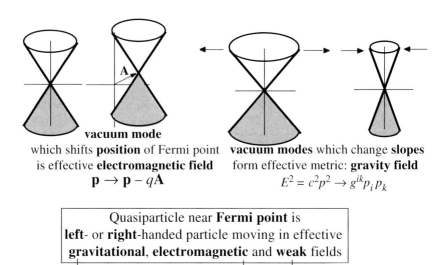

FIG. 8.6. Bosonic collective modes of the fermionic vacuum which belongs to the Fermi point universality class. The slow (low-energy) vacuum motion cannot destroy the topologically stable Fermi point, it can only shift the point (*top left*) and/or change its slopes (*top right*). The shift corresponds to the gauge field **A**, while the slopes ('speeds of light') form the metric tensor field g^{ik}. Both fields are dynamical, representing the collective modes of the fermionic vacuum with Fermi points. Such a collective motion of the vacuum is seen by an individual quasiparticle as gauge and gravity fields. Thus the chiral fermions, gauge fields and gravity appear in the low-energy corner together with physical laws: Lorentz and gauge invariance.

Thus in the vicinity of the Fermi point the massless quasiparticles are always described by the Lorentzian metric $g^{\mu\nu}$, even if the underlying Fermi system is not Lorentz invariant; superfluid ^{3}He-A is an example.

8.2.7 Collective modes of fermionic vacuum – electromagnetic and gravitational fields

The quantities $g^{\mu\nu}$ (or e_b^μ) and $p_\mu^{(a)}$, which enter the general fermionic spectrum at low energy in eqn (8.19), are dynamical variables, related to the bosonic collective modes of the fermionic vacuum. They play the role of an effective gravity and gauge fields, respectively (Fig. 8.6). In principle, there are many different collective modes of the vacuum, but these two modes are special. Let us consider the slow collective dynamics of the vacuum, and how it influences the quasiparticle spectrum. The dynamical perturbation with large wavelength and low frequency cannot destroy the topologically stable Fermi point. Thus under such a perturbation the general form of eqn (8.18) is preserved. The only thing that the perturbation can do is to shift locally the position of the Fermi point $p_\mu^{(a)}$ in momentum space and to deform locally the vierbein e_b^μ (which in particular includes slopes of the energy spectrum).

This means that the low-frequency collective modes in such Fermi systems are the propagating collective oscillations of the positions of the Fermi point and the propagating collective oscillations of the slopes at the Fermi point (Fig. 8.6). The former is felt by the right- or the left-handed quasiparticles as a dynamical gauge (electromagnetic) field, because the main effect of the electromagnetic field $A_\mu = (A_0, \mathbf{A})$ is just the dynamical change in the position of zero in the energy spectrum: in the simplest case, $(E - q_a A_0)^2 = c^2 (\mathbf{p} - q_a \mathbf{A})^2$.

The collective modes related to a local change of the vierbein e_b^μ correspond to the dynamical gravitational field $g^{\mu\nu}$. The quasiparticles feel the inverse tensor $g_{\mu\nu}$ as the metric of the effective space in which they move along the geodesic curves

$$ds^2 = g_{\mu\nu} dx^\mu dx^\nu . \tag{8.20}$$

Therefore, the collective modes related to the slopes play the part of a gravity field.

Thus near a Fermi point the quasiparticle is a chiral massless fermion moving in the effective dynamical electromagnetic and gravitational fields generated by the low-frequency collective motion of the vacuum.

8.2.8 *Fermi points and their physics are natural*

From the topological point of view the Standard Model and the Lorentz non-invariant ground state of ^{3}He-A belong to the same universality class of systems with topologically non-trivial Fermi points, though the underlying 'microscopic' physics can be essentially different. Pushing the analogy further, one may conclude that classical (and quantum) gravity, as well as electromagnetism and weak interactions, are not fundamental interactions. If the vacuum belongs to the universality class with Fermi points, then matter (chiral particles and gauge fields) and gravity (vierbein or metric field) inevitably appear together in the low-energy corner as collective fermionic and bosonic zero modes of the underlying system.

The emerging physics is natural because vacua with Fermi points are natural: they are topologically protected. If a pair of Fermi points with opposite topological charges exist, it is difficult to destroy them because of their topological stability: the only way is to annihilate the points with opposite charges. Vacua with Fermi points are not as abundant as vacua with Fermi surfaces, since the topological protection of the Fermi surface is more powerful. But they appear in most of the possible phases of spin-triplet superfluidity. They can naturally appear in semiconductors without any symmetry breaking (see Nielsen and Ninomiya (1983) where the Fermi point is referred to as a generic degeneracy point). Possible experimental realization of such Fermi points in semiconductors was discussed by Abrikosov (1998) and Abrikosov and Beneslavskii (1971). In

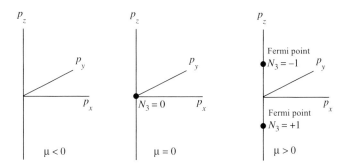

FIG. 8.7. Quantum phase transition between two axial vacua with the same symmetry but of different universality class. It occurs when the chemical potential μ in the BCS model in eqn (7.26) crosses zero value. At $\mu > 0$ the axial vacuum has two Fermi points which annihilate each other when $\mu = 0$. At $\mu < 0$ the Green function has no singularities and the axial quantum vacuum is fully gapped.

vacua with a fully gapped spectrum the Fermi points appear inside the cores of vortices (Fig. 23.4).

In quantum liquids the Fermi points always appear in pairs, so that the sum of topological charges N_3 of all the Fermi points is zero. This is similar to the observation made by Nielsen and Ninomiya (1981) for the Fermi points in RQFT on a lattice. In their case the vanishing of the total topological charge is required by the periodicity in momentum space. As a result, in addition to the chiral fermion living in the vicinity of the Fermi point at $\mathbf{p} = 0$, there must be a fermion with opposite chirality living in some point of momentum space far from the origin. This means the doubling of fermions.

Appearance of the pair of Fermi points in the initial topologically trivial vacuum (or their annihilation) represents the quantum phase transition at which the universality class of the fermionic spectrum changes. Such a transition occurs for instance in the axial vacuum when the interaction between the bare atoms increases and the BCS model transforms to the Bose–Einstein condensate of Cooper pairs: at some critical value of the interaction parameter λ in eqn (7.2) the Fermi points annihilate each other. See Fig. 8.7 where the same quantum phase transition occurs when the chemical potential μ in eqn (7.26) changes sign. The symmetry of the vacuum state is the same above and below this quantum phase transition.

Another important property of chiral quasiparticles in the vicinity of the Fermi point is that such a quasiparticle is 'a flat membrane moving in the orthogonal direction with the speed of light' ('t Hooft 1999); the position of the membrane and its velocity are well determined simultaneously. The fact that just these 'deterministic' quasiparticles emerge in the Standard Model may shed light on the fundamental concepts of quantum mechanics.

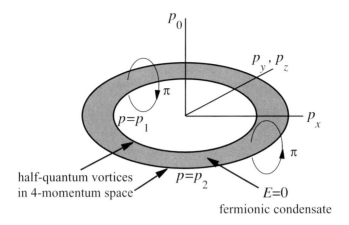

FIG. 8.8. Zeros of co-dimension 0. The vortex line in momentum space in Fig. 8.1 representing the Fermi surface – the manifold of zeros of co-dimension 1 – splits in two fractional vortices, say half-quantum vortices, at $p = p_1$ and $p = p_2$. The Green function has singularities on a whole band $p_1 < p < p_2$ between the fractional vortices. Such a band can be represented by the fermionic condensate suggested by Khodel and Shaginyan (1990) where the quasiparticle energy is zero, $E(p) = 0$, for all p within the band $p_1 < p < p_2$.

8.2.9 Manifolds of zeros in higher dimensions

The topological classification of universality classes of fermionic vacua in terms of their fermion zero modes can be extended to higher-dimensional space. The manifolds of zeros can be characterized by co-dimension, which is the dimension of **p**-space minus the dimension of the manifold of zeros:

(1) Co-dimension 1: The 2D Fermi surface in 3D **p**-space, described by the topological invariant N_1, has according to this definition co-dimension 1. The topologically stable manifolds of zeros with the same invariant N_1 can exist for any spatial dimension $D \geq 1$. Such Fermi hypersurfaces have dimension $D - 1$.

(2) Co-dimension 3: The Fermi points in 3D **p**-space described by the topological invariant N_3 have co-dimension 3. The same invariant N_3 describes in higher dimensions the topologically stable hypersurfaces with co-dimension 3. Since their dimension is $D - 3$, they can exist only for spatial dimensions $D \geq 3$.

(3) Co-dimension 5: Topologically stable hypersurfaces of dimension $D - 5$ are described by the higher-order topological invariant N_5. They can exist if the spatial dimension $D \geq 5$. The N_5 invariant for co-dimension 5 will be discussed in Sec. 22.2.1.

There is no invariant which can be constructed for the Green function singularities of even co-dimension. That is why the lines of nodes in 3D **p**-space (co-dimension 2) and also the point nodes in 2D **p**-space (also of co-dimension 2) are topologically unstable. They can be protected by the symmetry of the vacuum, but disappear under small deformations violating the symmetry. Zeros of co-dimension 0 have also been discussed. The homogeneous vacuum state with such a Fermi band in which all fermions have zero energy (the so-called fermionic condensate) was suggested by Khodel and Shaginyan (1990). The qauntum phase transition from the Fermi surface to fermionic condensate can be visualized as splitting of the vortex in momentum space in Fig. 8.1 representing the Fermi surface into fractional vortices in Fig. 8.8 connected by the vortex sheet – the fermionic condensate (Volovik 1991).

The manifold of co-dimension 0 can also be formed by fermion zero modes in the vortex core (Sec. 23.1.2).

9

EFFECTIVE QUANTUM ELECTRODYNAMICS IN ^{3}He-A

The new point compared to Bose superfluids where the effective gravity arises, is that in the fermionic vacuum there appear in addition chiral fermions and an effective electromagnetic field. One obtains all the ingredients of quantum electrodynamics (QED) with massless fermions. Let us start with the fermionic content of the theory.

9.1 Fermions

9.1.1 *Electric charge and chirality*

The Hamiltonian (7.57) for fermionic quasiparticles in ^{3}He-A is

$$\mathcal{H}_A = M(\mathbf{p})\check{\tau}^3 + c_\perp \sigma^\mu \hat{d}_\mu (\check{\tau}_1 \hat{\mathbf{m}} \cdot \mathbf{p} - \check{\tau}^2 \hat{\mathbf{n}} \cdot \mathbf{p}) \ . \tag{9.1}$$

First we consider a fixed unit vector $\hat{\mathbf{d}}$. Then the projection $S_z = (1/2)\sigma^\mu \hat{d}_\mu$ of the quasiparticle spin on the $\hat{\mathbf{d}}$ axis is a good quantum number, $S_z = \pm 1/2$. Let us recall that the conventional spin of the fermionic quasiparticle in ^{3}He-A comes from the nuclear spin of the bare ^{3}He atoms. It plays the role of isospin in emerging RQFT (see Sec. 9.1.5).

There are four different species a of quasiparticles in ^{3}He-A. In addition to the 'isospin' $S_{za} = \pm 1/2$, they are characterized by 'electric charge' $q_a = \pm 1$. As before, we assign the 'electric charge' $q_a = +1$ to the quasiparticle which lives in the vicinity of the Fermi points at $\mathbf{p} = +p_F \hat{\mathbf{l}}$. Correspondingly the quasiparticle living in the vicinity of the opposite node at $\mathbf{p} = -p_F \hat{\mathbf{l}}$ has the 'electric' charge $q_a = -1$. Then close to the Fermi points the Hamiltonian for these four quasiparticle species has the form

$$\mathcal{H}_a = e_b^{k(a)}(p_k - q_a A_k)\check{\tau}^b$$
$$= 2S_{za}c_\perp \left[\hat{\mathbf{m}} \cdot (\mathbf{p} - q_a \mathbf{A})\check{\tau}^1 - \hat{\mathbf{n}} \cdot (\mathbf{p} - q_a \mathbf{A})\check{\tau}^2\right] + q_a c_\parallel \hat{\mathbf{l}} \cdot (\mathbf{p} - q_a \mathbf{A})\check{\tau}^3, \tag{9.2}$$

$$\mathbf{A} = p_F \hat{\mathbf{l}} \ . \tag{9.3}$$

The chirality C_a of quasiparticles is determined by the sign of the determinant of matrices $e_b^{k(a)}$:

$$C_a = \text{sign } |e_b^{k(a)}| = -q_a \ . \tag{9.4}$$

Thus in the vicinity of the Fermi point at $\mathbf{p} = +p_F \hat{\mathbf{l}}$ there live two left-handed fermions with 'isotopic' spins $S_z = \pm 1/2$ and with 'electric' charge $q_a = +1$. Two right-handed fermions with 'isotopic' spins $S_z = \pm 1/2$ and charge $q_a = -1$ live near the opposite node at $\mathbf{p} = -p_F \hat{\mathbf{l}}$.

9.1.2 Topological invariant as generalization of chirality

For the chiral fermions in eqn (8.10) the momentum space invariant (8.15) coincides with the chirality of the fermion: $N_3 = C_a$, where $C_a = \pm 1$ is the chirality. In other words, for a single relativistic chiral particle the topological invariant N_3 is equivalent to the chirality. In the case of several species of relativistic particles, the invariant N_3 is simply the number of right-handed minus the number of left-handed fermions. However, this connection is valid only in the relativistic edge. The topological invariant is robust to any deformation, including those which violate Lorentz invariance. In the latter case the notion of chirality loses its meaning, while the topological invariant persists. This means that the topological description is far more general than the description in terms of chirality, which is valid only when Lorentz symmetry is obeyed. Actually the momentum space charge N_3 is the topological generalization of chirality.

According to eqn (8.18) we know that in the originally non-relativistic system, the effective Lorentz invariance gradually arises in the vicinity of the Fermi points with $N_3 = \pm 1$. Simultaneously the quasiparticles acquire chirality: they gradually become right-handed (left-handed) in the vicinity of the Fermi point with $N_3 = +1$ ($N_3 = -1$). Thus the topological index N_3 is not only the generalization of chirality to the Lorentz non-invariant case, but also the source of chirality and of Lorentz invariance in the low-energy limit.

Later in Sec. 12.4 we shall see that in the vicinity of the degenerate Fermi point, i.e. the Fermi point with $|N_3| > 1$, chirality of fermions is protected by the symmetry only. If the corresponding symmetry is violated the quasiparticles are no longer relativistic and chiral.

9.1.3 Effective metric viewed by quasiparticles

In eqn (9.2) the Hamiltonian is represented in terms of a dreibein, which gives the spatial part of the effective metric:

$$g_{(a)}^{ij} = \sum_{b=1}^{3} e_b^{i(a)} e_b^{j(a)} = c_\parallel^2 \hat{l}^i \hat{l}^j + c_\perp^2 (\delta^{ij} - \hat{l}^i \hat{l}^j) \ . \tag{9.5}$$

Note that all four fermions have the same metric tensor and thus the same 'speed of light'. This is the result of symmetries of the vacuum state connecting different fermionic species. The energy spectrum in the vicinity of each of the nodes

$$E_a^2(\mathbf{p}) = g^{ij}(p_i - q_a A_i)(p_j - q_a A_j). \tag{9.6}$$

demonstrates that the interaction of fermions with effective electromagnetic field depends on their electric charge, while their interaction with gravity is universal.

What is still missing in eqn 89.6 is the non-diagonal metric g^{0i} and the A_0 component of the electromagnetic field. Both of them are simulated by the superfluid velocity field $\mathbf{v}_s$ of the ^{3}He-A vacuum. Equation (9.6) is valid in the superfluid-comoving frame. The Hamiltonian and the energy spectrum in the environment frame are obtained by the Doppler shift:

$$\tilde{\mathcal{H}}_{(a)} = \mathcal{H}_{(a)} + \mathbf{p} \cdot \mathbf{v}_s = \mathcal{H}_{(a)} + (\mathbf{p} - q_a \mathbf{A}) \cdot \mathbf{v}_s + q_a p_F \hat{\mathbf{l}} \cdot \mathbf{v}_s \ , \tag{9.7}$$

$$\tilde{E}_a(\mathbf{p}) = E_a(\mathbf{p}) + (\mathbf{p} - q_a\mathbf{A}) \cdot \mathbf{v}_s + q_a p_F \hat{\mathbf{l}} \cdot \mathbf{v}_s . \quad (9.8)$$

This gives finally the fully relativistic Weyl equations for the spinor wave function of fermions

$$e_b^{\mu(a)}(p_\mu - q_a A_\mu)\check{\tau}^b \chi = 0 , \quad \check{\tau}^b = (1, \check{\tau}^1, \check{\tau}^2, \check{\tau}^3) , \quad (9.9)$$

$$g^{\mu\nu}(p_\mu - q_a A_\mu)(p_\nu - q_a A_\nu) = 0 , \quad (9.10)$$

where the components of the effective metric and electromagnetic fields acting on the fermionic quasiparticles are

$$g^{ij} = c_\parallel^2 \hat{l}^i \hat{l}^j + c_\perp^2 (\delta^{ij} - \hat{l}^i \hat{l}^j) - v_s^i v_s^j , \quad g^{00} = -1 , \quad g^{0i} = -v_s^i , \quad (9.11)$$

$$\sqrt{-g} = \frac{1}{c_\parallel c_\perp^2} , \quad (9.12)$$

$$g_{ij} = \frac{1}{c_\parallel^2}\hat{l}^i \hat{l}^j + \frac{1}{c_\perp^2}(\delta^{ij} - \hat{l}^i \hat{l}^j) , \quad g_{00} = -1 + g_{ij} v_s^i v_s^j , \quad g_{0i} = -g_{ij} v_s^j , \quad (9.13)$$

$$ds^2 = -dt^2 + g_{ij}(dx^i - v_s^i dt)(dx^j - v_s^j dt) , \quad (9.14)$$

$$A_0 = p_F \hat{\mathbf{l}} \cdot \mathbf{v}_s , \quad \mathbf{A} = p_F \hat{\mathbf{l}} . \quad (9.15)$$

9.1.4 Superfluid velocity in axial vacuum

Superfluid motion of the quantum vacuum in ^{3}He-A has very peculiar properties. They come from the fact that there is no phase factor in the order parameter $\hat{\mathbf{m}} + i\hat{\mathbf{n}}$ in eqn (7.25), which characterizes the axial vacuum. The phase factor is absorbed by the complex vector. That is why the old definition of superfluid velocity $\mathbf{v}_s = \frac{\hbar}{2m}\nabla\Phi$ is meaningless. The phase Φ becomes meaningful if the vector $\hat{\mathbf{l}} = \hat{\mathbf{m}} \times \hat{\mathbf{n}}$ is uniform in space, say $\hat{\mathbf{l}} = \hat{\mathbf{z}}$. Then one can write the order parameter as $\hat{\mathbf{m}} + i\hat{\mathbf{n}} = (\hat{\mathbf{x}} + i\hat{\mathbf{y}})e^{i\Phi}$. In this case global phase rotations $U(1)_N$ have the same effect on the order parameter as orbital rotations about axis $\hat{\mathbf{l}}$, so that the superfluid velocity can be expressed in terms of the orbital rotations: $\mathbf{v}_s = \frac{\hbar}{2m}\nabla\Phi = \frac{\hbar}{2m}\hat{m}^i \nabla \hat{n}^i$. Dropping the intermediate expression in terms of the phase Φ, which is valid only for homogeneous $\hat{\mathbf{l}}$, one obtains the equation for the superfluid velocity valid for any non-uniform $\hat{\mathbf{l}}$-field:

$$\mathbf{v}_s = \frac{\hbar}{2m}\hat{m}^i \nabla \hat{n}^i . \quad (9.16)$$

Since in our derivation we used the connection to the $U(1)_N$ group, the obtained superfluid velocity transforms properly under the Galilean transformation: $\mathbf{v}_s \to \mathbf{v}_s + \mathbf{u}$. It follows from eqn (9.16) that in contrast to the curl-free superfluid velocity in superfluid ^{4}He, the vorticity in ^{3}He-A can be continuous. It is expressed through an $\hat{\mathbf{l}}$-texture by the following relation:

$$\nabla \times \mathbf{v}_s = \frac{\hbar}{4m}e_{ijk}\hat{l}_i \nabla \hat{l}_j \times \nabla \hat{l}_k , \quad (9.17)$$

which was first found by Mermin and Ho (1976) and is called the Mermin–Ho relation. It reflects the properties of the symmetry-breaking pattern $SO(3)_\mathbf{L} \times$

$U(1)_N \to U(1)_{L_3-N/2}$, which connects orbital rotations and $U(1)_N$ transformations.

Since the triad $\hat{\mathbf{m}}$, $\hat{\mathbf{n}}$ and $\hat{\mathbf{l}}$ play the role of dreibein, the superfluid velocity is analogous to the torsion field in the tetrad formalism of gravity. It describes the space-dependent rotations of vectors $\hat{\mathbf{m}}$ and $\hat{\mathbf{n}}$ about the third vector of the triad, the $\hat{\mathbf{l}}$-vector.

9.1.5 Spin from isospin, isospin from spin

We have seen in eqn (9.1) that nuclear spin of ^{3}He atoms is not responsible for chirality of quasiparticles in superfluid ^{3}He-A. The chirality which appears in the low-energy limit is determined by the orientation of its Bogoliubov–Nambu spin $\check{\tau}^b$ with respect to the direction of the motion of a quasiparticle. Even for the originally 'spinless' fermions one obtains (after rescaling the 'speeds of light', rotating in $\check{\tau}^b$-space, and shifting the momentum) the standard isotropic Hamiltonian for the Weyl's chiral quasiparticles

$$\mathcal{H}_{(a)} = c C_a \check{\tau}^b p_b \;, \tag{9.18}$$

where $C_a = \pm 1$ is the emerging chirality. This Hamiltonian is invariant under rotations in the rescaled coordinate space (or rescaled momenta p_b) if they are accompanied by rotations of matrices $\check{\tau}^b$ in the Bogoliubov–Nambu space. The elements of this combined symmetry of the low-energy Hamiltonian form the $SO(3)$ group of rotations with the generators

$$J^b = L^b + \frac{1}{2}\check{\tau}^b \;, \quad b = (1,2,3) \;. \tag{9.19}$$

This combined $SO(3)$ group is effective; it arises only in the low-energy corner where it is part of the emerging effective Lorentz group. It has nothing to do with the $SO(3)_\mathbf{J}$ group of rotations in the underlying liquid helium which is spontaneously broken in ^{3}He-A. From eqn (9.19) it follows that it is the Bogoliubov–Nambu spin, which plays the role of the 'relativistic' spin for an inner observer living in the world of quasiparticles. While the inner observer believes that the spin is a consequence of the fundamental group of space rotations, or of the fundamental Lorentz group, an external observer finds that the spin is not fundamental, but emerges in the low-energy corner where it gives rise to the full group of relativistic rotations (or Lorentz group) as the effective combined symmetry group of the low-energy quasiparticle world.

Spin 1/2 naturally arising for the fermion zero modes of co-dimension 3 – the fermionic quasiparticles near the Fermi points – produces the connection between spin and statistic in the low-energy world: even if the original bare fermions are spinless, the fermion zero modes of quantum vacuum acquire spin 1/2 in the vicinity of the Fermi point.

On the other hand the conventional spin $S_z = \pm 1/2$ (the spin in the underlying physics of ^{3}He atoms), which leads to the degeneracy of the Fermi point in ^{3}He-A, plays the role of isotopic spin in the world of quasiparticles. The global

$SO(3)_S$ group of spin rotations (actually this is the $SU(2)$ group) is seen by quasiparticles as an isotopic group. It is responsible for the $SU(2)$ degeneracy of quasiparticles, but not for their chirality. We shall see later that it corresponds to the weak isospin in the Standard Model, so that in the low-energy corner the global $SU(2)$ group gradually becomes a local one.

Such interchange of spin and isospin degrees of freedom shows that the only origin of chirality of the (quasi)particle is the non-zero value of the topological invariant N_3 of the fermion zero modes of the quantum vacuum. What kind of spin is related to the chirality depends on the details of the matrix structure of the Green function in the vicinity of the Fermi point, i.e. what energy levels cross each other at the Fermi point. For example, the 'relativistic' spin can be induced near the conical point where two electronic bands in a crystal touch each other. In a sense, there is no principal difference between spin and isospin: by changing continuously the matrix structure of the Green function one can gradually convert isospin to spin, while the topological charge N_3 of the Fermi point remains invariant under such a rotation in spin–isospin space. In RQFT the conversion between spin and isospin degrees of freedom can occur due to fermion zero modes which appear in the inhomogeneous vacuum of a magnetic monopole (Jackiw and Rebbi 1976b).

9.1.6 Gauge invariance and general covariance in fermionic sector

It is not only the Lorentz invariance which arises in the low-energy corner. The Weyl equation (9.9) for the wave function of fermionic quasiparticles near the elementary Fermi point is gauge invariant and even obeys general covariance. For example, the local gauge transformation of the wave function of the fermionic quasiparticle, $\chi \to \chi e^{iq_a \alpha(\mathbf{r},t)}$, can be compensated by the shift of the 'electromagnetic' field $A_\mu \to A_\mu + \partial_\mu \alpha$. The same occurs with general coordinate transformations, which can be compensated by the deformations and rotations of the vierbein fields. These attributes of the electromagnetic (A_μ) and gravitational ($g^{\mu\nu}$) fields arise gradually and emergently when the Fermi point is approached. In this extreme limit, massless relativistic Weyl fermions have another symmetry – the invariance under the global scale transformation $g_{\mu\nu} \to \Omega^2 g_{\mu\nu}$.

9.2 Effective electromagnetic field

9.2.1 Why does QED arise in ^{3}He-A?

Will the low-energy collective modes – the effective gravity and electromagnetic fields – experience the same symmetries as fermions? If yes, we would obtain completely invariant RQFT emerging naturally at low energy. However, this is not as simple. We have already seen in Bose liquids that the low-energy quasiparticles are relativistic and that there is a dynamical metric field $g_{\mu\nu}$ which simulates gravity, but the gravity itself does not obey the Einstein equations. The same occurs for the effective gravity in ^{3}He-A. But the situation with the effective electromagnetic field is different: the leading term in the action for the electromagnetic field is fully relativistic.

The reason for such a difference is the following. The effective action for the collective modes is obtained by integration over the fermionic degrees of freedom. This principle was used by Sakharov (1967a) to obtain an effective gravity and by Zel'dovich (1967) to derive an effective electrodynamics, with both effective actions arising from quantum fluctuations of the fermionic vacuum. In quantum liquids all the fermionic degrees of freedom are known at least in principle for all energy scales. At any energy scale the bare fermions are certainly non-relativistic and do not obey any of the invariances which the low-energy quasiparticles enjoy. That is why one should not expect a priori that the dynamics of the collective vacuum modes will acquire some symmetries.

The alternative is to integrate over the dressed particles – the quasiparticles. If the main contribution to the integral comes from the low-energy corner, i.e. from the fermion zero modes of the quantum vacuum, then the use of the low-energy quasiparticles is justified, and the action obtained in this procedure will automatically obey the same symmetry as fermion zero modes. This is where the difference between the effective gravity and electromagnetic fields shows up. We know that the Newton constant has the dimension $[\hbar][E]^{-2}$, that is why the contribution to G^{-1} depends quadratically on the cut-off energy $G^{-1} \sim E_{\text{Planck}}^2/\hbar$, and thus it practically does not depend on the low-energy fermions. In case of the electromagnetic field the divergence of the dimensionless coupling constant is logarithmic, $\gamma^{-2} \sim \ln(E_{\text{Planck}}/E)$ (eqn (9.21)). Thus the smaller the energy E of gapless fermions, the larger is their contribution to the effective action. That is why in the leading logarithmic approximation, the action for the effective electromagnetic field is determined by the low-energy quasiparticles and thus it automatically acquires all the symmetries experienced by these quasiparticles: gauge invariance, general covariance and even conformal invariance.

Thus in terms of QED, the ^{3}He-A behaves as almost 'perfect' condensed matter, producing in a leading logarithmic approximation the general covariant, gauge and conformal invariant action for the effective electromagnetic field:

$$S_{em} = \int \frac{d^4x}{4\gamma^2\hbar} \sqrt{-g} g^{\mu\nu} g^{\alpha\beta} F_{\mu\alpha} F_{\nu\beta} \ . \quad (9.20)$$

Bearing in mind that the gauge field in the effective QED in ^{3}He-A comes from the shift of the 4-momentum p_μ, we must choose the same dimension for A_μ as that of the 4-momentum p_μ. Then γ becomes a dimensionless coupling 'constant'.

Since the effective action in eqn (9.20) is obtained by integration over momenta where fermions are mostly relativistic, it is covariant in terms of the same metric $g^{\mu\nu}$ as that acting on fermions. This is the reason why photons have the same speed as the maximum attainable speed for fermions. Moreover, since fermions are massless the action obeys conformal invariance: $g^{\mu\nu} \to \Omega^2 g^{\mu\nu}$ does not change the action.

Note that eqn (9.20) does not contain the speed of light c explicitly: it is hidden within the metric field, That is why it can be equally applied both to the Standard Model and to ^{3}He-A, where the speed of light is anisotropic as viewed by an external observer.

9.2.2 Running coupling constant

Both in QED with massless fermions (or fermions with a mass small compared to the cut-off energy scale) and in the effective QED in ^{3}He-A, the coupling constant is not really constant but 'runs', i.e. changes with the energy scales. The vacuum of QED can be considered as some kind of polarizable medium, where virtual pairs of fermions and anti-fermions screen the electric charge. If the fermions are massless this screening is perfect: the charge is completely screened by the polarized vacuum, whence the name zero-charge effect. Mathematically this means that the running coupling constant γ is logarithmically divergent. The ultraviolet cut-off is given by either the Planck or GUT energy scale. If q_a are dimensionless electric charges of chiral fermions in terms of, say, electric charge of the positron, then after integration over the virtual fermions one obtains eqn (9.20) with the following γ (see e.g. Terazawa et al. 1977; Akama et al. 1978):

$$\gamma^{-2} = \frac{1}{24\pi^2} \sum_a q_a^2 \ln\left(\frac{E_{\text{Planck}}^2}{\max(B, T^2, M^2)}\right) . \tag{9.21}$$

The infrared cut-off is provided by the temperature T, external frequency ω, the inverse size r of inhomogeneity, the magnetic field B itself, or, if the fermions are paired forming Dirac particles, by the Dirac mass M. Thus the infrared cut-off in QED is $\max(T^2, \omega^2, (\hbar c/r)^2, B, M^2)$. If $T = \omega = B = M = 0$, then the infrared cut-off is given by the distance r from the charge. This demonstrates that the effective electric charge of the body logarithmically decreases with the distance r from the body: $e^2 \sim e_0^2/\ln(rp_{\text{Planck}}/\hbar)$. At infinity the observed electric charge is zero.

Equation (9.21) is not complete (see the book by Weinberg 1995): (i) It gives only the logarithmic contribution, while the constant term is ignored. The constant term is important since, because of the mass of fermions, the logarithm is not very big. (ii) It misses the antiscreening contribution of charge bosons. That is why it cannot be applied to real QED, but it works well in ^{3}He-A, where the fermions are always massless and the contribution of bosons can be neglected.

9.2.3 Zero-charge effect in ^{3}He-A

Since ^{3}He-A belongs to the same universality class of the vacua with chiral fermions, the above equations can be applied to ^{3}He-A. There are $N_F = 4$ chiral fermionic species in ^{3}He-A with charges $q_a = \pm 1$ and 'isotopic' spin $S_{za} = \pm 1/2$. However, since we doubled the number of degrees of freedom by introducing particles and holes, we must divide the final result by two to compensate the double counting. That is why the effective number of chiral fermionic species in ^{3}He-A is $N_F = 2$. The ultraviolet cut-off of the logarithmically divergent coupling is provided by the gap amplitude $\Delta_0 = c_\perp p_F$, which is the second Planck energy scale in ^{3}He-A, eqn (7.31). It is the same energy scale which enters the vacuum energy density $N_F\sqrt{-g}\Delta_0^4/12\pi^2$ in eqn (7.32), and the Newton gravitational constant of the effective gravity which arises in ^{3}He-A: $G^{-1} = (N_F/9\pi)\Delta_0^2$ (see Sec. 10.5). As a result the logarithmically divergent running coupling constant in ^{3}He-A is

$$\gamma^{-2} = \frac{1}{12\pi^2} \ln\left(\frac{\Delta_0^2}{T^2}\right). \tag{9.22}$$

In ^{3}He-A, the effective metric $g^{\mu\nu}$, its determinant $\sqrt{-g}$ and the gauge field A_μ must be expressed in terms of the ^{3}He-A observables in eqns (9.11) and eqn (9.15). Substituting all these into eqn (9.20) one obtains the logarithmically divergent contribution to the Lagrangian for the $\hat{\mathbf{l}}$ field in ^{3}He-A:

$$\ln\left(\frac{\Delta_0^2}{T^2}\right) \frac{p_F^2}{24\pi^2 \hbar v_F} \left(c_\parallel^2 \left(\hat{\mathbf{l}} \times (\nabla \times \hat{\mathbf{l}})\right)^2 + c_\perp^2 \left(\hat{\mathbf{l}} \cdot (\nabla \times \hat{\mathbf{l}})\right)^2 - \left(\partial_t \hat{\mathbf{l}} + (\mathbf{v}_s \cdot \nabla)\hat{\mathbf{l}}\right)^2 \right). \tag{9.23}$$

This is what was obtained in a microscopic BCS theory of ^{3}He-A (see Chapter 10; Leggett and Takagi 1978; Dziarmaga 2002). The second term contains the parameter $c_\perp$, which is small in ^{3}He-A. That is why it is usually neglected compared to the regular (non-divergent) term with the same structure in eqn (10.41). At $T = 0$ the infrared cut-off instead of T^2 is given by the field B, which in our case is $c_\perp^2 p_F |\hat{\mathbf{l}} \times (\nabla \times \hat{\mathbf{l}})|$.

9.2.4 Light – orbital waves

There are many drawbacks of the effective QED in ^{3}He-A, which are consequences of the fact that the low-energy physics is formed mostly by fermions at the trans-Planckian scale where these fermions are not relativistic. Though eqn (9.23) was obtained from the completely symmetric eqn (9.20) for QED, the difference between QED and ^{3}He-A is rather large. For example, the vector field $\hat{\mathbf{l}}$ is an observable variable in ^{3}He-A. Only at low energy does it play the role of the vector potential of gauge field $\mathbf{A}$, which is not the physical observable for the inner observer because of the emerging gauge invariance. Moreover, the $\hat{\mathbf{l}}$-vector and the superfluid velocity $\mathbf{v}_s$ form both the gauge field ($\mathbf{A} = p_F \hat{\mathbf{l}}$ and $A_0 = p_F \mathbf{v}_s \cdot \hat{\mathbf{l}}$) and the metric field in eqn (9.11). The only case when the gauge and gravitational fields are not mixed is when $c_\parallel = c_\perp$, and $\mathbf{v}_s \perp \hat{\mathbf{l}}$. Also there are the non-logarithmic terms in action, which can be ascribed to both the gravitational and electromagnetic fields.

There are, however, situations where these fields are decoupled. One physically important case, where only the 'electromagnetic part' is relevant, will be discussed in Sec. 19.3. In this particular case we need the linearized equations in the background of the equilibrium state which is characterized by the homogeneous direction of the $\hat{\mathbf{l}}$-vector. The latter is fixed by the counterflow: $\hat{\mathbf{l}}_0 \parallel \mathbf{v}_n - \mathbf{v}_s$. The vector potential of the electromagnetic field is simulated by the small deviations of the $\hat{\mathbf{l}}$-vector from its equilibrium direction, $\mathbf{A} = p_F \delta\hat{\mathbf{l}}$, while the metric is determined by the equilibrium background $\hat{\mathbf{l}}_0$-vector. Since $\hat{\mathbf{l}}$ is a unit vector, its variation is $\delta\hat{\mathbf{l}} \perp \hat{\mathbf{l}}_0$. This corresponds to the gauge choice $A_z = 0$, if the z axis is chosen along the background orientation, $\hat{\mathbf{z}} = \hat{\mathbf{l}}_0$. In the considered case only the dependence on z and t is relevant; as a result $\hat{\mathbf{l}} \cdot (\nabla \times \hat{\mathbf{l}}) = \nabla \cdot \hat{\mathbf{l}} = 0$ and the dynamics of the field $\mathbf{A} = p_F \delta\hat{\mathbf{l}}$ is determined by the Lagrangian (9.23), which acquires the following form in the superfluid-comoving frame:

$$\frac{1}{2\gamma^2}\sqrt{-g}\left(\mathbf{B}^2 - \mathbf{E}^2\right) = \frac{p_F^2 v_F}{24\pi^2 \hbar}\ln\left(\frac{\Delta_0^2}{T^2}\right)\left((\partial_z \delta \hat{\mathbf{l}})^2 - \frac{1}{v_F^2}(\partial_t \delta \hat{\mathbf{l}})^2\right) . \quad (9.24)$$

There are two modes of transverse oscillations of $\mathbf{A} = p_F \delta \hat{\mathbf{l}}$ propagating along z, with the effective electric and magnetic fields (E_y, B_x) in one mode and (E_x, B_y) in the other. These effective electromagnetic waves are called orbital waves. 'Light' propagates along z with the velocity $v_F = c_\parallel$, i.e. with the same velocity as fermionic quasiparticles propagating in the same direction. Fermions impose their speed on light. Though in ^{3}He-A this happens only for the z-direction (^{3}He-A is not a perfect system for a complete simulation of RQFT), nevertheless this example demonstrates the mechanism leading to the same maximum attainable speed for fermions and bosons in an effective theory.

9.2.5 Does one need the symmetry breaking to obtain massless bosons?

Ironically the enhanced symmetry of orbital dynamics in ^{3}He-A characterized by the fully 'relativistic' equation (9.20) arises due to the spontaneous symmetry breaking into the ^{3}He-A state which leads to formation of the Fermi points at $\mathbf{p}^{(a)} = -C_a p_F \hat{\mathbf{l}}$, and at low energy to all phenomena following from the existence of the Fermi points. In other words, the low-energy symmetry emerges from the symmetry breaking, which occurs at the second Planck energy scale, at $T = T_c \sim \Delta_0 = E_{\text{Planck 2}}$ (eqn (7.31)). Below T_c the liquid becomes anisotropic with the anisotropy axis along the $\hat{\mathbf{l}}$-vector. The propagating oscillations of the anisotropy axis $\mathbf{A} = p_F \hat{\mathbf{l}}$ in eqn (9.24) are the Goldstone bosons arising in ^{3}He-A due to spontaneously broken symmetry. They are viewed by an inner observer as the propagating electromagnetic waves.

However, the spontaneous symmetry breaking is not a necessary condition for the Fermi points to exist. For instance, the Fermi points can naturally appear in semiconductors without any symmetry breaking (Nielsen and Ninomiya 1983; Abrikosov 1998). It is the collective dynamics of the fermionic vacuum with Fermi points which gives rise to the massless photon.

Moreover, one must avoid symmetry breaking in order to have a closer analogy with the Einstein theory of gravity. This is because the symmetry breaking is the main reason why the effective action for the metric $g^{\mu\nu}$ and gauge field A_ν is contaminated by non-covariant terms. The latter come from the gradients of Goldstone fields, and in most cases they dominate or are comparable to the natural Einstein and Maxwell terms in the effective action. An example is the $(1/2)m n \mathbf{v}_s^2$ term in eqn (4.6). It is always larger than the Einstein action for $g^{\mu\nu}$, which, when expressed through $\mathbf{v}_s$, contains the gradients of $\mathbf{v}_s$:

$$\sqrt{-g}\mathcal{R} = \frac{1}{c^3}\left(2\partial_t \nabla \cdot \mathbf{v}_s + \nabla^2(v_s^2)\right) . \quad (9.25)$$

This is why a condensed matter system, where the analogy with RQFT is realized in full, would be a quantum liquid in which Fermi points exist without symmetry breaking, and thus without superfluidity. In other words, our physical vacuum is not a superfluid liquid.

9.2.6 Are gauge equivalent states physically indistinguishable?

The discussed example of the effective QED arising in ^{3}He-A shows the main difference between effective and fundamental theories of QED. In a fundamental theory the gauge equivalent fields – the fields obtained from each other by the gauge transformation $A_\mu \to A_\mu + \nabla_\mu \alpha$ – are considered as physically indistinguishable. In an effective theory gauge invariance is not fundamental: it exists in the low-energy world of quasiparticles and is gradually violated at higher energy. That is why gauge equivalent states are physically different as viewed by an external (i.e. high-energy) observer. The homogeneous states with different directions of $\hat{\mathbf{l}}$ and thus with different homogeneous vector potentials $\mathbf{A} = p_F \hat{\mathbf{l}}$ are equivalent for an inner observer, but can be easily resolved by an external observer, who does not belong to the ^{3}He-A vacuum. The same can occur in gravity: if general relativity is an effective theory, general covariance must be violated at high energy. This means that metrics which are equivalent for us, i.e. the metrics which can be transformed to each other by general coordinate transformations, become physically distinguishable at high energy. We shall discuss this in Chapter 32.

Finally we note that, as we already discussed in Sec. 5.3, the speed of light c does not enter eqn (9.20) explicitly. It is absorbed by the metric tensor $g_{\mu\nu}$ and by the fine structure constant γ. According to eqn (9.22) γ is a dimensionless quantity determined by the number of chiral fermionic species N_F and by the Planck energy cut-off Δ_0. The factorization of $\gamma^2 = e^2/4\pi\hbar c$ in terms of the dimensionful parameters – electric charge e and speed of light c – would be artificial, especially if one must choose between two considerably different speeds of 'light', $c_\parallel$ and $c_\perp$. Such a factorization can be appropriate for an inner observer, for whom the 'speeds of light' in different directions are indistinguishable, but not for an external observer, for whom they are different.

9.3 Effective $SU(N)$ gauge fields

9.3.1 Local $SU(2)$ from double degeneracy

In ^{3}He-A each Fermi point is doubly degenerate owing to the ordinary spin σ of the ^{3}He atom (Sec. 9.1.5). These two species of fermions living in the vicinity of the Fermi point transform to each other by the discrete Z_2-symmetry operation P in eqn (7.56), which in turn originates from the global $SU(2)_S$ group of spin rotations in the normal liquid (Sec. 7.4.6). The total topological charge of the point, say at the north pole $\mathbf{p} = p_F \hat{\mathbf{l}}$, is $N_3 = -2$. This degeneracy, however, exists only in equilibrium. If the vacuum is perturbed the P-symmetry is violated, and the doubly degenerate Fermi point splits into two elementary Fermi points with $N_3 = -1$ each. The perturbed A-phase effectively becomes the axiplanar phase discussed in Sec. 7.4.7 (see Fig. 7.3), though it remains the A-phase in a full equilibrium. The positions of the split Fermi points, $p_F \hat{\mathbf{l}}_1$ and $p_F \hat{\mathbf{l}}_2$, can oscillate separately. This does not violate momentum space topology as the total topological charge of the two Fermi points is conserved in such oscillations: $N_3 = -1 - 1 = -2$.

The additional degrees of freedom of the low-energy collective dynamics of the vacuum, which are responsible for the separate motion of the Fermi points, have a direct analog in RQFT. These degrees of freedom are viewed by an inner observer as the local $SU(2)$ gauge field. The reason for this is the following.

The propagator describing the two fermionic species, each being a spinor in the Bogoliubov–Nambu space, is a 4×4 matrix. Let us consider a general form of the perturbations of the propagator. In principle one can perturb the vierbein separately for each of the two fermionic species. We ignore such degrees of freedom, since they appear to be massive. Then the general perturbation of the propagator for quasiparticles living near the north Fermi point, where its 'electric' charge $q = +1$, has the following form:

$$\mathcal{G}^{-1} = \check{\tau}^b e_b^\mu (p_\mu - qA_\mu - q\sigma_a W_\mu^a) \,, \quad a = (1,2,3) \,. \tag{9.26}$$

The last term contains the Pauli matrix for conventional spin and the collective variable W_μ^a which interacts with the spin. This new effective field is viewed by quasiparticles (and thus by an inner observer) as an $SU(2)$ gauge field – it is the analog of the weak field in the Standard Model. The ordinary spin of the ^{3}He atoms is viewed by an inner observer as the weak isospin.

This weak field W_μ^a is dynamical, since it represents some collective motion of the fermionic vacuum. The leading logarithmic contribution to the action for this Yang–Mills field can be obtained by integration over fermions:

$$S_{YM} = \int \frac{d^4 x}{4\gamma_W^2} \sqrt{-g} g^{\mu\nu} g^{\alpha\beta} F_{\mu\alpha}^a F_{\nu\beta}^a \,. \tag{9.27}$$

Calculations show that the coupling constant γ_W for the Yang–Mills field in ^{3}He-A in eqn (9.27) coincides with the coupling constant γ for the Abelian field in eqn (9.22), i.e. the 'weak' charge is also logarithmically screened by the fermionic vacuum. It experiences a zero-charge effect in ^{3}He-A instead of the asymptotic freedom in the Standard Model, where the antiscreening is produced by the dominating contribution from bosonic degrees of freedom of the vacuum. The reason for such a difference is that in ^{3}He-A the $SU(2)$ gauge bosons are not fundamental: they appear in the low-energy limit only. That is why their contribution to the vacuum polarization is relatively small compared to the fermionic contribution, and thus their antiscreening effect can be neglected in spite of the prevailing number of bosons. The same can in principle occur in the Standard Model above, say the GUT scale.

9.3.2 Role of discrete symmetries

The spin degrees of freedom come from the global $SU(2)_S$ group of spin rotations of the normal liquid. This group is spontaneously broken in superfluid ^{3}He-A, but not completely. The π rotation combined with either orbital or phase rotation form the important Z_2-symmetry group P of the ^{3}He-A vacuum in eqn (7.56). This discrete symmetry is the source of the degeneracy of the Fermi point in

^{3}He-A, which finally leads to the local $SU(2)$ symmetry group in the effective low-energy theory. Thus in ^{3}He-A we have the following chain:

$$SU(2)_{\text{global}} \to Z_2 \to SU(2)_{\text{local}} \ . \tag{9.28}$$

In principle, the first stage is not necessary: the simplest discrete Z_2 symmetry of the vacuum state can be at the origin of the local $SU(2)$ gauge field in the effective theory.

This implies that the higher local symmetry group of the quantum vacuum, such as $SU(4)$, can also arise from the discrete symmetry group, such as Z_4 or $Z_2 \times Z_2$. The higher discrete symmetry group leads to multiple degeneracy of the Fermi point, and thus to the multiplet of chiral fermionic species. The collective modes of such a vacuum contain effective gauge fields of the higher symmetry group. The discrete symmetry between fermions ensures that all the fermions of the multiplet have the same maximum attainable speed. The same speed will enter the effective action for the gauge bosons, which is obtained by integration over fermions. Thus, in principle, it is possible that the discrete symmetry can be at the origin of the degeneracy of the Fermi point in the Standard Model too, giving rise to its fermionic and bosonic content, and to the same Lorentz invariance for all fermions and bosons. Thus the Lorentz invariance gradually becomes the general principle of the low-energy physics, which in turn is the basis for the general covariance of the effective action.

The importance and possible decisive role of the discrete symmetries in RQFT was emphasized by Harari and Seiberg (1981), Adler (1999) and Peccei (1999). The relation between the discrete symmetry in the Standard Model and the topological invariant for the Fermi point will be discussed in Sec. 12.3.2.

9.3.3 Mass of W-bosons, flat directions and supersymmetry

In the effective theory the gauge invariance is approximate. It is violated due to the so-called non-renormalizable terms in the action, i.e. terms which do not diverge logarithmically. They come from 'Planckian' physics beyond the logarithmic approximation, i.e. from the fermions which are not close to the Fermi point. Due to such a term the $SU(2)$ gauge boson – the W-boson – acquires a mass in ^{3}He-A. This mass comes from the physics beyond the BCS model; as a result it is small compared to the typical gap of order Δ_0 in the spectrum of the other massive collective modes (see the book by Volovik 1992a).

As we know the BCS model for ^{3}He-A enjoys a hidden symmetry: the spin-up and spin-down components are independent. The energy of the axiplanar phase does not depend on the mutual orientation of the vectors $\hat{\mathbf{l}}_1$ and $\hat{\mathbf{l}}_2$ (Fig. 7.3 *top*). Since the mutual orientation is the variable corresponding to the W-boson, one obtains that the W-boson is the Goldstone boson and therefore massless. Thus the masslessness of the W-boson is a consequence of the flat direction in the vacuum energy functional obtained within the BCS model (Fig. 7.3 *bottom*), and the W-boson becomes massive when flatness is disturbed. Flat directions are popular in RQFT, especially in supersymmetric models (see e.g. Achúcarro et al. (2001) and references therein).

Probably the same hidden symmetry in the BCS model is at the basis of the supersymmetry between fermions and bosons revealed by Nambu (1985), who observed that in superconductors, in nuclear matter and in superfluid ^{3}He-B, if they are described within the BCS theory, there is a remarkable relation between the masses of the fermions and the masses of the bosons: the sum of the squares of the masses is equal. The same elements of the supersymmetry can be shown to exist in superfluid ^{3}He-A as well.

Consequences of the hidden symmetry in ^{3}He-A are discussed by Volovik and Khazan (1982), Novikov (1982), and in Sec. 5.15 of the book by Volovik (1992a).

9.3.4 Different metrics for different fermions. Dynamic restoration of Lorentz symmetry

In eqn (9.26) we did not take into account that in the axiplanar phase each of the two elementary Fermi points can have its own dynamical vierbein field. Though in equilibrium their vierbeins coincide due to internal symmetry, they can oscillate separately. As a result the number of collective modes could increase even more. In the BCS model for ^{3}He-A, which enjoys hidden symmetry, the spin-up and spin-down components are decoupled. As a result the two vierbeins are completely independent, and each of them acts on its own 'matter': fermions with given spin projection. Thus we have two non-interacting worlds: two 'matter' fields each interacting with their own gravitational field.

However, in the real liquid, beyond the BCS model, the coupling between different spin populations is restored. As a result the collective mode, which corresponds to the out-of-phase oscillations of the two vierbeins, acquires a mass. This means that in the low-energy limit the vierbeins are in the in-phase regime: i.e. all fermions have the same metric $g^{\mu\nu}$ in this limit.

A similar example has been discussed by Chadha and Nielsen (1983). They considered massless electrodynamics with a different metric (vierbein) for the left-handed and right-handed fermions. In this model Lorentz invariance is violated. They found that the two metrics converge to a single one as the energy is lowered. Thus in the low-energy corner the Lorentz invariance becomes better and better, and at the same time the number of independent massless bosonic modes decreases.

10
THREE LEVELS OF PHENOMENOLOGY OF SUPERFLUID ^{3}He

There are three levels of phenomenology of superfluid ^{3}He, determined by the energy scale of the variables and by the temperature range.

The first one is the Ginzburg–Landau level. It describes the behavior of the general order parameter field which is not necessarily close to the vacuum manifold. The order parameter is the analog of the Higgs field in the Standard Model. As distinct from the Higgs field in the Standard Model, which belongs to the spinor representation of the $SU(2)$ group, the Higgs field $e_{\mu i}$ in ^{3}He belongs to the product of vector representations ($S = 1$ and $L = 1$) of groups $SO(3)_\mathbf{S}$ and $SO(3)_\mathbf{L}$, respectively. At fixed i the elements $e_{\mu i}$ with $\mu = 1, 2, 3$ form a vector which rotates under the group $SO(3)_\mathbf{S}$ of spin rotations. Correspondingly, at fixed μ the same elements $e_{\mu i}$ with $i = 1, 2, 3$ form a vector which rotates under the group $SO(3)_\mathbf{L}$ of the orbital (coordinate) rotations.

The static order parameter field is described by the Ginzburg–Landau free energy functional, which is valid only in the vicinity of the temperature T_c of the phase transition into the broken-symmetry state, $T_c - T \ll T_c$. Extrema of the Ginzburg–Landau free energy determine the vacuum states: the true vacuum if it is a global minimum, or the false vacuum if it is a local minimum. In some rather rare cases the Ginzburg–Landau theory can be extended to incorporate the dynamics of the Higgs field. However, in most typical situations this dynamics, which originates from the microscopic fermionic degrees of freedom, is non-local and the time-dependent Ginzburg–Landau theory does not exist.

Another level of phenomenology is the statics and dynamics of the soft variables describing the 'vacuum' in the vicinity of a given vacuum manifold. This is an analog of elasticity theory in crystals; in superfluid ^{4}He this is the two-fluid hydrodynamics; and in superconductors and also in Fermi superfluids it is called the London limit: equations which are valid in this limit were first introduced for superconductivity by H. and F. London in 1935. This phenomenology is applicable in the whole range of the broken symmetry state, $0 < T < T_c$, if the considered length scale exceeds the characteristic microscopic length scale ξ – the coherence length.

Finally, in the low-temperature limit $T \ll T_c$ the analog of RQFT arises in ^{3}He-A as the third level of phenomenology.

10.1 Ginzburg–Landau level

10.1.1 Ginzburg–Landau free energy

The Ginzburg–Landau theory of superconductivity and also of superfluidity in Fermi systems is based on Landau's theory of second-order phase transition. Close to the transition temperature, $T_c - T \ll T_c$, the static behavior of the order parameter (the Higgs field) is determined by the Ginzburg–Landau free energy functional, which contains second- and fourth-order terms only. In our case the order parameter is the matrix $e_{\mu i}(\mathbf{r})$ in eqn (7.38). The general form of the Ginzburg–Landau functional must be consistent with the symmetry group $G = U(1)_N \times SO(3)_{\mathbf{L}} \times SO(3)_{\mathbf{S}}$ of the symmetric normal state of liquid ^{3}He. The condensation energy term, i.e. the free energy of the spatially uniform state measured from the energy of the normal Fermi liquid with unbroken symmetry, is given by

$$F_{\rm GL}^{\rm bulk} = -\alpha(T) e^*_{\mu i} e_{\mu i} + \beta_1 e^*_{\mu i} e^*_{\mu i} e_{\nu j} e_{\nu j} + \beta_2 e^*_{\mu i} e_{\mu i} e^*_{\nu j} e_{\nu j}$$
$$+ \beta_3 e^*_{\mu i} e^*_{\nu i} e_{\mu j} e_{\nu j} + \beta_4 e^*_{\mu i} e_{\nu i} e^*_{\nu j} e_{\mu j} + \beta_5 e^*_{\mu i} e_{\nu i} e_{\nu j} e^*_{\mu j} \ . \tag{10.1}$$

Here $\alpha(T) = \alpha'(1 - T/T_c)$ changes sign at the phase transition. The input parameters – coefficient α' of the second-order term, and the coefficients $\beta_1, \ldots, \beta_5$ of the fourth-order invariants – can be considered as phenomenological. In principle, they can be calculated from the exact microscopic Theory of Everything in eqn (3.2), but this requires serious numerical simulations and still has not been completed. In the BCS model of superfluid ^{3}He these parameters are expressed in terms of the 'fundamental constants' entering the BCS model, p_F, v_F (or the effective mass $m^* = p_F/v_f$), and $T_c \sim \Delta_0(T=0)$:

$$\alpha = \frac{1}{3} N(0)(1 - T/T_c) \ , \quad -2\beta_1 = \beta_2 = \beta_3 = \beta_4 = -\beta_5 = \frac{7N(0)\zeta(3)}{120(\pi T_c)^2} \ , \tag{10.2}$$

where $N(0) = m^* p_F/2\pi^2 \hbar^3$ is the density of fermionic states in normal ^{3}He for one spin projection at the Fermi level; $\zeta(3) \approx 1.202$.

The gradient terms in the Ginzburg–Landau free energy functional contain three more phenomenological parameters:

$$F_{\rm GL}^{\rm grad} = \gamma_1 \partial_i e_{\mu j} \partial_i e^*_{\mu j} + \gamma_2 \partial_i e_{\mu i} \partial_j e^*_{\mu j} + \gamma_3 \partial_i e_{\mu j} \partial_j e^*_{\mu i} \ . \tag{10.3}$$

In the BCS model they are expressed in terms of the same 'fundamental constants':

$$\gamma_1 = \gamma_2 = \gamma_3 = \gamma = 7\zeta(3) N(0) \frac{\hbar^2 v_F^2}{240(\pi T_c)^2} \equiv \frac{1}{3} N(0) \xi_0^2 \ , \tag{10.4}$$

where $\xi_0 \sim \hbar v_F/T_c$ enters the temperature-dependent coherence or healing length

$$\xi(T) = \left(\frac{\gamma}{\alpha}\right)^{1/2} = \frac{\xi_0}{\sqrt{1 - T/T_c}} \ . \tag{10.5}$$

The coherence length $\xi(T)$ characterizes the spatial extent of an inhomogeneity, at which the gradient energy becomes comparable to the bulk energy in the liquid: perturbations of the order parameter decay over this distance.

10.1.2 Vacuum states

If the system is smooth enough, i.e. if the size of the inhomogeneity regions is larger than $\xi(T)$, the gradient terms may be neglected compared to the condensation energy in eqn (10.1). Then the minimization of this bulk energy (10.1) determines the vacuum states – possible equilibrium phases of spin-triplet p-wave superfluidity. It is known from experiment that depending on pressure (which certainly influences the relations between the parameters β_i in eqn (10.1)) only two of the possible superfluid phases are realized in the absence of an external magnetic field. These are: ^{3}He-A with $e_{\mu i} = \Delta_0(T)\hat{d}_\mu(\hat{m}_i + i\hat{n}_i)$ in eqn (7.55); and ^{3}He-B with $e_{\mu i} = \Delta_0(T)R_{\mu i}e^{i\Phi}$ in eqn (7.47). The gap parameters $\Delta_0(T)$ in these two vacua are respectively

$$\Delta_0^2(T) = \frac{\alpha(T)}{4(\beta_2 + \beta_4 + \beta_5)}, \quad \text{A-phase}, \tag{10.6}$$

$$\Delta_0^2(T) = \frac{\alpha(T)}{2(\beta_3 + \beta_4 + \beta_5) + 6(\beta_1 + \beta_2)}, \quad \text{B-phase}. \tag{10.7}$$

10.2 London level in ^{3}He-A

10.2.1 London energy

In the London limit only such deformations are considered which leave the system in the vicinity of a given vacuum state. Here we are mostly interested in the ^{3}He-A vacuum. Because of the $SO(3)_L \times SO(3)_S$ symmetry of the bulk energy, the vacuum energy in the A-phase does not depend on the orientation of unit vectors $\hat{m}$, $\hat{n}$, $\hat{l}$, forming the triad, and on the unit vector $\hat{d}$ describing spin degrees of freedom of the vacuum. These vectors thus characterize the degenerate states of the A-phase vacuum. The space swept by these vectors forms the so-called vacuum manifold: each point of this manifold corresponds to the minimum of the bulk energy. These unit vectors thus represent the Goldstone modes – massless bosonic fields which appear due to symmetry breaking. For the slow (hydrodynamic) motion of the vacuum, only these soft Goldstone modes are excited. The energetics of these static Goldstone fields is given by the so-called London energy, which is the quadratic form of the gradients of the Goldstone fields. The London energy is only applicable if the length scale is larger than the coherence length ξ, such that the system is locally in the vacuum manifold.

Let us for simplicity ignore the spin degrees of freedom. The gradients of the remaining degrees of freedom – orbital variables $\hat{m}$, $\hat{n}$, $\hat{l}$ – can be conveniently expressed in terms of physical observables. The $\hat{l}$-vector is observable (for an external observer), since it determines the direction of the anisotropy axis in this anisotropic liquid. However, it is not so easy to detect the rotation angle of the other two vectors of the triad around the $\hat{l}$-vector, since according to the symmetry $U(1)_{L_z-N/2}$ of the ^{3}He-A vacuum, the orbital rotation about the $\hat{l}$-vector is equivalent to the global gauge transformation from $U(1)_N$ group. The observable variable is the gradient of the angle of rotation of $\hat{m}$ and $\hat{n}$ about $\hat{l}$. It is the superfluid velocity $\mathbf{v}_s = (\hbar/2m)\hat{m}_i\nabla\hat{n}_i$ in eqn (9.16). Since the three components of $\mathbf{v}_s$ are constrained by the Mermin–Ho relation eqn (9.17), the variables $\hat{\mathbf{l}}$ and $\mathbf{v}_s$ represent the same three degrees of freedom as the original triad. We assume that the normal component is in a global equilibrium, i.e. $\mathbf{v}_n = 0$ in the environment frame. Then the London energy density, which is quadratic in the gradients of the soft Goldstone variables, is written in the following general form, satisfying the global symmetry $U(1)_N \times SO(3)_L$ of the energy functional:

$$F_{\text{London}} = \frac{m}{2}n_{sik}v_s^i v_s^k + \frac{1}{2}\mathbf{v}_s \cdot \left(C\nabla \times \hat{\mathbf{l}} - C_0\hat{\mathbf{l}}(\hat{\mathbf{l}} \cdot (\nabla \times \hat{\mathbf{l}}))\right) \quad (10.8)$$

$$+ K_s(\nabla \cdot \hat{\mathbf{l}})^2 + K_t(\hat{\mathbf{l}} \cdot (\nabla \times \hat{\mathbf{l}}))^2 + K_b(\hat{\mathbf{l}} \times (\nabla \times \hat{\mathbf{l}}))^2. \quad (10.9)$$

10.2.2 *Particle current*

Compared to the simplest superfluids discussed in Part I, there are several complications related to the particle current or mass current:

(1) Superfluid ^{3}He-A is anisotropic with the uniaxial anisotropy along the $\hat{\mathbf{l}}$-vector; as a result the superfluid density is a tensor with uniaxial anisotropy:

$$n_{sik} = n_{s\|}\hat{l}_i\hat{l}_k + n_{s\perp}(\delta_{ik} - \hat{l}_i\hat{l}_k) . \quad (10.10)$$

(2) The direction $\hat{\mathbf{l}}$ of the anisotropy axis is the Goldstone part of the order parameter, and thus the London energy also contains gradients of $\hat{\mathbf{l}}$ in eqn (10.9). This gradient energy is the same as in liquid crystals with uniaxial anisotropy, the so-called nematic liquid crystals. The coefficients K_s, K_t and K_b describe the response of the system to *splay*-, *twist*- and *bend*- like deformations of the $\hat{\mathbf{l}}$-vector, respectively.

(3) The flow properties of ^{3}He-A are remarkably different. We have already seen that the vorticity of the flow velocity $\mathbf{v}_s$ of the superfluid vacuum is nonzero in the presence of an $\hat{\mathbf{l}}$-texture according to the Mermin–Ho relation (9.17). Also the particle current, in addition to the superfluid current carried by the superfluid component of the liquid, and the normal current carried by the normal component of the liquid, contains (as follows from eqn (10.8)) the orbital current carried by textures of the $\hat{\mathbf{l}}$-vector:

$$P_i = mn_{sik}v_s^k + mn_{nik}v_n^k + P_i\{\hat{\mathbf{l}}\} \,, \quad \mathbf{P}\{\hat{\mathbf{l}}\} = \frac{C}{2}\nabla \times \hat{\mathbf{l}} - \frac{C_0}{2}\hat{\mathbf{l}}(\hat{\mathbf{l}} \cdot (\nabla \times \hat{\mathbf{l}})) . \quad (10.11)$$

We recall that $n_{sik} + n_{nik} = n\delta_{ik}$ due to Galilean invariance, where $n \approx p_F^3/3\pi^2\hbar^3$ is the particle density of the liquid. The last (textural) term in the current density is allowed by symmetry and thus must be present. This current is related to the phenomenon of chiral anomaly in RQFT, which we shall discuss later on.

10.2.3 *Parameters of London energy*

In general all seven temperature-dependent coefficients in London energy, $n_{s\|}$, $n_{s\perp}$, C, C_0, K_s, K_t and K_b, must be considered as phenomenological. However, in the limit $T_c - T \ll T_c$, all these parameters are obtained from the higher-level theory – the Ginzburg–Landau free energy functional. The latter, in turn, can be derived using the more fundamental BCS theory. Using the BCS values (10.4) of parameters γ which enter the gradient energy (10.3) one obtains (see the book by Vollhardt and Wölfle (1990)):

$$n_{s\|} = \frac{1}{2}n_{s\perp} = 16\frac{m}{\hbar^2}K_s = 16\frac{m}{\hbar^2}K_t = \frac{16m}{3\hbar^2}K_b = \frac{C_0}{\hbar} = \frac{2C}{\hbar} = \frac{2}{5}\frac{m}{m^*}n\left(1 - \frac{T}{T_c}\right). \quad (10.12)$$

10.3 Low-temperature relativistic regime

10.3.1 *Energy and momentum of vacuum and 'matter'*

In the low-temperature limit, the system enters the regime where the 'matter' consists of a dilute gas of 'relativistic' quasiparticles interacting with the vacuum collective degrees of freedom, $\mathbf{v}_s$ and $\hat{\mathbf{l}}$. These vacuum variables produce the fields which are seen by quasiparticles as effective electromagnetic and gravity fields.

The particle current density $\mathbf{J}$ (or momentum density $\mathbf{P} = m\mathbf{J}$) and energy density of the system are given by the same eqn (6.3) as for a Bose liquid, modified to include the orbital degrees of freedom – the $\hat{\mathbf{l}}$-field:

$$E = \frac{m}{2}n\mathbf{v}_s^2 + \mathbf{P}\{\hat{\mathbf{l}}\} \cdot \mathbf{v}_s + \sum_a \int \frac{d^3p}{(2\pi\hbar)^3}\tilde{E}_a(\mathbf{p})f_a(\mathbf{p}) \quad (10.13)$$

$$+ K_s(\nabla \cdot \hat{\mathbf{l}})^2 + K_t(\hat{\mathbf{l}} \cdot (\nabla \times \hat{\mathbf{l}}))^2 + K_b(\hat{\mathbf{l}} \times (\nabla \times \hat{\mathbf{l}}))^2, \quad (10.14)$$

$$\mathbf{P} = mn\mathbf{v}_s + \mathbf{P}\{\hat{\mathbf{l}}\} + \mathbf{P}^M + \mathbf{P}^F, \quad (10.15)$$

$$\mathbf{P}^M = \sum_a \int \frac{d^3p}{(2\pi\hbar)^3}\left(\mathbf{p} - \mathbf{p}^{(a)}\right)f_a(\mathbf{p}), \quad (10.16)$$

$$\mathbf{P}^F = \sum_a \int \frac{d^3p}{(2\pi\hbar)^3}\mathbf{p}^{(a)}f_a(\mathbf{p}). \quad (10.17)$$

Here we excluded from consideration the variation of the particle density n and the Goldstone field $\hat{\mathbf{d}}$. The quasiparticle energy $\tilde{E}_a$ is given by eqn (9.8).

The momentum of the liquid contains four contributions which have different physical meaning. The momentum $\mathbf{P}\{\hat{\mathbf{l}}\}$ carried by the texture in eqn (10.11) is anomalous: in Sec. 19.2 we shall show that this current is determined by the chiral anomaly. The corresponding energy term $\mathbf{P}\{\hat{\mathbf{l}}\} \cdot \mathbf{v}_s$ in eqn (10.13) is the analog of the Chern–Simons term in the Standard Model (Sec. 19.2).

The momentum carried by quasiparticles, $\propto \sum_a \int \mathbf{p}f_a(\mathbf{p})$, is written as a sum of two terms. Such separation stems from the fact that the Fermi points $\mathbf{p}^{(a)}$ are not in the origin in the momentum space, i.e. in the equilibrium vacuum $\mathbf{p}^{(a)} \neq 0$. As a result the momentum $\mathbf{P}^M$ counted from the Fermi points corresponds to the momentum of 'matter' introduced for quasiparticles in Bose systems in eqn (5.14). The remaining part of the quasiparticle momentum $\mathbf{P}^F$ represents the fermionic charge which serves as an analog of the baryonic charge in the effect of chiral anomaly (Sec. 18.3). $\mathbf{P}^{\mathrm{F}}$ is directed along the $\hat{\mathbf{l}}$-vector, while $\mathbf{P}^{\mathrm{M}}$ becomes perpendicular to $\hat{\mathbf{l}}$ in the limit of low T.

Let us first consider the uniform state of the $\hat{\mathbf{l}}$-field, when the texture does not carry the momentum, and consider these two quasiparticle currents.

10.3.2 Chemical potential for quasiparticles

In the local thermodynamic equilibrium the distribution of quasiparticles is characterized by the local temperature T and by the local velocity $\mathbf{v}_{\mathrm{n}}$ of the normal component:

$$f_T^{(a)}(\mathbf{p}) = \frac{1}{\exp\frac{\tilde{E}_a - \mathbf{p}\cdot\mathbf{v}_{\mathrm{n}}}{T} + 1} = \frac{1}{\exp\frac{E_a - \mathbf{p}\cdot\mathbf{w}}{T} + 1} . \tag{10.18}$$

Here, as before, $\mathbf{w} = \mathbf{v}_{\mathrm{n}} - \mathbf{v}_{\mathrm{s}}$ is the counterflow velocity – the velocity of the 'matter' in the superfluid-comoving frame. Let us now introduce the following quantities:

$$\mu_a = q_a p_F (\hat{\mathbf{l}}\cdot\mathbf{w}) . \tag{10.19}$$

Then eqn (10.18) can be rewritten in the gauge invariant form

$$f_T^{(a)}(\mathbf{p}) = \frac{1}{\exp\frac{E_a - \mu_a - (\mathbf{p} - q_a \mathbf{A})\cdot\mathbf{w}}{T} + 1} . \tag{10.20}$$

From this equation it follows that parameters μ_a play the role of effective chemical potentials for relativistic quasiparticles living in the vicinity of two Fermi points. The counterflow thus produces the effective chemical potentials with opposite sign for quasiparticles of different chiralities $C_a = -q_a$.

The effective chemical potential is different from the true chemical potential μ of the original bare particles, ^{3}He atoms of the underlying liquid, which arises from the microscopic physics as a result of the conservation of the number of atoms related to the $U(1)_N$ symmetry. The effective chemical potential μ_a for quasiparticles appears only in the low-energy corner of the effective theory, i.e. in the vicinity of the a-th Fermi point. For the external observer, i.e. at the fundamental level, there is no conservation law for the quasiparticle number. But in the low-energy corner such a conservation law gradually emerges, and the quasiparticle number

$$N_a = \sum_{\mathbf{p}} f_a(\mathbf{p}) \tag{10.21}$$

is conserved better and better, as the energy is lowered. For an inner observer this is a true conservation law, which is, however, violated at the quantum level due to the chiral anomaly.

10.3.3 Double role of counterflow

Note that the counterflow velocity $\mathbf{w} = \mathbf{v}_{\mathrm{n}} - \mathbf{v}_{\mathrm{s}}$ enters the distribution function (10.20) through the chemical potential (10.19), and also explicitly. This reflects the separation of quasiparticle momentum, which is obtained by variation of energy over counterflow, into $\mathbf{P}^{\mathrm{F}}$ and $\mathbf{P}^{\mathrm{M}}$.

Let us first consider the case when the counterflow is orthogonal to the $\hat{\bf l}$-vector: ${\bf w} \perp \hat{\bf l}$. Then the effective chemical potentials $\mu_a = 0$ and only the current ${\bf P}^{\rm M}$ is induced. The explicit dependence of f_T on ${\bf w}$ does the same job as in the case of Bose superfluids. It leads to the modification of the effective temperature in eqn (5.23) for the thermodynamic potential. For the dilute gas of fermionic quasiparticles, the thermodynamic potential (the pressure) has the form

$$\Omega = \frac{7\pi^2}{360\hbar^3} N_F T_{\rm eff}^4 \sqrt{-g}, \quad T_{\rm eff} = \frac{T}{\sqrt{1-w^2}}, \quad w^2 = g_{ik} w^i w^k . \quad (10.22)$$

Here N_F is the number of chiral fermions ($N_F = 2$ for ^{3}He-A) and g_{ik} is the effective metric for quasiparticles in ^{3}He-A. Equation (10.22) is valid when $T_c \gg T \gg |\mu_a|$ and ${\bf w} \perp \hat{\bf l}$. From this equation one obtains the momentum density of 'matter' ${\bf P}^{\rm M}$ in eqn (10.16) expressed in terms of the transverse component of the normal density tensor:

$$P_i^{\rm M} = \left.\frac{d\Omega}{dw_\perp^i}\right|_{{\bf w} \to 0} = m n_{nik\perp} w^k , \quad (10.23)$$

$$m n_{nik\perp} = m n_{\rm n\perp}(\delta_{ik} - \hat{l}_i \hat{l}_k) , \quad m n_{\rm n\perp} = \frac{7\pi^2 T^4}{45\hbar^3 v_F c_\perp^4} . \quad (10.24)$$

10.3.4 Fermionic charge

There are effective fermionic charges which become conserved in the low-energy corner together with the quasiparticle number. We discuss here the most important fermionic charge, which for the inner observer is an analog of the baryon number in the Standard Model. The non-conservation of this charge is governed by the same Adler–Bell–Jackiw equation derived for the phenomenon of axial anomaly which is responsible for the non-conservation of the baryon number (see Chapter 18). Moreover, in the case of ^{3}He-A, the Adler–Bell–Jackiw equation for axial anomaly has been experimentally verified.

The relevant fermionic charge is the momentum carried by a quasiparticle just at the Fermi point. This momentum is non-zero since the Fermi points are situated away from the origin in the momentum space. For a quasiparticle at the Fermi point with chirality C_a this fermionic charge is

$${\bf p}^{(a)} = -C_a p_F \hat{\bf l} . \quad (10.25)$$

In the low-temperature limit, the momentum ${\bf p}$ of quasiparticles is close to ${\bf p}^{(a)}$ and thus the total fermionic charge carried by the gas of quasiparticles becomes indistinguishable from the current along the $\hat{\bf l}$-vector

$${\bf P}^{\rm F} = \sum_a {\bf p}^{(a)} N_a \approx \hat{\bf l} \left(\hat{\bf l} \cdot \sum_a \sum_{\bf p} {\bf p} f_a({\bf p}) \right) . \quad (10.26)$$

In equilibrium this fermionic charge is non-zero if the effective chemical potentials μ_a in eqn (10.19) are non-zero. This is clear: the momentum (or current) along $\hat{\mathbf{l}}$ carried by quasiparticles can be non-zero only if there is a counterflow along $\hat{\mathbf{l}}$. Let us first consider the case of small chemical potentials $|\mu_a| \ll T \ll \Delta_0$. The total fermionic charge stored in the heat bath of 'relativistic' massless quasiparticles in this temperature range is given by the relativistic equation for the particle number

$$\mathbf{P}^{\mathrm{F}} = \sum_a \mathbf{p}^{(a)} N_a = -\sum_a \mathbf{p}^{(a)} \mu_a \sum_{\mathbf{p}} \frac{\partial f_a}{\partial E} = \frac{V}{6\hbar^3}\sqrt{-g}T^2 \sum_a \mathbf{p}^{(a)} \mu_a \ . \quad (10.27)$$

As expected, the fermionic charge is proportional to the effective chemical potential. If the fermionic charges $\mathbf{p}^{(a)}$ and chemical potentials μ_a are substituted by the baryonic charges and chemical potentials of the Standard Model fermions, this equation would describe the baryonic charge stored in the heat bath of quarks.

In the case of ^{3}He-A, introducing the ^{3}He-A variables from eqns (10.25) and (10.19) one obtains

$$\sum_a \mathbf{p}^{(a)} \mu_a = 2p_F^2 \hat{\mathbf{l}}(\hat{\mathbf{l}} \cdot \mathbf{w}) \ . \quad (10.28)$$

Then eqn (10.27) gives the relation between the longitudinal momentum of quasiparticles and the longitudinal component $\hat{\mathbf{l}} \cdot \mathbf{w}$ of the counterflow velocity. By definition (5.20) their ratio represents the longitudinal part of the anisotropic normal density of the liquid

$$mn_{\mathrm{n}\|} = \frac{p_F^2}{3\hbar^3}\sqrt{-g}T^2 = \pi^2 m^* n \frac{T^2}{\Delta_0^2} \ . \quad (10.29)$$

It plays the part of the density of states of massless relativistic fermions

$$\sum_a \frac{dN_a}{d\mu_a} = \frac{mn_{\mathrm{n}\|}}{p_F^2} \ . \quad (10.30)$$

Comparing eqn (10.29) to eqn (10.24) for the transverse part of the normal component tensor, one finds that the longitudinal part is much larger at low T. This is a consequence of two physically different contributions to the quasiparticle momentum, and of the double role of the counterflow velocity $\mathbf{w}$ in the thermal distribution of relativistic quasiparticles in ^{3}He-A. In the case of a longitudinal flow, the counterflow velocity entering the effective chemical potential produces the fermionic charge $\mathbf{P}^{\mathrm{F}}$.

10.3.5 Normal component at zero temperature

Let us now consider the fermionic charge $\mathbf{P}^{\mathrm{F}}$ in the opposite limit, where T is small compared to the effective chemical potentials, $T \ll |\mu_a| \ll \Delta_0$. In this $T \to 0$ limit let us introduce the relevant thermodynamic potential which takes into account that the number of quasiparticles is conserved:

$$W = \sum_{\mathbf{p}} E(\mathbf{p}) f(\mathbf{p}) - \sum_a \mu_a N_a = -\frac{\sqrt{-g}}{24\pi^2 \hbar^3} \sum_a \mu_a^4 \ . \quad (10.31)$$

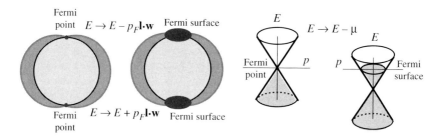

FIG. 10.1. Fermi surface is formed from the Fermi point at finite chemical potential of chiral fermions *right* or in the presence of counterflow in ^{3}He-A *left*.

In terms of the ^{3}He-A variables in eqns (10.19) and (10.13) this reads as

$$W = \sum_{\mathbf{p}} E(\mathbf{p})f(\mathbf{p}) - \mathbf{P}^{\mathrm{F}} \cdot \mathbf{w} \approx -\frac{m^* p_F^3}{12\pi^2 \hbar^3 c_\perp^2} \left(\hat{\mathbf{l}} \cdot \mathbf{w}\right)^4 . \tag{10.32}$$

This means that the energy of the counterflow along the $\hat{\mathbf{l}}$-vector simulates the energy stored in the system of chiral fermions with the non-zero chemical potentials μ_a. Variation of eqn (10.32) with respect to $\mathbf{w}$ gives the fermionic charge

$$\mathbf{P}^{\mathrm{F}}(T \to 0) = -\frac{dW}{d\mathbf{w}} = \frac{m^* p_F^3}{3\pi^2 \hbar^3 c_\perp^2} \hat{\mathbf{l}} \left(\hat{\mathbf{l}} \cdot \mathbf{w}\right)^3 . \tag{10.33}$$

Thus one arrives at an apparent paradox: quasiparticles carry a finite momentum even in the limit $T \to 0$ (Volovik and Mineev 1981; Muzikar and Rainer 1983; Nagai 1984; Volovik 1990a). The normal component density at $T \to 0$ is also non-zero:

$$mn_{\mathrm{n}\parallel}(T \to 0) = \frac{d(\mathbf{P}^{\mathrm{F}} \cdot \hat{\mathbf{l}})}{d(\mathbf{w} \cdot \hat{\mathbf{l}})} = \frac{p_F^2}{2\pi^2 \hbar^3} \sqrt{-g} \sum_a \mu_a^2 = \frac{m^* p_F^3}{\pi^2 \hbar^3 c_\perp^2} (\hat{\mathbf{l}} \cdot \mathbf{w})^2 . \tag{10.34}$$

This corresponds to the non-zero density of states of chiral fermions $\partial N_a/\partial \mu_a$ in the presence of non-zero chemical potentials.

10.3.6 *Fermi surface from Fermi point*

The situation at first glance is really paradoxical. At $T = 0$ there are no excitations, and thus one cannot determine the counterflow, since the velocity of the normal component $\mathbf{v}_{\mathrm{n}}$ has no meaning. Then what is the meaning of eqns (10.33) and (10.34)?

The situation becomes clear when we introduce the walls of a container, and thus the preferred environment frame which determines the velocity $\mathbf{v}_{\mathrm{n}}$ in equilibrium: $\mathbf{v}_{\mathrm{n}} = 0$ in the laboratory frame in a global equilibrium even if $T = 0$.

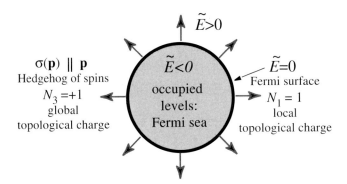

FIG. 10.2. Fermi surface with global topological charge $N_3 = +1$. When such a Fermi surface shrinks, it transforms to the Fermi point.

(If there are no boundaries, we must assume that there are impurities which form the preferred reference frame of the environment, or the temperature of the heat bath is not exactly zero, i.e. we consider the limit when $T \to 0$ but T is still large compared to, say, inverse relaxation time or other relevant energy scale.)

If there is a flow of vacuum with respect to the walls of the container, then, since the fermions are massless, the energy spectrum of quasiparticles measured in the environment frame $\tilde{E} = E(\mathbf{p}) - \mathbf{p}\cdot\mathbf{w}$ acquires negative values. The non-zero contribution to eqn (10.33) for the fermionic charge comes just from the fermionic quasiparticles which in a global equilibrium at $T = 0$ occupy the negative energy levels:

$$\mathbf{P}^{\mathrm{F}}(T \to 0) = \sum_{a,\mathbf{p}} \mathbf{p}^{(a)} \Theta(-\tilde{E}) \; , \qquad (10.35)$$

where $\Theta(x)$ is the step function: $\Theta(x > 0) = 1$ and $\Theta(x < 0) = 0$. As distinct from the quasiparticles in Bose superfluids discussed in Sec. 5.3 where the appearance of a state with negative energy $\tilde{E} = E(\mathbf{p}) - \mathbf{p}\cdot\mathbf{w} < 0$ is catastrophic, for the fermionic quasiparticles the catastrophe is avoided by the Pauli principle which restricts the number of fermions in one state by unity. When quasiparticles fill all the negative levels, the production of quasiparticles stops and the equilibrium state is reached.

This new equilibrium state of the vacuum contains the Fermi surface shown in Fig. 10.1 for the relativistic quasiparticles *left* and in ^{3}He-A *right*. In the relativistic description the energy states of massless fermions with $cp - \mu_a < 0$ are occupied at $T = 0$ if the chemical potential is non-zero and positive. The quasiparticles filling the negative energy levels form the solid Fermi sphere, with the Fermi surface determined by the equation $p_F = \mu_a/c$. For $\mu_a < 0$ the Fermi surface with $p_F = -\mu_a/c$ is formed by holes. In both cases the Fermi point is transformed to the Fermi surface. This Fermi surface, in addition to the momentum space invariant N_1 which characterizes it locally, also has the topological invariant $N_3 = \pm 1$ in eqn (8.15) which characterizes it globally: the surface σ_3 in eqn (8.15) now surrounds the whole Fermi sphere (Fig. 10.2). When such a Fermi surface shrinks it cannot disappear completely due to its non-zero global charge N_3, and a Fermi point will remain.

The non-zero density of the normal component at $T \to 0$ can also be calculated from the finite density of fermionic states $N(\omega) = 2 \sum_a \sum_{\mathbf{p}} \delta(\omega - \tilde{E}_a(\mathbf{p}))$ at $\omega = 0$, which appears as a consequence of the formation of the Fermi surface at $\hat{\mathbf{l}} \cdot \mathbf{w} \neq 0$:

$$N(0) = \frac{\sqrt{-g}}{\pi^2 \hbar^3} \sum_a \mu_a^2 \equiv \frac{p_F m^*}{\pi^2 \hbar^3 c_\perp^2} (\hat{\mathbf{l}} \cdot (\mathbf{v}_s - \mathbf{v}_n))^2 \ , \ mn_{n\|}(T \to 0) = p_F^2 N(0) \ .$$
(10.36)

As we shall see in Sec. 26.1.3, Fermi surface always appears in the supercritical regime where the flow velocity exceeds the Landau critical velocity. A similar formation of the Fermi surface in the presence of the counterflow occurs in high-temperature superconductors where the Landau critical velocity is zero and thus the counterflow is always supercritical. This leads to the non-zero density of states in the presence of the counterflow. The low-energy fermionic quasiparticles there are 2+1-dimensional massless Dirac fermions. Since the space dimension is reduced, the energy stored in the counterflow in eqn (10.32) has the power 3 instead of 4: $W \propto |\mathbf{v}_s - \mathbf{v}_n|^3$. The second derivative gives the non-analytic normal density at $T \to 0$ and the non-analytic density of states: $N(0) \propto |\mathbf{v}_s - \mathbf{v}_n|$.

In a mixed state of superconductor in applied magnetic field $\mathbf{B}$, the counterflow occurs around vortices, with the average counterflow velocity being determined by the applied field: $|\mathbf{v}_s - \mathbf{v}_n| \sim \sqrt{B}$. This causes the opening of the Fermi surface which gives the finite density of states, $N(0) \propto |\mathbf{v}_s - \mathbf{v}_n| \propto \sqrt{B}$ (Volovik 1993c). The non-analytic DOS as a function of magnetic field leads to the non-analytic dependence on B of the thermodynamic quantities in d-wave superconductors which has been observed experimentally (see Revaz et al. (1998) and references therein).

In superconductors all this occurs not only in the limit $T \to 0$, but even at exactly zero temperature, $T = 0$. This is because, even if there is no heat bath of thermal quasiparticles, the crystal itself provides the preferred reference frame: $\mathbf{v}_n$ is the velocity of the crystal.

10.4 Parameters of effective theory in London limit

10.4.1 Parameters of effective theory from BCS theory

In the low-T limit some of the temperature-dependent coefficients in the London energy, such as $n_{n\|}$, $n_{n\perp}$ and K_b, can be determined within the effective relativistic theory of chiral fermions interacting with gauge and gravity fields.

To obtain the temperature dependence of all the coefficients in the entire temperature range $0 \leq T < T_c$ one needs the microscopic theory, which is provided by the BCS model in conjunction with the Landau Fermi liquid theory. The gradient expansion within this scheme has been elaborated by Cross (1975). We are interested in the low-temperature region $T \ll T_c \sim \Delta_0$, which corresponds to low energies compared to the Planck energy $E_{\text{Planck 2}}$ where most of the analogies with RQFT are found. In this limit, according to Cross (1975), one obtains the following coefficients in the gradient energy in eqns (10.11) and (10.14):

$$C_0 - C = \frac{1}{2m} n_{s\|} \;,\; C = \frac{1}{2m} n_{s\perp} \frac{n_{s\|}^0}{n_{s\perp}^0} \;, \tag{10.37}$$

$$K_s = \frac{1}{32m^*} n_{s\perp}^0 \;, \tag{10.38}$$

$$K_t = \frac{1}{96m^*} \left(n_{s\perp}^0 + 4 n_{s\|}^0 + 3 \left(\frac{m^*}{m} - 1 \right) \frac{n_{s\|} n_{s\|}^0}{n} \right) \;, \tag{10.39}$$

$$K_b = \frac{1}{32m^*} \left(2 n_{s\|}^0 + \left(\frac{m^*}{m} - 1 \right) \frac{n_{s\|} n_{s\|}^0}{n} \right) + \text{Log} \;, \tag{10.40}$$

$$\text{Log} = \frac{1}{4m^*} n \int \frac{d\Omega}{4\pi} \frac{(\hat{\mathbf{l}} \cdot \hat{\mathbf{p}})^4}{(\hat{\mathbf{l}} \times \hat{\mathbf{p}})^2} \left[1 + 2 \int_0^\infty dM \frac{\partial f_T}{\partial E} \right] \;. \tag{10.41}$$

Here, as before, $m^* = p_F/v_F$ is the effective mass of quasiparticles in the normal Fermi liquid, while m is the bare mass of the ^{3}He atoms: these masses coincide only in the limit of weak interactions between the particles. The index 0 marks the bare (non-renormalized) values of the superfluid and normal densities, computed in the weak-interaction limit, where $m^* = m$:

$$n_{s\|}^0 = n - n_{n\|}^0 \;,\; n_{n\|}^0 = \pi^2 n \frac{T^2}{\Delta_0^2} = \frac{m}{m^*} n_{n\|} \;, \tag{10.42}$$

$$n_{s\perp}^0 = n - n_{n\perp}^0 \;,\; n_{n\perp}^0 = \frac{7\pi^4}{15} n \frac{T^4}{\Delta_0^4} = \frac{m}{m^*} n_{n\perp} \;. \tag{10.43}$$

The different power law for the momentum carried by quasiparticles along and transverse to $\hat{\mathbf{l}}$ shows that there is a different physics of longitudinal $\mathbf{P}^F$ and transverse $\mathbf{P}^M$ momenta of quasiparticles behind the scenes.

In eqn (10.41) Log is the term which diverges at $T \to 0$ as $\ln(\Delta_0/T)$.

10.4.2 Fundamental constants

In the simplest BCS theory, the London gradient energy is determined by five 'fundamental' parameters:

The 'speeds of light' $c_\| = v_F$ and $c_\perp$ characterize the 'relativistic' physics of the low-energy corner. These are the parameters of the quasiparticle energy spectrum in the relativistic low-energy corner below the first Planck scale, $E \ll E_{\text{Planck 1}} = \Delta_0^2/v_F p_F = m^* c_\perp^2$. Let us, however, repeat that the inner observer cannot resolve between the two velocities, $c_\|$ and $c_\perp$. For him (or her) the measured speed of light is fundamental. In particular, it does not depend on the direction of propagation.

The parameter p_F is the property of the higher level in the hierarchy of the energy scales – the Fermi liquid level (the related quantity is the quasiparticle mass in the Fermi liquid theory, $m^* = p_F/c_\|$). Thus the spectrum of quasiparticles in ^{3}He-A in the entire range $E \ll v_F p_F$ is determined by three parameters, $c_\| = v_F$, $c_\perp$ and p_F: $E^2 = M^2(p) + c_\perp^2 (\mathbf{p} \times \hat{\mathbf{l}})^2$, where $M(p) = v_F(p - p_F)$.

The bare mass m of ^{3}He is the parameter of the underlying microscopic physics of interacting 'indivisible' particles – ^{3}He atoms. This parameter does not enter the spectrum of quasiparticles in ^{3}He-A. However, the Galilean invariance of the underlying system of ^{3}He atoms requires that the kinetic energy of superflow at $T = 0$ must be $(1/2)mn\mathbf{v}_s^2$, i.e. the bare mass m must be incorporated into the BCS scheme to maintain Galilean invariance. This is achieved in the Landau theory of the Fermi liquid, where the dressing occurs due to quasiparticle interaction. In the simplified approach one can consider only that part of the interaction which is responsible for the renormalization of the mass and which restores Galilean invariance of the Fermi system. This is the current–current interaction with the Landau parameter $F_1 = 3(m^*/m - 1)$, containing the bare mass m. This is how the bare mass m enters the Fermi liquid and thus the BCS theory.

Together with $\hbar$ this gives five 'fundamental' constants: $\hbar, m, p_F, m^*, c_\perp$. But only one of them is really fundamental, the $\hbar$, since it is the same for all energy scales. The important combinations of these parameters are $c_\| = v_F = p_F/m^*$ and $\Delta_0 = p_F c_\perp$. There are two dimensionless parameters, m^*/m and $c_\|/c_\perp$, and now we can play with them: we would like to know how to 'improve' the liquid by changing its dimensionless parameters so that it would resemble closer the quantum vacuum of RQFT. In this way one could possibly understand the main features of the quantum vacuum.

10.5 How to improve quantum liquid

10.5.1 Limit of inert vacuum

Though ^{3}He-A and RQFT – the Standard Model – belong to the same universality class and thus have similar properties of the fermionic spectrum, we know that ^{3}He-A cannot serve as a perfect model for the quantum vacuum in RQFT. While the properties of the chiral fermions are well reproduced in ^{3}He-A, the effective action for bosonic gauge and gravity fields obtained by integration over fermionic degrees of freedom is contaminated by the terms which are absent in a fully relativistic system. This is because the integration over the vacuum fermions is not always concentrated in the region where their spectrum obeys the Lorentz invariance.

The missing detail, which is to have a generally covariant theory as an emergent phenomenon, is the mechanism which would confine the integration over fermions to the 'relativistic' region. The effective ultraviolet cut-off must be lower than the momenta at which the 'Lorentz' invariance is violated. It is reasonable to expect that strong interactions between the bare particles can provide such a natural cut-off, which is well within the 'relativistic' region. Unfortunately, at the moment there is no good model which can treat the system of strongly interacting particles. Instead, let us use the existing BCS model and try to 'correct' the ³He-A by moving the parameters of the theory in the direction of strong interactions but still within the BCS scheme, in such a way that the system more easily forgets its microscopic origin. The first step is to erase the memory of the microscopic parameter m – the bare mass. Thus the ratio between the bare mass and the mass m^* of the dressed particle, which enters the effective theory, must be either $m/m^* \to 0$ or $m/m^* \to \infty$.

In real liquid ³He the ratio m^*/m varies between about 3 and 6 at low and high pressure, respectively. However, we shall consider this ratio as a free parameter which one can adjust to make the system closer to the relativistic theories in the low-energy corner. As we discussed in Sec. 9.2.5, in the system with broken symmetry the Goldstone fields spoil the Einstein action. The effect of the symmetry breaking, i.e. the superfluidity, must therefore be suppressed. This happens in the limit $m \to \infty$, since the superfluid velocity is inversely proportional to m according to eqn (9.16): $\mathbf{v}_s \propto 1/m$. Because of the heavy mass m of atoms comprising the vacuum, the vacuum becomes inert, and the kinetic energy of superflow $(1/2)mn\mathbf{v}_s^2$ vanishes as $1/m$. The superfluidity effectively disappears, and at the same time the bare mass drops out of the physical results. Since the influence of the microscopic level on the effective theory of gauge field and gravity is weakened one may expect that the effective action for the collective modes would more closely resemble the covariant and gauge invariant limit of the Einstein–Maxwell action. So, let us consider the limit of the 'inert' vacuum, $m \gg m^*$.

10.5.2 *Effective action in inert vacuum*

In the limit $m \to \infty$ all the terms in the London energy related to superfluidity vanish since $v_s \propto 1/m$. In the remaining $\hat{\mathbf{l}}$-terms we take into account that in this limit according to eqn (10.42) one has $n_{s\parallel} = n$. As a result the London energy is substantially reduced:

$$F_{\text{London}}(m \to \infty) = \frac{1}{32m^*} n_{s\perp}^0 (\nabla \cdot \hat{\mathbf{l}})^2 \qquad (10.44)$$

$$+ \frac{1}{96m^*}(n_{s\perp}^0 + n_{s\parallel}^0)(\hat{\mathbf{l}} \cdot (\nabla \times \hat{\mathbf{l}}))^2 \qquad (10.45)$$

$$+ \left(\frac{1}{32m^*} n_{s\parallel}^0 + \text{Log}\right)(\hat{\mathbf{l}} \times (\nabla \times \hat{\mathbf{l}}))^2 \ . \qquad (10.46)$$

All three terms have a correspondence in QED and Einstein gravity. We have already seen that the bend term, i.e. $(\hat{\mathbf{l}} \times (\nabla \times \hat{\mathbf{l}}))^2$, is exactly the energy of the magnetic field in curved space in eqn (9.20) with the logarithmically diverging coupling constant – the coefficient (Log) in eqn (10.41). This coupling constant does not depend on the microscopic parameter m, which shows that it is within the responsibility of the effective field theory.

Let us now consider the twist term, i.e. $(\hat{\mathbf{l}} \cdot (\nabla \times \hat{\mathbf{l}}))^2$, and show that it corresponds to the Einstein action.

10.5.3 Einstein action in ^{3}He-A

The part of effective gravity which is simulated by the superfluid velocity field, vanishes in the limit of inert vacuum. The remaining part of the gravitational field is simulated by the inhomogeneity of the $\hat{\mathbf{l}}$-field, which plays the part of the 'Kasner axis' in the metric

$$g^{ij} = c_\|^2 \hat{l}^i \hat{l}^j + c_\perp^2 (\delta^{ij} - \hat{l}^i \hat{l}^j) \,,\; g^{00} = -1,\; g^{0i} = 0,\; \sqrt{-g} = \frac{1}{c_\| c_\perp^2}, \quad (10.47)$$

$$g_{ij} = \frac{1}{c_\|^2} \hat{l}^i \hat{l}^j + \frac{1}{c_\perp^2}(\delta^{ij} - \hat{l}^i \hat{l}^j),\; g_{00} = -1\,,\; g_{0i} = 0\,. \quad (10.48)$$

The curvature of space with this metric is caused by spatial rotations of the 'Kasner axis' $\hat{\mathbf{l}}$. For the stationary metric, $\partial_t \hat{\mathbf{l}} = 0$, one obtains that in terms of the $\hat{\mathbf{l}}$-field the Einstein action is

$$-\frac{1}{16\pi G} \int d^4 x \sqrt{-g} R = \frac{1}{32\pi G \Delta_0^2} \left(1 - \frac{c_\perp^2}{c_\|^2}\right)^2 \frac{p_F^3}{m^*} \int d^4 x (\hat{\mathbf{l}} \cdot (\nabla \times \hat{\mathbf{l}}))^2 \,. \quad (10.49)$$

It has the structure of the twist term in the gradient energy (10.45) obtained in a gradient expansion, which in the inert vacuum limit is

$$F_{\text{twist}} = \frac{1}{288} \left(\frac{2}{\pi^2} - \frac{T^2}{\Delta_0^2}\right) \frac{p_F^3}{\hbar m^*} \int d^3 x (\hat{\mathbf{l}} \cdot (\nabla \times \hat{\mathbf{l}}))^2 \,. \quad (10.50)$$

We thus can identify the twist term with the Einstein action. Neglecting the small anisotropy factor $c_\perp^2/c_\|^2$ one obtains that the Newton constant in the effective gravity of the 'improved' ^{3}He-A is:

$$G^{-1} = \frac{2}{9\pi \hbar} \Delta_0^2 - \frac{\pi}{9\hbar} T^2 \,. \quad (10.51)$$

10.5.4 Is G fundamental?

In the limit of the inert vacuum the action for the metric in eqn (10.49) expressed in terms of the $\hat{\mathbf{l}}$-vector has the general relativistic form with a temperature-dependent Newton constant G (10.51). This 'improved' 3He-A is still not the right model to provide the analogy on a full scale, because the $\hat{\mathbf{l}}$-vector enters both the gravity and electromagnetism. However, there has already been some progress on the way toward the right model of effective gravity, and one can draw some conclusions probably concerning real gravity.

The temperature-independent part of G certainly depends on the details of trans-Planckian physics: it contains the second Planck energy scale Δ_0. To compare this to G following from the Sakharov theory of induced gravity let us introduce the number of fermionic species N_F. Since the bosonic action for $\hat{\mathbf{l}}$ is obtained by integration over fermions it is proportional to N_F. That is why the general form for the parameter G in the modified A-phase with N_F species (for real ^{3}He-A, $N_F = 2$) is

$$G^{-1}(T=0) = \frac{N_F}{9\pi\hbar}\Delta_0^2 . \qquad (10.52)$$

This eqn (10.52) is similar to that obtained by Sakharov, where G^{-1} is the product of the square of the cut-off parameter and the number of fermion zero modes of quantum vacuum. The numerical factor depends on the cut-off procedure.

The temperature dependence of the Newton constant pretends to be more universal, since it does not depend on the microscopic parameters of the system. Moreover, it does not contain any 'fundamental constant'. Trans-Planckian physics of the modified ^{3}He-A thus suggests the following temperature dependence of the Newton constant in the vacuum with N_F Weyl fermions:

$$G^{-1}(T) - G^{-1}(T=0) = -\frac{\pi}{18\hbar}N_F T^2 . \qquad (10.53)$$

Since in the effective theory of gravity the Newton constant G depends on temperature, and thus on the energy scale at which the gravity is measured, G is not a fundamental constant. Its asymptotic value, $G(T=0, r=\infty)$, also cannot be considered as a fundamental constant according to Weinberg's (1983) criterion: it is derived from a more fundamental quantity – the second Planck energy scale.

If the temperature dependence of G in eqn (10.53) were applied to real cosmology, one would find that the Newton constant depends on the cosmological time t. In the radiation-dominated regime, when $T^2 \sim \hbar E_{\rm Planck}/t$, the correction to G decays as $\delta G/G \propto \hbar/(tE_{\rm Planck})$, while G approaches its asymptotic value dictated by the Planck energy scale. This is very different from the Dirac (1937, 1938) suggestion that the gravitational constant decays as $1/t$ and is small at present time simply because the universe is old. Thus the effective theory of gravity rules out the Dirac conjecture.

10.5.5 *Violation of gauge invariance*

Let us finally consider the splay term $(\nabla \cdot \hat{\bf l})^2$ in eqn (10.44). It has only fourth-th order temperature corrections, T^4/Δ_0^4. This term has a similar coefficient to the curvature term, but it is not contained in the Einstein action, since it cannot be written in covariant form. The structure of this term can, however, be obtained using the gauge field presentation of the $\hat{\bf l}$-vector, where ${\bf A} = p_F\hat{\bf l}$. It is known that a similar term can be obtained in the renormalization of QED to leading order of a $1/N$ expansion regularized by introducing a momentum cut-off (Sonoda 2000):

$$L'_{\rm QED} = \frac{1}{96\pi^2}(\partial_\mu A_\mu)^2 . \qquad (10.54)$$

This term violates gauge invariance and cannot appear in the dimensional regularization scheme (see the book by Weinberg 1995), but it can appear in the momentum cut-off procedure, since such a procedure violates gauge invariance.

When written in covariant form, eqn (10.54) can be applied to ^{3}He-A, where $\mathbf{A} = p_F \hat{\mathbf{l}}$ and $\sqrt{-g} = \text{constant}$:

$$L'_{\text{QED}} = \frac{1}{96\pi^2}\sqrt{-g}(\partial_\mu(g^{\mu\nu}A_\nu))^2 = \frac{1}{96\pi^2}\frac{p_F^2 c_\parallel^3}{c_\perp^2}(\nabla \cdot \hat{\mathbf{l}})^2 = \frac{c_\parallel^2}{c_\perp^2}\frac{1}{96\pi^2}\frac{p_F^3}{m^*}(\nabla \cdot \hat{\mathbf{l}})^2. \tag{10.55}$$

The last term is just the splay term in eqn (10.44) except for an extra big factor of the vacuum anisotropy $c_\parallel^2/c_\perp^2$: $F_{\text{London splay}} = (c_\perp^2/c_\parallel^2)L'_{\text{QED}}$. However, in the isotropic case, where $c_\parallel = c_\perp$, they exactly coincide. This suggests that the regularization provided by the 'trans-Planckian physics' of ^{3}He-A represents an anisotropic version of the momentum cut-off regularization of QED.

10.5.6 Origin of precision of symmetries in effective theory

There is, however, an open question in the above scheme as well as in the general approach to the problem of emergence of effective theories: it is necessary to explain the high precision of symmetries which we observe today. At the moment according to Kostelecky and Mewes (2001) the existing constraint on different Lorentz-violating coefficients is about a few parts in 10^{31}. Such accuracy of symmetries observed in nature certainly cannot be explained by logarithmic selections of terms in the effective action as suggested by Chadha and Nielsen (1983): the logarithm is too slow a function for that (Iliopoulus *et al.* 1980).

According to Bjorken (2001b) the effective RQFT with such high precision can only emerge if there is a small expansion parameter in the game, say about 10^{-15}. Bjorken relates this parameter to the ratio of electroweak and Planck scales. In the above scheme this can be the ratio m^*/m, but the more instructive development of Bjorken's idea would be the suggestion that the small parameter is the ratio of two different Planck scales: the lowest one provides the natural ultraviolet cut-off for integrals in momentum space. The second one (with higher energy) marks the energy above which the Lorentz symmetry is violated.

The ^{3}He-A analogy indicates that the non-covariant terms in the effective action appear due to integration over fermions far from the Fermi point, where the 'Lorentz' invariance is not obeyed. If the natural ultraviolet cut-off is very much below the scale where the Lorentz symmetry is violated, all the symmetries of the effective theory, including the gauge invariance and general covariance, will be protected to a high precision. On the other hand the ratio of the Planck scales can provide the origin of the other small numbers, such as electroweak energy with respect to the Planck scale, say, in the mechanism of reentrant violation of symmetry discussed in Sec. 12.4.

^{3}He-A does provide several different 'Planck' energy scales, but unfortunately the hierarchy of Planck scales is the opposite of what we need: the Lorentz symmetry is violated before the natural ultraviolet cut-off is reached. That is why ^{3}He-A is not a good example of emergent RQFT: the integration over fermions occurs mainly in the region of momenta where there is no symmetry, and as a result the effective action for bosonic fields is contaminated by non-covariant terms. In future efforts must be made to reverse the hierarchy of Planck scales.

11
MOMENTUM SPACE TOPOLOGY OF 2+1 SYSTEMS

11.1 Topological invariant for 2+1 systems

11.1.1 Universality classes for 2+1 systems

As distinct from the 3D systems, in the 2D case there is only one universality class of systems with gap nodes. This is the class of nodes in the quasiparticle energy spectrum with co-dimension 1, which is described by the non-trivial topological invariant N_1 in eqn (8.3). In 3D **p**-space this manifold forms the surface, the Fermi surface, while in 2D **p**-space (p_x, p_y) it is a line of zeros. The next non-trivial class with nodes has co-dimension 3 and thus simply cannot exist in 2D **p**-space.

However, it appears that the fully gapped systems, i.e. without singularities in the Green function, are not so dull in the 2D case: the **p**-space topology of their vacua can be non-trivial. The states with non-trivial topology can be obtained by a dimensional reduction of states with Fermi points in 3D systems. Such non-trivial vacua without singularities in **p**-space have a counterpart in **r**-space: these are the topologically non-trivial but non-singular configurations – textures or skyrmions – characterized by the third homotopy groups π_3 in 3D space, by the second homotopy group π_2 in 2D space, or by the relative homotopy groups (see Chapter 16; Skyrme (1961) considered non-singular solitons in high-energy physics as a model for baryons, hence the name). We discuss here **p**-space skyrmions – topologically non-trivial vacua with a fully non-singular Green function (Fig. 11.1).

For the 2D systems the non-trivial momentum space topology of the gapped vacua are of particular importance, because it gives rise to quantization of physical parameters (see Chapter 21). These are: 2D electron systems exhibiting the quantum Hall effect (Kohmoto 1985; Ishikawa and Matsuyama 1986, 1987; Matsuyama 1987); thin films of ^{3}He-A (Volovik and Yakovenko 1989, 1997; and Sec. 9 of the book by Volovik 1992a); 2D (or layered) superconductors with broken time reversal symmetry (Volovik 1997a); fermions living in the 2+1 world within a domain wall etc. Topological quantization of physical parameters, such as the Hall conductance, is possible only for dissipationless systems (Kohmoto 1985). The fully gapped systems at $T = 0$ satisfy this condition.

11.1.2 Invariant for fully gapped systems

The gapped ground states (vacua) in 2D systems or in quasi-2D thin films are characterized by the invariant obtained by dimensional reduction from the topological invariant for the Fermi point in eqn (8.15):

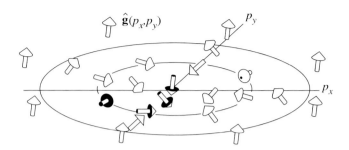

FIG. 11.1. Skyrmion in **p**-space with momentum space topological charge $\tilde{N}_3 = -1$. It describes topologically non-trivial vacua in 2+1 systems with a fully non-singular Green function.

$$\tilde{N}_3 = \frac{1}{24\pi^2} e_{\mu\nu\lambda} \, \mathbf{tr} \int d^2p \, dp_0 \, \mathcal{G}\partial_{p_\mu}\mathcal{G}^{-1}\mathcal{G}\partial_{p_\nu}\mathcal{G}^{-1}\mathcal{G}\partial_{p_\lambda}\mathcal{G}^{-1} \, . \tag{11.1}$$

The integral is now over the entire 3-momentum space $p_\mu = (p_0, p_x, p_y)$. (If a crystalline system is considered the integration over (p_x, p_y) is bounded by the Brillouin zone; we shall use it in Sec. 21.2.1) The integrand is determined everywhere in the 3-momentum space since the system is fully gapped and thus the Green function is nowhere singular.

In Sec. 12.3.1 we shall discuss the 4-momentum topological invariants protected by symmetry. The dimensional reduction of these invariants to the 3-momentum space leads to other topological invariants in addition to eqn (11.1), and thus to other fermionic charges characterizing the ground state of 2+1 systems (see e.g. Yakovenko 1989 and Sec. 21.2.3).

Further dimensional reduction determines the topology of edge states: the fermion zero modes, which appear on the surface of a 2D system or within a domain wall separating domains with different values of $\tilde{N}_3$ (see Chapter 22).

11.2 2+1 systems with non-trivial p-space topology

11.2.1 p-space skyrmion in p-wave state

An example of the 2D system with non-trivial $\tilde{N}_3$ is the crystal layer of the chiral p-wave superconductor – the 2D analog of ^{3}He-A, where both time reversal symmetry and reflection symmetry are spontaneously broken. Current belief holds that such a superconducting state occurs in the tetragonal Sr$_2$RuO$_4$ material (Rice 1998; Ishida et al. 1998). Because of the interaction with crystal fields, the $\hat{\mathbf{l}}$-vector is normal to the layers, $\hat{\mathbf{l}} = \pm\hat{\mathbf{z}}$. Let us consider fermionic quasiparticles with a given projection of the ordinary spin. Then the Bogoliubov–Nambu Hamiltonian for the 2+1 fermions living in the layer is actually the same as for a 3+1 system with a Fermi point in eqn (8.12) with the exception that there is no dependence on the momentum p_z along the third dimension.

$$\mathcal{H} = \check{\tau}^b g_b(\mathbf{p}) \;,\; g_3 = \frac{p_x^2 + p_y^2}{2m^*} - \mu \;,\; g_1 = cp_x \;,\; g_2 = \mp cp_y \;,\; c = \frac{\Delta_0}{p_F} \, . \tag{11.2}$$

The sign of g_2 depends on the orientation of the $\hat{\mathbf{l}}$-vector. Due to suppression of a third dimension the quasiparticle energy spectrum $E(\mathbf{p})$,

$$E^2(\mathbf{p}) = \mathcal{H}^2 = \mathbf{g}^2(\mathbf{p}) = \left(\frac{p^2}{2m^*} - \mu\right)^2 + c^2 p^2 , \qquad (11.3)$$

is fully gapped for both negative and positive chemical potential μ, which is counted from the bottom of the band. If μ is positive and large, $\Delta_0 \ll \mu$, the gap in the energy spectrum coincides with Δ_0, i.e. $\min(E(\mathbf{p})) \approx \Delta_0$:

$$m^* c^2 \ll \mu : \qquad E^2 \approx v_F^2 (p - p_F)^2 + \Delta_0^2 . \qquad (11.4)$$

If μ is negative and large, $\Delta_0 \ll |\mu|$, the gap is determined by μ, i.e. $\min(E(\mathbf{p})) \approx |\mu|$.

The special case is when the chemical potential crosses the bottom of the band, i.e. when μ crosses zero. At the moment of crossing the quasiparticle energy spectrum becomes gapless: it is zero at the point $p_x = p_y = 0$. The point $\mu = p_x = p_y = 0$ represents the singularity in the Green function: it is the hedgehog in 3D space (μ, p_x, p_y). This case will be discussed later in Sec. 11.4. Close to the crossing point, where $|\mu| \ll \Delta_0$, the Bogoliubov–Nambu Hamiltonian transforms to the 2+1 Dirac Hamiltonian, with the minimum of the energy spectrum being at $p = 0$:

$$m^* c^2 \gg |\mu| : \qquad E^2 \approx \mu^2 + c^2(p_x^2 + p_y^2) . \qquad (11.5)$$

In the case of a simple 2×2 Hamiltonian, the topological invariant $\tilde{N}_3$ in eqn (11.1) can be expressed in terms of the unit vector field $\hat{\mathbf{g}}(\mathbf{p}) = \mathbf{g}/|\mathbf{g}|$, just in the same way as in eqn (8.13) for the topology of Fermi points in 3D momentum space:

$$\tilde{N}_3 = \frac{1}{4\pi} \int dp_x dp_y \, \hat{\mathbf{g}} \cdot \left(\frac{\partial \hat{\mathbf{g}}}{\partial p_x} \times \frac{\partial \hat{\mathbf{g}}}{\partial p_y}\right) . \qquad (11.6)$$

Since at infinity the unit vector field $\hat{\mathbf{g}}$ has the same value, $\hat{\mathbf{g}}_{p \to \infty} \to (0, 0, 1)$, the 2-momentum space (p_x, p_y) becomes isomoprhic to the compact S^2 sphere. The function $\hat{\mathbf{g}}(\mathbf{p})$ realizes the mapping of the S^2 sphere to the S^2 sphere, $\hat{\mathbf{g}} \cdot \hat{\mathbf{g}} = 1$, described by the second homotopy group π_2. If $\mu > 0$, the winding number of such a mapping is $\tilde{N}_3 = 1$ or $\tilde{N}_3 = -1$ depending on the orientation of the $\hat{\mathbf{l}}$-vector. In the case $\tilde{N}_3 = -1$ the $\hat{\mathbf{g}}(\mathbf{p})$-field forms a skyrmion in Fig. 11.1. For $\mu < 0$ one has $\tilde{N}_3 = 0$. The zero value of the chemical potential μ marks the border between quantum vacua with different momentum space topological charge $\tilde{N}_3$, the quantum phase transition.

11.2.2 Topological invariant and broken time reversal symmetry

It is important that the necessary condition for a non-zero value of $\tilde{N}_3$ in a condensed matter system is a broken time reversal symmetry. The non-zero integrand in eqn (11.6) is possible only if all three components of vector $\hat{\mathbf{g}}$ are non-zero. This includes the component $g_2 \neq 0$, which is in front of the imaginary

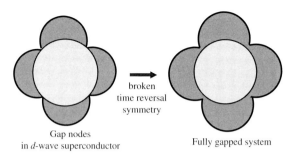

FIG. 11.2. Zeros of co-dimension 2 are topologically unstable. Nodal lines in 3D superconductor or point nodes in 2D superconductor disappear when perturbation violating time reversal symmetry is introduced.

matrix $\check{\tau}^2$, and as a result one has $\mathcal{H}^* \neq \mathcal{H}$. The time reversal operation changes the sign of g_2 and thus the sign of the topological charge: $T\tilde{N}_3 = -\tilde{N}_3$. This is the reason why systems with $\tilde{N}_3 \neq 0$ typically exhibit ferromagnetic behavior. An example is provided by ^{3}He-A with orbital ferromagnetism of the Cooper pairs along the vector $\hat{\mathbf{l}}$ (Leggett 1977b; Vollhardt and Wölfle 1990).

11.2.3 *d-wave states*

It is believed that in the cuprate high-temperature superconductors the Cooper pairs are in the *d*-wave state with $g_1 = d_{x^2-y^2}(p_x^2 - p_y^2)$ and $g_2 = 0$. Such a state has four point nodes at $p_x = \pm p_F$, $p_y = \pm p_F$. These nodes have co-dimension 2 and thus are topologically unstable (Fig. 11.2). Perturbations which destroy the nodes are those which make g_2 non-zero. Simultaneously these perturbations break time reversal symmetry: a *d*-wave superconductor with broken time reversal symmetry is fully gapped. For the popular model with $g_2 = d_{xy} p_x p_y$ the Bogoliubov–Nambu Hamiltonian for the 2+1 fermions living in the atomic layer is

$$\mathcal{H} = \check{\tau}^3 \frac{p_x^2 + p_y^2 - p_F^2}{2m^*} + \check{\tau}^1 d_{x^2-y^2}(p_x^2 - p_y^2) + \check{\tau}^2 d_{xy} p_x p_y \ . \tag{11.7}$$

For $\mu > 0$ the topological invariant in eqn (11.6) is $\tilde{N}_3 = \pm 2$ for any small value of the amplitude d_{xy}; the sign of $\tilde{N}_3$ is determined by the sign of $d_{xy}/d_{x^2-y^2}$ (Volovik 1997a). If both spin components are taken into account one has $\tilde{N}_3 = \pm 4$. And again $\tilde{N}_3 = 0$ for $\mu < 0$ with the quantum phase transition at $\mu = 0$.

11.3 Fermi point as diabolical point and Berry phase

11.3.1 *Families (generations) of fermions in 2+1 systems*

In thin films, in addition to spin indices, the Green function matrix $\mathcal{G}$ can contain the indices of the transverse levels, which come from the quantization of motion along the normal to the film (see the book by Volovik 1992a). In periodic systems (2D crystals) in addition the band indices appear. In these cases the simple

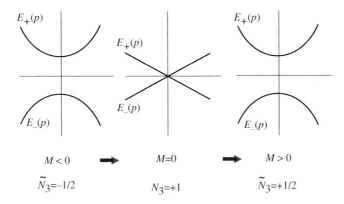

FIG. 11.3. The Fermi point in (p_x, p_y, M) space describes a quantum transition between the two vacuum states with different topological charge $\tilde{N}_3$, when the parameter M crosses the critical value. In this particular example M is the mass of the Dirac particle in 2+1 spacetime, which changes sign after the transition, while $\tilde{N}_3$ changes from $-1/2$ to $+1/2$. This Fermi point is a diabolical point at which two branches of the spectrum can touch each other.

equation (11.6) for topological invariant is not applicable, and one must use the more general eqn (11.1) in terms of the Green function.

Quasiparticles on different transverse levels represent different families of fermions with the same properties. This would correspond to generations of fermions in the Standard Model, if our 3+1 world is embedded within a soliton wall (brane) in higher-dimensional spacetime. The transverse quantization of fermions living inside the brane demonstrates one of the possible routes to the solution of the family problem: why each fermion is replicated three (or maybe more) times, i.e. why in addition to, say, the electron we also have the muon and the tau lepton, which exhibit the same electric and other charges.

11.3.2 Diabolical points

For the periodic systems the fermionic spectrum is described by the same invariant in eqn (11.1), but in this case the $dp_x dp_y$-integral is over the Brillouin zone. The Brillouin zone is a compact space which is topologically equivalent to a 2-torus T^2. The invariant $\tilde{N}_3$ can be extended to characterize the topological property of a given band $E_n(p_x, p_y)$, if it is not overlapping with the other bands (Avron et al. 1983).

Let us now introduce some external parameter M, which we can change to regulate the band structure. This can be, for instance, the chemical potential μ whose position can cross different bands. Now we have the 3D space of parameters, (p_x, p_y, M), which mark the energy levels $E_n(p_x, p_y, M)$. The general properties of the crossing of different branches of an energy spectrum in the space of external parameters were investigated by Von Neumann and Wigner (1929). From their analysis it follows that two bands can touch each other at isolated

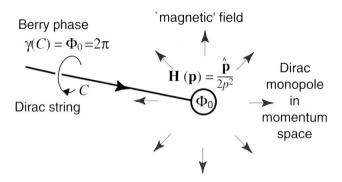

FIG. 11.4. The Fermi point as a Dirac magnetic monopole in 3-momentum space. The geometric Berry phase acquired by a chiral fermion after it circumvents an infinitesimal contour C around the Dirac string in 3-momentum **p**-space is $\gamma(C) = 2\pi$. The Dirac string carries the 'magnetic' flux $\Phi_0 = 2\pi$ to the monopole, from which the flux radially propagates outwards.

points, say (p_{x0}, p_{y0}, M_0). Typically different branches of the energy spectrum, if they have the same symmetry, repel each other. But in 3D space of parameters a contact point is possible, which has the following properties. After M crosses M_0 the bands again become isolated, but the topological charge $\tilde{N}_3$ of the vacuum changes.

In the case of a chiral particle with the Hamiltonian $\mathcal{H} = c(\check{\tau}^1 p_x + \check{\tau}^2 p_y + \check{\tau}^3 M)$, the two relevant branches are branches with positive and negative square root, $E_n = \pm c\sqrt{p_x^2 + p_y^2 + M^2}$. The parameter M here plays the role of the mass M of the 2+1 Dirac particle, or the momentum p_z of the 3+1 Weyl fermion. The two branches contact each other when M crosses zero (Fig. 11.3). For the 2+1 system, the topological charge of the 2+1 Dirac vacuum $\tilde{N}_3$ changes from $\tilde{N}_3 = -1/2$ to $\tilde{N}_3 = +1/2$ after crossing. The point $(p_x = 0, p_y = 0, M = 0)$ at which the reconnection takes place is a diabolical (or conical) point – an exceptional point in the energy spectrum at which two different energy levels with the same symmetry can touch each other. In our case the point $(p_x = 0, p_y = 0, M = 0)$ forms a Fermi point in 3D space – the hedgehog with topological charge $N_3 = +1$. Thus Fermi points with the topological charge N_3 are diabolical points in the spectrum. This again shows the abundance of Fermi points in physics.

The fractional charge $\tilde{N}_3 = \pm 1/2$ appears because the Hamiltonian of the 2+1 Dirac particle is pathological: it is linear in momentum **p** everywhere in 2-momentum space. A non-linear correction restores the integer values of $\tilde{N}_3$ on both sides of this quantum phase transition between two quantum vacua (see Sec. 11.4.2).

11.3.3 Berry phase and magnetic monopole in **p**-space

The diabolical or conical point also represents the Dirac magnetic monopole related to the Berry (1984) phase. Let us consider this in an example of the

Fermi point arising in Hamiltonian $\mathcal{H} = \check{\tau}^b g_b(\mathbf{p})$ in 3-momentum space. Let us adiabatically change the momentum $\mathbf{p}$ of the quasiparticle; then the unit vector $\hat{\mathbf{g}} = \mathbf{g}/|\mathbf{g}|$ moves along some path on its unit sphere. If the path C is closed, the wave function of the chiral quasiparticle acquires the geometrical phase factor $\gamma(C)$, called the Berry phase. It is half an area of the surface $S(C)$ enclosed by the closed contour C: $\gamma(C) = S(C)/2$. The Berry phase can be expressed in terms of 'magnetic flux' through the surface $S(C)$:

$$\gamma(C) = \int_{S(C)} d\mathbf{S} \cdot \mathbf{H}(\mathbf{p}) \ , \quad H_i(\mathbf{p}) = \frac{1}{4|\mathbf{g}|^3} e_{ijk} \, \mathbf{g} \cdot \left(\frac{\partial \mathbf{g}}{\partial p_j} \times \frac{\partial \mathbf{g}}{\partial p_k} \right) . \quad (11.8)$$

The 'magnetic' field providing this flux is determined in 3-momentum space. The analog of such a topological magnetic field in $\mathbf{r}$-space will be discussed in Sec. 21.1.1 (see eqn (21.2)).

This $\mathbf{p}$-space magnetic field has a singularity – a magnetic monopole – at the Fermi point:

$$\frac{\partial}{\partial p_i} H_i(\mathbf{p}) = 2\pi \delta(\mathbf{p}) . \quad (11.9)$$

Because of the 'magnetic' monopole in momentum space, the eigenfunctions of the Weyl Hamiltonian cannot be defined globally for all $\mathbf{p}$. In any determination of the eigenfunctions, one cannot continue them to all $\mathbf{p}$: one always finds a 'Dirac string' in momentum space (see Fig. 11.4), a vortex line in $\mathbf{p}$-space emerging from the monopole, on which the solution is ill defined; along an infinitely small path around such a line the solution acquires the Berry phase $\gamma(C) = 2\pi$.

11.4 Quantum phase transitions

11.4.1 *Quantum phase transition as change of momentum space topology*

Figure 11.3 demonstrates the phase transition between two vacuum states of a 2+1 system. Such a transition, occurring at $T = 0$, is an example of a quantum phase transition which is accompanied by a change of the topological quantum number of the quantum vacuum. On both sides of the transition the quasiparticle spectrum is fully gapped and the vacuum is characterized by the topological charge $\tilde{N}_3$. At the transition point the spectrum becomes gapless and the invariant $\tilde{N}_3$ is ill defined. In the extended 3D space (M, p_x, p_y) the node in the spectrum represents the Fermi point characterized by the topological charge N_3. The relation between the charges of the two vacua and the charge of the Fermi point is evident:

$$N_3 = \tilde{N}_3(M > 0) - \tilde{N}_3(M < 0) . \quad (11.10)$$

Let us consider quantum phase transitions which occur if we change some parameters of the system. For the chiral p-wave 2+1 superconductor in eqn (11.2), let us choose as the parameter M the chemical potential μ. At $\mu = 0$ the quantum phase transition occurs between the vacuum states with $\tilde{N}_3 = 0$ at $\mu < 0$ and $\tilde{N}_3 = +1$ at $\mu = 0$. The intermediate state between these two fully gapped vacua at $\mu = 0$ is gapless: $E_{\mu=0}(\mathbf{p} = 0) = 0$. In the extended 3D space

(μ, p_x, p_y), the intermediate state represents the Fermi point with $N_3 = +1$ in agreement with eqn (11.10).

In the d-wave superconductor in eqn (11.7) we can choose as the parameter M the amplitude d_{xy} of the order parameter. At $\mu > 0$ the quantum phase transition occurs when d_{xy} changes sign. When d_{xy} crosses zero the invariant in eqn (11.6) changes from $\tilde{N}_3 = -2$ to $\tilde{N}_3 = +2$. The intermediate state with $d_{xy} = 0$ is also gapless, and in the extended 3D space (d_{xy}, p_x, p_y) it represents Fermi point(s). It can be for example one degenerate Fermi point with $N_3 = +4$, or four elementary Fermi points each with $N_3 = +1$.

The quantum phase transition between the states with different $\tilde{N}_3$ can also be achieved by choosing as the parameter M the inverse effective mass m^* in eqn (11.2). The transition between states with $\tilde{N}_3 \neq 0$ and $\tilde{N}_3 = 0$ occurs when at fixed $\mu \neq 0$ the inverse mass $1/m^*$ crosses zero. Such a transition has a special property since the intermediate state with $1/m^* = 0$ is not gapless. Let us consider this example in more detail.

11.4.2 *Dirac vacuum is marginal*

In the chiral p-wave superconductor in eqn (11.2), the quasiparticles in intermediate state at $1/m^* = 0$ are Dirac 2+1 particles with mass $M = -\mu$ (see eqn (11.5)). This intermediate state is fully gapped everywhere, and the topological invariant in eqn (11.6) for this gapped Dirac vacuum is well defined. But it has the fractional value $\tilde{N}_3 = +1/2$, i.e. the intermediate value between two integer invariants on two sides of the quantum transition, $\tilde{N}_3 = +1$ and $\tilde{N}_3 = 0$. This happens because momentum space is not compact in this intermediate state: the unit vector $\hat{\mathbf{g}}$ does not approach the same value at infinity, but instead forms the 2D hedgehog.

The fractional topology, $\tilde{N}_3 = 1/2$, of the intermediate Dirac state demonstrates the marginal behavior of the vacuum of 2+1 Dirac fermions. The physical properties of the vacuum, which are related to the topological quantum numbers in momentum space (see Chapter 21), are not well defined for the Dirac vacuum. They crucially depend on how the Dirac spectrum is modified at high energy: toward $\tilde{N}_3 = 0$ or toward $\tilde{N}_3 = 1$. This modification is provided by the quadratic term $p^2/2m^*$ which is non-zero on both sides of the transition; as a result $\tilde{N}_3(1/m^* < 0) = 0$ and $\tilde{N}_3(1/m^* > 0) = +1$.

Thus the intermediate state at the point of the quantum phase transition between two regular, fully gapped vacua with integer topological charges $\tilde{N}_3$ in 2+1 systems is either gapless or marginal. The Dirac vacuum in 2+1 systems serves as the marginal intermediate state.

12

MOMENTUM SPACE TOPOLOGY PROTECTED BY SYMMETRY

12.1 Momentum space topology of planar phase

12.1.1 *Topology protected by discrete symmetry*

Let us consider first the **p**-space topology of the planar state of p-wave superfluid in eqn (7.61). We know that in the model of the independent spin components, the planar phase and ^{3}He-A differ only by their spin structure. That is why in this model the planar phase has point nodes at the same points $\mathbf{p} = \pm p_F \hat{\mathbf{l}}$ of the **p**-space as the A-phase. The only difference is that the Fermi points for quasiparticles with spin projection $S_z = 1/2$ and $S_z = -1/2$ have the same topological charges in the A-phase, $N_3(S_z = 1/2) = N_3(S_z = -1/2) = \pm 1$, and opposite topological charges in the planar phase, $N_3(S_z = 1/2) = -N_3(S_z = -1/2) = \pm 1$. That is why, while in the A-phase the total charge of the Fermi point is finite, $N_3(S_z = 1/2) + N_3(S_z = -1/2) = \pm 2$, in the planar one the total charge $N_3(S_z = 1/2) + N_3(S_z = -1/2) = 0$ for each of the two Fermi points.

The zero value of the invariant does not support the singularity of the Green function. Thus the question arises: Do the nodes at $\mathbf{p} = \pm p_F \hat{\mathbf{l}}$ survive if instead of the above model we consider the 'real' planar phase, where the interaction between the particles with different spin projection is non-zero?

The answer depends on the symmetry of the vacuum state. Let us consider the homogeneous vacuum of the planar phase whose order parameter is, say,

$$e_{\mu i} = \Delta_0 (\hat{x}_\mu \hat{x}_i + \hat{y}_\mu \hat{y}_i) . \tag{12.1}$$

The Bogoliubov–Nambu Hamiltonian (7.62) for quasiparticles in this state,

$$\mathcal{H}_{\text{planar}} = M(\mathbf{p})\check{\tau}^3 + c_\perp \check{\tau}^1 (\sigma_x p_x + \sigma_y p_y) \approx q_a c_\| \check{\tau}^3 (p_z - q_a p_F) + c_\perp \check{\tau}^1 (\sigma_x p_x + \sigma_y p_y) , \tag{12.2}$$

is equivalent to the Hamiltonian $\mathcal{H}_{\text{Dirac}} = \alpha^i (p_i - qA_i)$ for Dirac fermions with zero mass, where α^i are Dirac matrices. In other words, in the vicinity of the Fermi point, say at $\mathbf{p} = p_F \hat{\mathbf{z}}$, there are the right-handed and the left-handed fermions which are not mixed. The absence of mixing (zero Dirac mass) is provided by the discrete symmetry P of the vacuum in eqn (7.66). For the fermionic Hamiltonian this symmetry is

$$\text{P} = \tau_3 \sigma_z , \quad \text{PP} = 1 , \quad \text{P}\mathcal{H}\text{P} = \mathcal{H}\text{P} . \tag{12.3}$$

Since $\mathcal{H}$ and P commute, the same occurs with the Green function: $[\mathcal{G}, \text{P}] = 0$. This allows us to construct in addition to N_3 the following topological invariant:

$$N_3(\mathrm{P}) = \frac{1}{24\pi^2} e_{\mu\nu\lambda\gamma} \, \mathbf{tr} \left(\mathrm{P} \int_\sigma dS^\gamma \, \mathcal{G}\partial_{p_\mu}\mathcal{G}^{-1}\mathcal{G}\partial_{p_\nu}\mathcal{G}^{-1}\mathcal{G}\partial_{p_\lambda}\mathcal{G}^{-1} \right) . \quad (12.4)$$

It can be shown that $N_3(\mathrm{P})$ is robust to any perturbations of the Green function, unless they violate the commutation relation $[\mathcal{G}, \mathrm{P}] = 0$. In other words, eqn (12.4) is invariant under perturbations conserving the symmetry P. This is the topological invariant protected by symmetry.

For the Fermi points in the planar state one obtains $N_3(\mathrm{P}) = N_3(S_z = 1/2) - N_3(S_z = -1/2) = \pm 2$. The non-zero value of this invariant shows that, though the total conventional topological charge of the Fermi point in the planar state is zero, $N_3 = N_3(S_z = 1/2) + N_3(S_z = -1/2) = 0$, the Fermi point is robust to any interactions which do not violate P.

In addition, because of the discrete symmetry P, the topological charge $N_3(\mathrm{P}) = \pm 2$ of the degenerate Fermi point is equally distributed between the fermionic species. This ensures that each quasiparticle has an elementary (unit) topological charge: $N_3(S_z = 1/2) = -N_3(S_z = -1/2) = \pm 1$. This is important since only such quasiparticles that are characterized by the elementary charge $N_3 = \pm 1$ acquire the relativistic energy spectrum in the low-energy corner. In ^{3}He-A the analogous discrete symmetry P ensures that the topological charge $N_3 = \pm 2$ of the doubly degenerate Fermi point is equally distributed between two fermions, and thus is responsible for the Lorentz invariance in the vicinity of the Fermi point. Thus the discrete symmetry is the necessary element for the development of the effective RQFT at low energy.

12.1.2 Dirac mass from violation of discrete symmetry

If the symmetry P is violated, either by the external field or by some extra interaction, or by inhomogeneity of the order parameter, $N_3(\mathrm{P})$ in eqn (12.4) ceases to be the invariant. As a result the Fermi points with the charge $N_3 = 0$ are not protected by symmetry and will be destroyed: the quasiparticles will acquire gap (mass). Let us consider how this occurs in the planar state.

First, we recall that in the planar state, the symmetry P in eqn (7.66) is the combination of a discrete gauge transformation from the $U(1)$ group and of spin rotation by π around the z axis. Applying eqn (7.66) to the Bogoliubov–Nambu Hamiltonian (7.62), for which the generator of the $U(1)_N$ gauge rotation is τ_3, one obtains $\mathrm{P} = e^{-\pi i \tau_3/2} e^{\pi i \sigma_z/2} = \tau_3 \sigma_z$ in eqn (12.3). Thus the symmetry P of the planar state includes the element of the original $SO(3)_\mathbf{S}$ symmetry group of liquid ^{3}He above the superfluid transition, which was spontaneously broken in the planar phase.

However, even in normal liquid ^{3}He, the symmetry $SO(3)_\mathbf{S}$ under spin rotations is an approximate symmetry violated by the spin–orbit interaction. This means that at a deep trans-Planckian level, the symmetry P is also approximate, and thus it cannot fully protect the topologically trivial Fermi point. The Fermi point must disappear leading to a small Dirac mass for relativistic quasiparticles proportional to the symmetry-violating interaction. In our case the Dirac mass is expressed in terms of the spin–orbit interaction:

$$M \sim \frac{E_D^2}{E_{\text{Planck 2}}} \quad , \quad E_D = \frac{\hbar c_\parallel}{\xi_D} \ . \tag{12.5}$$

Here E_D is the energy scale characterizing the spin–orbit coupling (see eqn (16.1)), which comes from the dipole–dipole interaction of ^{3}He atoms; the corresponding length scale called the dipole length is typically $\xi_D \sim 10^{-3}$ cm.

This symmetry-violating spin–orbit coupling also generates masses for the 'gauge bosons' in the planar state and in ^{3}He-A: see eqn (19.23) for the mass of a 'hyperphoton' in ^{3}He-A.

This example demonstrates three important roles of discrete symmetry: (i) it establishes stability of the Fermi point; (ii) it ensures the emergence of relativistic chiral fermions and thus RQFT at low energy; (iii) the violation of the discrete symmetry at a deep fundamental level is the source of a small Dirac mass of relativistic fermions.

The analogous discrete symmetry P is also important for the quasiparticle spectrum in the vicinity of the degenerate Fermi point with $N_3 = 2$ in ^{3}He-A. We know that the quasiparticle spectrum becomes relativistic in the vicinity of the Fermi point with elementary topological charge, $N_3 = \pm 1$. In the case of $N_3 = 2$ there is no such rule. The role of the discrete symmetry P in eqn (7.56) is again to ensure that in the vicinity of the $N_3 = 2$ Fermi point, the spectrum behaves as if there are two quasiparticle species each with the elementary charge $N_3 = 1$. Though P is violated by the spin–orbit interaction, the Fermi point cannot disappear since its topological charge is non-zero, $N_3 = 2$. However, because of the violation of P the quasiparticle spectrum is modified and again becomes non-relativistic, but now at a very low energy. This reentrant violation of Lorentz symmetry will be discussed in Sec. 12.4.

Now we proceed to the Standard Model and show that the massless chiral fermions there are also described by the Fermi points protected by discrete symmetry. When this symmetry is violated at low energy, all the fermions acquire mass.

12.2 Quarks and leptons

The Standard Model of particle physics is the greatest achievement in the physics of second half of the 20th century. It is in excellent agreement with experimental observation even today. It is a common view now that the Standard Model is an effective theory, which describes the physics well below the Planck and GUT scales, 10^{19} GeV and 10^{15} GeV respectively. At temperatures above the electroweak scale, i.e. at $T > 10^2$–10^3 GeV, the elementary particles – quasiparticles if the theory is effective – are massless chiral fermions. This means that the vacuum of the Standard Model belongs to the universality class of Fermi points. At lower temperature fermions acquire masses in almost the same way as electrons in metals gain the gap below the phase transition into the superconducting state (see Sec. 7.4.5). Since the fermionic and bosonic contest of the Standard Model is much bigger than that in conventional superconductors, an effective theory of such a transition (or crossover) to the 'superconducting' state contains many

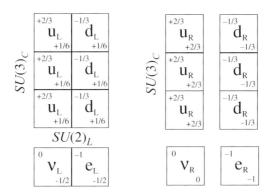

FIG. 12.1. First family of quarks and leptons. Number in bottom right corner is the hypercharge – the charge of $U(1)_Y$ group. Electric charge $Q = Y + T_{L3}$ is shown in the top left corner.

parameters which must be considered as phenomenological. Here we discuss the Standard Model above the electroweak energy scale, when fermions are massless and chiral.

12.2.1 Fermions in Standard Model

In the Standard Model of electroweak and strong interactions each family of quarks and leptons contains eight left-handed and eight right-handed fermions (Fig. 12.1). We assume here that the right-handed neutrino is present, as follows from the Kamiokande experiments. Experimentally, there are three families (generations) of fermions, $N_g = 3$. However, the larger number of families does not contradict observations if Dirac masses of fermions in extra families are large enough.

In the Standard Model fermions transform under the gauge group G(213) = $SU(2)_L \times U(1)_Y \times SU(3)_C$ of weak, hypercharge and strong interactions respectively. In addition there are two global charges: baryonic B and leptonic L. Quarks, u and d, appear in three colors, i.e. they are triplets under the color group $SU(3)_C$ of quantum chromodynamics (QCD). They have $B = 1/3$ and $L = 0$. Leptons e and ν are colorless, i.e. they are $SU(3)_C$-singlets. They have $B = 0$ and $L = 1$. There is a pronounced asymmetry between left fermions, which are $SU(2)_L$-doublets (their weak isospin is $T_L = 1/2$), and right fermions, which are all weak singlets ($T_L = 0$). The group $SU(2)_L$ thus transforms only left fermions, hence the index L.

At low energy below about 200 GeV (the electroweak scale) the $SU(2)_L \times U(1)_Y$ group of electroweak interactions is violated so that only its subgroup $U(1)_Q$ is left, where $Q = Y + T_{L3}$ is the electric charge. This is the group of quantum electrodynamics (QED). Thus the group G(213) is broken into its subgroup G(13) = $U(1)_Q \times SU(3)_C$. Figure 12.1 shows the hypercharge Y_a of fermions (bottom right corner) and also the electric charge $Q = Y + T_{L3}$ (top left corner).

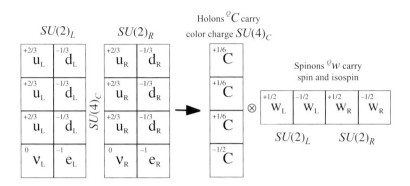

FIG. 12.2. *Left*: Standard Model fermions organized in the G(224) group. *Right*: Slave-boson description of Standard Model fermions. Spinons are fermions with spin and isospin, while the $SU(4)$ color charge is carried by slave bosons – holons. The number in the top left corner is electric charge Q.

12.2.2 *Unification of quarks and leptons*

According to Fig. 12.1, charges of Standard Model fermions are rather clumsily distributed between the fermions. It looks very improbable that the fundamental theory can have such diverse charges. There certainly must be a more fundamental theory, the Grand Unification Theory (GUT), where such charges elegantly arise from a simpler construction. The idea of Grand Unification is supported by the observation that three running coupling constants of the G(213) group, when extrapolated to high energy, meet each other at an energy about 10^{15}–10^{16} GeV. This suggests that above this scale the groups of weak, hypercharge and strong interactions are unified into one big symmetry group, such as $SO(10)$, with a single coupling constant. Let us remind ourselves, however, that from the condensed matter point of view even the GUT remains the effective sub-Planckian theory.

There is another group, the subgroup of $SO(10)$, which also unites in a very simple way all $16 N_g$ fermions with their diverse hypercharges and electric charges. This is a type of Pati–Salam model (Pati and Salam 1973, 1974; Foot et al. 1991) with the symmetry group G(224) = $SU(2)_L \times SU(2)_R \times SU(4)_C$. This group G(224) is the minimal subgroup of the more popular $SO(10)$ group which preserves all its important properties (Pati 2000; 2002). It naturally arises in the compactification scheme (Shafi and Tavartkiladze 2001).

Though the G(224) group does not represent GUT, since it has two coupling constants instead of a single one, it has many advantages when compared to the $SO(10)$ group. In particular, the $SO(10)$ group organizes fermions in a multiplet which contains both matter and antimatter. This does not happen with the G(224) group which does not mix matter and antimatter. What is most important is that this group allows the correct definition of the momentum space topological charge.

The G(224) group organizes all 16 fermions of one generation of matter into left and right baryon–lepton octets (Fig. 12.2 left). Here the $SU(3)_C$ color group of QCD is extended to the $SU(4)_C$ color group by introducing as a charge the difference between baryon and lepton charges $B - L$. Now the quarks and leptons are united into quartets of the $SU(4)_C$ group, with leptons treated as a fourth color. Quarks and leptons can thus transform to each other, so that the baryon and lepton charges are not conserved separately, but only in combination $B - L$. The non-conservation of baryon charge is an important element in modern theories of the origin of the baryonic asymmetry of our Universe (see Sec. 18.2). At present B and L are conserved with a high precision, since the mutual transformation of quarks and leptons is highly suppressed at low energy. That is why, though the decay of the proton is allowed in the GUT scheme, the proton lifetime estimated using the GUT scheme is about 10^{33}–10^{34} years. The discovery of proton decay would be direct proof of the unification of quarks and leptons.

The $SU(2)_R$ group for the right particles is added to make the nature left–right symmetric: at this more fundamental level the parity is conserved. All the charges of 16 fermions are collected in the following table:

Fermion	T_{L3}	T_{R3}	$B-L$	$\to Y$	$\to Q$
$u_L(3)$	$+\frac{1}{2}$	0	$\frac{1}{3}$	$\frac{1}{6}$	$\frac{2}{3}$
$u_R(3)$	0	$+\frac{1}{2}$	$\frac{1}{3}$	$\frac{2}{3}$	$\frac{2}{3}$
$d_L(3)$	$-\frac{1}{2}$	0	$\frac{1}{3}$	$\frac{1}{6}$	$-\frac{1}{3}$
$d_R(3)$	0	$-\frac{1}{2}$	$\frac{1}{3}$	$-\frac{1}{3}$	$-\frac{1}{3}$
ν_L	$+\frac{1}{2}$	0	-1	$-\frac{1}{2}$	0
ν_R	0	$+\frac{1}{2}$	-1	0	0
e_L	$-\frac{1}{2}$	0	-1	$-\frac{1}{2}$	-1
e_R	0	$-\frac{1}{2}$	-1	-1	-1

(12.6)

When the energy is reduced the G(224) group transforms to the intermediate subgroup G(213) of the weak, hypercharge and strong interactions with the hypercharge given by

$$Y = \frac{1}{2}(B - L) + T_{R3} \ . \tag{12.7}$$

This equation naturally reproduces all the diversity of the hypercharges $Y_{(a)}$ of fermions in Fig. 12.1. When the energy is reduced further the electroweak–strong subgroup G(213) is reduced to the G(13) = $U(1)_Q \times SU(3)_C$ group of electromagnetic and strong interactions. The electric charge Q of the $U(1)_Q$ group is left–right symmetric:

$$Q = \frac{1}{2}(B - L) + T_{R3} + T_{L3} \ . \tag{12.8}$$

12.2.3 Spinons and holons

Another advantage of the G(224) group is that all 16 chiral fermions of one generation can be considered as composite objects being the product Cw of four C-bosons and four w-fermions in Fig. 12.2 *right* (Terazawa 1999).

The number in the top left corner shows the electric charge Q of C-bosons and w-fermions. This scheme is similar to the slave-boson approach in condensed matter, where the particle (electron) is considered as a product of the spinon and holon. Spinons in condensed matter are fermions which carry electronic spin, while holons are 'slave' bosons which carry its electric charge (see e.g. Marchetti *et al.* (1996) and references therein and Sec. 8.1.7).

In the scheme demonstrated in Fig. 12.2 *right* four 'holons' C have zero spin and zero isospins, but they carry the color charge of the $SU(4)_C$ group; their $B - L$ charges of the $SU(4)_C$ group are $(\frac{1}{3}, \frac{1}{3}, \frac{1}{3}, -1)$. Correspondingly their electric charges in eqn (12.8) are $Q = (B - L)/2 = (\frac{1}{6}, \frac{1}{6}, \frac{1}{6}, -\frac{1}{2})$.

The 'spinons' w are the $SU(4)_C$ singlets, but they carry spin and weak isospins. Left and right spinons form doublets of $SU(2)_L$ and $SU(2)_R$ groups respectively. Since their $B - L$ charge is zero, the electric charges of spinons according to eqn (12.8) are $Q = T_{L3} + T_{R3} = \pm 1/2$.

12.3 Momentum space topology of Standard Model

In the case of a single chiral fermion, the massless (gapless) character of its energy spectrum, $E = cp$, is protected by the momentum space topological invariant N_3 of the Fermi point at $\mathbf{p} = 0$. However, the Standard Model has an equal number of left and right fermions, so the Fermi point there is marginal: the total topological charge N_3 of the Fermi point in eqn (8.15) is zero, if the trace is over all the fermionic species. Thus the topological mechanism of mass protection does not work and, in principle, an arbitrarily small interaction between the fermions can provide the Dirac masses for all eight pairs of fermions. This indicates that the vacuum of the Standard Model is marginal in the same way as the planar state of superfluid ^{3}He as was discussed in Sec. 12.1.

In an example of the planar phase, we have seen that in the fermionic systems with marginal Fermi points, the mass (gap) does not appear if the vacuum has the proper symmetry element. The same situation occurs for the Fermi points of the Standard Model. Here also the momentum space topological invariants protected by symmetry can be introduced. These invariants are robust under symmetric perturbations, and provide the protection against the mass if the relevant symmetry is exact.

In the Standard Model the protection against the mass is provided by both discrete and continuous symmetries. Let us first consider the relevant continuous symmetries. They form the electroweak group $G(12) = U(1)_Y \times SU(2)_L$ generated by the hypercharge and by the weak isospin respectively. The topological invariants protected by symmetry are the functions of parameters of these symmetry groups. If these symmetries are violated all the fermions acquire masses. However, we shall see that instead of the G(12) group it is enough to have one

12.3.1 Generating function for topological invariants constrained by symmetry

Following Volovik (2000a) let us introduce the matrix $\mathcal{N}$ whose trace gives the invariant N_3 in eqn (8.15):

$$\mathcal{N} = \frac{1}{24\pi^2} e_{\mu\nu\lambda\gamma} \int_\sigma dS^\gamma \, \mathcal{G}\partial_{p_\mu}\mathcal{G}^{-1}\mathcal{G}\partial_{p_\nu}\mathcal{G}^{-1}\mathcal{G}\partial_{p_\lambda}\mathcal{G}^{-1} , \qquad (12.9)$$

where, as before, the integral is about the Fermi point in the 4-momentum space. Let us consider the expression

$$\mathbf{tr}\,(\mathcal{N}Y) , \qquad (12.10)$$

where Y is the generator of the $U(1)_Y$ group, the hypercharge matrix. It is clear that eqn (12.10) is robust to any perturbation of the Green function which does not violate the $U(1)_Y$ symmetry, since in this case the hypercharge matrix Y commutes with the Green function: $[Y, \mathcal{G}] = 0$. The same occurs with any power of Y, i.e. $\mathbf{tr}\,(\mathcal{N}Y^n)$ is also invariant under symmetric deformations. That is why one can introduce the generating function for all the topological invariants containing powers of the hypercharge

$$\mathbf{tr}\left(e^{i\theta_Y\,Y}\mathcal{N}\right) . \qquad (12.11)$$

All the powers $\mathbf{tr}(\mathcal{N}Y^n)$, which are topological invariants, can be obtained by differentiating eqn (12.11) over the group parameter θ_Y. Since the above parameter-dependent invariant is robust to interactions between the fermions, it can be calculated for the non-interacting particles. In the latter case the matrix $\mathcal{N}$ is diagonal and its eigenvalues coincide with chirality $C_a = +1$ and $C_a = -1$ for right and left fermions correspondingly. The trace of the matrix $\mathcal{N}$ over the given irreducible fermionic representation of the gauge group is (with minus sign) the symbol $N_{(y/2,\underline{a},I_W)}$ introduced by Froggatt and Nielsen (1999). In their notation $y/2(=Y)$, $\underline{a}$, and I_W denote hypercharge, color representation and the weak isospin correspondingly.

For the Standard Model with hypercharges for 16 fermions given in Fig. 12.1 one has the following generating function:

$$\mathbf{tr}\left(e^{i\theta_Y\,Y}\mathcal{N}\right) = \sum_a C_a e^{i\theta_Y\,Y_a} = 2\left(\cos\frac{\theta_Y}{2} - 1\right)\left(3e^{i\theta_Y/6} + e^{-i\theta_Y/2}\right) . \qquad (12.12)$$

The factorized form of the generating function reflects the composite spinon–holon representation of fermions and directly follows from Fig. 12.2 *right*. Since $Y = \frac{1}{2}(B - L) + T_{R3}$ one has

$$\mathbf{tr}\left(e^{i\theta_Y\,Y}\mathcal{N}\right) = \mathbf{tr}\left(e^{i\theta_Y\,T_{R3}}\mathcal{N}_{sp}\right)\,\mathbf{tr}\left(e^{i\theta_Y\,(B-L)/2}\right) \qquad (12.13)$$

$$= 2\left(\cos\frac{\theta_Y}{2} - 1\right)\left(3e^{i\theta_Y/6} + e^{-i\theta_Y/2}\right) .$$

Here $\mathcal{N}_{sp}$ is the matrix $\mathcal{N}$ for spinons.

In addition to the hypercharge the weak charge is also conserved in the Standard Model above the electroweak transition. The generating function for the topological invariants which contain the powers of both the hypercharge Y and the weak isospin T_{L3} also has the factorized form

$$\mathbf{tr}\left(e^{i\theta_W T_{L3}}e^{i\theta_Y Y}\mathcal{N}\right) = \mathbf{tr}\left(e^{i\theta_W T_{L3}}e^{i\theta_Y T_{R3}}\mathcal{N}_{sp}\right) \mathbf{tr}\left(e^{i\theta_Y (B-L)/2}\right) \quad (12.14)$$

$$= 2\left(\cos\frac{\theta_Y}{2} - \cos\frac{\theta_W}{2}\right)\left(3e^{i\theta_Y/6} + e^{-i\theta_Y/2}\right). \quad (12.15)$$

The non-zero result of eqn (12.15) shows that the Green function is singular at the Fermi point $\mathbf{p} = 0$ and $p_0 = 0$, which means that at least some fermions must be massless if either of the symmetries, $U(1)_Y$ or $SU(2)_L$, is exact.

12.3.2 *Discrete symmetry and massless fermions*

Choosing the parameters $\theta_Y = 0$ and $\theta_W = 2\pi$ one obtains the maximally possible value of the generating function:

$$\mathbf{tr}\left(\mathrm{P}\mathcal{N}\right) = 16 \;,\; \mathrm{P} = e^{2\pi i T_{L3}}\;. \quad (12.16)$$

This implies 16-fold degeneracy of the Fermi point which provides the existence of 16 massless fermions. Thus all 16 fermions of one generation are massless above the electroweak scale 200 GeV. This also shows that as in the case of the planar phase, it is the discrete symmetry group, the Z_2 group $\mathrm{P} = e^{2\pi i T_{L3}}$, which is responsible for the protection against the mass (mass protection).

Since $\mathrm{P} = 1$ for all right-handed fermions, which are $SU(2)_L$ singlets, and $\mathrm{P} = -1$ for all left-handed fermions, whose isospins $T_{L3} = \pm 1/2$, one obtains that in the relativistic limit P coincides with chirality C_a. That is why $\mathbf{tr}\left(\mathrm{P}\mathcal{N}\right) = \sum_a P_a C_a = \sum_a C_a C_a = 16$. So, in this limit the discrete symmetry P is equivalent to the chiral symmetry, the γ^5 symmetry which protects the Dirac fermions from the masses. If the γ^5 symmetry is obeyed, i.e. it commutes with the Dirac Hamiltonian, $\gamma^5 \mathcal{H} \gamma^5 = \mathcal{H}$, then the Dirac fermion has no mass. This is consistent with the non-zero value of the topological invariant: it is easy to check that for the massless Dirac fermion one has $\mathbf{tr}\left(\gamma^5 \mathcal{N}\right) = 2$.

This connection between topology and mass protection looks trivial in the relativistic case, where the absence of mass due to γ^5 symmetry can be directly obtained from the Dirac Hamiltonian. However, it is not so trivial if the interaction is introduced, or if the Lorentz and other symmetries are violated at high energy, so that the Dirac equation is no longer applicable, and even the chirality ceases to be a good quantum number. In particular, transitions between the fermions with different chirality are possible at high energy, see Sec. 29.3.1 for an example. In such a general case it is only the symmetry-protected topological charge, such as eqn (12.16), that gives the information on the gap (mass) protection of fermionic quasiparticles.

12.3.3 *Relation to chiral anomaly*

The momentum space topological invariants are related to the axial anomaly in fermionic systems. In particular, the charges related to the local gauge group cannot be created from vacuum, and the condition for that is the vanishing of some of these invariants (see Chapter 18):

$$\mathbf{tr}\,(Y\mathcal{N}) = \mathbf{tr}\,(Y^3\mathcal{N}) = \mathbf{tr}\,((T_{L3})^2 Y \mathcal{N}) = \mathbf{tr}\,(Y^2 T_{L3} \mathcal{N}) = \ldots = 0 \,. \quad (12.17)$$

From the form of the generating function in eqn (12.15) it really follows that all these invariants are zero, though in this equation it is not assumed that the groups $U(1)_Y$ and $SU(2)_L$ are local.

12.3.4 Trivial topology below electroweak transition and massive fermions

When the electroweak symmetry $U(1)_Y \times SU(2)_L$ is violated to $U(1)_Q$, the only remaining charge – the electric charge $Q = Y + T_{L3}$ – produces a zero value for the whole generating function according to eqn (12.15):

$$\mathbf{tr}\,\left(e^{i\theta_Q Q}\mathcal{N}\right) = \mathbf{tr}\,\left(e^{i\theta_Q Y} e^{i\theta_Q T_{L3}} \mathcal{N}\right) = 0 \,. \quad (12.18)$$

The zero value of the topological invariants implies that even if the singularity in the Green function exists in the Fermi point it can be washed out by interaction. Thus, if the electroweak symmetry-breaking scheme is applicable to the Standard Model, each elementary fermion in our world must have a mass below the electroweak transition temperature. The mass would also occur if the Standard Model is such an effective theory that its electroweak symmetry is not exact at the fundamental level.

What is the reason for such a symmetry-breaking pattern, and, in particular, for such choice of electric charge Q? Why had nature not chosen the more natural symmetry breaking, such as $U(1)_Y \times SU(2)_L \to U(1)_Y$, $U(1)_Y \times SU(2)_L \to SU(2)_L$ or $U(1)_Y \times SU(2)_L \to U(1)_Y \times U(1)_{T_3}$? The possible reason is provided by eqn (12.15), according to which the nullification of all the momentum space topological invariants occurs only if the symmetry-breaking scheme $U(1)_Y \times SU(2)_L \to U(1)_Q$ takes place with the charge $Q = \pm Y \pm T_{L3}$. Only in such cases does the topological mechanism for the mass protection disappear. This can shed light on the origin of the electroweak transition. It is possible that the elimination of the mass protection is the only goal of the transition. This is similar to the Peierls transition in condensed matter: the formation of mass (gap) is not the consequence but the cause of the transition. It is energetically favorable to have masses of quasiparticles, since this leads to a decrease of the energy of the fermionic vacuum. Formation of the condensate of top quarks, which generates the heavy mass of the top quark, could be a relevant scenario for that (see the review by Tait 1999).

Another hint of why it is the charge Q which is the only remnant charge in the low-energy limit, is that in the G(224) model the electric charge $Q = \frac{1}{2}(B - L) + T_{L3} + T_{R3}$ is left–right symmetric. For any left–right symmetric charge Q the topological invariant $\mathbf{tr}\,\left(e^{i\theta_Q Q}\mathcal{N}\right) = 0$. Such remnant charge does not prevent the formation of mass, and thus there is no reason to violate the $U(1)_Q$ symmetry.

12.4 Reentrant violation of special relativity

12.4.1 Discrete symmetry in ^{3}He-A

Now let us consider the peculiarities of the quasiparticle energy spectrum in the case when the Fermi point has multiple topological charge $|N_3| > 1$ (Volovik 2001a). This happens in the ^{3}He-A vacuum described by eqn (7.54), where the topological charges of Fermi points are $N_3 = \pm 2$. Let us discuss the Fermi point, say, at $\mathbf{p} = +p_F \hat{\mathbf{l}}$, which has $N_3 = -2$. In the model with two independent spin projections, $S_z = +1/2$ and $S_z = -1/2$, we have actually two independent flavors of fermions, each with the topological charge $N_3 = -1$. Both flavors have a $\mathbf{p}$-space singularity situated at the same point in $\mathbf{p}$-space. Near this doubly degenerate Fermi point, two flavors of left-handed 'relativistic' quasiparticles are described by the following Bogoliubov–Nambu Hamiltonian:

$$\mathcal{H} = c_\| \tilde{p}_z \check{\tau}^3 + c_\perp \sigma^z (\check{\tau}^1 p_x - \check{\tau}^2 p_y) \ , \quad \tilde{p}_z = p_z - p_F \ . \tag{12.19}$$

The question is whether this relativistic physics survives if the interaction between two populations, $S_z = +1/2$ and $S_z = -1/2$, is introduced. Of course, since the charge of the Fermi point is non-zero, $N_3 = -2$, the singularity in the Green function will persist, but there is no guarantee that quasiparticles would necessarily have the relativistic spectrum at low energy. Again, the answer depends on the existence of the discrete symmetry of the vacuum. The ^{3}He-A vacuum has the proper discrete symmetry P in eqn (7.56) which couples the two flavors and forces them to have identical elementary topological charges $N_3 = -1$, and thus their spectrum is 'relativistic'. This $\mathbf{P} = U_{x2}^S e^{i\pi}$ in eqn (7.56) is the combined symmetry: it is the element U_{x2}^S of the $SO(3)_\mathbf{S}$ group, the π rotation of spins about, say, axis x, which is supplemented by the gauge rotation $e^{i\pi}$ from the $U(1)_N$ group. Applying this to the Bogoliubov–Nambu Hamiltonian (12.19), for which the generator of the $U(1)_N$ gauge rotation is $\check{\tau}^3$, one obtains $\mathbf{P} = e^{\pi i \check{\tau}^3/2} e^{\pi i \sigma^x/2} = -\check{\tau}^3 \sigma^x$, and $[\mathbf{P}, \mathcal{H}] = 0$.

12.4.2 Violation of discrete symmetry

The P symmetry came from the 'fundamental' microscopic physics at an energy well above the first Planck scale $E_{\text{Planck 1}} = m^* c_\perp^2$ at which the spectrum becomes non-linear and thus the Lorentz invariance is violated. However, from the trans-Planckian physics of the ^{3}He atoms we know that symmetry P is not exact in ^{3}He. Due to the spin–orbit coupling the symmetry group of spin rotations $SO(3)_\mathbf{S}$ is no longer the exact symmetry of the normal liquid, and this in particular concerns the spin rotation U_{x2}^S which enters P in eqn (7.56). Since the P symmetry was instrumental for establishing special relativity in the low-energy corner, its violation must lead to violation of the Lorentz invariance and also to mixing of the two fermionic flavors at the very low energy determined by this tiny spin–orbit coupling.

Let us consider how this reentrant violation of the Lorentz symmetry at low energy happens in ^{3}He-A. Due to the symmetry violating spin–orbit coupling the symmetry group of ^{3}He-A, $\mathbf{H}_\mathbf{A} = U(1)_{L_z - N/2} \times U(1)_{S_z}$ is also not exact. The exact symmetry now is the combined symmetry constructed from the sum of two generators: $U(1)_{J_z-N/2}$, where $J_z = S_z + L_z$ is the generator of the simultaneous rotations of spins and orbital degrees of freedom. The spin–orbit coupling does not destroy the symmetry of the normal liquid under combined rotations. The order parameter in eqn (7.54) acquires a small correction consistent with the $U(1)_{J_z-N/2}$ symmetry:

$$e_{\mu i} = \Delta_0 \hat{z}_\mu (\hat{x}_i + i\hat{y}_i) + \alpha \Delta_0 (\hat{x}_\mu + i\hat{y}_\mu) \hat{z}_i \ . \tag{12.20}$$

The first term corresponds to a Cooper pair state with $L_z = 1/2$ per atom and $S_z = 0$, while the second one is a small admixture of the state with $S_z = 1/2$ per atom and $L_z = 0$. Both components have $J_z = 1/2$ per atom and thus must be present in the order parameter. Due to the second term the order parameter is not symmetric under the P operation. The small parameter $\alpha \sim \xi^2/\xi_D^2 = E_D^2/E_{\text{Planck 2}}^2 \sim 10^{-5}$ is the relative strength of the spin–orbit coupling (see eqn (12.5)).

The Bogoliubov–Nambu Hamiltonian for fermionic quasiparticles in such a vacuum is now modified as compared to that in the pure vacuum state with $L_z = 1/2$ and $S_z = 0$ in eqn (7.57):

$$\mathcal{H}_A = c_\| \tilde{p}_z \check{\tau}^3 + c_\perp \sigma^z (\check{\tau}^1 p_x - \check{\tau}^2 p_y) + \alpha c_\perp p_z (\sigma^x \check{\tau}^1 - \sigma^y \check{\tau}^2) \ . \tag{12.21}$$

12.4.3 Violation of 'Lorentz invariance' at low energy

Diagonalization of the Hamiltonian (12.21) shows that the small correction due to spin–orbit coupling gives rise to the following splitting of the energy spectrum in the low-energy corner:

$$E_\pm^2 = c_\|^2 \tilde{p}_z^2 + c_\perp^2 \left(\alpha |p_z| \pm \sqrt{\alpha^2 p_z^2 + p_\perp^2} \right)^2 \ . \tag{12.22}$$

In ^{3}He-A p_z is close to p_F, so one can put $\alpha|p_z| = \alpha p_F$. Then the $+$ and $-$ branches give the gapped and gapless spectra respectively. For $p_\perp \ll \alpha p_F$ one has

$$E_+^2 \approx c_\|^2 \tilde{p}_z^2 + \tilde{c}_\perp^2 p_\perp^2 + \tilde{m}^2 \tilde{c}_\perp^4 \ , \tag{12.23}$$

$$E_-^2 \approx c_\|^2 \tilde{p}_z^2 + \frac{p_\perp^4}{4\tilde{m}^2} \ , \tag{12.24}$$

$$\tilde{c}_\perp = \sqrt{2} c_\perp \ , \ \tilde{m} c_\perp^2 = \alpha p_F c_\perp \sim \frac{E_D^2}{E_{\text{Planck 2}}} \ , \ p_\perp \ll \tilde{m} c_\perp \ . \tag{12.25}$$

In this ultra-low-energy corner the gapped branch of the spectrum in eqn (12.23) is relativistic, but with different speed of light $\tilde{c}_\perp$. The gapless branch in eqn (12.24) is relativistic in one direction $E = c_\| |\tilde{p}_z|$, and is non-relativistic, $E = p_\perp^2/2\tilde{m}$, for the motion in the transverse direction.

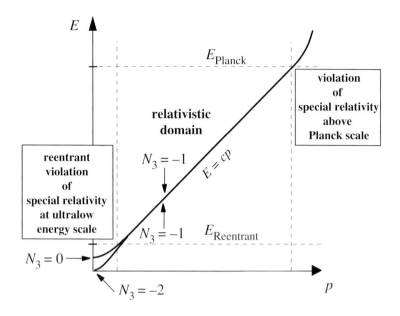

FIG. 12.3. Low-energy memory of the high-energy non-symmetric physics

12.4.4 Momentum space topology of exotic fermions

What is important here is that the reentrant violation of Lorentz invariance caused by the violation of discrete symmetry is generic and thus can occur in the effective RQFT such as the Standard Model. This is because of the topological properties of the spectrum: the mixing of the two fermionic flavors occurs with the redistribution of the topological charge $N_3 = -2$ between the two fermions. In the relativistic domain each of two fermions has the topological charge $N_3 = -1$. It is easy to check that in the ultra-low-energy corner this is not the case. While the total topological charge, $N_3 = -2$, must be conserved, it is now redistributed between the fermions in the following manner: the massive fermion (with energy E_+) acquires the trivial topological charge $N_3 = 0$ (that is why it becomes massive), while another one (with energy E_-) has double topological charge $N_3 = -2$ (see Fig. 12.3). It is important that only one species of the massless fermions now has $N_3 = -2$. Thus it cannot split into two fermions each with $N_3 = -1$. This exotic fermion with $N_3 = -2$ is gapless because of the non-zero value of the topological charge, but the energy spectrum of such a fermion is not linear. That is why it cannot be described in relativistic terms.

In the same way as the $N_3 = \pm 1$ fermions are necessarily relativistic and chiral in the low-energy corner, the fermions with higher $|N_3|$ are in general non-relativistic, unless they are protected by the discrete symmetry. The momentum space topology which induces the special relativity if $|N_3| = 1$ becomes incompatible with the relativistic invariance if $|N_3| > 1$ and the corresponding discrete symmetry is not exact. Thus the Lorentz symmetry, if it is effective, may be violated at very low energy. Properties of the fermionic systems with multiple zeros, $|N_3| > 1$, including the axial anomaly in its non-relativistic version, were discussed by Voloshev and Konyshev (1988).

The energy scale at which the non-relativistic splitting of the energy spectrum occurs is $E_{\text{Reentrant}} = E_D^2/E_{\text{Planck 2}}$ which is much less than the second and even the first Planck scale in ^{3}He-A, $E_{\text{Planck 1}} = m^* c_\perp^2$. Thus the relativistic region for the ^{3}He-A fermions, $E_{\text{Reentrant}} \ll E \ll E_{\text{Planck 1}}$, is sandwiched by the non-relativistic regions at the high and low energies.

12.4.5 Application to RQFT

The above example of ^{3}He-A shows that the special relativity in the low-energy corner is produced by the combined effect of the degenerate Fermi point and the discrete symmetry between the fermionic species. If the discrete symmetry is approximate, then in the ultra-low-energy corner the redistribution of the momentum space topological charges occurs between the fermions with the appearance of exotic fermions with higher topological charge $|N_3| > 1$. This topological transition leads to strong modification of the energy spectrum which again becomes essentially non-relativistic, but now at the ultra-low energies.

In principle, such a topological transition with the appearance of the exotic fermions with $N_3 = \pm 2$ can occur in the relativistic theories too, if these theories are effective. In the effective theory the Lorentz invariance (and thus the special relativity) appears in the low-energy corner as an emergent phenomenon, while it can be violated at high energy approaching the Planck scale. At low energy, the fermions are chiral and relativistic, if there is a symmetry between the flavors of the fermions. If such symmetry is violated, spontaneously or due to the fundamental physics well above the Planck scale, then in the extreme low-energy limit the asymmetry between the fermionic flavors becomes important, and the system starts to remember its high-energy non-relativistic origin. The rearrangement of the topological charges N_3 between the fermionic species occurs and the special relativity disappears again. This can serve as a low-energy window for the trans-Planckian physics.

This scenario can be applied to the massless neutrinos. The violation of the horizontal symmetry between the left-handed neutrino flavors can lead to the reentrant violation of Lorentz invariance at the very low energy. If the neutrinos remain massless at such an ultra-low-energy scale, then below this scale, two flavors – electronic and muonic left-handed neutrinos each with $N_3 = -1$ – hybridize and produce the $N_3 = 0$ fermion with the gap and the exotic gapless $N_3 = -2$ fermion with the essentially non-linear non-relativistic spectrum. This is another example of violation of the special relativity, which can also give rise to the neutrino oscillations. The previously considered effect of the violation of the special relativity on neutrino oscillations was related by Coleman and Glashow (1997) and Glashow *et al.* (1997) to the different speeds of light for different neutrino flavors. equivalence principle: different flavors are differently coupled to gravity (Majumdar *et al.* 2001; Gago *et al.* 2001).

Part III

Topological defects

13
TOPOLOGICAL CLASSIFICATION OF DEFECTS

We have seen that the effective metric and effective gauge fields are simulated in superfluids by the inhomogeneity of the superfluid vacuum. In superfluids many inhomogeneous configurations of the vacuum are stable and thus can be experimentally investigated in detail, since they are protected by **r**-space topology. In particular, the effect of the chiral anomaly, which will be discussed in Chapter 18, has been verified using such topologically stable objects as vortex-skyrmions in ^{3}He-A and quantized vortices in ^{3}He-B. Other topological objects can produce non-trivial effective metrics. In addition many topological defects have almost direct analogs in some RQFT. That is why we discuss here topologically stable objects in both phases of ^{3}He.

13.1 Defects and homotopy groups

The space-dependent configurations of the order parameter can be distributed into large classes, defined by their distinct topological invariants, or topological 'charges'. Due to conservation of topological charge, configurations from a given topological class cannot be continuously transformed into a configuration belonging to a different class, while continuous deformation between configurations within the given class is allowed by topology. The homogeneous state has zero topological charge (in **r**-space) and therefore configurations with non-zero charge are topologically stable: they cannot dissolve into the uniform vacuum state in a continuous manner.

In the broken-symmetry phase below T_c the system acquires a set R of degenerate equilibrium states. This set is the same in the entire temperature range $0 \leq T < T_c$, unless there is an additional symmetry breaking. That is why this set is called the vacuum manifold, referring to the $T = 0$ case. The space-dependent configurations of the order parameter define a mapping of the relevant part X of the coordinate space **r** (or space time $x^\mu = (\mathbf{r}, t)$) into the vacuum manifold R of the degenerate states. The relevant part of space (or of spacetime) depends on which type of configurations we are interested in.

In the case of vortices – and other linear defects – the relevant subspace X is a circle S^1 that encloses the defect line. Then the topological classes are defined by the elements of the first (fundamental) homotopy group, $\pi_1(\mathrm{R})$. They describe the classes of continuous mappings $S^1 \to \mathrm{R}$ of the closed circle S^1 around the defect line to the closed contour in the vacuum manifold.

In the same manner the point defects in 3D space (or the defects of co-dimension 3 in multi-dimensional space) are determined by classes of mapping $S^2 \to \mathrm{R}$ of a closed surface S^2 embracing the defect point into the vacuum manifold. These topological classes form the second homotopy group $\pi_2(R)$.

The non-singular point-like topological objects in 3D space – skyrmions – are determined by classes of mapping $S^3 \to R$, where S^3 is the compactified 3D space. The latter is obtained if the order parameter at infinity is homogeneous and thus the whole infinity is equivalent to one point. The topological classes form the third homotopy group $\pi_3(R)$. Such skyrmions are observed in liquid crystals (see the book by Kleman and Lavrentovich 2003). In RQFT with non-Abelian gauge fields this topological charge marks topologically different homogeneous vacua.

The same group $\pi_3(R)$ determines the point defects in spacetime – instantons (Belavin *et al.* 1975), and defects of co-dimension 4 in multi-dimensional space. Here the closed surface S^3 is around the point in 3+1 spacetime. The instanton represents the process of transition between two vacuum states with different topological charge (in RQFT), or the process of nucleation of the non-singular point-like topological object (in condensed matter). Usually in the literature it is assumed that the instanton is the quantum tunneling through the energy barrier. But in condensed matter this can be the classical process without any energy barrier.

The relevant subspaces X can be more complicated in the case of configurations described by the relative homotopy groups, such as solitons (Sec. 16.1.1); defects on surface (boojums) (Sec. 17.3); etc. The subspace is not closed, but has boundary ∂X. Due to, say, boundary conditions, the mapping $X \to R$ is accompanied by the mapping $\partial X \to \tilde{R}$, where $\tilde{R}$ is the subspace of the vacuum manifold R constrained by the boundary conditions.

13.1.1 Vacuum manifold

In the simplest cases the vacuum manifold R is fully determined by two groups, G and H. The first one is the symmetry group of the disordered state above T_c. This symmetry G is broken in the ordered state below T_c. Typically this group is broken only partially, which means that the vacuum remains invariant under some subgroup H of G, called the residual symmetry. Then the vacuum manifold R is the so-called coset space G/H.

This can be viewed in the following way. Let A_0 be the order parameter in some chosen vacuum state. Then all the degenerate vacuum states in the vacuum manifold can be obtained by symmetry transformations, i.e. through the action on A_0 of all the elements g of the group G: $A = gA_0$. However, A_0 is invariant under the action of the elements h of the residual symmetry group H: $A_0 = hA_0$. Thus all the elements h must be identified with unity, so that the actual space of the degenerate vacuum states – the vacuum manifold – is the factor space

$$R = G/H . \tag{13.1}$$

13.1.2 Symmetry G of physical laws in ^{3}He

For the superfluid ^{3}He the relevant group G is the symmetry of the Theory of Everything of this quantum liquids, i.e. the symmetry of the Hamiltonian in eqn

(3.2) which contains all the symmetries allowed in non-relativistic condensed matter. These include all the symmetries of the physical laws in non-relativistic physics, except for the Galilean invariance which is broken by the liquid: liquid has a preferred reference frame, where it is stationary. The normal liquid ^{3}He above the critical phase transition temperature T_c has the same symmetry G which thus contains the following subgroups.

The group of solid rotations of coordinate space, which we denoted as $SO(3)_\mathbf{L}$ to separate it from the spin rotations.

The spin rotations forming the group $SO(3)_\mathbf{S}$ can be considered as a separate symmetry operation if one neglects the spin–orbit interaction: the magnetic dipole interaction between the nuclear spins is about five orders of magnitude smaller in comparison to the energies characterizing the superfluid transition.

The group $U(1)_N$ of the *global* gauge transformations, which stems from the conservation of the particle number N for the ^{3}He atoms. $U(1)$ is an exact symmetry if one neglects extremely rare processes of the excitations or ionization of the atom, as well as the transformation of ^{3}He nuclei under external radiation.

In addition there is the translational symmetry, in which we are not interested at the moment: it is not broken in the superfluid state but becomes broken under rotation. We also ignore discrete symmetries for the moment, the space and time inversion, P and T.

Thus the continuous symmetries, whose breaking is relevant for the topological classification of the defects in the superfluid phases of ^{3}He, form the symmetry group

$$G = SO(3)_\mathbf{L} \times SO(3)_\mathbf{S} \times U(1)_N \ . \tag{13.2}$$

The broken-symmetry states of spin-triplet p-wave superfluidity below T_c are characterized by the order parameter, the 3×3 matrix $\mathbf{A} = e_{\alpha i}$, which transforms as a vector under a spin rotation for given orbital index (i), and as a vector under an orbital rotation for given spin index (α). The transformation of $e_{\alpha i}$ under the action of the elements of the group G (see e.g. eqn (7.47)) can be written in the following symbolic form:

$$\mathbf{GA} = e^{2i\alpha} R_\mathbf{S} \mathbf{A} R_\mathbf{L}^{-1} \ , \tag{13.3}$$

where α is the parameter of the *global* gauge transformation; $R_\mathbf{S}$ and $R_\mathbf{L}$ are matrices of spin and orbital rotations.

13.1.3 *Symmetry breaking in ^{3}He-B*

In ^{3}He-B the symmetry G in eqn (13.2) is broken in the following way: the $U(1)$ group is broken completely, while the product of two other groups breaks down to the diagonal subgroup:

$$G = SO(3)_\mathbf{L} \times SO(3)_\mathbf{S} \times U(1)_N \to H_\mathbf{B} = SO(3)_\mathbf{S+L} \ . \tag{13.4}$$

The residual symmetry group $H_\mathbf{B}$ of ^{3}He-B is the symmetry under simultaneous rotations of spin and orbital spaces.

For the triplet p-wave pairing, this is the only possible superfluid phase which has an isotropic gap in the quasiparticle spectrum and thus no gap nodes. The order parameter is isotropic: in the simplest realization it has the form

$$e^{(0)}_{\alpha i} = \Delta_0 \delta_{\alpha i} \ . \tag{13.5}$$

All other degenerate states of the B-phase vacuum in eqn (7.47) are obtained by the action of the symmetry group G:

$$(\mathbf{G A}^{(0)})_{\alpha i} = \Delta_0 e^{i\Phi} R_{\alpha i} \ . \tag{13.6}$$

Here Φ is the phase of the order parameter, which manifests the breaking of the $U(1)_N$ group. $R_{\alpha i}$ is the real 3×3 matrix obtained from the original unit matrix in eqn (13.5) by spin and orbital rotations: $R_{\alpha i} = (R_\mathbf{S} R_\mathbf{L}^{-1})_{\alpha i}$. Matrices $R_{\alpha i}$ characterizing the vacuum states span the group $SO(3)$ of rotations. They can be expressed in terms of the direction $\hat{\mathbf{n}}$ of the rotation axis, and angle θ of rotation about this axis:

$$R_{\alpha i}(\hat{\mathbf{n}}, \theta) = (1 - \cos\theta)\delta_{\alpha i} + \hat{n}_\alpha \hat{n}_i \cos\theta - \varepsilon_{\alpha i k}\hat{n}_k \sin\theta \ . \tag{13.7}$$

The vacuum manifold of ^{3}He-B is thus the product of the $SO(3)$-space of matrices and of the $U(1)$-space of the phase Φ: $R_B = SO(3) \times U(1)$, which reflects the general relation in eqn (13.1),

$$R_B = G/H_B = SO(3) \times U(1) \ . \tag{13.8}$$

^{3}He-B has four Goldstone bosons: one propagating mode of the phase Φ (sound) and three propagating modes of the matrix $R_{\alpha i}$ (spin waves). The number of Goldstone bosons corresponds to the dimension of the vacuum manifold R_B. This is a general rule which, however, can be violated in the presence of hidden symmetry discussed in Sec. 9.3.3.

13.2 Analogous 'superfluid' phases in high-energy physics

There are several models in high-energy physics with a similar pattern of symmetry breaking. Each of the discussed groups G contains $U(1)$ and the product of two similar groups ($SU(2) \times SU(2)$, $SU(3) \times SU(3)$, etc.). In the phase transition the $U(1)$ group is broken completely, while the product of two other groups breaks down to the diagonal subgroup.

13.2.1 Chiral superfluidity in QCD

The analogy between the chiral phase transition in QCD and the superfluid transition to ^{3}He-B has been discussed in the book by Vollhardt and Wölfle (1990), pages 172–173. In both systems the rotational symmetry is doubled when small terms in the Lagrangian are omitted. In ^{3}He, the exact symmetry is the rotational symmetry $SO(3)_\mathbf{J}$, where $\mathbf{J} = \mathbf{L} + \mathbf{S}$. Since the spin–orbit interaction is very small, the group $SO(3)_\mathbf{J}$ can be extended to the group $SO(3)_\mathbf{L} \times SO(3)_\mathbf{S}$

of separate orbital and spin rotations, which is an approximate symmetry. After the phase transition to the broken-symmetry state the approximate extended symmetry is spontaneously broken back to $SO(3)_\mathbf{J}$. But at the end of this closed route $SO(3)_\mathbf{J} \to SO(3)_\mathbf{L} \times SO(3)_\mathbf{S} \to SO(3)_\mathbf{J}$ one obtains: (i) the gap in the fermionic spectrum; (ii) Goldstone and/or pseudo-Goldstone bosons with small gap dictated by the spin–orbit interaction (see the book by Weinberg 1995); and (iii) topological defects – solitons terminating on strings (Sec. 14.1.5). The source of all these is the small parameter $E_D^2/E_{\text{Planck}}^2$, the relative magnitude of the spin–orbit interaction.

Exactly the same thing occurs in the effective theory of nuclear forces arising in the low-energy limit of QCD. There exists an 'exact' global group $SU(2)$ – the symmetry of nuclear forces with respect to the interchange of the proton and neutron – called the isotopic spin symmetry (not to be confused with the local group $SU(2)_L$ of weak interactions). The proton and neutron form the isodoublet of this global group. It appears that this symmetry can be extended to the approximate symmetry which is the product of two global groups $SU(2)_\mathbf{L} \times SU(2)_\mathbf{R}$ of separate isorotations of left and right quarks. The reason for that is that the 'bare' masses of u and d quarks ($m_u \sim 4$ MeV and $m_d \sim 7$ MeV) are small compared to the QCD scale of 100–200 MeV. If the masses are neglected, one obtains two isodoublets of chiral quarks, left (u_L, d_L) and right (u_R, d_R), which can be independently transformed by the $SO(3)_\mathbf{L}$ and $SO(3)_\mathbf{R}$ groups.

As in ^{3}He-B, the symmetry-breaking scheme in low-energy QCD also contains the global $U(1)$ symmetry appropriate for the massless quarks: $(u_L, d_L) \to (u_L, d_L)e^{i\alpha}$, $(u_R, d_R) \to (u_R, d_R)e^{-i\alpha}$. This chiral symmetry, denoted $U(1)_A$, is approximate and is violated by the chiral anomaly. This is somewhat similar to ^{3}He, where strictly speaking the $U(1)_N$ symmetry is approximate, since the number of ^{3}He atoms is not conserved because of the possibility of chemical and nuclear reactions. Thus in this model the extended approximate group of low-energy QCD is the global group (see the book by Weinberg 1995)

$$G = SU(2)_\mathbf{L} \times SU(2)_\mathbf{R} \times U(1)_A . \quad (13.9)$$

It is assumed that in the chiral phase transition, which occurs at $T_c \sim 100$–200 MeV, the $U(1)_A$ symmetry is broken completely, while the $SU(2)_\mathbf{L} \times SU(2)_\mathbf{R}$ symmetry is broken back to their diagonal subgroup $SU(2)_\mathbf{L+R}$. This fully reproduces the symmetry-breaking pattern in ^{3}He-B:

$$SU(2)_\mathbf{L} \times SU(2)_\mathbf{R} \times U(1)_A \to SU(2)_\mathbf{L+R} . \quad (13.10)$$

This symmetry breaking gives rise to pseudo-Goldstone bosons – pions, and topological defects – strings and domain walls terminating on strings. Mixing the left and right quarks by the order parameter generates the quark masses, which are larger than their original masses generated in the electroweak transition and ignored in this theory.

The quark–antiquark chiral condensates in the state with chiral superfluidity, $\langle u_L \bar{u}_R \rangle$, $\langle d_L \bar{d}_R \rangle$, $\langle u_L \bar{d}_R \rangle$, $\langle d_L \bar{u}_R \rangle$, form the 2×2 matrix order parameter. Its simplest form and the general form obtained by symmetry transformations are

$$\mathbf{A}^{(0)} = \Delta \tau^0 , \quad (13.11)$$
$$\mathbf{A} = g\mathbf{A}^{(0)} = \left(\sigma\tau^0 + i\pi_b \cdot \tau^b\right) e^{i\eta} , \quad \sigma^2 + \vec{\pi}^2 = \Delta^2 \quad (13.12)$$

Here τ^0 and τ^b with $b = (1, 2, 3)$ are the Pauli matrices in isospin space; Δ is the magnitude of the order parameter, which determines the dressed mass of u and d quarks; and η is the phase of the chiral condensate: $\eta \to \eta + 2\alpha$ under transformations from the chiral symmetry group $U(1)_A$. In the hypothetical ideal case, when the initial symmetry G is exact, the situation is similar to ^{3}He-B with four Goldstone bosons: the η mode – the so-called η' meson – is analogous to sound waves in ^{3}He-B; and three pions $\vec{\pi}$, analogs of three spin waves in ^{3}He-B. Since the original symmetry G is approximate, the Goldstone bosons are massive. The masses of these pseudo-Goldstone modes are small, since they are determined by small violations of G, except for the η' meson, whose mass is bigger due to chiral anomaly. This is similar to one of the spin-wave modes in ^{3}He-B which has a small gap due to the spin–orbit interaction violating G.

13.2.2 *Chiral superfluidity in QCD with three flavors*

The extended $SU(3)_\text{Flavur}$ model of the chiral QCD transition includes three quark flavors u, d and s, assuming that the mass of the strange quark s is also small enough. Now $SU(3)_\mathbf{L} \times SU(3)_\mathbf{R}$ is spontaneously broken to the diagonal subgroup $SU(3)_\text{Flavor}$:

$$SU(3)_\mathbf{L} \times SU(3)_\mathbf{R} \to SU(3)_\text{Flavor} . \quad (13.13)$$

13.2.3 *Color superfluidity in QCD*

The similar symmetry-breaking pattern for the color superfluidity of the quark condensate $\langle qq \rangle$ in dense baryonic matter was discussed by Ying (1998), Alford *et al.* (1998) and Wilczek (1998). As distinct from the $\langle q\bar{q} \rangle$ condensates discussed above the Cooper pairs $\langle qq \rangle$ carry color, whence the name color superconductivity. The original approximate symmetry group above the superfluid phase transition can be, for example,

$$G = SU(3)_C \times SU(3)_\text{Flavor} \times U(1)_B , \quad (13.14)$$

where $SU(3)_C$ is the local group of QCD, and $U(1)_B$ corresponds to the conservation of the baryon charge. In the symmetry-breaking scheme

$$SU(3)_C \times SU(3)_\text{Flavor} \times U(1)_B \to SU(3)_{C+F} , \quad (13.15)$$

the breaking of the baryonic $U(1)_B$ group manifests the superfluidity of the baryon charge in the baryonic quark matter, while the breaking of the color group of QCD manifests the superfluidity of the color. In the more extended theories, eqn (13.13) and eqn (13.15) are combined: $SU(3)_\mathbf{L} \times SU(3)_\mathbf{R} \times SU(3)_C \to SU(3)_{C+\mathbf{L}+\mathbf{R}}$.

14
VORTICES IN ^{3}He-B

14.1 Topology of ^{3}He-B defects
14.1.1 Fundamental homotopy group for ^{3}He-B defects

The homotopy group describing the linear topological defects in ^{3}He-B is

$$\pi_1(G/H_B) = \pi_1(U(1)) + \pi_1(SO(3)) = Z + Z_2 \ . \tag{14.1}$$

Here Z is the group of integers, and Z_2 contains two elements, 1 and 0, with the summation law $1+1=0$. These elements describe the defects with singular core, which means that inside this core of coherence length ξ the order parameter is no longer in the vacuum manifold of ^{3}He-B. Such a core is also called the hard core to distiguish it from the smooth or soft cores of continuous structures. Since $\pi_2(G/H_B) = 0$, in ^{3}He-B there are no topologically stable hard-core point defects – hedgehogs. The hedgehogs with non-singular (soft) core will be discussed later in Sec. 14.2.5.

14.1.2 Mass vortex vs axion string

The group Z of integers in eqn (14.1) describes the conventional singular vortices with integer winding number n_1 of the phase Φ of the order parameter (13.6) around the vortex core. The simplest realization of such vortices is $\Phi(\mathbf{r}) = n_1\phi$, where ϕ is an azimuthal angle in the cylindrical coordinate frame. In superfluid ^{3}He-B, the superfluid velocity $\mathbf{v}_s = (\hbar/2m)\nabla\Phi$ characterizes the superfluidity of mass carried by ^{3}He atoms: the mass flow of the superfluid vacuum is $\mathbf{P} = mn\mathbf{v}_s$. Thus the vortices of the Z group have circulating mass flow around the core and are called the mass vortices. The mass vortex carries the quantized circulation $\kappa = \oint d\mathbf{r} \cdot \mathbf{v}_s = n_1\kappa_0$, with $\kappa_0 = \pi\hbar/m$. Mass vortices with $n_1 = 1$ form a regular array in the rotating vessel (see the cluster of mass vortices in Fig. 14.1). Within the cluster the average superfluid velocity obeys the solid-body rotation $\langle\mathbf{v}_s\rangle = \mathbf{v}_n = \mathbf{\Omega}\times\mathbf{r}$, where $\mathbf{\Omega}$ is the angular velocity of rotation. The areal density of circulation quanta in the cluster has a value

$$n_v = \frac{1}{S\kappa_0}\int_S d\mathbf{S}\cdot(\nabla\times\mathbf{v}_s) = \frac{2\Omega}{\kappa_0} \ , \tag{14.2}$$

In the chiral condensate phase of QCD, the mass vortices are equivalent to the η'-vortices or axion strings around which the supercurrent of chiral charge is circulating (see e.g. Zhang et al. 1998) Defect formation during the non-equilibrium phase transition into the state with the broken chiral symmetry was discussed by Balachandran and Digal (2002). This is analogous to the mechanism of vortex formation in ^{3}He-B which will be discussed in Chapter 28.

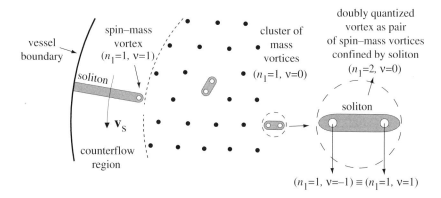

FIG. 14.1. Vortices in rotating ^{3}He-B. Mass vortices form a regular structure like Abrikosov vortices in an applied magnetic field. If the number of vortices is less than equilibrium number for given rotation velocity, vortices are collected in the vortex cluster. Within the cluster the average superfluid velocity $\langle \mathbf{v}_s \rangle = \mathbf{v}_n$. On the periphery there is a region void of vortices – the counterflow region, where $\langle \mathbf{v}_s \rangle \neq \mathbf{v}_n$. Spin–mass vortices with $n_1 = 1$, $\nu = 1$ can be created and stabilized in the rotating vessel. The confining potential produced by the soliton wall is compensated by logarithmic repulsion of vortices forming the vortex pair – the doubly quantized vortex with $n_1 = 2$, $\nu = 0$ inside the cluster. A single spin–mass vortex is stabilized at the periphery of the cluster by the combined effect of soliton tension and Magnus force in eqn (18.26).

14.1.3 Spin vortices vs pion strings

The Z_2 group in ^{3}He-B describes the singular spin vortices, with the summation rule $1 + 1 = 0$ for the topological charge ν. This summation rule means that unlike vortices of the Z group, the Z_2-vortex coincides with its antivortex, in other words it is unoriented. In the simplest realization of spin vortices, the orthogonal matrix of the B-phase vacuum (13.6) has the following form far from the vortex core:

$$R_{\alpha i}(\phi) = (1 - \cos\theta(\phi))\hat{z}_\alpha \hat{z}_i + \delta_{\alpha i}\cos\theta(\phi) - \varepsilon_{\alpha ik}\hat{z}_k \sin\theta(\phi) \quad (14.3)$$

$$= \begin{pmatrix} 1 & 0 & 0 \\ 0 & \cos\theta(\phi) & -\sin\theta(\phi) \\ 0 & \sin\theta(\phi) & \cos\theta(\phi) \end{pmatrix}. \quad (14.4)$$

Here $\theta(\phi) = \nu\phi$ with integer ν. Around such a vortex there is a circulation of the spin current $\propto \nabla\theta$, hence the name spin vortex. The topologically stable vortex corresponds to $\nu = 1$, i.e. to 2π rotation around the string axis. The rule $1 + 1 = 0$ means that the spin vortex with winding number $\nu = 2$ (i.e. with 4π rotation around the string), or with any even ν, is topologically unstable and can be continuously unwound. Spin vortices with odd ν can continuously transform to the $\nu = 1$ spin vortex.

In the chiral condensate, where the groups $SU(2)$ substitute the groups $SO(3)$ of ^{3}He, the homotopy group $\pi_1(G/H) = \pi_1(U(1)) = Z$ supports the topological stability only for the η'-vortices. There is no additional Z_2 group. This is because in the $SO(3)$ group the 2π rotation is the identity transformation, while in the $SU(2)$ group the identity transformation is the 4π rotation. But the string with 4π rotation around the core is equivalent to the topologically unstable $\nu = 2$ spin vortex in ^{3}He-B and can be continuously unwound. This means that any pion vortex, with any winding of the $\vec{\pi}$-field around the core, is topologically unstable. The topologically unstable pion strings, nevertheless, can survive under special conditions, see e.g. Zhang et al. (1998). In the simplest realization of the pion string the order parameter in eqn (13.12) is

$$\pi_1 = \pi_2 = 0 \,, \quad \sigma + i\pi_3 = \frac{\Delta(r)}{\sqrt{2}} e^{i\phi} \,. \tag{14.5}$$

Such a solution can be locally stable due to its symmetry, but this stability is rather ephemeral since it is not supported by the topology.

14.1.4 Casimir force between spin and mass vortices and composite defect

Mass vortices are stabilized by rotation of the liquid forming the vortex cluster, and thus can be investigated experimentally. This is not the case for spin vortices: there is no such external field as rotation in superfluids, or magnetic field in superconductors, which can regulate the position of the spin vortex in the sample. However, the spin vortex has been observed because of its two unique properties: one of them is that the spin vortex can be pinned by the mass vortex.

Mass and spin vortices do not interact significantly – they 'live in different worlds', since they are described by non-interacting Φ and θ fields, respectively. Let us for simplicity fix the Goldstone variable $\hat{\mathbf{n}}$ in eqn (13.7), leaving only two Goldstone fields Φ and θ. Then from the gradient energy in eqn (10.3) one obtains the following London energy density for the remaining Goldstone fields:

$$E_{\text{London}} = n_s \frac{\hbar^2}{8m} (\nabla \Phi)^2 + n_s^{spin} \frac{\hbar^2}{8m} (\nabla \theta)^2 \,. \tag{14.6}$$

Here n_s^{spin} is spin rigidity, which enters the spin current $\frac{\hbar^2}{4m} n_s^{spin} \nabla \theta$.

Let us consider two rectilinear defects of length L: the mass vortex with charge n_1 (i.e. $\Phi = n_1 \phi$) and the spin vortex with charge ν (i.e. $\theta = \nu\phi$; since $\hat{\mathbf{n}}$ is fixed the spin vortex in eqn (14.4) can have any integer winding number). Equation (14.6) contains no interaction term between the two defects, so the total energy of two defects is simply the sum of two energies:

$$E(n_1, \nu) = E(n_1) + E(\nu) = L \left(n_1^2 n_s + \nu^2 n_s^{spin} \right) \frac{\pi \hbar^2}{4m} \ln \frac{R_0}{\xi} \,. \tag{14.7}$$

Here R_0 is the external (infrared) cut-off of the logarithmically divergent integral, which is given by the size of the vessel, and the coherence length ξ is the core size which provides the ultraviolet cut-off. The London energy is valid only for scales above ξ, while in the core the deformations with the scale of order ξ drive the system out of the vacuum manifold of the B-phase.

The most surprising property of the energy (14.7) is that it does not depend on the distance between the vortices. It is very similar to the case when two particles are charged, but their charges correspond to different gauge fields. Another analog corresponds to electrons of two kinds: one is the conventional electron, while the other one belongs to mirror matter, which can exist if the parity is an unbroken symmetry of nature (Silagadze 2001; Foot and Silagadze 2001). The electron and mirror electron interact gravitationally, and also through the Casimir effect.

In our case it is the Casimir effect that is important. Each string disturbs the vacuum, and this disturbance influences another string. Such Casimir interaction can be described by the higher-order gradient term $\sim n_s(\hbar^2\xi^2/m)(\nabla\Phi)^2(\nabla\theta)^2$. Then the Casimir interaction and the Casimir force between spin and mass vortices, parallel to each other, are

$$E_{\text{Casimir}}(R) = E(n_1,\nu) - E(n_1) - E(\nu) \sim -n_1^2\nu^2 L \frac{n_s\hbar^2\xi^2}{mR^2} \,, \qquad (14.8)$$

$$F_{\text{Casimir}}(R) = -\partial_R E_{\text{Casimir}} \sim -n_1^2\nu^2 L \frac{E_F^2}{\Delta_0^2}\frac{\hbar v_F}{R^3} \,, \qquad (14.9)$$

where R is the distance between the vortex lines. If one disregards the difference between Planck scales in superfluid ^{3}He, i.e. assumes that $v_F = c$ in eqn (7.48), one obtains $F_{\text{Casimir}} \sim -n_1^2\nu^2 L\hbar c/R^3$. This is reminiscent of the Casimir force between two conducting plates, $F_{\text{Casimir}} \sim -L^2\hbar c/R^4$, where L^2 is the area of the plates.

For comparison, the Casimir force between two point defects (hedgehogs or global monopoles) 'living in different worlds', i.e. described by different fields, would be $F_{\text{Casimir}}(R) \propto -\hbar c/R^2$, which has the same R-dependence as the gravitational attraction between two point particles.

The Casimir attraction between spin and mass vortices is small at large distances, but becomes essential when R is of order of the core size ξ. It is energetically preferable for the two defects to form a common core: according to Thuneberg (1987a) by trapping the spin vortex on a mass vortex the combined core energy is reduced. As a result the two strings form a composite linear defect – the spin–mass vortex – characterized by two non-zero topological quantum numbers, $n_1 = 1$ and $\nu = 1$. Because of the mass-vortex constituent of this composite defect, it is influenced by the Magnus force in eqn (18.26). It is stabilized in a rotating vessel due to another peculiar property of the spin-vortex constituent.

14.1.5 *Spin vortex as string terminating soliton*

Since both the $SO(3)_{\mathbf{S}}$ part of the group G in ^{3}He and the global $SU(2)$ or $SU(3)$ groups of chiral symmetries in QCD are approximate, some or all of the

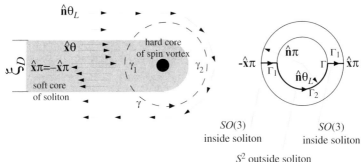

FIG. 14.2. Spin vortex as the termination line of soliton in ^{3}He-B. *Left*: The field of the vector $\hat{\mathbf{n}}\theta$, whose direction shows the axis of rotation, and the magnitude – the angle of rotation. In the core of the soliton the angle θ deviates from its 'magic' value $\theta_L \approx 104°$. *Right*: The $SO(3)$ space of the vector $\hat{\mathbf{n}}\theta$. Solid line Γ is the contour drawn by the vector $\hat{\mathbf{n}}\theta$ when circling around the spin vortex along the contour γ. Γ is closed since vectors $\hat{\mathbf{n}}\theta$ and $-\hat{\mathbf{n}}\theta$ are identical in the $SO(3)$ space of rotations.

Goldstone bosons in these systems have a mass. In ^{3}He-B the spin–orbit (dipole–dipole) energy explicitly depends on the degeneracy parameter θ in eqn (14.3):

$$F_D = g_D \left(\cos\theta + \frac{1}{4}\right)^2. \qquad (14.10)$$

It fixes the value of θ at the 'magic' Leggett angle $\theta_L \approx 104°$, and as a result the θ-boson – the spin-wave mode in which θ oscillates – acquires the mass. The spin–orbital energy reduces the vacuum manifold $SO(3)$ of the matrix $R_{\alpha i}$ in eqn (13.7) to the spherical surface $S^2 = SO(3)_\mathbf{J}/SO(2)_{J_3}$ of unit vector $\hat{\mathbf{n}}$. The parameter g_D is connected with the dipole length ξ_D in eqn (12.5) as $g_D \sim n_s \hbar^2 / m \xi_D^2 \sim n_s M$, where M is the Dirac mass of chiral fermions in the planar state in eqn (12.5) induced by the spin–orbit interaction.

The 'pion' mass term in eqn (14.10) essentially modifies the structure of the spin vortex: the region where θ deviates from the Leggett angle shrinks to the soliton – the planar object terminating on the spin vortex (Fig. 14.2 *left*). The closed path γ around the spin vortex in Fig. 14.2 *left* is mapped to the contour Γ in $SO(3)$ space of the vector $\hat{\mathbf{n}}\theta$, whose direction shows the axis of rotation, and the magnitude – the angle of rotation (Fig. 14.2 *right*). This contour Γ is closed since the pair of points $\hat{\mathbf{n}}\pi$ and $-\hat{\mathbf{n}}\pi$ are identical: they correspond to the same matrix $R_{\alpha i}$ of the ^{3}He-B order parameter in eqn (13.7). The contour Γ cannot be shrunk to a point by smooth perturbations of the system; this demonstrates the topological stability of the spin vortex with $\nu = 1$.

Let us now discuss the topological stability of the soliton. On the arc γ_2 of the contour γ in the region outside the soliton, the system is in the vacuum manifold S^2 determined by the minimum of eqn (14.10). This arc is thus mapped to the arc Γ_2 of Γ along the spherical surface S^2 of radius θ_L. On the arc γ_1 crossing the soliton, the order parameter leaves the S^2 sphere and varies in the larger $SO(3)$ space. For a given field configuration, γ_1 maps to the horizontal segment Γ_1. Though the segment Γ_1 is not closed, the mapping $\gamma_1 \to \Gamma_1$ is non-trivial: Γ_1 cannot shrink to a point because of the requirement that the segment Γ_1 must terminate on the sphere S^2 of the vacuum manifold outside the soliton. As a result the soliton is stable. This is an illustration of the non-triviality of the relative homotopy group $\pi_1(SO(3), S^2) = Z_2$, which provides the topological stability of solitons in ^{3}He-B. More on topological solitons and on the relative homotopy groups will be given in Sec. 16.1.2.

The topologically similar phenomenon of cosmic walls bounded by cosmic strings was discussed by Hindmarsh and Kibble (1995) (see also Chapter 17).

14.1.6 Topological confinement of spin–mass vortices

In Sec. 14.1.4 we found that spin and mass vortices attract each other by Casimir forces and form a composite defect – the spin–mass vortex with $n_1 = 1$ and $\nu = 1$. Since this object contains the spin vortex, according to the results of the previous section 14.1.5 it has a solitonic tail. Such a topological confinement of the linear defect (the mass vortex) and the planar defect (the soliton) allows us to stabilize the composite object under rotation.

Two configurations containing the spin–mass vortex–soliton were experimentally observed by Kondo et al. (1992). In the first configuration (Fig. 14.1 bottom) two spin–mass vortices form the vortex pair, with the confining potential produced by the tension of the soliton wall between them. The confining potential is proportional to the distance R between the spin–mass vortices. On the other hand there is a logarithmic interaction of the vortex 'charges': the $n_1^{(1)}$ charge of the first vortex interacts with the $n_1^{(2)}$ of the second one, and the same logarithmic interaction exists between the ν charges of vortices. In the simplest case of fixed variable $\hat{\mathbf{n}}$ the total interaction between the two spin–mass vortices is

$$E(n_1^{(1)}, \nu^{(1)}; n_1^{(2)}, \nu^{(2)}) - E(n_1^{(1)}, \nu^{(1)}) - E(n_1^{(2)}, \nu^{(2)})$$
$$= \beta L R - L \left(n_1^{(1)} n_1^{(2)} n_s + \nu^{(1)} \nu^{(2)} n_s^{spin} \right) \frac{\pi \hbar^2}{2m} \ln \frac{R}{\xi} \, , \quad \beta \sim g_D \xi_D \, . (14.11)$$

In the experimentally observed vortex pairs, the spin–mass vortices have charges $(n_1^{(1)}, \nu^{(1)}) = (1, 1)$ and $(n_1^{(2)}, \nu^{(2)}) = (1, -1)$. Since $n_s > n_s^{spin}$ the repulsion of like charges n_1 prevails over the attraction of unlike charges ν. The overall logarithmic repulsion of charges and the linear attractive potential due to the soliton produce the equilibrium distance R between spin–mass vortices in the molecule, which is about several ξ_D. The molecule as a whole has topological charges $n_1 = 2$ and $\nu = 0$, which means that it represents the doubly quantized mass vortex. As a vortex, in the rotating sample it finds its position among the other vortices within the vortex cluster (Fig. 14.1 top).

In the other experimental realization of the composite defect (Fig. 14.1 left), the position of the spin–mass vortex is stabilized at the edge of the vortex cluster, while the second end of the soliton is attached to the wall of the vessel. In this case the Magnus force acting on the mass-vortex part of the object pushes the vortex toward the cluster, and thus compensates the soliton tension which attracts the vortex to the wall of the container. The soliton cannot be unpinned from the wall of the vessel, since this requires nucleation of a singularity – spin or spin–mass vortex. In superfluid ^{3}He, processes forming singular defects are highly suppressed because of huge energy barriers as compared to temperature (see Sec. 26.3.2).

14.2 Symmetry of defects

14.2.1 Topology of defects vs symmetry of defects

The first NMR measurement on rotating ^{3}He-B by Ikkala et al. (1982) revealed a first order phase transition in the vortex core structure. Other transitions related to the change of the structure (and even topology) of the individual topological defects have been identified in ^{3}He-A since then, but the B-phase transition remains the most prominent one. The first-order phase transition line in Fig. 14.3 separates two regions in the phase diagram. In both regions the vortex with the lowest energy has the same winding number $n_1 = 1$, i.e. the same topology. But the symmetry of the core of the vortex is different in the two regions. This was later experimentally verified by Kondo et al. (1991) who observed the Goldstone mode in the core of the vortex in the low-T part of the phase diagram. The Goldstone boson arises from the breaking of continuous symmetry; in a given case it is the axial symmetry which is spontaneously broken in the core: the core becomes anisotropic and the direction of the anisotropy axis is the soft Goldstone variable. This discovery illustrated that in ^{3}He-B the mass vortices have a complex core structure, unlike in superfluid ^{4}He or conventional s-state superconductors, where the superfluid order parameter amplitude goes to zero on approaching the center of the vortex core.

This observation required fine consideration of the vortex core structure. The resulting general scheme of defect classification in the ordered media, which includes the symmetry of defects, has three major steps. The first step is the symmetry classification of possible homogeneous ordered media in terms of the broken symmetry. This actually corresponds to the enlisting of the possible subgroups H of the symmetry group G. For instance, to enumerate all the possible crystal and liquid-crystal states, it is sufficient to find all the subgroups of the Euclidean group. In the case of superconductivity, where in addition the electromagnetic $U(1)_Q$ group is broken, the classification of the superconductivity classes in crystals was carried out by Volovik and Gor'kov (1985).

At the second stage, for a given broken-symmetry state, i.e. for given symmetry H, the topological classification of defects is made using the topological properties of the vacuum manifold R = G/H. This topological classification distributes configurations into big classes. Such a classification is too general and does not exhibit much information on the distribution of the order parameter outside or inside the core. Within a given topological class one can find many dif-

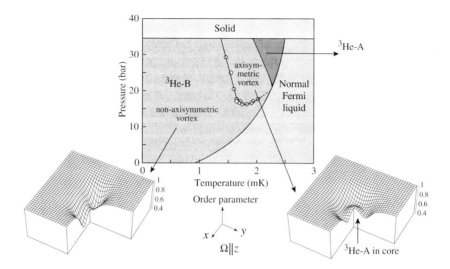

FIG. 14.3. *Top*: Experimental phase diagram of vortex core states in ^{3}He-B. Both vortices have winding number $n_1 = 1$, but the low-T vortex has broken axisymmetry in the core which was verified by observation by Kondo *et al.* (1991) of the Goldstone mode related to broken axial symmetry. *Bottom*: Normalized amplitude of the order parameter in the core of symmetric (*right*) and asymmetric (*left*) vortices according to calculations by Thuneberg (1986) in the Ginzburg–Landau region. Note that the order parameter is nowhere zero. Schematic illustration of the core structure can be also found in Fig. 14.5.

ferent solutions, and even the phase transitions between different configurations with the same topological charge as in Fig. 14.3.

Such a phase transition between the defects can also be described in terms of the symmetry breaking, and here it is the symmetry of the defect that is important. The third step is thus the fine classification of defects in terms of the symmetry group.

To get an idea of how the symmetry classification of defects works, let us consider two examples: symmetry of the vortex in superfluid ^{4}He or ^{3}He-B (see the review by Salomaa and Volovik 1987); and symmetry of hedgehogs in ferromagnets or, which is similar, the symmetry of the 't Hooft–Polyakov magnetic monopoles in the $SU(2)$ theory (Wilkinson and Goldhaber 1977). The symmetry classification of disclination lines in liquid crystals has been considered in detail by Balinskii *et al.* (1984).

14.2.2 *Symmetry of hedgehogs and monopoles*

In the homogeneous phase transition of a paramagnetic material to the ferromagnetic state with spontaneous magnetization $\mathbf{M} = M\hat{\mathbf{m}}$, the $G = SO(3)_\mathbf{S}$ symmetry of global spin rotations is broken to its subgroup $H = SO(2) \equiv U(1)$

of rotations around the direction $\hat{\mathbf{m}}$ of spontaneous magnetization. The vacuum manifold R = G/H = S^2 is the spherical surface of unit vector $\hat{\mathbf{m}}$. It has non–trivial second homotopy group $\pi_2(S^2) = Z$ describing hedgehogs with integer topological charges n_2:

$$n_2 = \frac{1}{8\pi} e^{ijk} \int_{\sigma_2} dS_k \, \hat{\mathbf{m}} \cdot \left(\frac{\partial \hat{\mathbf{m}}}{\partial x^i} \times \frac{\partial \hat{\mathbf{m}}}{\partial x^j} \right) . \qquad (14.12)$$

It is the winding number of mapping of the sphere σ_2 around the singular point in **r**-space to the 2-sphere of unit vector $\hat{\mathbf{m}}$.

In the simplest model of the 't Hooft–Polyakov magnetic monopoles, one has an analogous scheme of symmetry breaking G = $SU(2) \to$ H = $U(1)_Q$. This leads to the stable hedgehogs. Since the $SU(2)$ group under consideration is local, the hedgehog represents topologically stable magnetic monopoles in the emerging $U(1)_Q$ electrodynamics.

To discuss the possible symmetries of hedgehogs (or magnetic monopoles) we must consider the phase transition from the paramagnetic state to a ferromagnetic state in which a single hedgehog (or magnetic monopole) is present. In the presence of the space-dependent texture the space itself becomes inhomogeneous and anisotropic, i.e. the group $SO(3)_\mathbf{L}$ of the rotations in the coordinate space is broken by the hedgehog/monopole. Thus both the group $SO(3)_\mathbf{S}$ of rotations in 'isotopic' space and the group $SO(3)_\mathbf{L}$ of rotations in coordinate space are broken in the state with the defect.

What is left? This can be seen from the simplest ansatz for the hedgehog with $n_2 = 1$ far from the core, where $M = M_0$ and $\hat{\mathbf{m}}(\mathbf{r}) = \hat{\mathbf{r}}$. This configuration has the symmetry group $SO(3)_{\mathbf{L}+\mathbf{S}}$. Thus in the presence of the simplest hedgehog/monopole the symmetry-breaking scheme is

$$G' = SO(3)_\mathbf{L} \times SO(3)_\mathbf{S} \to SO(3)_{\mathbf{S}+\mathbf{L}} . \qquad (14.13)$$

$SO(3)_{\mathbf{L}+\mathbf{S}}$ represents the maximum possible symmetry of the asymptote of the order parameter far from the core of the hedgehog/monopole. Since this symmetry is the maximum possible, it can be extended to the core region. In other words, among all the possible solutions of the corresponding Euler–Lagrange equations there always exists the solution with the most symmetric configuration of the core. This spherically symmetric solution for the magnetization vector $\mathbf{M}(\mathbf{r})$ has the form $\mathbf{M}(\mathbf{r}) = M(r)\hat{\mathbf{r}}$ with $M(r = 0) = 0$ and $M(r = \infty) = M_0$. Other degenerate solutions are obtained by rotation in spin space (in isotopic spin space in the case of the $SU(2)$ 't Hooft–Polyakov magnetic monopole). Though the most symmetric solutions are always the stationary points of the energy functional, they do not necessarily represent the local minimum of the energy. They can be the saddle point, and in this case the energy can be reduced by spontaneous breaking of the maximal symmetry $SO(3)_{\mathbf{L}+\mathbf{S}}$. In nematic liquid crystals, in some region of external parameters of the system, the symmetric hedgehog becomes unstable. It loses the spherical symmetry and transforms, for example, to a small disclination loop with the same global charge $n_2 = 1$ (see

Lubensky et al. (1997); the book by Mineev (1998); and the book by Kleman and Lavrentovich (2003)). Similarly, the $\hat{\mathbf{d}}$-hedgehog in ^{3}He-A (Sec. 15.2.1) can be unstable toward the loop of the Alice string – the half-quantum vortex in Secs 15.3.1 and 15.3.2. Also the breaking of the maximum possible symmetry group can occur in the core of magnetic monopoles (Axenides et al. 1998).

14.2.3 Spherically symmetric objects in superfluid ^{3}He

In superfluid ^{3}He the relevant symmetry group G′ for the point defects coincides with the group G in eqn (13.2): G′ = G. This is because the group $SO(3)_L$ of coordinate rotations is already included in G. The general structure of the spherically symmetric object in superfluid ^{3}He, i.e. the object with the symmetry $SO(3)_{L+S}$, is (up to a constant rotation in spin space)

$$e_{\alpha i}(\mathbf{r}) = a(r)(\delta_{\alpha i} - \hat{r}_\alpha \hat{r}_i) + b(r)\hat{r}_\alpha \hat{r}_i - c(r)\varepsilon_{\alpha i k}\hat{r}_k , \qquad (14.14)$$

where $a(r)$, $b(r)$ and $c(r)$ are arbitrary functions of radial coordinate. However, neither the A-phase nor the B-phase have stable hedgehogs with such symmetry. ^{3}He-A is too anisotropic: one cannot construct the spherically symmetric object which has the A-phase vacuum far from the core. One can construct the spherically symmetric object in the isotropic ^{3}He-B vacuum, but such an object is topologically unstable in ^{3}He-B since the second homotopy group is trivial there, $\pi_2(R_B) = 0$. However, the spin–orbit interaction modifies the B-phase topology at large distances and gives rise to the topologically stable hedgehog with the soft core of the dipole legth ξ_D (Sec. 14.2.5). The spherically symmetric hedgehog with the hard core can exist in the planar phase. The core matter there has a very peculiar property, so let us discuss this object.

14.2.4 Enhanced superfluidity in the core of hedgehog. Generation of Dirac mass in the core

The order parameter in the planar phase of superfluid ^{3}He in eqn (7.61) is consistent with the spherically symmetric ansatz in eqn (14.14), if we choose the following asymptotes far from the core: $a(r = \infty) = \Delta_0$, $b(r = \infty) = c(r = \infty) = 0$. Then one has $e_{\alpha i}(\mathbf{r} \to \infty) = \Delta_0(\delta_{\alpha i} - \hat{r}_\alpha \hat{r}_i)$, which represents the spherically symmetric hedgehog in the planar phase. It has two discrete symmetries, space inversion P and time inversion T. In the most symmetric hedgehog these symmetries persist in the core. Because of P symmetry the function $c(r) = 0$ everywhere in the core, and due to T symmetry the other two functions are real: they represent the transverse and longitudinal gaps in the quasiparticle spectrum, $a(r) = \Delta_\perp(r)$ and $b(r) = \Delta_\parallel(r)$ (Fig. 14.4 bottom center).

The evolution of the gap in the quasiparticle spectrum in the hard core of coherence length ξ is shown in Fig. 14.4. In the center of the spherically symmetric core the two gaps must coincide to prevent the discontinuity in the order parameter: $a(r = 0) = b(r = 0)$. This means that the isotropic superfluid phase – ^{3}He-B – arises in the center of the hedgehog (Fig. 14.4 bottom left). The superfluid order parameter $e_{\alpha i}$ is thus nowhere zero throughout the core: the core

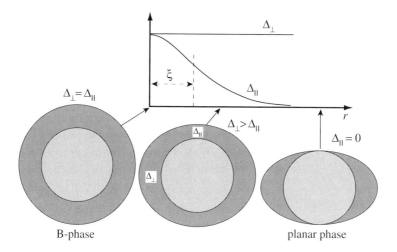

FIG. 14.4. *Top*: Schematic evolution of the order parameter in the hard core of the hedgehog in planar phase of superfluid ^{3}He. In the center of the hedgehog the isotropic B-phase appears. *Bottom*: Parameters $a(r) = \Delta_\perp(r)$ and $b(r) = \Delta_\parallel(r)$ in eqn (14.14) are transverse and longitudinal gaps in the quasiparticle spectrum. Far from the core one has a pure planar state with $\Delta_\parallel = 0$ and thus with gap nodes – the marginal Fermi points with $N_3 = 0$ – on poles. In the core the gap $\Delta_\parallel$ appears and fermions acquire the Dirac mass. Finally the order parameter develops to the isotropic ^{3}He-B in the center of the hedgehog.

matter is superfluid. This is typical for defects in systems with multi-component order parameter. It is not advantageous to nullify all the components of the order parameter simultaneously, since below T_c the system prefers the superfluid state.

Far from the core of the hedgehog, i.e. in a pure planar phase with $\Delta_\parallel = 0$, $\Delta_\perp = \Delta_0$ (Fig. 14.4 *bottom right*), the energy spectrum has two Fermi points in the directions parallel and antiparallel to the radius vector, $\mathbf{p}^{(a)} = \pm p_F \hat{\mathbf{r}}$. Quasiparticles in the vicinity of nodes are massless 'Dirac' fermions (see Sec. 12.1). Since the Fermi points in the planar phase are marginal, perturbations caused by the inhomogeneity of the order parameter in the core destroy the Fermi points, and fermions become massive. The gap – the Dirac mass $M = \Delta_\parallel$ in eqn (7.64) – gradually appears in the core region. Deeper in the core the order parameter develops into the isotropic ^{3}He-B in the center of the hedgehog, and the fully gapped fermionic spectrum becomes isotropic. This is just the opposite of the common wisdom according to which the massive fermions in the bulk become massless in the core of defect. The system here naturally prefers to have the isotropic superfluid state in the core with massive fermions, rather than the isotropic normal state with massless fermions.

14.2.5 Continuous hedgehog in B-phase

The spherically symmetric object (obeying eqn (14.14)) can be stabilized in ^{3}He-B by the spin–orbit interaction in eqn (14.10), which tries to fix the angle θ in the rotation matrix $R_{\alpha i}(\hat{\mathbf{n}}, \theta)$ in eqn (13.7) at the magic value $\cos\theta_L = -1/4$ (the Leggett angle). The axis of rotation $\hat{\mathbf{n}}$ is not fixed by this interaction, that is why the vacuum manifold reduced by the spin–orbit interaction becomes $\tilde{R} = S^2 \times U(1)$, where S^2 is the sphere of unit vector $\hat{\mathbf{n}}$. It has a non-trivial second homotopy group $\pi_2(\tilde{R}) = Z$, and thus there are hedgehogs in the field of rotation axis $\hat{\mathbf{n}}$. The most symmetric hedgehog with the winding number $\tilde{n}_2 = 1$ has $\hat{\mathbf{n}} = \hat{\mathbf{r}}$ and the following structure of the core consistent with eqn (14.14):

$$R_{\alpha i}(\mathbf{r}) = \delta_{\alpha i}\cos\theta(r) + (1-\cos\theta(r))\hat{r}_\alpha \hat{r}_i - \varepsilon_{\alpha i k}\hat{r}_k \sin\theta(r) \ . \tag{14.15}$$

It has the soft core within which the spin–orbit interaction is not saturated, and θ deviates from its magic value. At the origin one has $\theta(r=0) = 0$, which means that the sphere S^2 shriks to the point, and the symmetry broken in the bulk is restored in the core in agreement with common wisdom. The size of the soft core of this hedgehog is on order of the dipole length ξ_D, i.e.it has the same size as the thickness of the soliton in Fig. 14.2.

14.2.6 Symmetry of vortices: continuous symmetry

From the above examples we can see that the symmetry classification of defects is based on the symmetry-breaking scheme of the transition from the normal state to the ordered state with a given type of defect (point, line or wall). The initial large symmetry group G is extended to the group $G' = G \times SO(3)_L$ to include the space rotation. One must look for those subgroups of the group G' which are consistent with (i) the geometry of defect; (ii) the symmetry group H of the superfluid phase far from the core; and (iii) the topological charge of the defect. For each maximum symmetry subgroup there is a stationary solution for the order parameter everywhere in space including the core of defect. The phase transition in the core can occur as the spontaneous breaking of one of the maximal symmetries, or as the first-order transitions between core states corresponding to different maximum symmetry subgroups.

Let us now turn to the symmetry of linear defects. We shall start with vortices in superfluid ^{4}He, whose order parameter is a complex scalar $\Psi = |\Psi|e^{i\Phi}$. In the homogeneous phase transition of normal liquid ^{4}He to the superfluid state, the $U(1)_N$ symmetry is broken completely, while the system remains invariant under space rotations $SO(3)_L$. The symmetry-breaking pattern is different if one considers the transition from the normal liquid ^{4}He to the superfluid state with one vortex line. In this case the group $SO(3)_L$ must be broken too, since the direction of the vortex line appears as the axis of spontaneous anisotropy. One can easily find out what the rest symmetry of the system is by inspection of the asymptote of the order parameter Ψ far from the vortex with winding number n_1:

$$\Psi(\rho \to \infty, \phi) \propto e^{in_1\phi} \ . \tag{14.16}$$

One finds that the symmetry-breaking scheme is now

$$G' = U(1)_N \times SO(3)_\mathbf{L} \to H' = U(1)_Q \ . \tag{14.17}$$

Here the remaining symmetry $U(1)_Q$ is the symmetry of the order parameter in eqn (14.16). It is rotation by angle θ, which transforms $\phi \to \phi + \theta$, accompanied by the global phase rotation $\Phi \to \Phi + \alpha$, with $\alpha = -n_1\theta$. The generator of such $U(1)_Q$ transformations is

$$Q = L_z - n_1 N \ . \tag{14.18}$$

Vortices with the symmetry $U(1)_Q$ are axisymmetric, since according to this symmetry the modulus of the order parameter $|\Psi|$ does not depend on ϕ: from the symmetry condition $Q|\Psi| = 0$ it follows that $\partial_\phi|\Psi| = iL_z|\Psi| = i(L_z - n_1 N)|\Psi| = iQ|\Psi| = 0$.

For the s-wave superconductors, as well as for superfluid ^{3}He, the order parameter corresponds to the vev of the annihilation operator of two particles. Thus the corresponding scalar order parameter in s-wave superconductors, Ψ, transforms under (global) gauge transformation generated by N as $\Psi \to \Psi e^{2i\alpha}$. That is why the generator Q of the axial symmetry in eqn (14.18) must be modified to

$$Q = L_z - \frac{n_1}{2} N \ . \tag{14.19}$$

The phase transition into the vortex state has the same pattern of the symmetry breaking as the phase transition from normal ^{3}He to the homogeneous ^{3}He-A in eqn (7.53), if we are interested only in the orbital part of the order parameter in ^{3}He-A. Thus the homogeneous ^{3}He-A corresponds to the vortex in s-wave superfluid with winding number $n_1 = 1$, while the $\hat{\mathbf{l}}$-vector of ^{3}He-A corresponds to the direction of the vortex axis. This suggests also that the total angular momentum of both systems, the homogeneous ^{3}He-A and $n_1 = 1$ vortex in s-wave superfluid (or in ^{3}He-B), is the same, with $\hbar/2$ per particle (see the discussion of the orbital angular momentum in ^{3}He-A in Sec. 20.2).

The same symmetry-breaking scheme occurs in electroweak transitions, where the generator Q is the electric charge (Secs 12.2 and 15.1.2).

14.2.7 Symmetry of vortices: discrete symmetry

There are also two discrete symmetries of the order parameter in eqn (14.16). One of them is the parity P, which transforms $\phi \to \phi + \pi$ (for odd n_1 this operation must be accompanied by gauge transformation from $U(1)_N$). The time reversal operation T transforms the order parameter to its complex conjugate, i.e. the phase Φ changes sign: $T\Phi = -\Phi$. This means that the time reversal symmetry is broken by the vortex. However, the combined symmetry TU_2 is retained, where U_2 is rotation by π about the axis perpendicular to the vortex axis. Since $U_2\phi = -\phi$ this compensates the complex conjugation.

The symmetries $U(1)_Q \times P \times TU_2$ comprise the maximum possible symmetry of the vortex with winding number n_1. This means that one can always find the solution where this symmetry is extended to the core region. This does not mean

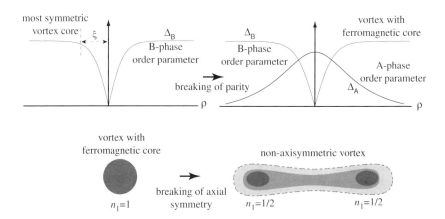

FIG. 14.5. Schematic illustration of the core structure of B-phase vortices with $n_1 = 1$. The most symmetric vortex core state with vanishing order parameter (*top left*) is never realized in ^{3}He-B. The vortex with the A-phase core (*top right*) is realized in the high-T part of the phase diagram in Fig. 14.3. Fermi points arising in the core of this vortex are shown in Fig. 23.4 *bottom*. The vortex with asymmetric core (*bottom*), which can be represented as a pair of half-quantum vortices, is realized in the low-T part of the phase diagram in Fig. 14.3.

that the solution with the symmetric core is the local minimum of the energy: it can be the saddle point and the energy can be reduced by spontaneous breaking of P, TU$_2$ or even continuous Q-symmetry.

14.3 Broken symmetry in B-phase vortex core

14.3.1 *Most symmetric vortex and its instability*

The simplest possible vortex with $n_1 = 1$ in ^{3}He-B is, of course, the most symmetric mass vortex, i.e. the vortex whose order parameter distribution has the same maximum symmetry as the asymptote:

$$\mathrm{H}_{\max} = U(1)_Q \times \mathrm{P} \times \mathrm{TU}_2 \ , \quad \mathrm{Q} = \mathrm{J}_z - \frac{n_1}{2}\mathrm{N} \ , \ \mathrm{J}_z = \mathrm{L}_z + \mathrm{S}_z \ . \quad (14.20)$$

For the maximum symmetric vortex with $n_1 = 1$, the symmetry requires that all the order parameter components become zero on the vortex axis as is illustrated in Fig. 14.5 *top left*; actually the most symmetric vortex in ^{3}He-B is somewhat more complicated than in the figure: it has three non-zero components, but all of them behave in the same manner as in Fig. 14.5 *top left*; the details can be found in the review by Salomaa and Volovik (1987).

Though the maximum symmetric vortex necessarily represents the solution of, say, Ginzburg–Landau equations, it is never realized in ^{3}He-B. The corresponding solution represents the saddle point of thr Ginzburg–Landau free energy functional, rather than a local minimum. It is energetically favorable to break the symmetry in order to escape the vanishing of the superfluidity in the vortex core. This is actually a typical situation for superfluids/superconductors with a multi-component order parameter: the superfluid/superconductor does not tolerate a full suppression of the superfluid fraction in the core, if there is a possibility to escape this by filling the core with other components of the order parameter.

In ^{3}He-B there are two structures of the vortex with the same asymptote of the order parameter as the maximum symmetric vortex, but with broken symmetry in the vortex core. These are the vortex with ferromagnetic core in the higher-T region of the phase diagram in Fig. 14.3 and the non-axisymmetric vortex in the lower-T region.

14.3.2 *Ferromagnetic core with broken parity*

In the region of the ^{3}He-B phase diagram, which has a border line with ^{3}He-A, the neighborhood (proximity) of the ^{3}He-A is felt: the core of the $n_1 = 1$ vortex there is filled by the A-phase order parameter components (Fig. 14.5 *top right*). This vortex structure is doubly degenerate: the vortex with opposite sign of the A-phase order parameter in the core has the same energy. This reflects the broken discrete Z_2 symmetry. In a given case parities P and TU_2 are broken in the core of the ^{3}He-B vortex while their combination PTU_2 persists (see details in the review paper by Salomaa and Volovik 1987).

The core of this ^{3}He-B vortex displays the orbital ferromagnetism of the A-phase component, and in addition there is a ferromagnetic polarization of spins in the core (Salomaa and Volovik 1983). This core ferromagnetism was observed by Hakonen *et al.* (1983*b*) in NMR experiments as the gyromagnetism of the rotating liquid mediated by quantized vortices: rotation of the cryostat produces the cluster of vortices with ferromagnetically oriented spin magnetic momenta of their cores. This net magnetic moment interacts with the external magnetic field, which leads to the dependence of the NMR line shape on the sense of rotation with respect to the magnetic field.

Similar, but antiferromagnetic, cores have been discussed for vortices in high-T_c superconductors within the popular $SO(5)$ model for the coexistence of superconductivity and antiferromagnetism (for the $SO(5)$ model see e.g. Mortensen *et al.* (2000) and references therein). It has been shown by Arovas *et al.* (1997) and Alama *et al.* (1999) that in the Ginzburg–Landau regime, in certain regions of the parameter values, a solution corresponding to the conventional vortex core is unstable with respect to that with the antiferromagnetic core.

14.3.3 *Double-core vortex as Witten superconducting string*

On the other side of the phase transition line in Fig. 14.3, the B-phase vortices have non-axisymmetric cores, i.e. the axial $U(1)_Q$ symmetry in eqn (14.20) is spontaneously broken in the core (Fig. 14.5 *bottom right*). This was demonstrated in experiments by Kondo *et al.* (1991) and calculations in the Ginzburg–Landau region by Volovik and Salomaa (1985), Thuneberg (1986) and Salomaa and Volovik (1986). The core with broken rotational symmetry can be considered as a pair of half-quantum vortices, connected by a non-topological soliton wall (Thuneberg 1986, 1987b; Salomaa and Volovik 1986, 1988, 1989). The separation of the half-quantum vortices increases with decreasing pressure and thus the double-core structure is most pronounced at zero pressure (Volovik 1990b). This is similar to the half-quantum vortices in momentum space in Fig. 8.8 connected by the non-topological fermionic condensate.

Related phenomena are also possible in superconductors. The splitting of the vortex core into a pair of half-quantum vortices confined by the non-topological soliton has been discussed in heavy-fermionic superconductors by Luck'yanchuk and Zhitomirsky (1995) and Zhitomirsky (1995). In fact a vortex core splitting may have been observed in high-T_c superconductors (Hoogenboom et al. 2000). The observed double core was interpreted as tunneling of a vortex between two neighboring sites in the potential wells created by impurities. However, the phenomenon can also be explained in terms of vortex core splitting.

In the physics of cosmic strings, an analogous breaking of continuous symmetry in the core was first discussed by Witten (1985), who considered the spontaneous breaking of the electromagnetic gauge symmetry $U(1)_Q$. Since the same symmetry group is broken the condensed matter superconductors, one can say that in the core of the cosmic string there appears the superconductivity of the electric charges, hence the name 'superconducting cosmic strings'. For the closed string loop one can have quantization of magnetic flux inside the loop provided by the electric supercurrent along the string. In a sense, a superconducting cosmic string is analogous to a closed superconducting wire, where the phase of the order parameter Φ changes by $2\pi n_1$. The instability toward the breaking of the $U(1)_Q$ symmetry in the string core can be triggered, for example, by fermion zero modes in the core of cosmic strings as was found by Naculich (1995). According to Makhlin and Volovik (1995) the same mechanism can take place in condensed matter vortices too.

14.3.4 Vorton – closed loop of the vortex with twisted core

For the ^{3}He-B vortices, the spontaneous breaking of the $U(1)_Q$ symmetry (eqn (14.20)) in the core leads to the Goldstone bosons – the mode in which the degeneracy parameter, the axis of anisotropy of the vortex core, is oscillating. The vibrational Goldstone mode is excited by a special type of B-phase NMR mode, called homogeneously precessing domain, in which the magnetization of the ^{3}He-B precesses coherently. In experiments by Kondo et al. (1991) it was possible also to rotate the core around its axis with constant angular velocity, and in addition, since the core was pinned on the top and the bottom of the container, it was possible even to screw the core (Fig. 14.6). Since the Goldstone field α – the angle of the anisotropy axis – has a gradient $\nabla\alpha$ along the string, such a twisted core corresponds to the Witten superconducting string with the supercurrent along the core.

If the vortex line with the twisted core is closed, one obtains the analog of the string loop with the quantized supercurrent along the loop. Such a closed cosmic string, if it can be made energetically stable, is called a vorton (see Carter

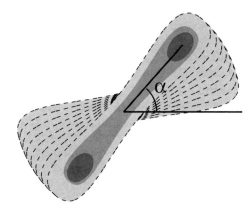

FIG. 14.6. Twisted core of non-axisymmetric vortex in ^{3}He-B. The gradient of the Goldstone field $\nabla\alpha$ along the string corresponds to the superconducting current along the superconducting cosmic string. The closed loop of the twisted vortex is called the vorton.

and Davis 2000). The stability can be provided by conservation of the winding number n of the phase of the superconducting order parameter along the loop. When the loop is shrinking due to the string tension, the conservation of n leads to an increase of supercurrent and its energy. This opposes the string tension and can even stabilize the loop if the parameters of the system are favorable. Vortons can have cosmological implications, since they can survive after creation during the cosmological phase transition.

Estimates show that in ^{3}He-B vortons cannot be stabilized.

15

SYMMETRY BREAKING IN ³He-A AND SINGULAR VORTICES

15.1 ³He-A and analogous phases in high-energy physics

15.1.1 Broken symmetry

In the phase transition from the normal liquid ³He to the A-phase of ³He, the symmetry G in eqn (13.2) is broken to the following subgroup:

$$H_A = U(1)_{S_z} \times U(1)_{L_z - \frac{1}{2}N} \times Z_2 . \tag{15.1}$$

We recall that the vacuum state corresponding to this residual symmetry is

$$e^{(0)}_{\alpha i} = \Delta_A \hat{z}_\alpha (\hat{x}_i + i\hat{y}_i) , \tag{15.2}$$

while the general form of the degenerate states, obtained by action of the symmetry group G, is

$$e_{\alpha i} = \Delta_0 \hat{d}_\alpha (\hat{m}_i + i\hat{n}_i) . \tag{15.3}$$

The $SO(3)_\mathbf{S}$ group of spin rotations is broken to its $U(1)_{S_z}$ subgroup: these are spin rotations about axis $\hat{\mathbf{d}}$, which do not change the order parameter. The other two groups are broken in the way it happens in the Standard Model. We also explicitly introduce the discrete Z_2 symmetry P (7.56), which plays the role of the parity in the effective theory. This symmetry is important for the classification of the topological defects.

15.1.2 Connection with electroweak phase transition

The group of orbital rotations and the global $U(1)_N$ group are broken to their diagonal subgroup

$$SO(3)_\mathbf{L} \times U(1)_N \to U(1)_{L_z - \frac{1}{2}N} . \tag{15.4}$$

The vacuum state in eqn (15.3) remains invariant under the orbital rotation $SO(2)_{L_z}$ about axis $\hat{\mathbf{l}} = \hat{\mathbf{m}} \times \hat{\mathbf{n}}$ if it is accompanied by the proper global transformation from the $U(1)_N$ group. The $\hat{\mathbf{l}}$-vector marks the direction of the spontaneous orbital momentum $\langle \text{vac}|\mathbf{L}|\text{vac}\rangle$ in the broken-symmetry state.

A very similar symmetry-breaking pattern occurs in the Standard Model of the electroweak interactions

$$SU(2)_L \times U(1)_Y \to U(1)_Q , \tag{15.5}$$

where $SU(2)_L$ is the group of the weak isotopic rotations with the generator **T**; Y is the hypercharge; and $Q = T_3 + Y$ is the generator of electric charge in the remaining electromagnetic symmetry (see Sec. 12.2; note that in standard notations the hypercharge Y is two times larger and thus the electric charge has the form $Q = T_3 + (1/2)Y$, which makes the analogy even closer). In this analogy the $\hat{\mathbf{l}}$-vector corresponds to the direction of the spontaneous isotopic spin $\langle \text{vac}|\mathbf{T}|\text{vac}\rangle$ in the broken-symmetry state.

The Higgs field in the electroweak model, which is the counterpart of the orbital vector $\hat{\mathbf{m}} + i\hat{\mathbf{n}}$ of the order parameter in ^{3}He-A, is the spinor

$$\Phi_{\text{ew}} = \begin{pmatrix} \phi_1 \\ \phi_2 \end{pmatrix}, \tag{15.6}$$

which is normalized by $\Phi_{\text{ew}}^{\dagger}\Phi_{\text{ew}} = |\phi_1|^2 + |\phi_2|^2 = \eta_{\text{ew}}^2/2$ in the vacuum where $\eta_{\text{ew}} \sim 250$ GeV. The amplitude of the order parameter is 18 orders of magnitude larger than the corresponding $\Delta_0 \sim 10^{-7}$ eV in ^{3}He-A. The $\hat{\mathbf{l}}$-vector, i.e. the direction of spontaneous isotopic spin $\langle \text{vac}|\mathbf{T}|\text{vac}\rangle$, is given by

$$\hat{\mathbf{l}}_{\text{ew}} = -\frac{\Phi_{\text{ew}}^{\dagger}\vec{\tau}\Phi_{\text{ew}}}{\Phi_{\text{ew}}^{\dagger}\Phi_{\text{ew}}}, \tag{15.7}$$

where $\vec{\tau}$ are the Pauli matrices in the isotopic space. Isotopic rotation about $\hat{\mathbf{l}}_{\text{ew}}$, together with the proper $U(1)_Y$ transformation induced by the hypercharge operator, is the remaining electromagnetic symmetry $U(1)_{T_3+Y/2}$ of the electroweak vacuum in the broken-symmetry state. This can be seen using the simplest representation of the order parameter, in which only the component with the isospin projection $T_3 = -1/2$ on the $\hat{\mathbf{l}}_{\text{ew}}$-vector is present:

$$\Phi_{\text{ew}} = \begin{pmatrix} 0 \\ \phi_2 \end{pmatrix}. \tag{15.8}$$

This form can be obtained from the general eqn (15.6) by $SU(2)$ rotations. It is invariant under $U(1)_Y$ gauge transformation $\Phi \to \Phi e^{i\alpha}$, if it is accompanied by rotation $\Phi \to \Phi e^{2iT_3\alpha}$.

This spinor Higgs field is also very similar to the order parameter of 2-component Bose–Einstein condensate in laser-manipulated traps (see Sec. 16.2.2).

15.1.3 Discrete symmetry and vacuum manifold

The residual Z_2 symmetry P in eqn (15.1) plays an important role in ^{3}He-A. It is this symmetry which couples two fermionic species living in the vicinity of the doubly degenerate Fermi point and gives rise to the effective $SU(2)$ gauge field in Sec. 9.3.1. In the world of topological defects, the P symmetry leads to exotic half-quantum vortices. Let us recall that according to eqn (7.56), P is the symmetry of the vacuum under spin rotation by π about an axis perpendicular to $\hat{\mathbf{d}}$ combined with the orbital rotation about $\hat{\mathbf{l}}$ by π.

The vacuum manifold of ^{3}He-A can be found as the factor space

$$R_A = G/H_A = \left(SO(3) \times S^2\right)/Z_2 \ . \tag{15.9}$$

It consists of the sphere S^2 of unit vector $\hat{\mathbf{d}}$, and of the space $SO(3)$ of solid rotations of $\hat{\mathbf{m}}$ and $\hat{\mathbf{n}}$ vectors. The symmetry P identifies on $S^2 \times SO(3)$ the pairs of points $\hat{\mathbf{d}}$, $\hat{\mathbf{m}} + i\hat{\mathbf{n}}$ and $-\hat{\mathbf{d}}$, $-(\hat{\mathbf{m}} + i\hat{\mathbf{n}})$.

15.2 Singular defects in ^{3}He-A

15.2.1 *Hedgehog in ^{3}He-A and 't Hooft–Polyakov magnetic monopole*

The A-phase manifold in eqn (15.9) has non-trivial homotopy groups

$$\pi_1(G/H_A) = \pi_1\left((SO(3) \times S^2)/Z_2\right) = Z_4 \ , \ \pi_2(G/H_A) = Z \ . \tag{15.10}$$

Now the second homotopy group is non-trivial. It comes from the sphere S^2 of unit vector $\hat{\mathbf{d}}$. The winding number of the mapping of the spherical surface σ_2 around the hedgehog to S^2 provides integer topological charge n_2 in eqn (14.12) now describing topologically stable hedgehogs in the field of unit vector $\hat{\mathbf{d}}$:

$$n_2 = \frac{1}{8\pi} e^{ijk} \int_{\sigma_2} dS_k \ \hat{\mathbf{d}} \cdot \left(\frac{\partial \hat{\mathbf{d}}}{\partial x^i} \times \frac{\partial \hat{\mathbf{d}}}{\partial x^j}\right) \ . \tag{15.11}$$

The simplest hedgehogs with the winding number $n_2 = \pm 1$ are given by radial distributions: $\hat{\mathbf{d}} = \pm \hat{\mathbf{r}}$. Note that the P symmetry transforms $\hat{\mathbf{d}}$ to $-\hat{\mathbf{d}}$, and thus in the P-symmetric vacuum point defects with $n_2 = \pm 1$ are equivalent. Indeed they can be transformed to each other by circling around the Alice string (see below, Sec. 15.3.2).

Let us recall now that there are two types of analogies between ^{3}He-A and the Standard Model; they correspond to the GUT and anti-GUT schemes. The anti-GUT analogy is based on **p**-space topology, where the common property is the existence of Fermi points leading to RQFT in the low-energy corner of ^{3}He-A. In this analogy, the $\hat{\mathbf{d}}$-vector plays the role of the quantization axis for the weak isospin of quasiparticles. Quasiparticles in eqn (9.1) with 'isospin' projection $1/2(\sigma^\mu \hat{d}_\mu) = +1/2$ and $-1/2$ correspond to the left neutrino and left electron respectively (Sec. 9.1.5). Thus the $\hat{\mathbf{d}}$-hedgehog presents the ^{3}He-A realization of the original 't Hooft–Polyakov magnetic monopole arising within the symmetry-breaking scheme $SU(2) \to U(1)$. This monopole has no Dirac string: the $\hat{\mathbf{d}}$-hedgehog is an isolated point object (Fig. 17.2 *bottom left*).

Another type of analogy is based on the GUT scheme. It exploits the similarity in broken-symmetry patterns, $SU(2)_L \times U(1)_Y \to U(1)_Q$ in the Standard Model and $SO(3)_L \times U(1)_N \to U(1)_{L_z-N/2}$ in the orbital part of ^{3}He-A, which leads to similar **r**-space topology of defects. In this analogy, it is now the $\hat{\mathbf{l}}$-vector which corresponds to the quantization axis $\hat{\mathbf{l}}_{\rm ew}$ of the weak isospin in the Standard Model. There are no isolated point defects in the $\hat{\mathbf{l}}$-field, since the corresponding second homotopy group is trivial: $\pi_2(SU(2)_L \times U(1)_Y/U(1)_Q) = \pi_2(SO(3)_L \times U(1)_N/U(1)_{L_z-N/2}) = 0$. The hedgehog in the $\hat{\mathbf{l}}$-field is always the termination point of the physical string playing the part of the Dirac strings. In the Standard Model it is called the electroweak magnetic monopole (Fig. 17.2 bottom right). In Sec. 17.1 we shall discuss different types of monopoles.

15.2.2 Pure mass vortices

The fundamental homotopy group $\pi_1(G/H_A)$ contains four elements, i.e. there are four topologically distinct classes of linear defects in ^{3}He-A. Each can be described by the topological charge which takes only four values, chosen to be 0, $\pm 1/2$ and 1 with summation modulo 2 (i.e. 1+1=0). We denote this charge as n_1, according to the circulation winding number of the simplest vortex representatives of these topological classes. Defects with $n_1 = \pm 1/2$ represent the fractional vortices – half-quantum vortices (Volovik and Mineev 1976b) also called the Alice strings (Sec. 15.3.1 below).

Let us start with the simplest representatives of classes $n_1 = 1$ and $n_1 = 2 \equiv 0$. Let us first fix the vector $\hat{\mathbf{l}}$ along the axis of the vortex, and also fix the $\hat{\mathbf{d}}$-vector. Then the only remaining degree of freedom is the rotation of $\hat{\mathbf{m}}$ and $\hat{\mathbf{n}}$ around $\hat{\mathbf{l}}$ which is equivalent to the phase rotation of group $U(1)_N$. Under this constraint the defects are pure mass vortices with integer winding number n_1 and with the following structure of the orbital part of the order parameter outside the core:

$$\hat{\mathbf{m}} + i\hat{\mathbf{n}} = e^{in_1\phi}(\hat{\mathbf{x}} + i\hat{\mathbf{y}}) \ . \tag{15.12}$$

The circulation of superfluid velocity around such a vortex is $\oint d\mathbf{r} \cdot \mathbf{v}_s = \kappa_0 n_1$, where the superfluid velocity $\mathbf{v}_s = (\kappa_0/2\pi)\hat{m}^i\vec{\nabla}\hat{n}^i = (n_1\kappa_0/2\pi)\vec{\nabla}\phi$, and the elementary circulation quantum is $\kappa_0 = 2\pi\hbar/(2m)$. We must keep in mind that circulation is quantized here only because we have chosen the fixed orientation of the $\hat{\mathbf{l}}$-vector: $\hat{\mathbf{l}} = \hat{\mathbf{z}}$. For the general distribution of the $\hat{\mathbf{l}}$-field, the vorticity can be continuous according to the Mermin–Ho relation in eqn (9.17), and thus circulation is not quantized in general.

Now let us allow the $\hat{\mathbf{l}}$-field to vary. Then one finds that for vortices with even n_1 in eqn (15.12) the singularity in their cores can be continuously washed out. This is because they all belong to the same class as homogeneous state $n_1 = 0$. As a result one obtains the continuous textures of the $\hat{\mathbf{l}}$-field with distributed vorticity, which we shall discuss later on in Chapter 16.

Mass vortices with odd n_1 in eqn (15.12) all belong to the same class as the mass vortex with $n_1 = 1$. They can continuously transform to each other, but they always have the singular core, where the order parameter leaves the vacuum manifold of ^{3}He-A. The simplest of the strings of this class is the pure mass vortex with $n_1 = 1$ circulation quanta. However, it is not the defect with the lowest energy within the class $n_1 = 1$. Let us consider other defects of the class $n_1 = 1$, the disclinations.

15.2.3 Disclination – antigravitating string

The linear defect of the class $n_1 = 1$, which will be discussed in Sec. 30.2 as an analog of an antigravitating cosmic string, has the following structure (Fig. 30.1). Around this defect one of the two orbital vectors, say $\hat{\mathbf{n}}$, remains constant, $\hat{\mathbf{n}} = \hat{\mathbf{z}}$, while the $\hat{\mathbf{m}}$-vector is rotating by 2π:

$$\hat{\mathbf{m}}(\mathbf{r}) = \hat{\phi} \, , \; \hat{\mathbf{l}}(\mathbf{r}) = \hat{\mathbf{m}}(\mathbf{r}) \times \hat{\mathbf{n}} = \hat{\rho} \, . \qquad (15.13)$$

Here, as before, z, ρ, ϕ are the cylindrical coordinates with the axis $\hat{\mathbf{z}}$ being along the defect line. There is no circulation of superfluid velocity around this defect, since $\mathbf{v}_s = (\kappa_0/2\pi)\hat{m}^i\vec{\nabla}\hat{n}^i = 0$, while the field of anisotropy axis $\hat{\mathbf{l}}$ is reminiscent of defects in nematic liquid crystals, where such defect lines are called disclinations. Equation (15.13) describes radial disclination, i.e. with radial distribution of the anisotropy axis. Tangential disclination, with $\hat{\mathbf{m}}(\mathbf{r}) = -\hat{\rho}$ and $\hat{\mathbf{l}}(\mathbf{r}) = \hat{\phi}$, is also the configuration which belongs to the class $n_1 = 1$ of linear defects.

The linear defect which has minimal energy within the class $n_1 = 1$ and was observed experimentally by Parts *et al.* (1995*a*) has a more complicated onion structure. It is the mass vortex at large distances, the vortex texture in the soft core and approaches the structure of the disclination near the hard core.

15.2.4 Singular doubly quantized vortex vs electroweak Z-string

The pure vortex with phase winding $n_1 = 2$ (as well as with any other even n_1) belongs to the topologically trivial class: it is topologically unstable and can be continuously unwound. This vortex can be compared to the electroweak strings. The electroweak vacuum manifold in the broken-symmetry state

$$\mathbf{R}_{\text{ew}} = (SU(2)_L \times U(1)_Y)/U(1)_{T_3+Y} = SU(2) \, , \qquad (15.14)$$

does not support any topologically non-trivial linear defects, since it has trivial fundamental homotopy group: $\pi_1(\mathbf{R}_{\text{ew}}) = 0$. However, there exist solutions of the Yang–Mills–Higgs equation which describe non-topological vortices with integer winding number. These are Nielsen–Olesen (1973) strings. An example of such a string in the Standard Model is the Z-strings in Fig. 15.1, in which the isotopic spin $\hat{\mathbf{l}}_{\text{ew}}$ is uniform and directed along the string axis:

$$\Phi_{\text{ew}}(\mathbf{r}) = f(\rho) \begin{pmatrix} 0 \\ e^{i\phi} \end{pmatrix} \, . \qquad (15.15)$$

The order parameter exhibits the 4π rotation in isotopic space when circling around the core, and this is equivalent to the 4π winding around the singular vortices in ^{3}He-A with $n_1 = 2$. Both 4π defects, the Z-string and the mass vortex with $n_1 = 2$ in ^{3}He-A, are topologically unstable. They can either completely or partially transform to a continuous texture. In case of partial transformation one obtains a piece of defect line terminating on a point defect – the electroweak monopole which we shall discuss in Sec. 17.1. In spite of the topological instability, the Z-string and the $n_1 = 2$ vortex can be stabilized energetically under favorable conditions.

Another realization of the topologically unstable defect in ^{3}He-A is the 4π disclination in the $\hat{\mathbf{l}}$ field:

$$e_{\alpha i} = \Delta_A \hat{z}_\alpha \left(\hat{z}_i + i(\hat{x}_i \cos 2\phi + \hat{y}_i \sin 2\phi) \right) \; , \; \hat{l}_i = \hat{y}_i \cos 2\phi - \hat{x}_i \sin 2\phi \; . \quad (15.16)$$

Its $\hat{\mathbf{l}}$-field corresponds to the W-string solution in the electroweak model (Volovik and Vachaspati 1996).

There is, however, an important difference between vortices in ^{3}He-A and in the Standard Model, which follows from the fact that the relevant symmetry groups are global in ^{3}He-A and local in the Standard Model. That is why the gauge field is instrumental for the structure of strings in the Standard Model, in the same way as the coupling with magnetic field is important for Abrikosov vortices in superconductors. Let us now compare the Nielsen–Olesen string and Abrikosov vortex.

15.2.5 Nielsen–Olesen string vs Abrikosov vortex

Superconductivity is the superfluidity of electric charge which interacts with the electromagnetic field – the $U(1)_Q$ gauge field. The gauge field screens the electric current due to the Meissner effect. In superconductors, the corresponding vortex with $n_1 = 1$ is the Abrikosov vortex (Abrikosov 1957). It is shown in Fig. 15.1 *top right* for a type-2 superconductor where the screening length of magnetic field called the London penetration length is larger than the coherence length determining the core size: $\lambda > \xi$. The magnetic flux carried by Abrikisov vortex is concentrated in the region $\rho \sim \lambda$. At $\rho > \lambda$ the electric current $\mathbf{j} = en_s \left(\mathbf{v}_s - \frac{e}{mc} \mathbf{A} \right)$ circulating around the vortex decays exponentially. Then the total magnetic flux trapped by the Abrikosov vortex with n_1 winding number is

$$\Phi = \int d\mathbf{S} \cdot \mathbf{B} = \oint d\mathbf{r} \cdot \mathbf{A} = \frac{mc}{e} \oint d\mathbf{r} \cdot \mathbf{v}_s = n_1 \frac{hc}{2e} \equiv \frac{n_1}{2} \Phi_0 \; . \quad (15.17)$$

Here we introduce the flux quantum $\Phi_0 = hc/e$ which coincides with the magnetic flux from the Dirac magnetic monopole; the conventional Abrikosov vortex with $n_1 = 1$ traps half of this quantum.

The same structure (Fig. 15.1 *top left*) with two coaxial cores of dimensions ξ and λ characterizes the local cosmic strings – Nielsen–Olesen vortices (Nielsen and Olesen 1973). The penetration length of the corresponding gauge field, say, of the hypermagnetic field, is determined by the inverse mass of the corresponding gauge boson, say the Z-boson: $\lambda = \hbar c / M_Z$. The healing length of the order parameter – the Higgs field – is determined by the inverse mass of the Higgs boson: $\xi = \hbar c / M_H$.

In both structures the order parameter is suppressed in the core of size ξ. In superconductors, the gap in the spectrum of Bogoliubov quasiparticles is proportional to the order parameter and thus vanishes on the vortex axis. This leads to fermion zero modes – bound states of quasiparticles in the potential well produced by the gap profile in the core of a vortex (Fig. 15.1 *bottom*) which will be discussed in Chapter 23. In the Standard Model, masses of quarks and leptons

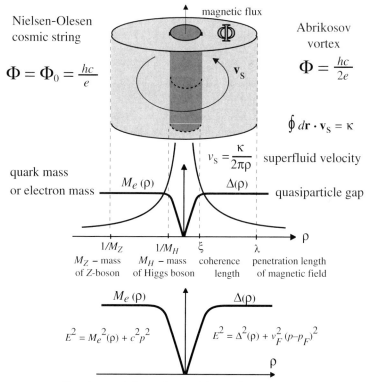

FIG. 15.1. Singular vortex and cosmic string. *Top*: The Abrikosov vortex in a superconductor (Abrikosov 1957) is the analog of the Nielsen–Olesen (1973) cosmic string. The role of the penetration length λ is played by the inverse mass of the Z-boson. If $\lambda \gg \xi$, the core size, within the region of dimension λ the Abrikosov vortex has the same distribution of the current as the vortex in neutral superfluids, such as ^{3}He-B, where the circulation of the superfluid velocity is quantized. At $\rho \gg \lambda$ the current is screened by the magnetic field concentrated in the region $\rho < \lambda$. *Bottom*: Masses of quarks and the gap of quasiparticles in superconductors, are suppressed in the vortex core. The core serves as a potential well for fermions which are bound in the vortex forming fermion zero modes.

are also proportional to the order parameter (Higgs field ϕ_2 in eqn (15.15)), and thus also vanish on the string axis leading to fermion zero modes in the core of the cosmic string.

Thus the physics of vortices is in many respect the same for different superfluids, whose non-dissipative supercurrents carry mass, spin and electric charge in condensed matter systems, and chiral, color, baryonic, hyper- and other charges in high-energy physics.

15.3 Fractional vorticity and fractional flux

15.3.1 Half-quantum vortex in ^{3}He-A

The asymptotic form of the order parameter in vortices with fractional circulation numbers $n_1 = \pm 1/2$ (or, simply, half-quantum vortices) is given by

$$e_{\alpha j} = \Delta_0 \, e^{\pm i\phi/2} \hat{d}_\alpha (\hat{x}_j + i\hat{y}_j) \;, \quad \hat{\mathbf{d}} = \hat{\mathbf{x}} \cos \frac{\phi}{2} + \hat{\mathbf{y}} \sin \frac{\phi}{2}. \tag{15.18}$$

On circumnavigating such a vortex, the vector $\hat{\mathbf{d}}$ changes its direction to the opposite (Fig. 15.2 *left*). The change of sign of the order parameter in eqn (15.18) is compensated by the winding of the phase around the string by π or $-\pi$. The latter means that the winding number of the vortex is $n_1 = \pm 1/2$, i.e. the circulation $\oint \mathbf{v}_s \cdot d\mathbf{r} = n_1 \pi \hbar/m$ of the superfluid velocity around this string is one-half of the circulation quantum $\kappa_0 = \pi \hbar/m$. It is the half-quantum vortex.

In a sense, the half-quantum vortex is the composite defect: the $n_1 = 1/2$ mass vortex is accompanied by the $\nu = 1/2$ vortex in the $\hat{\mathbf{d}}$-field. Together they form the spin–mass vortex similar to that in ^{3}He-B (Sec. 14.1.4), with one important difference. In ^{3}He-B, the mass vortex with $n_1 = 1$ and the spin vortex with $\nu = 1$ 'live in two different worlds' and interact through the Casimir forces. In ^{3}He-A, the $n_1 = 1/2$ mass vortex and $\nu = 1/2$ spin vortex also 'live in two different worlds': they are described by non-interacting fields $(\hat{\mathbf{m}} + i\hat{\mathbf{n}})$ and $\hat{\mathbf{d}}$ correspondingly. If $\hat{\mathbf{m}} + i\hat{\mathbf{n}} = (\hat{\mathbf{x}} + i\hat{\mathbf{y}})e^{i\Phi}$ and $\hat{\mathbf{d}} = \hat{\mathbf{x}} \cos \theta + \hat{\mathbf{y}} \sin \theta$, the gradient energy is the sum of the energies of Φ- and θ-fields in the same manner as in ^{3}He-B:

$$E_{\text{London}} = \frac{\hbar^2}{8m} \left(n_s (\nabla \Phi)^2 + n_s^{spin} (\nabla \theta)^2 \right) . \tag{15.19}$$

But, the important distinction from ^{3}He-B is that here the two defects are confined topologically: half-quantum vortices in Φ- and θ-fields cannot exist as separate isolated entities. The continuity of the vacuum order parameter around the defect requires that they must always have a common core.

The **p**-space analogs of fractional defects with the common core have been discussed in secs 8.1.7 and 12.2.3 where a (quasi)particle was considered as a product of the spinon and holon.

As the spin–mass vortex in ^{3}He-B (Sec. 14.1.5), the half-quantum spin–mass vortex in ^{3}He-A becomes the termination line of the topological Z_2-solitons when the symmetry violating spin–orbit interaction is turned on (see Sec. 16.1.1, Fig. 16.1). This property will allow identification of this object in future experiments.

15.3.2 Alice string

The $n_1 = 1/2$ vortex in Sec. 15.3.1 is the counterpart of Alice strings considered in particle physics by Schwarz (1982). The motion of the 'matter' around such a string has unexpected consequences. In nematic liquid crystals, the 'monopole' (the hedgehog) that moves around the $n_1 = 1/2$ disclination line transforms to the anti-monopole; the same occurs for the ^{3}He-A monopoles: the hedgehog

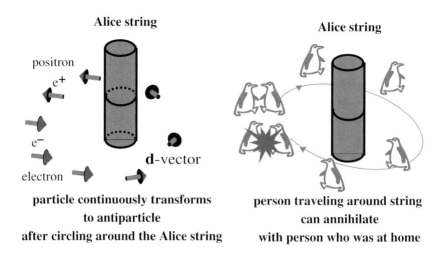

FIG. 15.2. Alice string in ^{3}He-A and in some RQFT. *Left*: The axis $\hat{\mathbf{d}}$ determining quantization of spin, charge or other quantum number changes direction to the opposite around the Alice string. This means that if a particle slowly moves around a half-quantum vortex, it flips its spin, charge, parity, etc., depending on the type of Alice string. *Right*: A person traveling around such an Alice string can annihilate with the person who did not follow this topologically non-trivial route.

in the $\hat{\mathbf{d}}$-field with the topological charge $n_2 = +1$ in eqn (15.11) transforms to the anti-hedgehog with $n_2 = -1$ (Volovik and Mineev 1977). Similarly, in some models of RQFT, a particle that moves around an Alice string (Fig. 15.2) continuously flips its charge, or parity, or enters the 'shadow' world (Schwarz 1982; Schwarz and Tyupkin 1982; Silagadze 2001; Foot and Silagadze 2001).

Let us consider such a flip of the quantum number in an example of a macroscopic body with a spin magnetization $\mathbf{M}$ immersed in ^{3}He-A. Since $\hat{\mathbf{d}}$ is the vector of magnetic anisotropy in ^{3}He-A, the magnetization of the object is aligned with $\hat{\mathbf{d}}$ due to the interaction term $-(\mathbf{M} \cdot \hat{\mathbf{d}})^2$. When this body slowly moves around a half-quantum vortex, its magnetization follows the direction of the anisotropy axis. Finally, after the body makes the complete circle, it will find that its magnetization is reversed with respect to the magnetization of those objects which did not follow the topologically non-trivial route. In this example, the $\hat{\mathbf{d}}$-vector is the spin quantization axis, that is why the object flips its $U(1)_{S_z}$ charge – the spin. In the same manner the electric or the topological charges are reversed around the corresponding Alice string.

There are several non-trivial quantum mechanical phenomena related to the flip of a quantum number around an Alice string. In particular, this leads to the Aharonov–Bohm effect experienced by some collective modes in the presence of a half-quantum vortex (Khazan 1985; Salomaa and Volovik 1987; Davis and Martin 1994).

Using the closed loop of the corresponding Alice string, one can produce the baryonic charge (or other charge) from the vacuum by creating the baryon–antibaryon pair and forcing the antibaryon to move through the loop. In this way the antibaryon transforms to the baryon, and one gains the double baryonic charge from the vacuum. In this process the loop of the Alice string acquires the opposite charge distributed along the string – the so-called Cheshire charge (Alford et al. 1990).

15.3.3 Fractional flux in chiral superconductor

In ^{3}He-A the fractional vorticity is still to be observed. However, its discussion extended to unconventional superconductivity led to predictions of half-quantum vortices in superconductors (Geshkenbein et al. 1987); and finally such a vortex was discovered by Kirtley et al. (1996). It was topologically pinned by the intersection line of three grain boundary planes in a thin film of a cuprate superconductor, YBa$_2$Cu$_3$O$_{7-\delta}$ (see Sec. 15.3.4).

In unconventional superconductors, the $U(1)_Q$ gauge symmetry is broken together with some symmetry of the underlying crystal, which is why the crystalline structure of the superconductor becomes important. Let us first start with the axial or chiral superconductor, whose order parameter structure is similar to that in ^{3}He-A. The possible candidate is the superconductivity in Sr$_2$RuO$_4$, whose crystal structure has tetragonal symmetry. In the simplest representation, the ^{3}He-A order parameter – the off-diagonal element in eqn (7.57) – adapted to the crystals with tetragonal symmetry has the following form:

$$\Delta(\mathbf{p}) = \Delta_0 \, (\hat{\mathbf{d}} \cdot \sigma) \, (\sin \mathbf{p} \cdot \mathbf{a} + i \, \sin \mathbf{p} \cdot \mathbf{b}) \, e^{i\theta} \, . \tag{15.20}$$

Here θ is the phase of the order parameter (here we use θ instead of Φ to distinguish the order parameter phase from the magnetic flux Φ); $\mathbf{a}$ and $\mathbf{b}$ are the elementary vectors of the crystal lattice within the layer. When $|\mathbf{p} \cdot \mathbf{a}|/\hbar \ll 1$ and $|\mathbf{p} \cdot \mathbf{b}|/\hbar \ll 1$, the order parameter acquires the familiar form applicable to liquids with triplet p-wave pairing: $\Delta(\mathbf{p}) = e_{\mu i} p^i \sigma_\mu$, with $e_{\mu i} \propto \hat{d}_\mu(\hat{a}_i + i\hat{b}_i)$.

Vortices with fractional winding number n_1 can be constructed in two ways. The traditional way discussed for liquid ^{3}He-A is applicable to superconductors when the $\hat{\mathbf{d}}$-vector is not strongly fixed by the crystal fields, and is flexible enough. Then one obtains the analog of $n_1 = 1/2$ vortex in eqn (15.18): after circling around this Alice string, $\hat{\mathbf{d}} \to -\hat{\mathbf{d}}$, while the phase of the order parameter $\theta \to \theta + \pi$. According to eqn (15.17) such a vortex traps the magnetic flux $\Phi = hc/4e$, which is one-half of the flux trapped by a conventional Abrikosov vortex having $n_1 = 1$.

In another scenario in Fig. 15.3, the crystalline properties of the chiral superconductor are exploited. Twisting the crystal axes $\mathbf{a}$ and $\mathbf{b}$ in the closed wire of a tetragonal superconductor, one obtains an analog of the Möbius strip geometry (Volovik 2000b). The closed loop traps the fractional flux, if it is twisted by an angle $\pi/2$ before gluing the ends. Since the local orientation of the crystal lattice continuously changes by $\pi/2$ around the loop, axes $\mathbf{a}$ and $\mathbf{b}$ transform to each

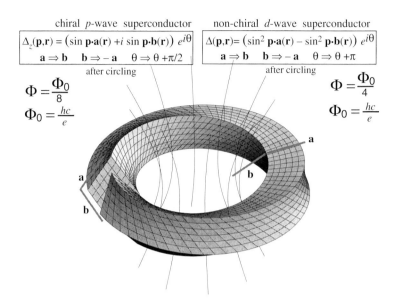

FIG. 15.3. Fractional flux in unconventional superconductors trapped by a topologically nontrivial loop. The twisted loop of a crystalline wire with tetragonal symmetry represents the disclination in crystal and has a geometry of 1/2 of the Möbius strip.

other after circling along the loop, and thus the order parameter is multiplied by i. The single-valuedness of the order parameter in eqn (15.20) requires that this change must be compensated by a change of the phase θ by $\pi/2$. Thus in the ground state, the phase winding around the twisted loop is $\pi/2$, and the circulation trapped by the loop is $n_1 = 1/4$ of the circulation quantum.

Applying eqn (15.17) for the magnetic flux in terms of the winding number, one would find that the loop of the chiral p-wave superconductor in the ground state traps a quarter of the magnetic flux $\Phi_0/2$ of the conventional Abrikosov vortex, i.e. $\Phi = \Phi_0/8$. However, this is not exactly so. Because of the breaking of time reversal symmetry in chiral crystalline superconductors, persistent electric current in the loop arises not only due to the order parameter phase θ but also due to deformations of the crystal axes (Volovik and Gor'kov 1984):

$$\mathbf{j} = en_s \left(\mathbf{v}_s - \frac{e}{mc}\mathbf{A} \right) + eC\hat{a}_i \nabla \hat{b}_i \ , \quad \mathbf{v}_s = \frac{\hbar}{2m}\nabla\theta \ . \qquad (15.21)$$

The parameter C is non-zero if the time reversal symmetry T is broken, and this modifies eqn (15.17). Now the condition of no electric current in the wire, $\mathbf{j} = 0$, leads to the following magnetic flux trapped by the loop:

$$\Phi = \frac{1-\tilde{C}}{8}\Phi_0 = \frac{1-\tilde{C}}{8}\frac{hc}{e} \ , \qquad (15.22)$$

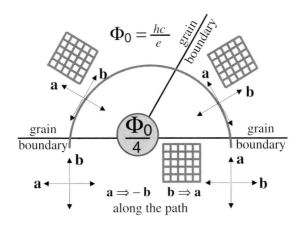

FIG. 15.4. Fractional flux is topologically trapped by intersection of three grain boundaries, according to experiments by Kirtley *et al.* (1996) in high-temperature cuprate superconductor.

where $\tilde{C} = 2mC/\hbar n_s$. If the underlying crystal lattice has hexagonal symmetry, and the wire is twisted by $\pi/3$, the trapped flux will be $\Phi = (1 - \tilde{C})\Phi_0/12$. In the limit case of $C = 0$, eqn (15.17) is restored, and the flux trapped by the loop becomes a quarter or a sixth of the conventional flux quantum in superconductors.

The supercurrent due to the deformation of the crystal lattice in eqn (15.21) can be considered as coming from the A-phase orbital vectors $\hat{\mathbf{m}}$ and $\hat{\mathbf{n}}$ trapped by crystal fields, $\hat{\mathbf{m}} = \hat{\mathbf{a}}$ and $\hat{\mathbf{n}} = \hat{\mathbf{b}}$, while the $\hat{\mathbf{l}}$-vector is trapped along the normal to the crystal layers. If the superconducting state is liquid, such as hypothetical electrically charged ^{3}He-A, its current is $\mathbf{j} = en_s\left(\mathbf{v}_s - \frac{e}{mc}\mathbf{A}\right)$, where $\mathbf{v}_s = \frac{\hbar}{2m}(\nabla\theta + \hat{m}_i\nabla\hat{n}_i)$. This corresponds to $\tilde{C} = 1$ in eqn (15.22), and thus if the $\hat{\mathbf{l}}$-vector is along the loop there is no trapped flux in the ground state of the liquid chiral superconductor. The fractional flux is trapped by the wire only because of the underlying crystal structure.

15.3.4 Half-quantum flux in d-wave superconductor

In the non-chiral (i.e. with conserved time reversal symmetry) spin-singlet superconductor in layered cuprate oxides, the order parameter can be represented by

$$\Delta(\mathbf{p}) = \Delta_0\left(\sin^2 \mathbf{p} \cdot \mathbf{a} - \sin^2 \mathbf{p} \cdot \mathbf{b}\right)e^{i\theta} . \quad (15.23)$$

In the liquid superconductors, or when $|\mathbf{p}\cdot\mathbf{a}|/\hbar \ll 1$ and $|\mathbf{p}\cdot\mathbf{b}|/\hbar \ll 1$, the order parameter acquires the more familiar d-wave form $\Delta(\mathbf{p}) = d_{x^2-y^2}(p_x^2 - p_y^2)$. The same twisted loop of superconducting wire in Fig. 15.3, with $\mathbf{a} \to \mathbf{b}$ and $\mathbf{b} \to -\mathbf{a}$ after circling, leads to the change of sign of the order parameter after circumnavigating along the loop. This must be compensated by a change of phase θ by π. As a result one finds that in the ground state the loop traps $n_1 = 1/2$ or $n_1 = -1/2$ of circulation quantum. The ground state has thus two-fold degeneracy, with the magnetic flux trapped by the loop being exactly $\pm\Phi_0/4$. This is because the time reversal symmetry is not broken in cuprate superconductors, and thus the crystal deformations do not produce an additional supercurrent in eqn (15.21): the parameter $C = 0$. Note that the observation of fractional flux different from $\Phi_0/4$, or $\Phi_0/2$, would indicate breaking of the time reversal symmetry (Sigrist *et al.* 1989, 1995; Volovik and Gor'kov 1984).

The same reasoning gives rise to the $\pm\Phi_0/4$ flux attached to the tricrystal line (Kirtley *et al.* 1996): circling around this line one finds that the crystal axes transform in the same way as in the twisted wire: $\mathbf{a} \to \mathbf{b}$ and $\mathbf{b} \to -\mathbf{a}$ (Fig. 15.4). This is an example of the topological interaction of the defects. In superfluid ^{3}He the half-quantum vortex is topologically attached to the disclination in the $\hat{\mathbf{d}}$-field with winding number $\nu = 1/2$. The latter is the termination line of the soliton, as we shall see in Sec. 16.1.1. The fractional vortex observed by Kirtley *et al.* (1996) is topologically coupled with the junction line of three grain boundaries. In the case of the twisted wire in Fig. 15.3, the fractional vortex is attached to the linear topological defect of the crystal lattice – disclination in the field of crystal axes. The empty space inside the loop thus represents the common core of the $n_1 = 1/2$ vortex and of the disclination in crystal with the winding number $\nu = 1/4$.

We must also mention the vortex sheet with fractional vortices, which has been predicted to exist in chiral superconductors by Sigrist *et al.* (1989) before the experimental identification of the vortex sheet in ^{3}He-A by Parts *et al.* (1994*b*) (see more on the vortex sheet in Sec. 16.3). The object in chiral superconductors which traps vorticity is the domain wall separating domains with opposite orientations of the $\hat{\mathbf{l}}$-vector discussed by Volovik and Gor'kov (1985). Such a domain wall has a kink – the Bloch line – which represents the vortex with the winding number $n_1 = 1/2$. When there are many trapped fractional vortices (kinks), they form the vortex sheet analogous to that in ^{3}He-A (see Sec. 16.3.4).

The half-quantum vortex – the Alice string – has also been suggested by Leonhardt and Volovik (2000) to exist in Bose–Einstein condensates with a hyperfine spin $F = 1$.

16
CONTINUOUS STRUCTURES

16.1 Hierarchy of energy scales and relative homotopy group

When several distinct energy scales are involved the vacuum symmetry is different for different length scales: the larger the length scale, the more the symmetry is reduced. The interplay of topologies on different length scales gives rise to many different types of topological defects, which are described by relative homotopy groups (Mineev and Volovik 1978). Here we discuss the continuous structures generated by this group, and in Chapter 17 the combinations of topological defects of different co-dimensions will be discussed.

16.1.1 Soliton from half-quantum vortex

Let us consider what happens with the half-quantum vortex (Sec. 15.3.1), when the spin–orbit interaction between $\hat{\mathbf{l}}$- and $\hat{\mathbf{d}}$- fields is turned on:

$$F_D = -g_D \left(\hat{\mathbf{l}} \cdot \hat{\mathbf{d}}\right)^2 . \tag{16.1}$$

The parameter g_D is connected with the dipole length ξ_D in eqn (12.5) as $g_D \sim n_s \hbar^2/m\xi_D^2$. It is also instructive to express it in terms of the characteristic energy E_D and Planck energy scale:

$$g_D \sim \sqrt{-g} E_D^2 E_{\text{Planck 2}}^2 . \tag{16.2}$$

This interaction requires that $\hat{\mathbf{l}}$ and $\hat{\mathbf{d}}$ must be aligned ('dipole-locked'): $\hat{\mathbf{l}} \cdot \hat{\mathbf{d}} = +1$ or $\hat{\mathbf{l}} \cdot \hat{\mathbf{d}} = -1$. But in the presence of a half-quantum vortex it is absolutely impossible to saturate the spin–orbit energy everywhere: the $\hat{\mathbf{d}}$-vector flips its direction around the Alice string, while the $\hat{\mathbf{l}}$-vector does not. That is why $\hat{\mathbf{l}}$ cannot follow $\hat{\mathbf{d}}$ everywhere, and the alignment ('dipole-locking') must be violated somewhere.

As in the case of the spin vortex in ^{3}He-B disturbed by the spin–orbit interaction in Sec. 14.1.5, the region where the vacuum energy (16.1) is not saturated forms a soliton in Fig. 16.1 terminated by the half-quantum vortex; for walls terminated by strings see Kibble (2000), Volovik and Mineev (1977), and Mineev and Volovik (1978). The opposite end of the soliton may be on a second Alice string, or it may be anchored to the wall of the container as in Fig. 14.1.

The thickness of the soliton wall is determined by the competition of the spin–orbit interaction and the gradient energy and is of order of the dipole length $\xi_D \sim 10^{-3}$ cm. The same occurs for solitons in ^{3}He-B (see Sec. 14.1.5). Solitons often appear in ^{3}He-A after cool-down into the superfluid state. Its existence

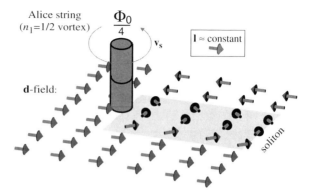

FIG. 16.1. Half-quantum vortex generates the $\hat{\mathbf{d}}$-soliton in ^{3}He-A. The arrows indicate the local direction of the vector $\hat{\mathbf{d}}$.

is displayed in cw NMR experiments as a special satellite peak in the NMR absorption as a function of excitation frequency.

How many topologically distinct solitons do we have in ^{3}He-A? First, since the $\hat{\mathbf{d}}$ texture in eqn (15.18) is the same for both half-quantum vortices, with $n_1 = +1/2$ and $n_1 = -1/2$, they give rise to the same soliton. Second, if we consider the pair of half-quantum vortices, then, on any path surrounding both strings, the $\hat{\mathbf{d}}$-field becomes single-valued. Thus outside the pair of any half-quantum vortices, the $\hat{\mathbf{d}}$-field loses its Alice behavior, so that the $\hat{\mathbf{l}}$-vector can follow $\hat{\mathbf{d}}$ saturating the vacuum spin–orbital energy. This demonstrates that any two solitons can kill each other, i.e. the summation law for the topological charges of solitons is $1 + 1 = 0$. A soliton equals its anti-soliton. Thus there is only one topologically stable soliton described by the non-trivial element of the Z_2 group.

This also means that the soliton in the $\hat{\mathbf{d}}$-field with constant $\hat{\mathbf{l}}$-vector in Fig. 16.1, and the soliton in the $\hat{\mathbf{l}}$-field with constant $\hat{\mathbf{d}}$-vector in Fig. 16.2, belong to the same topological class: they can be continuously transformed to each other. Typically the real solitons are neither $\hat{\mathbf{d}}$-solitons, nor $\hat{\mathbf{l}}$-solitons: both $\hat{\mathbf{d}}$ and $\hat{\mathbf{l}}$ change across the soliton with $\hat{\mathbf{l}} = \hat{\mathbf{d}}$ on one side of the soliton and $\hat{\mathbf{l}} = -\hat{\mathbf{d}}$ on the other.

These topological properties of the soliton can be obtained by direct calculations of the relative homotopy group.

16.1.2 *Relative homotopy group for soliton*

At first glance the $\hat{\mathbf{l}}$-soliton in Fig. 16.2 is equivalent to the domain wall in ferromagnets. The $\hat{\mathbf{l}}$-vector is a ferromagnetic vector, because the time reversal operation reverses its direction, $T\hat{\mathbf{l}} = -\hat{\mathbf{l}}$. The $\hat{\mathbf{l}}$-vector shows the direction of the orbital magnetism of Cooper pairs. On the other hand, the $\hat{\mathbf{d}}$-vector shows the direction of the 'easy axis' of the magnetic anisotropy in eqn (16.1). Thus the soliton wall separates domains with opposite orientations of the magnetization

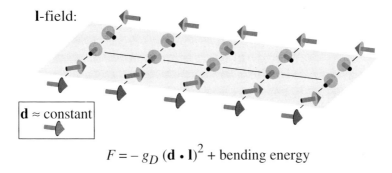

FIG. 16.2. The $\hat{\mathbf{l}}$-soliton in ^{3}He-A. The arrows indicate the local direction of the vector $\hat{\mathbf{l}}$. The soliton in the $\hat{\mathbf{l}}$ field belongs to the same topological class as the $\hat{\mathbf{d}}$-soliton in Fig. 16.1.

along the easy axis: with $\hat{\mathbf{l}} = \hat{\mathbf{d}}$ on one side and $\hat{\mathbf{l}} = -\hat{\mathbf{d}}$ on the other one.

Nevertherless the discussed soliton is not the domain wall and it is described by essentially different topology.

The vacuum manifold R in ferromagnets with an easy axis consists of two points only: $\mathbf{M} = \pm M_0 \hat{\mathbf{z}}$, if $\hat{\mathbf{z}}$ is the direction of an easy axis. That is why $\mathrm{R} = Z_2$, and the only possible non-trivial topology comes from the zeroth homotopy set $\pi_0(\mathrm{R}) = Z_2$, which shows the number of disconnected pieces of the vacuum manifold. Since vacua are disconnected, the domain wall cannot terminate inside the sample: it must either stretch 'from horizon to horizon' (i.e. from one boundary of the system to another) or form a closed surface. The same is valid for the Fermi surface in **p**-space (and also for the topologically stable vortex line in **r**-space): it must be either infinite or form a closed surface (a closed loop).

On the contrary, the vacua on two sides of the soliton belong to the same piece of the vacuum manifold. One can go from one vacuum to the other continuously, simply by going around the Alice string. This means that it is possible to drill a hole in the soliton, an operation which is absolutely impossible for the domain walls. The hole is bounded by the topological line defect – the string loop (Fig. 16.3). In our case it is the Alice string which can terminate the soliton, while for the ^{3}He-B soliton it is the spin vortex or spin–mass vortex.

This is the reason why the solitons are topological objects – their topology is determined by the topology of the string on which the soliton plane can terminate. But this topology is essentially different from the π_0 topology of the domain wall. In strict mathematical terms solitons correspond to the non-trivial elements of the relative homotopy group. The relevant relative homotopy group for the soliton in ^{3}He-A is

$$\pi_1(G/H_A, \tilde{G}/\tilde{H}_A) \ . \tag{16.3}$$

This group describes the classes of the following mapping of the path C crossing the soliton in Fig. 16.3 to the order parameter space. The part of the path within

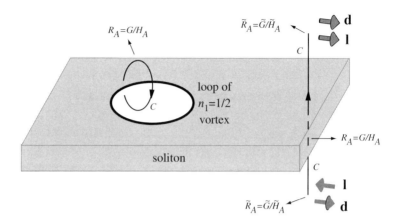

FIG. 16.3. A topological soliton can terminate on a topologically stable string. The soliton in ^{3}He-A can terminate on a half-quantum vortex. The hole in the soliton is bounded by the loop of the Alice string.

the soliton where there is no dipole locking is mapped to the vacuum manifold of the A-phase, $R_A = G/H_A = (SO(3) \times S^2)/Z_2$.

The end parts of the path are far from the soliton, where the spin–orbit interaction is saturated. These parts must be mapped to the order parameter space $\tilde{R}_A = \tilde{G}/\tilde{H}_A$, which is restricted by the dipole locking of $\hat{\mathbf{l}}$- and $\hat{\mathbf{d}}$- vectors. Since orbital and spin degrees of freedom are coupled at large distance, the large-distance symmetry group of the physical laws is $\tilde{G} = SO(3)_J \times U(1)_N$; its subgroup representing the symmetry group of the vacuum state is $\tilde{H}_A = U(1)$; and the vacuum manifold is $\tilde{R}_A = \tilde{G}/\tilde{H}_A = SO(3)$. This is the space of the solid rotations of the triad $\hat{\mathbf{m}}$, $\hat{\mathbf{n}}$ and $\hat{\mathbf{l}}$, while $\hat{\mathbf{d}}$ is dipole-locked with $\hat{\mathbf{l}}$. Formal calculations give

$$\pi_1(R_A, \tilde{R}_A) = \pi_1((SO(3) \times S^2)/Z_2, SO(3)) = Z_2 \ . \tag{16.4}$$

The same result can be obtained by the following consideration. The path C can be deformed and closed so that it encircles the half-quantum vortex (Fig. 16.3). The mapping of the closed path C to the vacuum manifold G/H_A gives rise to the homotopy group $\pi_1(G/H_A) = Z_4$. Two elements of this group correspond to two half-quantum vortices, with $n_1 = +1/2$ and $n_1 = -1/2$. Since both Alice strings give rise to the same soliton, the group describing the soliton is $Z_4/Z_2 = Z_2$.

16.1.3 How to destroy topological solitons and why singularities are not easily created in ^{3}He

The essential difference in the topological properties of the domain walls, described by the symmetry group $\pi_0(R)$, and the soliton walls, described by the relative homotopy group $\pi_1(R, \tilde{R})$, results in different energy barriers separating them from topologically trivial configurations. To destroy the domain wall one must restore the broken symmetry in the half-space on the left or on the right of the wall. The corresponding energy barrier is thus proportional to L^3, where L is the size of the system.

To destroy the soliton it is sufficient to drill a hole and form a vortex loop of the size on the order of the thickness of the soliton. After that the surface tension of the wall exceeds the linear tension of the string and it becomes energetically favorable for the string to grow and eat the soliton. Thus the corresponding energy barrier is the energy required for the nucleation of the vortex loop of the size ξ_D. This energy is finite: it does not depend on the size of the system. In the same way, nucleation of the loop of spin vortex of the size ξ_D can destroy the soliton in ^{3}He-B.

However, for solitons in both phases of ^{3}He the energy barrier – the energy of the optimal size of string loop – normalized to the transition temperature is $E_{\text{barrier}}/T_c \sim p_F^2 \xi_D \xi/\hbar^2 \sim 10^8$. Though the barrier is finite, it is huge. That is why topological solitons in both superfluid phases of ^{3}He are extremely stable. The reason for this is that the characteristic lengths ξ and ξ_D, which determine the core size of string and soliton respectively, are both big in superfluid ^{3}He compared to the interatomic space $a_0 \sim \hbar/p_F$. Singularities are not easily created in superfluid ^{3}He.

16.2 Continuous vortices, skyrmions and merons

16.2.1 *Skyrmion – Anderson–Toulouse–Chechetkin vortex texture*

According to the Landau picture of superfluidity, the superfluid flow is potential: its velocity $\mathbf{v}_s$ is curl-free: $\nabla \times \mathbf{v}_s$. Later Onsager (1949) and Feynman (1955) found that this statement must be generalized: $\nabla \times \mathbf{v}_s \neq 0$ at singular lines, the quantized vortices, around which the phase of the order parameter winds by $2\pi n_1$. The discovery of superfluid ^{3}He-A further weakened the rule: the non-singular vorticity can be produced by the regular texture of the order parameter according to the Mermin–Ho relation in eqn (9.17).

Let us consider the structure of the continuous vortex in its simplest axisymmetric form, as it was first discussed by Chechetkin (1976) and Anderson and Toulouse (1977) (ATC vortex, Fig. 16.4). It has the following distribution of the $\hat{\mathbf{l}}$-field:

$$\hat{\mathbf{l}}(\rho, \phi) = \hat{\mathbf{z}} \cos \eta(\rho) + \hat{\rho} \sin \eta(\rho) . \quad (16.5)$$

Here $\hat{\mathbf{z}}$, $\hat{\rho}$ and $\hat{\phi}$ are unit vectors of the cylindrical coordinate system; $\eta(\rho)$ changes from $\eta(0) = 0$ to $\eta(\infty) = \pi$. Such $\hat{\mathbf{l}}$-texture forms the so-called soft core of the vortex, since it is the region of texture which contains a non-zero vorticity of superfluid velocity in eqn (9.16):

$$\mathbf{v}_s(\rho, \phi) = \frac{\hbar}{2m\rho}[1 - \cos\eta(\rho)] \hat{\phi} \;\; , \;\; \nabla \times \mathbf{v}_s = \frac{\hbar}{2m} \sin\eta \, \partial_\rho \eta \, \hat{\mathbf{z}} \; . \quad (16.6)$$

In comparison to a more familiar singular vortex, the continuous vortex has a regular superfluid velocity field $\mathbf{v}_s$, with no singularity on the vortex axis.

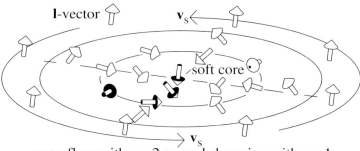

superflow with $n_1=2$ around skyrmion with $n_2=1$

FIG. 16.4. Continuous vortex-skyrmion. The arrows indicate the local direction of the order parameter vector $\hat{\mathbf{l}}$. In ^{3}He-A the winding number of the Anderson–Toulouse–Chechetkin vortex is $n_1 = 2$.

However, the circulation of the superfluid velocity about the soft core is still quantized: $\kappa = \oint d\mathbf{r} \cdot \mathbf{v}_s = 2\kappa_0$. This is twice the conventional circulation quantum number in the pair-correlated system, $\kappa_0 = \pi\hbar/m$, i.e. far from the soft core this object is viewed as the vortex with the winding number $n_1 = 2$.

Quantization of circulation for continuous vortex is related to the topology of the $\hat{\mathbf{l}}$-field. By following the $\hat{\mathbf{l}}$-field in the cross-section of the vortex texture, it is noted that all possible 4π directions of the $\hat{\mathbf{l}}$-vector on its unit sphere are present here. Such a 4π topology of the unit vector orientations in 2D is known as a *skyrmion*. This topology ensures two quanta of circulation $n_1 = 2$ according to the Mermin–Ho relation in eqn (9.17):

$$\oint d\mathbf{r} \cdot \mathbf{v}_s = \int d\mathbf{S} \cdot (\nabla \times \mathbf{v}_s) = \frac{\hbar}{2m} \int dx\, dy\, \hat{\mathbf{l}} \cdot \left(\frac{\partial \hat{\mathbf{l}}}{\partial x} \times \frac{\partial \hat{\mathbf{l}}}{\partial y} \right) = 4\pi \frac{\hbar}{2m} \ . \quad (16.7)$$

In the general case the winding number n_1 can be related to the degree n_2 of the mapping of the cross-section of the soft core to the sphere S^2 of unit vector $\hat{\mathbf{l}}$:

$$n_1 = \frac{m}{\pi\hbar} \oint d\mathbf{r} \cdot \mathbf{v}_s = \frac{1}{2\pi} \int dx\, dy\, \hat{\mathbf{l}} \cdot \left(\frac{\partial \hat{\mathbf{l}}}{\partial x} \times \frac{\partial \hat{\mathbf{l}}}{\partial y} \right) = 2n_2 \ . \quad (16.8)$$

This relation demonstrates the topological structure of this object. Since n_1 is even, this configuration belongs to the trivial class of topological defects described by the first homotopy group in the classification scheme of Sec. 15.2. That is why there is no singularity: any singularity which is not supported by topology can be continuously dissolved, so that finally the system everywhere will be in the vacuum manifold of ^{3}He-A. However, the continuous configurations are not necessarily topologically trivial: they are described by a finer topological scheme utilizing the relative homotopy group. In a given case it is the group $\pi_2(SO(3), U(1))$ which describes the following hierarchical mapping: the disk – the cross-section of the soft core – is mapped to the $SO(3)$ manifold of the

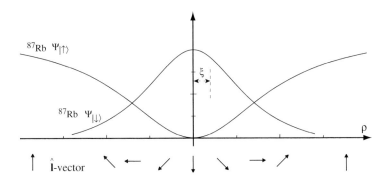

FIG. 16.5. Continuous vortex-skyrmion with $n_1 = n_2 = 1$ in two component Bose condensate. The 'isotopic spin' – the $\hat{\mathbf{l}}$-vector (shown by arrows) – sweeps the 4π solid angle in the soft core. Far from the core the vector $\hat{\mathbf{l}}$ is up, which means that only $\Psi_\uparrow$ component is present; this component has winding number $n_1 = 1$. Near the origin the vector $\hat{\mathbf{l}}$ is down which means that the core region is mainly filled by $\Psi_\downarrow$ component of the Bose condensate.

triad $\hat{\mathbf{m}}$, $\hat{\mathbf{n}}$ and $\hat{\mathbf{l}}$, while the boundary of the disk is mapped to the subspace $U(1) = SO(2)$ – the vacuum manifold which is left after $\hat{\mathbf{l}}$ is fixed outside the soft core.

The relation (16.8) also reflects the interplay of the **r**-space and **p**-space topologies, since the $\hat{\mathbf{l}}$-vector shows the position of the Fermi point in momentum space. It is a particular case of the general rule in eqn (23.32), which will be discussed in Sec. 23.3: in the Anderson–Toulouse–Chechetkin texture in eqn (16.5) the Fermi point with $N_3 = +2$ sweeps a 4π angle, while the Fermi point with $N_3 = -2$ sweeps a -4π solid angle, and as a result the winding number of the vortex is $n_1 = 2$.

16.2.2 Continuous vortices in spinor condensates

If the order parameter is a spinor (the 'half of vector'), as occurs in 2-component Bose condensates and in the Standard Model of electroweak interactions (eqn (15.6)), the corresponding skyrmions – the continuous vortices and the continuous cosmic strings (Achúcarro and Vachaspati 2000) – have a two times smaller winding number for the phase of the condensate, $n_1 = n_2$. Let us illustrate this in an example of a mixture of two Bose condensates in a laser-manipulated trap.

In the experiment one starts with a single Bose condensate which is denoted as the $|\uparrow\rangle$ component. Within this component a pure $n_1 = 1$ phase vortex with singular core is created. Next the hard core of the vortex is filled with the second component in the $|\downarrow\rangle$ state. As a result the core expands and becomes essentially larger than the coherence length ξ, and a vortex-skyrmion with continuous structure is obtained. Such a skyrmion has recently been observed with ^{87}Rb atoms by Matthews *et al.* (1999). It can be represented in terms of the $\hat{\mathbf{l}}$-vector which is constructed from the two components of the order parameter as in eqn

(15.7) for the direction of isotopic spin in the Standard Model

$$\begin{pmatrix} \Psi_\uparrow \\ \Psi_\downarrow \end{pmatrix} = |\Psi_\uparrow(\infty)| \begin{pmatrix} e^{i\phi} \cos \frac{\beta(\rho)}{2} \\ \sin \frac{\beta(\rho)}{2} \end{pmatrix} , \; \hat{\mathbf{l}} = (\sin \beta \cos \phi, -\sin \beta \sin \phi, \cos \beta) \; . \quad (16.9)$$

The polar angle $\beta(\rho)$ of the $\hat{\mathbf{l}}$-vector changes from 0 at infinity, where $\hat{\mathbf{l}} = \hat{\mathbf{z}}$ and only the $|\uparrow\rangle$ component is present, to $\beta(0) = \pi$ on the axis, where $\hat{\mathbf{l}} = -\hat{\mathbf{z}}$ and only the $|\downarrow\rangle$ component is present. Thus the vector $\hat{\mathbf{l}}$ sweeps the whole unit sphere, $n_2 = 1$, while the vortex winding number is also $n_1 = 1$. As distinct from the vector order parameter, where $n_1 = 2n_2$, for the spinor order parameter the skyrmion winding number n_2 and vortex winding number n_1 are equal: $n_2 = n_1$.

16.2.3 *Continuous vortex as a pair of merons*

The NMR measurements by Hakonen *et al.* (1983*a*) had already in 1982 provided an indication for the existence of skyrmions – the continuous 4π lines – in rotating ^{3}He-A. From that time many different types of continuous vorticity have been identified and their properties have been investigated in detail. This allowed us to use the continuous vortices for 'cosmological' experiments. In particular, the investigation of the dynamics of these vortices by Bevan *et al.* (1997*b*) presented the first demonstration of the condensed matter analog of the axial anomaly in RQFT, which is believed to be responsible for the present excess of matter over antimatter (see Sec. 18.4).

Another cosmological phenomenon related to nucleation of skyrmions has been simulated by Ruutu *et al.* (1996*b*, 1997). The onset of helical instability, which triggers formation of these vortices, is described by the same equations and actually represents the same physics as the helical instability of the bath of right-handed electrons toward formation of the helical hypermagnetic field discussed by Joyce and Shaposhnikov (1997) and Giovannini and Shaposhnikov (1998). This experiment thus supported the Joyce–Shaposhnikov scenario of formation of the primordial cosmological magnetic field (see Chapter 20).

This is why it is worthwhile to look at the real structure of these objects.

The typical structure of the doubly quantized continuous vortex line in the applied magnetic field needed to perform the NMR measurements (see Blaauwgeers *et al.* (2000) and references therein), is shown in Fig. 16.6. The topology of the $\hat{\mathbf{l}}$-vector and vorticity remain the same as in the axisymmetric ATC vortex in Fig. 16.4: all 4π directions of the $\hat{\mathbf{l}}$-vector are present here, which ensures two quanta of circulation $n_1 = 2$ trapped by skyrmions. However, the applied magnetic field changes the topology of the $\hat{\mathbf{d}}$-field and the symmetry of the $\hat{\mathbf{l}}$-field. The most important difference from the ATC vortex is that the magnetic field keeps the $\hat{\mathbf{d}}$-vector everywhere in the plane perpendicular to the magnetic field. Thus the $\hat{\mathbf{d}}$-vector cannot follow $\hat{\mathbf{l}}$: it is dipole-unlocked from $\hat{\mathbf{l}}$ in the region of size of dipole length $\xi_D \sim 10^{-3}$ cm. The dipole-unlocked region represents the soft core of the vortex, which serves as the potential well trapping the spin-wave modes. The standing spin waves localized inside the soft core are excited by NMR giving rise to the well-defined satellite peak in the NMR spectrum in Fig. 16.6. The

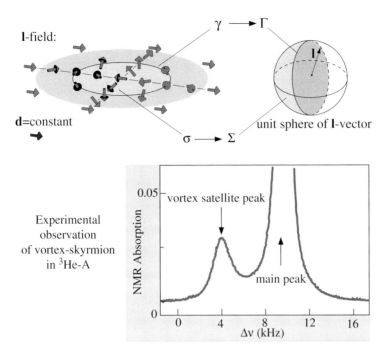

FIG. 16.6. *Top left*: An $n_1 = 2$ continuous vortex in ^{3}He-A. The arrows indicate the local direction of the order parameter vector $\hat{\mathbf{l}}$. Under experimental conditions the direction of the $\hat{\mathbf{l}}$-vector in the bulk liquid far from the soft core is kept in the plane perpendicular to the applied magnetic field. *Top right*: Illustration of the mapping of the space (x, y) to the unit sphere of the $\hat{\mathbf{l}}$-vector, and of continuous vorticity. Circulation of superfluid velocity along the contour γ is expressed in terms of the area $\Sigma(\Gamma)$ bounded by its image – the contour Γ, i.e. one has $\int_\sigma d\mathbf{S} \cdot (\nabla \times \mathbf{v}_s) = \oint_\gamma d\mathbf{r} \cdot \mathbf{v}_s = (\hbar/2m)\Sigma(\Gamma)$. In the whole cross-section of the Anderson–Toulouse–Chechetkin vortex, the $\hat{\mathbf{l}}$-vector covers the whole 4π sphere within the soft core. As a result there is 4π winding of the phase of the order parameter around the soft core, which corresponds to $n_1 = 2$ quanta of anticlockwise circulation. *Bottom*: The NMR absorption in the characteristic vortex satellite originates from the soft core where the $\hat{\mathbf{l}}$ orientation deviates from the homogeneous alignment in the bulk. Each soft core contributes equally to the intensity of the satellite peak and gives a practical tool for measuring the number of vortices.

position of the peak shows the type of the skyrmion, while the intensity of the peak gives the information on the number of skyrmions of this type. In 2000 the single-vortex sensitivity was reached and the quantization number $n_1 = 2$ of the vortex-skyrmion was verified directly by Blaauwgeers *et al.* (2000) [60].

The magnetic field also disturbs the axisymmetric structure of the ATC vortex: due to the spin–orbit coupling, far from the core the $\hat{\mathbf{l}}$-vector must be kept

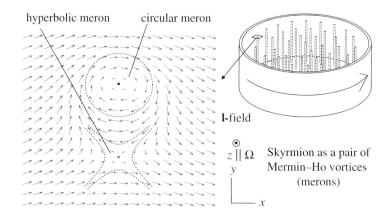

FIG. 16.7. Continuous vortex in applied magnetic field as a pair of merons. In each meron the $\hat{\mathbf{l}}$-vector covers half of the unit sphere. Merons serve as potential wells for spin-wave modes excited in NMR experiments.

in the plane perpendicular to the direction of the field, and this violates the axisymmetry. In Fig. 16.7 the projection of the $\hat{\mathbf{l}}$ to the plane perpendicular to the vortex line is shown, while $\hat{\mathbf{d}} = \hat{\mathbf{x}}$ everywhere. The vortex-skyrmion is divided into a pair of *merons* (Callan *et al.* 1977; Fateev *et al.* 1979; Steel and Negele 2000). $\mu\epsilon\rho o\sigma$ means fraction (Callan *et al.* 1977). In the ^{3}He literature such merons are known as Mermin–Ho vortices. In the complete skyrmion, the $\hat{\mathbf{l}}$-vector sweeps the whole unit sphere while each meron, or Mermin–Ho vortex, covers only the orientations in one hemisphere and therefore carries one quantum of vorticity, $n_1 = 1$. The meron covering the northern hemisphere forms a circular 2π Mermin–Ho vortex, while the meron covering the southern hemisphere is the hyperbolic 2π Mermin–Ho vortex. The division of a skyrmion into two merons is not artificial. The centers of the merons correspond to minima in the potential for spin waves. Also, approaching the transition to ^{3}He-A_1, the merons become well separated from each other, i.e. the distance between them grows faster than their size (Volovik and Kharadze 1990).

Skyrmions and merons are popular structures in physics. For instance, the model of nucleons as solitons was proposed 40 years ago by Skyrme (1961) (see the latest review on skyrmions by Gisiger and Paranjape 1998). Merons in QCD were suggested by Steel and Negele (2000) to produce the color confinement. The double-quantum vortex in the form of a pair of merons similar to that in ^{3}He-A was also discussed in the quantum Hall effect where it is formed by pseudospin orientations in the magnetic structure (see e.g. Girvin 2000), etc. Various schemes have recently been discussed (Ho 1998; Isoshima *et al.* 2000; Marzlin *et al.* 2000) by which a meron can be created in a Bose condensate formed within a 3-component $F = 1$ manifold.

16.2.4 Semilocal string and continuous vortex in superconductors

In the chiral superconductor (in superconductors with A-phase-like order parameter) the $\hat{\mathbf{l}}$-vector is fixed by the crystal lattice, and thus the continuous vorticity does not exist there. The continuous vortices become possible if the electromagnetic $U(1)_Q$ group is mixed not with orbital rotations but with the spin rotations. In principle the spin–orbit coupling between the electronic spins and the crystal lattice can be small, so that the spin rotation group $SO(3)_\mathbf{S}$ can be almost exact. Then the following symmetry-breaking scheme is possible:

$$G = SO(3)_\mathbf{S} \times U(1)_Q \to H = U(1)_{S_z - Q/2} . \qquad (16.10)$$

The spin part of the order parameter corresponding to this symmetry breaking is

$$\mathbf{e} = \Delta_0 \left(\hat{\mathbf{d}}' + i\hat{\mathbf{d}}'' \right) , \ \hat{\mathbf{d}}' \cdot \hat{\mathbf{d}}'' = 0 , \ |\hat{\mathbf{d}}'| = |\hat{\mathbf{d}}''| = 1 , \ \hat{\mathbf{d}}' \times \hat{\mathbf{d}}'' = \hat{\mathbf{s}} . \qquad (16.11)$$

Here $\hat{\mathbf{s}}$ is the orientation of spin of the Cooper pairs, which now plays the same role as the axial $\hat{\mathbf{l}}$-vector: the Mermin–Ho relation for continuous vorticity is obtained by substitution of $\hat{\mathbf{s}}$ instead of the orbital momentum $\hat{\mathbf{l}}$ in eqn (9.17). Thus in the analog of the ATC vortex in superconductors it is the spin $\hat{\mathbf{s}}$ which covers a solid angle of 4π in the soft core of the vortex with $n_1 = 2$. Such vortex a has been discussed by Burlachkov and Kopnin (1987).

The main difference from the corresponding scheme in superfluids where the group $U(1)_N$ is global, is that in superconductors this group is local: it is the gauge group $U(1)_Q$ of QED. As for the group $SO(3)_\mathbf{S}$ it is global in both superfluids and superconductors. Thus in the symmetry-breaking scheme in eqn (16.10) both local and global groups are broken and become mixed. In particle physics the topological defects resulting from breaking of the combination of the global and local groups are called semilocal defects (Preskill 1992; Achúcarro and Vachaspati 2000). Thus the Burlachkov–Kopnin (1987) vortex in superconductors represents the semilocal string in condensed matter. In the Standard Model one can obtain the 'semilocal' limit by using such a ratio of running couplings that the $SU(2)_L$ symmetry becomes effectively global. Semilocal electroweak defects in such a limit case have been discussed by Vachaspati and Achúcarro (1991).

In 'semilocal' superconductors the magnetic field can penetrate to the bulk of the superconductor due to $\hat{\mathbf{s}}$-texture, and one can construct the analog of the magnetic monopole in Sec. 17.1.5 using the $\hat{\mathbf{s}}$-hedgehog instead of the $\hat{\mathbf{l}}$-hedgehog.

16.2.5 Topological transition between continuous vortices

The continuous vortex-skyrmion is characterized by another topological invariant, which describes the topology of the unit vector $\hat{\mathbf{d}}$ of the axis of spontaneous magnetic anisotropy

$$n_2\{\hat{\mathbf{d}}\} = \frac{1}{4\pi} \int dx \, dy \, \hat{\mathbf{d}} \cdot \left(\frac{\partial \hat{\mathbf{d}}}{\partial x} \times \frac{\partial \hat{\mathbf{d}}}{\partial y} \right) . \qquad (16.12)$$

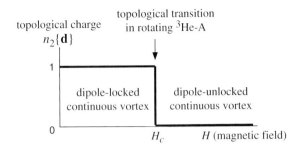

FIG. 16.8. Topological transition between continuous vortices. The vector field $\hat{\mathbf{d}}$ in the continuous vortex lattice is characterized by integer topological charge $n_2\{\hat{\mathbf{d}}\}$ – the degree of mapping of the isolated vortex core (or the elementary cell of the vortex lattice – torus) onto the sphere of the unit vector $\hat{\mathbf{d}}$.

This integer topological charge $n_2\{\hat{\mathbf{d}}\}$ is the degree of mapping of the cross-section of the isolated vortex core (or the elementary cell of the vortex lattice – the torus) onto the sphere of the unit vector $\hat{\mathbf{d}}$. An integer charge $n_2\{\hat{\mathbf{d}}\}$ can change only abruptly, which leads to the observed first-order topological phase transition in rotating ^{3}He-A when the magnetic field changes (Fig. 16.8). (Pecola et al. 1990) At zero or low field the $\hat{\mathbf{d}}$-vector is dipole-locked with $\hat{\mathbf{l}}$, and thus has the same winding $n_2\{\hat{\mathbf{d}}\} = n_2\{\hat{\mathbf{l}}\} = 1$. This vortex is the double skyrmion: both in $\hat{\mathbf{l}}$- and $\hat{\mathbf{d}}$-fields. At large field the $\hat{\mathbf{d}}$-field is kept in the plane perpendicular to the field and thus the vortex with the lowest energy has a dipole-unlocked core. In this vortex the $\hat{\mathbf{l}}$-vector topology remains intact, $n_2\{\hat{\mathbf{l}}\} = 1$, while the $\hat{\mathbf{d}}$-vector topology becomes trivial, $n_2\{\hat{\mathbf{d}}\} = 1$.

The double-skyrmion vortex with $n_2\{\hat{\mathbf{d}}\} = n_2\{\hat{\mathbf{l}}\} = 1$ can exist at high field as a metastable object, and this allows us to investigate this vortex in NMR experiments. The position of the satellite peak in the NMR line in Fig. 17.1 produced by this vortex is much closer to the main peak than the dipole-unlocked vortex (Parts et al. 1995a). This is why these two topologically different skyrmions can be observed simultaneously in the same NMR experiment.

16.3 Vortex sheet

16.3.1 Kink on a soliton as a meron-like vortex

Solitons and vortices, in both phases of superfluid ^{3}He, are related topologically. In A-phase the half-quantum vortex is the termination line of the soliton (Fig. 16.1), in the B-phase the soliton can terminate on a spin vortex or on a spin–mass vortex (Sec. 14.1.5). Here we discuss another topological interaction between these planar and linear objects: the topological defect of the soliton matter – a kink within the soliton – represents the continuous vortex with $n_1 = 1$. Such vortices bounded to the soliton core produce the vortex sheet, which appears

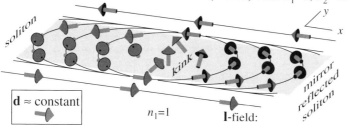
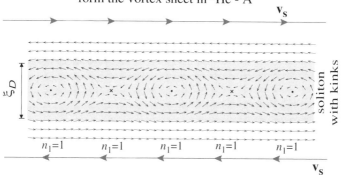

FIG. 16.9. Kink on the soliton is continuous vortex with $n_1 = 1$ (Mermin–Ho vortex or meron). The kink can live only within the soliton, as the Bloch line within the Bloch wall in magnets (see the books by Chen 1977 and Malozemoff and Slonczewski 1979). The chain of alternating circular and hyperbolic merons, with the same circulation $n_1 = 1$, forms the vortex sheet.

in the rotating vessel instead of the conventional $n_1 = 2$ continuous vortex-skyrmions (Parts *et al.* 1994*b*; Heinilä and Volovik 1995).

A kink in the soliton structure is the building block of the vortex sheet. Within the $\hat{\mathbf{l}}$-soliton in Fig. 16.2 the parity is broken. As a result there are two degenerate soliton structures (Fig. 16.9 *top*):

$$\hat{\mathbf{l}}(y) = \hat{\mathbf{x}} \cos \alpha(y) \pm \hat{\mathbf{y}} \sin \alpha(y) \;, \tag{16.13}$$

where $\alpha(-\infty) = 0$, $\alpha(+\infty) = \pi$. These two structures transform to each other by parity transformation. The domains with different degenerate structures can be separated by the domain wall (line), which is often called the kink. This kink has no singularity in the $\hat{\mathbf{l}}$-field and is equivalent to the Bloch line within the Bloch wall in magnets.

In ^{3}He-A the meron-like $\hat{\mathbf{l}}$-texture of the kink is the same as in the Mermin–Ho vortex: the $\hat{\mathbf{l}}$-vector sweeps the hemisphere and thus carries the $n_1 = \pm 1$ units of circulation quanta. Such continuous vortices with $n_1 = \pm 1$ can live only

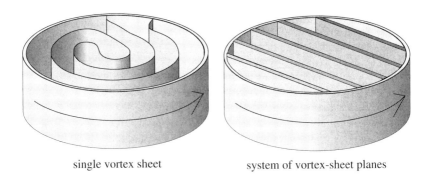

single vortex sheet   system of vortex-sheet planes

FIG. 16.10. Single and multiple vortex sheets in rotating vessel.

within the soliton. Outside the soliton they can live only in pairs forming the regular continuous vortices with $n_1 = \pm 2$.

16.3.2 Formation of a vortex sheet

A vortex sheet is the chain of alternating circular and hyperbolic kinks with the same orientation of circulation, say, all with $n_1 = 1$ (Fig. 16.9 *bottom*).

If there are no solitons in the vessel, then rotating the vessel results in the ordinary array of $n_1 = 2$ continuous vortices. However, if the rotation is started when there is a soliton plane parallel to the rotation axis, even at very slow acceleration of rotation the vortex sheet starts to grow. The new vorticity in the form of the kinks enters the soliton at the line of contact with the cylindrical cell wall. This is because the measured critical velocity of creation of a new kink within the soliton is lower compared to the measured critical velocity needed for nucleation of a conventional isolated vortex with $n_1 = 2$. That is why the vortex-sheet state can be grown in spite of its larger energy compared to the regular vortex state. The difference in critical velocities needed for nucleation of different structures is actually a very important factor which allows manipulation of textures, and even creation of the vortex state with a prescribed ratio between vortices with different structures.

16.3.3 Vortex sheet in rotating superfluid

When growing, the vortex sheet uniformly fills the rotating container by folding as illustrated in Fig. 16.10 *left*. Locally the folded sheet corresponds to a configuration with equidistant soliton planes. That is why we can consider an ideal system of planes shown in Fig. 16.10 *right*, which is, however, not so easy to reach in experiment.

Let us find b, the equilibrium distance between the planes of the vortex sheet, in a container rotating with $\mathbf{\Omega} \parallel \hat{\mathbf{z}}$. It is determined by the competition of the surface tension σ of the soliton and the kinetic energy of the counterflow $\mathbf{w} = \mathbf{v}_n - \mathbf{v}_s$ outside the sheet. The motion of the normal component, which corresponds to the vorticity $\nabla \times \mathbf{v}_n = 2\Omega\hat{\mathbf{z}}$, can be represented by the shear flow

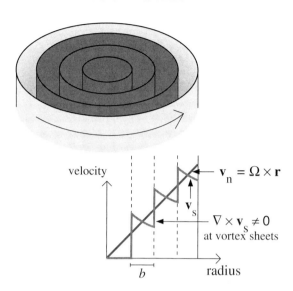

FIG. 16.11. Vortex-sheet array suggested by Landau and Lifshitz (1955) as the ground state of rotating superfluid ^{4}He.

$\mathbf{v}_n = -2\Omega y \hat{\mathbf{x}}$ parallel to the planes. In the gap between nearest planes, the vortex-free superfluid velocity $\mathbf{v}_s$ is constant and equals the average $\mathbf{v}_n$ to minimize the counterflow. Thus the velocity jump across the is $\Delta \mathbf{v}_s = 2\Omega b \hat{\mathbf{x}}$. The counterflow energy per volume is $(1/b) \int \frac{1}{2}\rho_{s\|} \mathbf{w}^2 dy = \frac{1}{6}\rho_{s\|}\Omega^2 b^2$, where $\rho_{s\|} = m n_{s\|}$ is the superfluid density for the flow along $\hat{\mathbf{l}}$. The surface energy per volume equals σ/b. Minimizing the sum of two contributions with respect to b one obtains

$$b = \left(\frac{3\sigma}{\rho_{s\|}\Omega^2}\right)^{1/3}. \tag{16.14}$$

This gives obout 0.3–0.4 mm at $\Omega = 1$ rad s^{-1} under the conditions of the experiment.

Equation (16.14) is the anisotropic version of the result obtained by Landau and Lifshitz (1955) for isotropic superfluid ^{4}He, where $\rho_{s\|} = \rho_s$. Before the concept of quantized vortices had been generally accepted, Landau and Lifshitz assumed that the system of coaxial cylindrical vortex sheets in Fig. 16.11 is the proper arrangement of vorticity in superfluid ^{4}He under rotation (historically the vortex sheet in superfluid ^{4}He was discussed even earlier by Onsager (unpublished) and London (1946). Though it turned out that in superfluid ^{4}He the vortex sheet is topologically and energetically unstable toward break-up into separated quantized vortex lines, the calculation made by Landau and Lifshitz (1955) happened to be exactly to the point for the topologically stable vortex sheet in ^{3}He-A.

The distance b between the sheets has been measured by Parts et al. (1994a). In addition to the vortex-sheet satellite peak in the NMR spectrum caused by

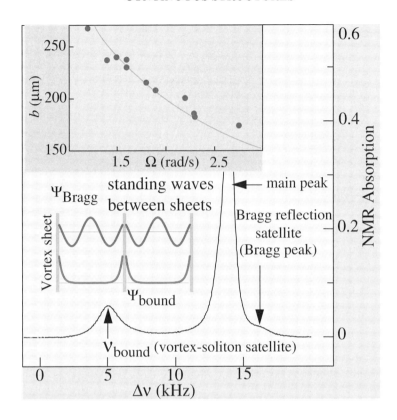

FIG. 16.12. Measurement of the distance between the planes of the vortex sheet using the Bragg reflection of the spin waves from the sheet (after Parts et al. 1994a).

the spin waves bound to the soliton (Fig. 16.12 *bottom left*), they resolved a small peak caused by the Bragg reflection of the spin waves from the equidistant sheet planes (Fig. 16.12 *bottom right*). The position of the Bragg peak as a function of Ω gives $b(\Omega)$ (Fig. 16.12 *top left*), which is in quantitative agreement with the Landau–Lifshitz equation (16.14).

The areal density of circulation quanta has the solid-body value $n_v = 2\Omega/\kappa_0$, as in the case of an array of singly quantized vortices (see eqn (14.2)): the vortex sheet also mimics the solid-body rotation of superfluid vacuum, $\langle \mathbf{v}_s \rangle = \mathbf{\Omega} \times \mathbf{r}$. This means that the length of the vortex sheet per two circulation quanta is $p = \kappa_0/(b\Omega)$, which is the periodicity of the order parameter structure in Fig. 16.9 *bottom* ($p \approx 180\,\mu\text{m}$ at $\Omega = 1\,\text{rad s}^{-1}$). The NMR absorption in the vortex-sheet satellite is proportional to the total volume of the sheet which in turn is proportional to $1/b \propto \Omega^{2/3}$. This non-linear dependence of the satellite absorption on rotation velocity is also one of the experimental signatures of the vortex sheet.

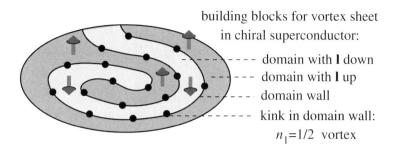

FIG. 16.13. Vortex sheet in chiral superconductor.

16.3.4 *Vortex sheet in superconductor*

The topologically stable vortex sheet has been discussed for chiral superconductors by Sigrist *et al.* (1989, 1995) and Sigrist and Agterberg (1999) (Fig. 16.13). Similar to superfluid ^{3}He-A, the vorticity is trapped in a wall separating two domains with opposite orientations of the $\hat{\bf l}$-vector. But, unlike the case of ^{3}He-A, this domain wall is not a continuous soliton described by the relative homotopy group: it is a singular defect described by the group π_0, as domain walls in ferromagnets. That is why the trapped kink is not a continuous meron but a singular vortex. Its winding is also two times smaller than that of the singular vortex in bulk superconductor: it has the fractional winding number $n_1 = 1/2$. If there are many trapped fractional vortices, then they form a vortex sheet, which as suggested by Sigrist and Agterberg (1999) can be responsible for the peculiarities in the flux-flow dynamics in the low-temperature phase of the heavy-fermionic superconductor UPt_3.

17

MONOPOLES AND BOOJUMS

17.1 Monopoles terminating strings

17.1.1 *Composite defects*

Composite defects exist in RQFT and in continuous media, if a hierarchy of energy scales with different symmetries is present. Examples are strings terminating on monopoles and walls bounded by strings. Many quantum field theories predict heavy objects of this kind that could appear only during symmetry-breaking phase transitions at an early stage in the expanding Universe (see reviews by Hindmarsh and Kibble (1995) and Vilenkin and Shellard (1994)). Various roles have been envisaged for them. For example, domain walls bounded by strings have been suggested by Ben-Menahem and Cooper (1992) as a possible mechanism for baryogenesis. Composite defects also provide a mechanism for avoiding the monopole overabundance problem as was suggested by Langacker and Pi (1980). On the interaction of the topological defects of different dimensionalities, which can be applicable to the magnetic monopole problem, see also Sec. 17.3.8.

In high-energy physics it is generally assumed that the simplest process for producing a composite defect is a two-stage symmetry breaking, realized in two successive phase transitions which are far apart in energy (Kibble 2000). An example of successive transitions in GUTs is $SO(10) \to G(224) \to G(213) \to G(13)$.

In condensed matter physics, composite defects are known to result even from a single phase transition, provided that at least two distinct energy scales are involved, such that the symmetry at large lengths can become reduced (Mineev and Volovik 1978) (see Sec. 16.1). Examples are the spin–mass vortex in superfluid ^{3}He-B (Sec. 14.1.5) and the half-quantum vortex in ^{3}He-A (Sec. 16.1.1), both serving as the termination line of a soliton. These composite defects result not from the second phase transition but from the fact that the original unbroken symmetry G is approximate. The original small interaction wich violates the symmetry G imposes the second length scale. At distances larger than this scale, the vacuum manifold is reduced by the interaction and linear defects become composite. In our case of superfluid ^{3}He, the symmetry violating interaction is the spin–orbit coupling.

Now we consider the other composite objects which appear as a result of two scales: monopoles terminating strings.

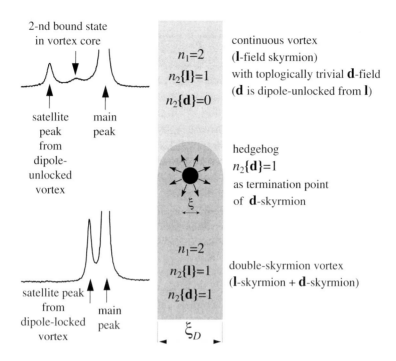

FIG. 17.1. d-hedgehog as an interface separating continuous vortices with different π_2 topological charge $n_2\{\hat{\mathbf{d}}\}$ of the $\hat{\mathbf{d}}$-field. The charge of the $\hat{\mathbf{l}}$-field is the same on both sides, $n_2\{\hat{\mathbf{l}}\} = 1$. The hedgehog can serve as a mediator of the observed phase transition between topologically different skyrmions discussed in Sec. 16.2.5. *Left*: NMR signatures of the corresponding vortex-skyrmions. Satellite peaks signify NMR absorption due to excitation of spin-wave modes localized inside the skyrmions, whose eigenfrequencies depend on orientations of $\hat{\mathbf{d}}$- and $\hat{\mathbf{l}}$-vectors.

17.1.2 Hedgehog and continuous vortices

In Sec. 16.2.5 we discussed the observed topological phase transition (Pecola *et al.* 1990) between different continuous vortex-skyrmions. The two continuous structures differ by the topological charge $n_2\{\hat{\mathbf{d}}\}$ in the $\hat{\mathbf{d}}$ field in eqn (16.12). The topological charge $n_2\{\hat{\mathbf{d}}\}$ comes from the homotopy group π_2 which describes the mapping of the cross-section of the soft core of the vortex to the unit sphere of unit vector $\hat{\mathbf{d}}$. The same topological charge in eqn (15.11) is carried by the point defect – the hedgehog in the $\hat{\mathbf{d}}$-field (Sec. 15.2.1). That is why the hedgehog can serve as a mediator of the topological transition between the single-skyrmion and double-skyrmion vortices which occurs continuously when the hedgehog moves along the vortex line from one 'horizon' to the other (Fig. 17.1).

This is an example of the topological interaction between the linear objects and point defects described by the same topological charge. In our case the n_2

charge of the hedgehog comes from the π_2 homotopy group of the A-phase, eqn (15.10). The topological charge n_2 of the continuous vortex-skyrmion below the hedgehog comes from the relative homotopy group. The relevant relative homotopy group is (compare with eqn (16.3))

$$\pi_2(\mathrm{G}/\mathrm{H_A}, \tilde{\mathrm{G}}/\tilde{\mathrm{H}}_\mathrm{A}) = \pi_2\left((SO(3) \times S^2)/Z_2, SO(3)\right) = \pi_2(S^2) = Z \ . \quad (17.1)$$

Here the cross-section of the skyrmion is mapped to the vacuum manifold $\mathrm{G}/\mathrm{H_A}$ which takes place at short distances below the dipole length ξ_D. At distances above the dipole length ξ_D, the vacuum manifold is restricted by the spin–orbit interaction, and thus the boundary of the disk is mapped to the restricted manifold $\tilde{\mathrm{G}}/\tilde{\mathrm{H}}_\mathrm{A}$.

In particle physics the point defects are usually associated with magnetic monopoles. Let us consider different types of monopoles with and without attached strings, and their analogs in ^{3}He and superconductors.

17.1.3 Dirac magnetic monopoles

Magnetic monopoles do not exist in classical electromagnetism. The Maxwell equations show that the magnetic field is divergenceless, $\nabla \cdot \mathbf{B} = 0$, which implies that the magnetic flux through any closed surface is zero: $\oint_S d\mathbf{S} \cdot \mathbf{B} = 0$. If one tries to construct the monopole solution $\mathbf{B} = g\mathbf{r}/r^3$, the condition that magnetic field is non-divergent requires that magnetic flux $\Phi = 4\pi g$ from the monopole must be accompanied by an equal singular flux supplied to the monopole by an attached Dirac string. QED, however, can be successfully modified to include magnetic monopoles. Dirac (1931) showed that the string emanating from a magnetic monopole (Fig. 17.2 *top left*) becomes invisible for electrons if the magnetic flux of the monopole is quantized in terms of the elementary magnetic flux:

$$\Phi = 4\pi g = n\Phi_0 \ , \ \Phi_0 = \frac{hc}{e} \ , \quad (17.2)$$

where e is the charge of the electron. When an electron circles around the string its wave function is multiplied by $\exp\left(ie \oint d\mathbf{r} \cdot \mathbf{A}/\hbar c\right) = \exp\left(ie\Phi/\hbar\right)$. If the magnetic charge g of the monopole is quantized according to eqn (17.2), the wave function does not change after circling, and the flux tube is invisible.

17.1.4 't Hooft–Polyakov monopole

It was shown by 't Hooft, and Polyakov (both in 1974) [172, 350], that a magnetic monopole with quantization of the magnetic charge according to eqn (17.2) can really occur as a physical object if the $U(1)_Q$ group of electromagnetism is a part of the higher gauge group $SU(2)$. The magnetic flux of a monopole in terms of the elementary magnetic flux coincides with the topological charge n_2 of the hedgehog in the isospin vector field $\hat{\mathbf{d}}$ (Fig. 17.2 *bottom left*): this is the quantity which remains constant under any smooth deformation of the quantum fields. Such monopoles do not appear in the electroweak symmetry breaking transition (eqn (15.5)), since the π_2 group is trivial there. But they can appear in GUT theories, where all interactions are united by, say, the $SU(5)$ group.

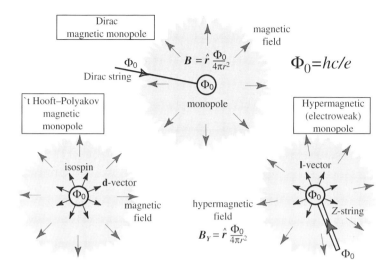

FIG. 17.2. Monopoles in high-energy physics. *Bottom left*: The 't Hooft–Polyakov magnetic monopole. In the anti-GUT analogy dictated by the similarity of **p**-space topology, the role of the 't Hooft–Polyakov magnetic monopole in ^{3}He-A is played by the hedgehog in the $\hat{\mathbf{d}}$-field in Sec. 15.2.1. *Bottom right*: The electroweak monopole. The monopole is the termination line of the string, which is physical in contrast to the Dirac string (*top*). In the GUT analogy which exploits the similarity in the symmetry-breaking pattern and **r**-space topology, the role of the electroweak monopole in ^{3}He-A is played by the hedgehog in the $\hat{\mathbf{l}}$-field (Fig. 17.3 *top left*).

In ^{3}He-A this type of magnetic monopole is reproduced by the hedgehog in the $\hat{\mathbf{d}}$-vector (Fig. 17.1). According to the **p**-space topology of Fermi points, the $\hat{\mathbf{d}}$-vector is felt by 'relativistic' quasiparticles in the vicinity of the Fermi point as the direction of the isotopic spin (see discussion in Sec. 15.2.1).

17.1.5 Nexus

An exotic composite defect in ^{3}He-A which bears some features of magnetic monopoles is the singular $n_1 = 2$ vortex terminated by the hedgehog in the $\hat{\mathbf{l}}$-field discussed by Blaha (1976) and Volovik and Mineev (1976a) (Fig. 17.3 *top left*). The $n_1 = 2$ vortex can terminate in the bulk ^{3}He-A because the $n_1 = 2$ vortices belong to the trivial element of the homotopy group. The superfluid velocity around such a monopole has the same form as the vector potential **A** of the electromagnetic field near the Dirac monopole

$$\mathbf{v}_s = \frac{\hbar}{2mr}\hat{\phi}\cot\frac{\theta}{2} \ , \ \nabla \times \mathbf{v}_s = \frac{\hbar}{2m}\frac{\hat{\mathbf{r}}}{r^2} \ . \tag{17.3}$$

It has a singularity at $\theta = \pi$, i.e. on the line of singular $n_1 = 2$ vortex emanating from the hedgehog. This type of monopole in ^{3}He-A has a physical 'Dirac string'.

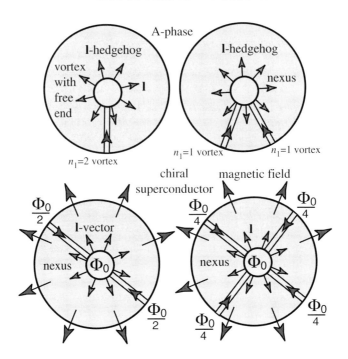

FIG. 17.3. *Top left*: Doubly quantized singular vortex with free end in ^{3}He-A. The termination point is the hedgehog in the $\hat{\mathbf{l}}$-field. *Top right*: Doubly quantized singular vortex splits into two $n_1 = 1$ vortices terminating on hedgehog. *Bottom left*: The same as in *top right* but in a chiral superconductor with the same order parameter as in ^{3}He-A. Each vortex with $n_1 = 1$ is an Abrikosov vortex which carries the magnetic flux $\Phi_0/2$ to the nexus. The nexus represents the magnetic pole with emanating flux Φ_0. *Bottom right*: The magnetic flux to the nexus is supplied by four half-quantum vortices.

The vorticity $\nabla \times \mathbf{v}_s$ which plays the role of magnetic field $\mathbf{B}$ is a conserved quantity: the continuous vorticity radially emanating from the hedgehog is compensated by the quantized vorticity entering the hedgehog along the hard core of the string. The Mermin–Ho relation in (9.17) must be modified to include the singular vorticity concentrated in the singular vortex:

$$\nabla \times \mathbf{v}_s = \frac{\hbar}{4m} e_{ijk} \hat{l}_i \nabla \hat{l}_j \times \nabla \hat{l}_k + \frac{2\pi\hbar}{m} \Theta(-z)\delta(x)\delta(y) \ . \quad (17.4)$$

The singular $n_1 = 2$ vortex can split into two topologically stable vortices with $n_1 = 1$ (Fig. 17.3 *top right*), or even into four half-quantum vortices, each with $n_1 = +1/2$ winding number (see Fig. 17.3 *bottom right* in the case of the chiral superconductor with the same order parameter as in ^{3}He-A). These vortices thus meet each other at one point – the hedgehog in the $\hat{\mathbf{l}}$-field. Such a composite object, which is reminiscent of the Dirac magnetic monopole with one or several physical Dirac strings, is called a nexus in relativistic theories (Cornwall 1999). In electroweak theory the nexus is represented by a hypermagnetic (or electroweak) monopole which is the termination point of the Z-string (Fig. 17.2 *bottom*); the monopole–anti-monopole pair can be connected by the Z-string (Nambu 1977); such configuration is called a dumbbell.

17.1.6 Nexus in chiral superconductors

Figure 17.3 *bottom* shows nexuses in chiral superconductors (Volovik 2000b). In a chiral superconductor the semilocal group is broken (Sec. 16.2.4); as a result the Meissner effect is not complete because of the $\hat{\mathbf{l}}$-texture. Magnetic flux is not necessarily concentrated in the tubes – Abrikosov vortices (Abrikosov 1957) – but can propagate radially from the hedgehog according to the Mermin–Ho relation eqn (9.17) extended to the electrically charged superfluids. The condition that the electric current $\mathbf{j} = en_s(\mathbf{v}_s - (e/m)\mathbf{A})$ is zero in the bulk of the superconductor gives for the vector potential according to eqn (17.3)

$$\mathbf{A} = \frac{m}{e}\mathbf{v}_s = \frac{\hbar}{2er}\hat{\phi}\cot\frac{\theta}{2}. \qquad (17.5)$$

This is exactly the vector potential of the Dirac monopole, i.e. the $\hat{\mathbf{l}}$-hedhehog in Fig. 17.3 *top* acquires in superconductors a magnetic charge $g = hc/4\pi e$. The magnetic flux $\Phi_0 = hc/e$ emanating radially from the hedgehog (nexus) is compensated by the flux supplied by the doubly quantized Abrikosov vortex ($n_1 = 2$) which plays the part of a Dirac string.

The flux Φ_0 can be supplied to the hedgehog by two conventional Abrikosov vortices, each having winding number $n_1 = 1$ and thus the flux $\Phi = \Phi_0/2$ (Fig. 17.3 *bottom left*) or by four half-quantum vortices, each having fractional winding number $n_1 = 1/2$ and thus the fractional flux $\Phi = \Phi_0/4$ (Fig. 17.3 *bottom right*).

17.1.7 Cosmic magnetic monopole inside superconductors

Since magnetic monopoles in GUT and the monopole-like defects in superconductors involve the quantized magnetic flux, there is a topological interaction between these topological objects. Let us imagine that the GUT magnetic monopole enters a conventional superconductor. If the monopole has unit Dirac magnetic flux Φ_0, it produces two Abrikosov vortices, each carrying $\Phi_0/2$ flux away from the magnetic monopole (Fig. 17.4 *top left*). This demonstrates the topological confinement of linear and point defects.

In chiral superconductors, a cosmic monopole with elementary magnetic charge can produce four half-quantum Abrikosov vortices (Fig. 17.4 *top right*). The other ends of the vortices can terminate on the nexus. Thus Abrikosov flux lines produce the topological confinement between the GUT magnetic monopole and the nexus of the chiral superconductor. Figure 17.4 (*bottom left*) shows the cosmic defect and defect of superconductor combined by the doubly quantized ($n_2 = 2$) Abrikosov vortex. This leads to the attraction between the objects, the

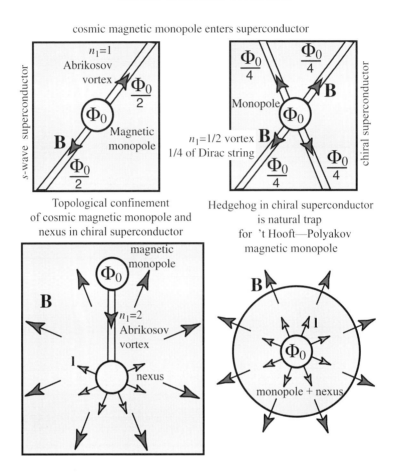

FIG. 17.4. Interaction of cosmic magnetic monopole with topological defects in superconductors.

final result of which is that the GUT magnetic monopole will find its equilibrium position in the core of the hedgehog in the $\hat{\mathbf{l}}$-vector field forming a purely point-like composite defect (Fig. 17.4 *bottom right*). The nexus in chiral superconductors thus provides a natural trap for the massive cosmic magnetic monopole: if one tries to separate the monopole from the hedgehog, one must supply the energy needed to create the segment(s) of the Abrikosov flux line(s) confining the hedgehog and the monopole.

17.2 Defects at surfaces

17.2.1 *Boojum and relative homotopy group*

Boojum is the point defect which can live only on the surface of the ordered medium. This name was coined by Mermin (1977) who discussed boojums in ^{3}He-A [300]. In ^{3}He-A boojums are always present on the surface of a rotating

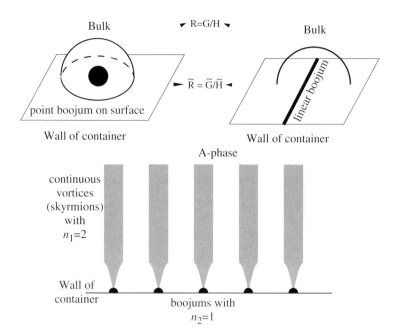

FIG. 17.5. *Top left*: Topological charge of point boojums is determined by relative homotopy group. The relevant subspace is a hemisphere, whose boundary – the circumference – is on the surface of the system. Such a hemisphere is mapped to the vacuum manifold R, while its boundary is mapped to the subspace R̃ of the vacuum manifold R constrained by the boundary conditions. *Bottom*: Doubly quantized continuous vortices in rotating vessel terminate on boojums. *Top right*: Classification of linear boojums – strings living on the surface of the system. The relevant subspace is a cemicircle, whose boundaries – two points – are on the surface of the system.

vessel as termination points of continuous vortex-skyrmions (Fig. 17.5 *bottom*; they are present at the interface separating ^{3}He-A and ^{3}He-B (see Figs 17.6 and 17.9 *top* below). Boojums on the surface of nematic liquid crystals are discussed in the book by Kleman and Lavrentovich (2003).

Boojums are described by the relative homotopy group $\pi_2(R, \tilde{R})$. Here the subspace R̃ is the vacuum manifold on the surface of the system, which is reduced due to the boundary conditions restricting the freedom of the order parameter (Volovik 1978; Trebin and Kutka 1995). Now the relevant part of the coordinate space to be mapped to the vacuum space is a hemisphere in Fig. 17.5 *top left*, whose boundary – the circumference – is on the surface of the system. Such a hemisphere surrounds the point defect living on the surface. The hemisphere is in the bulk liquid and thus is mapped to the vacuum manifold R. The boundary of the hemisphere is on the surface of the container and thus is mapped to the subspace $\tilde{R}$ of the vacuum manifold R, which is constrained by the boundary conditions.

In principle, the relative homotopy group $\pi_2(R,\tilde{R})$ describes two types of point defects: (i) hedgehogs which came from the bulk but did not disappear on the surface because of boundary conditions; and (ii) boojums which live only on the surface and cannot move to the bulk liquid. The topological description, which resolves between these two topologically different types of point defects on the surface, can be found in Volovik (1978).

Accordingly, linear defects on the surface of the system are desribed by the relative homotopy group $\pi_1(R,\tilde{R})$ (Fig. 17.5 *top right*). The semicircle surrounding the defect line is mapped to the vacuum manifold R of the bulk liquid, while the end points, which are on the surface, are mapped to the vacuum submanifold $\tilde{R}$ constrained by the boundary conditions. This group also includes two types of defects: (i) strings which came from the bulk but did not disappear on the surface because of boundary conditions (an example is the half-quantum vortex of the A-phase which survives on the AB-interface in Sec. 17.3.9); and (ii) linear boojums – the strings which live only on the surface and cannot move to the bulk liquid.

17.2.2 Boojum in ^{3}He-A

In the case of ^{3}He-A the boundary conditions restrict the orientation of the $\hat{\mathbf{l}}$-vector: $\hat{\mathbf{l}} = \pm\hat{\mathbf{s}}$, where $\hat{\mathbf{s}}$ is the normal to the surface. As a result the restricted vacuum manifold on the surface is

$$\tilde{R} = S^2 \times U(1) \times Z_2 \ . \tag{17.6}$$

Here S^2 is the sphere of the unit vector $\hat{\mathbf{d}}$: it is not disturbed by the boundary (we neglect here the spin–orbit interaction); $U(1)$ is the space of phase rotation group: $\hat{\mathbf{m}}$ and $\hat{\mathbf{n}}$ are parallel to the surface and can freely rotate about $\hat{\mathbf{l}}$, since such rotation corresponds to the change of the phase of the order parameter; Z_2 marks two possible directions of the $\hat{\mathbf{l}}$-vector on the surface: $\hat{\mathbf{l}} = \pm\hat{\mathbf{s}}$.

The relative homotopy group describing the point defects on the surface is

$$\pi_2(R_A, \tilde{R}_A) = \pi_2\left((S^2 \times SO(3))/Z_2, S^2 \times U(1) \times Z_2\right) = Z \ . \tag{17.7}$$

A more detailed inspection of this group shows that the group Z of integers refers to the topological charge n_2 of the boojums (see below), while the point defects which come from the bulk liquid – hedgehogs in the $\hat{\mathbf{d}}$-field – can be continuously destroyed on the boundary. The $\hat{\mathbf{d}}$-field belongs to the spin part of the order parameter and thus is not influenced by the surface: The $\hat{\mathbf{d}}$-hedgehog simply penetrates through the surface to the 'shadow world behind the wall' (this shadow world can be constructed by mirror reflection without violation of the boundary condition for the $\hat{\mathbf{l}}$-vector).

The boojum is a singular defect: its hard core has size of order coherence length ξ. This defect has a double topological nature, bulk and surface. On the

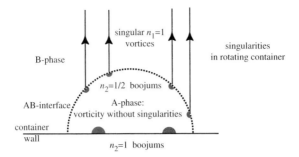

FIG. 17.6. Singularities in superfluid ^{3}He in a rotating container: singular B-phase vortices, AB-interface, A-phase boojums on the surface of the container and A-phase boojums at the interface. Boojums on the surface of container (Mermin 1977) have the surface winding number $n_1 = 2$ and the bulk winding number $n_2 = 1$. Boojums on the AB-interface have two times smaller winding numbers: $n_1 = 1$ and $n_2 = 1/2$.

one hand, it is a point defect for the bulk $\hat{\mathbf{l}}$-field. The integral charge n_2 is the degree of the mapping of the hemisphere, which is S^2 since $\hat{\mathbf{l}}$ is fixed on the boundary, to the sphere S^2 of unit vector $\hat{\mathbf{l}}$ in the bulk liquid. On the other hand, if one considers the order parameter field on the surface where $\hat{\mathbf{l}}$ is fixed, one finds that it is the point defect in the $\hat{\mathbf{m}} + i\hat{\mathbf{n}}$ field. Since rotations of these vectors about $\hat{\mathbf{l}}$ are equivalent to phase rotations, this defect is a point vortex. The winding number of vectors $\hat{\mathbf{m}}$ and $\hat{\mathbf{n}}$ along the closed path on the surface around the boojum, which determines the circulation number of superfluid velocity, is $n_1 = 2n_2$.

The relation between the two winding numbers is exactly the same as for continuous vortex-skyrmions in eqn (16.8), where the $\hat{\mathbf{l}}$-texture gives rise to the winding number of the vortex with $n_1 = 2n_2$. This is the reason why the continuous vortex with $n_1 = 2$ in the bulk liquid terminates on the singular vortex with $n_1 = 2$ on the surface – the boojum (Fig. 17.5 *bottom*). Boojums on the top and bottom walls of the container are the necessary attributes of the rotating state in ^{3}He-A.

17.3 Defects on interface between different vacua

17.3.1 *Classification of defects in presence of interface*

The phase boundary between two superfluid vacua, ^{3}He-A and ^{3}He-B, is the 2D object which is under extensive experimental investigation for many reason (see e.g. Blaauwgeers *et al.* 2002)

The classification of defects in the presence of the interface between A- and B-phases (Volovik 1990c; Misirpashaev 1991) must give answers to the following questions: What is the fate of defects which come to the interface from the bulk liquid? Do they survive or disappear? Can they propagate through the interface or do they have a termination point on the interface? Are there special defects which live only at the interface? The latter are shown in Fig. 17.6. Boojums at the AB-interface do exist, but they have different topological charges than the boojums living on the surface of ^{3}He-A.

The existence and topological stability of boojums are determined by the topological matching rules across the interface, which in turn are defined by boundary conditions, combined with the topology of the A- and B-phases in the bulk liquid. The boundary conditions in turn depend on the internal symmetry of the order parameter in the core of the interface. Thus we must start with the symmetry classes of interfaces.

17.3.2 Symmetry classes of interfaces

The symmetry classification of possible interfaces is similar to the symmetry classification of the structures of the core of defects – hedgehogs/monopoles and vortices/cosmic strings – discussed in Sec. 14.2.6. We must consider the symmetry-breaking scheme of the transition from the normal ^{3}He with symmetry G to the superfluid state having the extended object in the form of a plane wall.

The are only two maximum symmetry subgroups of G which are consistent with the geometry of defects, and with the groups H$_A$ and H$_B$ of the superfluid phases A and B far from the interface. Both of them contain only discrete elements:

$$H_{AB1} = (1, U_{2x}T, U_{2z}P, U_{2y}TP) \qquad (17.8)$$

and

$$H_{AB2} = (1, U_{2x}, U_{2y}PT, U_{2z}PT). \qquad (17.9)$$

Here U_{2x}, U_{2y} and U_{2z} denote rotations through π about $\hat{\mathbf{x}}$, $\hat{\mathbf{y}}$ and $\hat{\mathbf{z}}$, with the axis $\hat{\mathbf{x}}$ along the normal $\mathbf{s}$ to the AB-interface. In each of these two subgroups there is a stationary solution for the order parameter everywhere in space including the interior of the interface. In the solution with symmetry H$_{AB2}$ the orbital unit vector of the A-phase anisotropy is oriented along the normal to the interface, $\hat{\mathbf{l}} = \pm\hat{\mathbf{x}}$. In the solution with symmetry H$_{AB1}$ the $\hat{\mathbf{l}}$-vector is in the plane of the interface, say $\hat{\mathbf{l}} = \hat{\mathbf{z}}$.

The solution with minimum interface energy belongs to the class H$_{AB1}$. It has the following asymptotes for the order parameter on both sides of the wall:

$$e^0_{\alpha i}(x=-\infty) = \Delta_A \hat{x}_\alpha(\hat{x}_i + i\hat{y}_i), \quad e^0_{\alpha i}(x=+\infty) = \Delta_B \delta_{\alpha i}. \qquad (17.10)$$

For the maximally symmetric interface the symmetry H$_{AB1}$ of the asymptote persists everywhere throughout the interface, i.e. at $-\infty < x < +\infty$.

17.3.3 Vacuum manifold for interface

Equation (17.10) represents only one of the possible degenerate states of the wall in the class H$_{AB1}$, the other states with the same energy being obtained by the symmetry operations. These are the symmetry operations from the group G which do not change the orientation of the interface

$$G_{AB} = U(1)_N \times SO(2)_{L_x} \times SO(3)_S \times PU_{2y} \times T. \qquad (17.11)$$

Here $SO(2)_{L_x}$ is the group of orbital rotations about the normal to the interface. Applying these symmetry transformations to the particular solution (17.10) one obtains all the possible orientations of the vacuum parameters in both phases across the interface (the unit vectors $\hat{\mathbf{d}}$, $\hat{\mathbf{m}}$, $\hat{\mathbf{n}}$ and $\hat{\mathbf{l}}$ on the A-phase side, and the phase Φ and the rotation matrix $R_{\alpha i}$ on the B-phase side):

$$e_{\alpha i} = \Delta_{\rm B} e^{i\Phi} R_{\alpha i}(\theta, \hat{\mathbf{n}}), \quad x = +\infty \;, \quad (17.12)$$

$$(\hat{\mathbf{m}} + i\hat{\mathbf{n}})_i = e^{i\Phi} R^L_{ik}(\alpha, \hat{\mathbf{x}})(\hat{x}_k + i\hat{y}_k), \quad \hat{d}_\alpha = R_{\alpha\beta}(\theta, \hat{\mathbf{n}})\hat{x}_\beta, \quad x = -\infty \;. \quad (17.13)$$

Here R^L is the matrix of orbital rotations from the group $SO(2)_{L_x}$. Equation (17.12) and eqn (17.13) represent the mutual boundary conditions for the two vacua across the interface, which can be written in general form introducing the unit vector $\hat{\mathbf{s}}$ along the normal to the boundary:

$$\hat{\mathbf{s}} \cdot \hat{\mathbf{l}} = 0 \;, \quad \hat{\mathbf{s}} \cdot (\hat{\mathbf{m}} + i\hat{\mathbf{n}}) = e^{i\Phi} \;, \quad \hat{d}_\alpha = R_{\alpha i}(\theta, \hat{\mathbf{n}})\hat{s}_i \;. \quad (17.14)$$

The vacuum manifold – the space of degenerate states – which describes all possible mutual orientations of the order parameter on both sides of the wall of class $\rm H_{AB1}$ is

$$\rm R_{AB1} = G_{AB}/H_{AB1} = \mathit{U}(1)_{\mathit{N}} \times \mathit{SO}(2)_{\mathit{L_x}} \times \mathit{SO}(3)_{S} \;. \quad (17.15)$$

17.3.4 Topological charges of linear defects

The vacuum manifold in eqn (17.15) is rather peculiar: there is no one-to-one correspondence between the orientations of the vacua across the interface. It follows, for example, that if we fix the orientation of the B-phase vacuum, the order parameter matrix $R_{\alpha i}$ and the phase Φ, there is still freedom on the A-phase side. While the spin vector $\hat{\mathbf{d}}$ and the phase Φ (the angle of rotation of vectors $\hat{\mathbf{m}}$ and $\hat{\mathbf{n}}$ around $\mathbf{l}$) become fixed by the B-phase, $\hat{d}_\mu = R_{\mu i}\hat{s}^i$ and $\Phi_{\rm A} = \Phi_{\rm B}$, the direction of the $\hat{\mathbf{l}}$ in the plane – the angle α in eqn (17.13) – remains arbitrary. In other words, each point in the B-phase vacuum produces a $U(1)$ manifold of vacuum states on the A-phase side.

On the other hand, if one fixes the orientation of the A-phase degeneracy parameters Φ, $\hat{\mathbf{l}}$ and $\hat{\mathbf{d}}$, then on the B-phase side the phase Φ will be fixed, but there is still some freedom in the orientation of the B-phase order parameter matrix $R_{\alpha i}$: the angle θ of rotations about axis $\hat{\mathbf{n}} = \hat{\mathbf{s}}$ is arbitrary. Thus each point in the A-phase vacuum also produces a $U(1)$ manifold of states on the B-phase side.

These degrees of freedom lead to a variety of possible defects on the AB-interface. The general classification of such defects in terms of the relative homotopy groups has been elaborated by Misirpashaev (1991) and Trebin and Kutka (1995). Let us first consider the defects which have singular points on the interface as termination or crossing points of the linear defects in the bulk. At these points the structure of the interface is violated: the mutual boundary conditions for the degeneracy parameters are satisfied everywhere on the AB-interface except these points.

These points can be described by the π_1 homotopy group

$$\pi_1(\mathrm{R_{AB1}}) = Z \times Z \times Z_2 . \qquad (17.16)$$

This group describes the mapping of the contour lying on the interface and surrounding the defect to the vacuum manifold of the interface in eqn (17.15). According to eqn (17.16), there are three topological charges – three integer numbers N_Φ, N_l and N_R – which characterize the topologically stable defects related to the interface.

The integer charge N_Φ comes from the subspace $U(1)_N$ of $\mathrm{R_{AB1}}$ in eqn (17.15). It is the winding number of the phase of the order parameter on the interface. In other words, this is the number of the circulation quanta of the superfluid velocity along the contour on the AB-interface which embraces the defect, i.e. $N_\Phi = n_1$. Since the phases Φ at the interface are the same for both vacua, $\Phi_A = \Phi_B$, the circulation number n_1 of superfluid velocity is conserved across the interface.

Integer N_l comes from the subspace $SO(2)_{L_x}$. It is the winding number for the $\hat{\mathbf{l}}$-vector on the interface around a defect. Let us recall that at the interface one has $\hat{\mathbf{l}} \perp \hat{\mathbf{s}}$ and thus the vacuum manifold of the $\hat{\mathbf{l}}$-vector is the circumference $SO(2)_{L_x}$.

The topological quantum number N_R comes from the space of spin rotations $SO(3)_\mathbf{S}$ in the vacuum manifold $\mathrm{R_{AB1}}$ in eqn (17.15). Since the spin–orbit interaction is neglected, spin rotations of the order parameter do not change the energy of the interface. The integer N_R takes only values 0 and 1, since $\pi_1(SO(3)_\mathbf{S}) = Z_2$. Corresponding defects are related to spin vortices.

17.3.5 Strings across AB-interface

All the linear defects intersecting the interface or terminating on the interface can be classified in terms of the 3-vector with integer-valued components $\vec{N} = (N_\Phi, N_l, N_R)$. Among these defects one can find:

(1) End points of the linear singularities in the bulk A-phase without propagation of the singularity into the B-phase.

(2) End points of the linear singularities in the bulk B-phase without propagation into the A-phase. The end of the linear B-phase defect sometimes gives rise to the special point singularity on the A-phase side, which is similar to boojums – the point defects which live on the surface of the ordered medium.

(3) Points of the intersection of the linear defects with the AB-interface, i.e. defects which propagate into the bulk liquid on both sides of the wall.

Let us start with three elementary defects, i.e. described by only one non-zero component of the $\vec{N}$-vector (Fig. 17.7):

$\vec{N} = (0, 0, 1)$. This elementary defect has non-zero winding $N_R = 1$ of the angle θ in the B-phase. It represents the end point of the B-phase spin vortex without any singularity on the A-phase side (Fig. 17.7 *top left*). This is possible because the angle θ is not fixed by the A-phase and thus can make winding even if the A-phase order parameter remains constant.

DEFECTS ON INTERFACE BETWEEN DIFFERENT VACUA

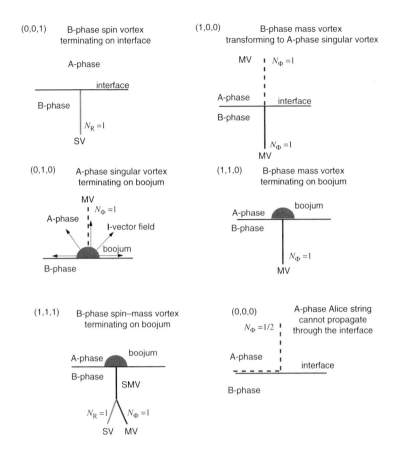

FIG. 17.7. Defects at and across the AB-interface. MV, SV and SMV denote mass vortex, spin vortex and spin–mass vortex respectively.

$\vec{N} = (1,0,0)$. This elementary defect is the B-phase mass vortex with 2π winding of the phase Φ, which transforms to the singular A-phase mass vortex with the same winding number: $N_\Phi \equiv n_1 = 1$ (Fig. 17.7 *top right*).

$\vec{N} = (0,1,0)$ (Fig. 17.7 *middle left*). This elementary defect is a disclination in the $\hat{\mathbf{l}}$-field with winding number $N_l = 1$ which does not propagate into the bulk B-phase, since at a given fixed value of the B-phase orientation the orientation of the $\hat{\mathbf{l}}$ can be arbitrary. However, it should propagate into the A-phase either in terms of the same disclination or in terms of the vortex, since in the A-phase the disclination and the vortex belong to the same topological class $n_1 = 1$ and may transform to each other. If the singularity propagates as a vortex, the end of the A-phase vortex is the boojum in the $\hat{\mathbf{l}}$-field with $n_2 = 1/2$.

All other defects can be constructed from the elementary ones. The important ones for us are the following defects, which can appear in the rotating vessel.

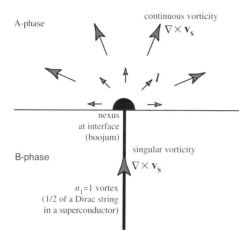

FIG. 17.8. Boojum as nexus. Since in the analogy with gravity $\nabla \times \mathbf{v}_s$ plays the role of the gravimagnetic field, this defect plays the role of gravimagnetic monopole.

17.3.6 Boojum as nexus

$\vec{N} = (1,1,0)$ (Fig. 17.7 *middle right*). This is a sum of the defects $(1,0,0) + (0,1,0)$. The B-phase vortex will persist but the two singularities in the A-phase will be annihilated due to the famous sum rule for the linear defects in the bulk A-phase: $N_\Phi + N_l = 1+1 = 0$. So only the boojum is left on the A-phase side and we come to the Dirac monopole structure: B-phase $n_1 = 1$ vortex terminating on the A-phase boojum $\hat{\mathbf{l}} = \hat{\mathbf{r}}$ (Fig. 17.8). As distinct from the boojums on the surface of the container, where the boundary conditions on the $\hat{\mathbf{l}}$-vector are different ($\hat{\mathbf{l}}$ is parallel to the normal to the surface), the boojum at the interface has $n_1 = 1$ circulation quantum (Fig. 17.6). Since $\nabla \times \mathbf{v}_s$ plays the role of the gravimagnetic field, this boojum is equivalent to the gravimagnetic monopole discussed by Lynden-Bell and Nouri-Zonoz (1998) (see Sec. 20.1.2 and Chapter 31 on the gravimagnetic field).

And finally the sum of three elementary defects, $(1,1,1) = (1,0,0) + (0,1,0) + (0,0,1)$, in Fig. 17.7 *bottom left* is the A-phase Mermin–Ho vortex, which propagates across the AB-interface forming the composite linear defect in the Φ- and $R_{\mu i}$- fields – the spin–mass B-phase vortex discussed in Sec. 14.1.4. The composite spin–mass vortices were experimentally resolved by Kondo *et al.* (1992) just after the AB-interface with continuous vorticity on the A-phase side traversed the rotating vessel. This implies that $(1,1,1)$ or $(1,-1,1)$ defects of the interface were the intermediate objects in formation of the spin–mass vortex.

17.3.7 AB-interface in rotating vessel

The vortex-full rotating state in the presence of the AB-interface is shown in Fig. 17.9 *top*. Both phases contain an equilibrium number of vortices at given

DEFECTS ON INTERFACE BETWEEN DIFFERENT VACUA 227

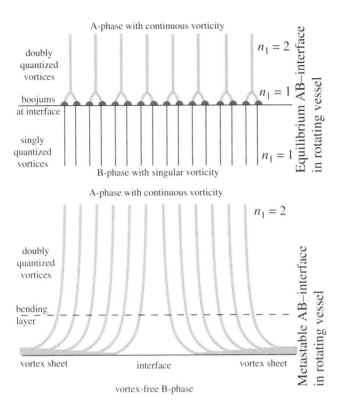

FIG. 17.9. *Top*: In an equilibrium rotating state, a continuous vortex-skyrmion in the A-phase with winding number $n_1 = 2$, approaching the interface, splits into Mermin–Ho vortices, radial and hyperbolic, each with $n_1 = 1$. Each Mermin–Ho vortex gives rise to a singular point defect on the AB-interface – a boojum. The boojum is the termination point of a singular 2π ($n_1 = 1$) vortex in the B-phase. Thus the continuous vorticity crosses the interface transforming to singular vorticity. However, singularities are not easily created in superfluid ^{3}He. That is why typically the non-equilibrium state arises (*bottom*), in which the continuous A-phase vorticity does not propagate through the interface. Instead it is accumulated at the A-phase side of the interface forming the vortex layer with a high density of vortices separating the vortex-free ^{3}He-B and the ^{3}He-A with equilibrium vorticity.

angular velocity Ω of rotation. Let us recall (see eqn (14.2)) that the areal density of vortices n_v is determined by the condition that in equilibrium the superfluid component performs on average the solid-body rotation, i.e. $\langle \mathbf{v}_s \rangle = \mathbf{\Omega} \times \mathbf{r}$, or $\langle \nabla \times \mathbf{v}_s \rangle = 2\mathbf{\Omega}$. Since $\langle \nabla \times \mathbf{v}_s \rangle = n_1 \kappa_0 n_v$, where $\kappa_0 = \pi\hbar/m$, one obtains

$$n_v = \frac{2m\Omega}{n_1 \pi \hbar} . \qquad (17.17)$$

This demonstrates that the number of doubly quantized vortex-skyrmions in the A-phase is two times smaller than the number of singular singly quantized B-phase vortices.

The vorticity is conserved when crossing the interface. When approaching the interface the continuous $n_1 = 2$ A-phase vortex splits into two Mermin–Ho vortices, radial and hyperbolic, each terminating on a boojum with $n_1 = 2n_2 = 1$. The quantized circulation around the boojum continues to the B-phase side in terms of $n_1 = 1$ singular vortices. In terms of the point singularities on the AB-interface these processes correspond to two defects: $\vec{N} = (1,1,0)$ and $\vec{N} = (1,-1,0)$, which are radial and hyperbolic boojums at the AB-interface, respectively.

However, we know that singularities are not easily created in superfluid ^{3}He: the energy barrier is too high, typically of 6–9 orders of magnitude bigger than the temperature (Sec. 16.1.3). When the vessel is accelerated from rest, the A-phase vortex-skyrmions are created even at a low velocity of rotation, while the critical velocity for nucleation of the singular vortices in ^{3}He-B are is higher (Sec. 26.3.3). Thus one may construct the state with vortex-full A-phase and vortex-free B-phase (Blaauwgeers *et al.* 2002). The A-phase thus rotates as a solid body, while the B-phase remains stationary in the inertial frame (Fig. 17.9 *bottom*). The A-phase continuous vorticity does not propagate through the interface; instead it is accumulated at the interface forming the vortex layer (the vortex sheet) with a high density of continuous vorticity. Thus two superfluids slide along each other with a tangential discontinuity of the superflow velocity at the interface. This tangential discontinuity is ideal – there is no viscosity in the motion of superfluids, so that such a state can persist for ever.

Vorticity starts to propagate into the B-phase only after the critical velocity of the Kelvin–Helmholtz instability of the tangential discontinuity is reached, in the process of development of this shear-flow instability (see Chapter 27), and thus one superfluid (^{3}He-A) spins up another one (^{3}He-B).

17.3.8 AB-interface and monopole 'erasure'

The interaction of cosmic defects of different dimensionalities can be important as a possible way of soving the magnetic monopole problem. The GUT magnetic monopoles could have been formed in the early Universe if the temperature had crossed the phase transition point at which the GUT symmetry of strong, weak and electromagnetic interactions was spontaneously broken. They appear according to Kibble's (1976) scenario of formation of topological defects in the process of non-equilibrium phase transition, which we shall discuss in Chapter 28. Since these monopoles are heavy, their energy density at the time of nucleosynthesis would by several orders of magnitude exceed the energy density of matter, which strongly contradicts the existence of the present Universe (see review by Vilenkin and Shellard 1994).

The possible solution of the cosmological puzzle of overabundance of monopoles was suggested by Dvali *et al.* (1998) and Pogosian and Vachaspati (2000). In the suggested mechanism the interaction of the defects of different dimensions

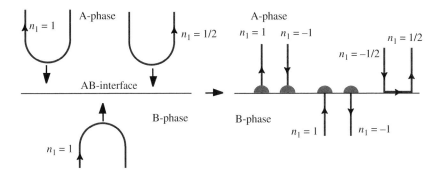

FIG. 17.10. All topological defects are destroyed by the AB-interface except for the Alice string. When a singular vortex or other defect (except for the half-quantum vortex) approaches the interface from either side (*left*), it can be cut to pieces by the interface (*right*). The half-quantum vortex – the Alice string – survives at the interface.

leads to 'defect erasure': monopoles can be swept away by the topological defects of higher dimension, the domain walls, which then subsequently decay.

The impenetrability of continuous A-phase vorticity through the interface in Fig. 17.9 *bottom*, as well as the impenetrability of singular defects through the interface from the B-phase to A-phase in Fig. 17.9 *top*, serve as a condensed matter illustration of the monopole 'erasure' mechanism. In our case the role of the monopoles is played by both vortices, singular and continuous.

In the situation in Fig. 17.9 *bottom*, continuous A-phase vortex-skyrmions cannot propagate through the domain wall – the interface between two different phases of ^{3}He. They are swept away when the interface moves toward the A-phase.

In the situation in Fig. 17.9 *top* the singular defects of the B-phase terminate on boojums and do not propagate to the A-phase side where vortex-skyrmions have no singularities. If one starts to move the interface toward the B-phase one finds that the singularities are erased from the experimental cell by the interface.

17.3.9 *Alice string at interface*

As was found by Misirpashaev (1991), the Alice string (half-quantum vortex) is the only singular defect which is topologically stable in the presence of the AB-interface (Fig. 17.7 *bottom right*). The other linear defects, when they approach and touch the interface, can smoothly annihilate leaving the termination points (see Fig. 17.10). The point defects – the A-phase hedgehog in the $\hat{\mathbf{d}}$-field – are erased completely when they collide with the interface. This is another illustration of the monopole erasure. The Alice string cannot propagate through the interface, but when it comes from the bulk A-phase (Fig. 17.10 *left*) to the interface it will survive there (Fig. 17.10 *right*).

Whether the Alice string is repelled from the interface or attracted by the

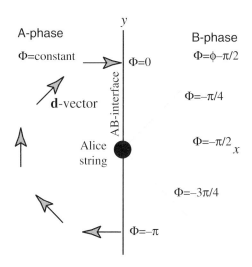

FIG. 17.11. Simplest realization of the Alice string at the interface. On the A-phase side it is the spin vortex in the $\hat{\mathbf{d}}$-field with $\nu = 1/2$, while on the B-phase side it is the mass vortex with $n_1 = 1/2$.

interface and lives there, is not determined by topology, but by the energetics of vortices on different sides of the interface. Let us recall that in the bulk A-phase, the Alice string is the combination of the mass vortex with $n_1 = 1/2$ and spin vortex with $\nu = 1/2$. This allows us to construct the simplest representation of the Alice string lying at the interface:

$$e_{\alpha i}(x < 0) = \Delta_A \left(\hat{x}_\alpha \sin \phi - \hat{y}_\alpha \cos \phi\right) (\hat{x}_i + i\hat{y}_i) , \tag{17.18}$$
$$e_{\alpha i}(x > 0) = -i\Delta_B \delta_{\alpha i} e^{i\phi} . \tag{17.19}$$

Here the vortex line and the $\hat{\mathbf{l}}$-vector are along the z axis. In this construction, the change of sign of the order parameter due to the reorientation of the $\hat{\mathbf{d}}$-vector on the A-phase side is compensated by the change of the phase Φ on the B-phase side. As a result the mass-current part of the string is on the B-phase side of the interface, where the phase $\Phi_B = \phi - \pi/2$ changes from $-\pi$ to 0, while the spin-current part of the string is on the A-phase side, i.e. at $x < 0$ (Fig. 17.11). In the real Alicie string the spin and mass currents are present in both phases due to the conservation law for the mass and spin:

$$e_{\alpha i A} = \Delta_A \left(\hat{x}_\alpha \sin((1-a)\phi) - \hat{y}_\alpha \cos((1-a)\phi)\right)(\hat{x}_i + i\hat{y}_i)e^{ib\phi}, \tag{17.20}$$
$$e_{\alpha i B} = -i\Delta_B R_{\alpha i}(\hat{z}, a\phi)e^{i(1-b)\phi}. \tag{17.21}$$

Parameters a and b of the string can be determined from the minimization of the London energy. In the BCS model at $T = 0$, eqns (14.6) and (15.19) give the following energy densities in the A- and B-phases:

$$E(x<0) = n\frac{\hbar^2}{8m}\left((\nabla\Phi)^2 + (\nabla\alpha)^2\right) \ , \ \Phi = b\phi \ , \ \alpha = (1-a)\phi \ , \quad (17.22)$$

$$E(x>0) = n\frac{\hbar^2}{8m}\left((\nabla\Phi)^2 + \frac{4}{5}(\nabla\alpha)^2\right) \ , \ \Phi = (1-b)\phi \ , \ \alpha = a\phi \ . \quad (17.23)$$

Minimization of the energy $\int dx E$ is equivalent to application of the conservation law for the mass and spin currents across the interface, $dE_B/d(\nabla\Phi) = dE_A/d(\nabla\Phi)$ and $dE_B/d(\nabla\alpha) = dE_A/d(\nabla\alpha)$, which gives $b = 1/2$ and $a = 5/9$. Thus the energy of the Alice string of length L attached to the interface is

$$E_{\text{Alice AB}} = nL\frac{\pi\hbar^2}{8m}\frac{17}{18}\ln\frac{R}{\xi} \ . \quad (17.24)$$

The energy of the Alice string in a bulk A-phase is somewhat higher:

$$E_{\text{Alice A}} = nL\frac{\pi\hbar^2}{8m}\ln\frac{R}{\xi} \ . \quad (17.25)$$

That is why in this simplified case the Alice string prefers to live at the interface for energetical reasons. The topology allows for that. All the other linear defects, even if they are topologically stable in bulk vacua, can be eaten by the interface (Fig. 17.10). This does not mean that they will always be destroyed by the interface: the non-topological energy barriers can prevent the destruction.

Part IV

Anomalies of chiral vacuum

18

ANOMALOUS NON-CONSERVATION OF FERMIONIC CHARGE

In this part we discuss physical phenomena in the vacuum with Fermi points. The non-trivial topology in the momentum space leads to anomalies produced by the massless chiral fermions in the presence of collective fields. In ^{3}He-A this gives rise to the anomalous mass current; non-conservation of the linear momentum of superflow at $T = 0$; the paradox of the orbital angular momentum, etc. All these phenomena are of the same origin as the chiral anomaly in RQFT (Adler 1969; Bell and Jackiw 1969).

As distinct from the pair production from the vacuum, which conserves the fermionic charge, in the chiral anomaly phenomenon the fermionic charge is nucleated from the vacuum one by one. This is a property of the vacuum of massless chiral fermions, which leads to a number of anomalies in the effective action. The advantage of ^{3}He-A in simulating these anomalies is that this system is complete: not only is the 'relativistic' infrared regime known, but also the behavior in the ultraviolet 'non-relativistic' (or 'trans-Planckian') range is calculable, at least in principle, within the BCS scheme. Since there is no need for a cut-off, all subtle issues of the anomaly can be resolved on physical grounds.

Whenever gapless fermions are present (due to the Fermi point or Fermi surface), and/or gapless collective modes, the measured quantities depend on the correct order of imposing limits. It is necessary to resolve which parameters of the system tend to zero faster in a given physical situation: temperature T; external frequency ω; inverse relaxation time $1/\tau$; inverse observational time; inverse volume; the distance ω_0 between the energy levels of fermions; or others. All this is very important for the $T \to 0$ limit, where the relaxation time τ is formally infinite.

We have already seen such ambiguity in the example discussed in Sec. 10.3.5, where the density of the normal component of the liquid is different in the limit $T \to 0$ and at exactly zero temperature. An example of the crucial difference between the results obtained using different limiting procedures is also provided by the so-called 'angular momentum paradox' in ^{3}He-A, which is also related to the anomaly: the density of the orbital momentum of the fluid at $T = 0$ differs by several orders of magnitude, depending on whether the limit is taken while lefting $\omega\tau \to 0$ or $\omega\tau \to \infty$. The same situation occurs for the Kopnin or spectral-flow force acting on vortices, which comes from the direct analog of the chiral anomaly in the vortex core (see Chapter 25), and for quantization of physical parameters in 2+1 systems (Sec. 21.2.1), which is also related to the anomaly.

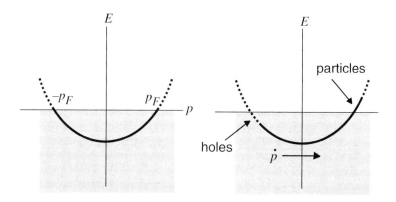

FIG. 18.1. Flow of the vacuum in 1D **p**-space under external force.

In many cases different regularization schemes lead to essentially different results. In some cases this does not mean that one scheme is better than another: each scheme can simply reflect the proper physical situation in its most extreme manifestation.

18.1 Chiral anomaly

18.1.1 *Pumping the charge from the vacuum*

The nucleation of the fermionic charge from the vacuum can be visualized in an example of the 1D energy spectrum $E(p) = (p^2 - p_F^2)/2m$ (Fig. 18.1). In the initial vacuum state all the negative energy levels are occupied by quasiparticles. Now let us apply an external force $F > 0$ acting on the particles according to the Newton law $\dot{p} = F$. Particles in the momentum space start to move to the right, they cross the zero-energy level at $p = p_F$ and become part of the positive energy world, i.e. become quasiparticles representing the matter. Thus, an external field pumps the liquid from the vacuum to the world. The number of quasiparticles which appear from the vacuum near $+p_F$ is $\dot{n} = F/2\pi\hbar$ (per unit time per unit length). On the other hand, the same number of holes appear near the left Fermi surface at $p = -p_F$, which also represent matter. If we assign the fermionic charge $B = +1$ to the fermions near $+p_F$ and the charge $B = -1$ to the fermions near $-p_F$, we find that the total charge B produced by the external force per unit time per unit length is $\dot{B} = 2F/2\pi\hbar$.

The same actually occurs in 3D space in the vacuum with Fermi points: under external fields the energy levels flow from the vacuum through the Fermi point to the positive energy world. Since the flux of the levels in momentum space is conserved, the rate of production of quasiparticles can be equally calculated in the infrared or in the ultraviolet limits. In ^{3}He-A it was calculated in both regimes: in the infrared regime below the first 'Planck' scale, where the chiral quasiparticles obey all the 'relativistic' symmetries; and in a more traditional approach utilizing the so-called quasiclassical method, which is applicable only above the 'Planck' scale, i.e. in the highly non-relativistic ultraviolet regime. Of course, the

CHIRAL ANOMALY

Chiral particles in magnetic field **B**

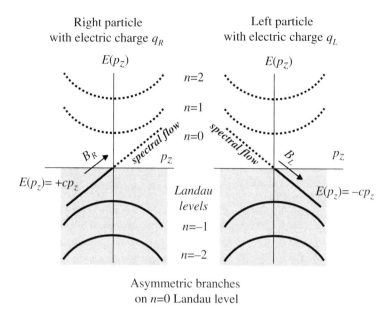

Asymmetric branches on $n=0$ Landau level

FIG. 18.2. Spectrum of massless right-handed and left-handed particles with electric charges q_R and q_L, respectively, in a magnetic field **B** along z; the thick lines show the occupied negative energy states. Particles are also characterized by some other fermionic charges, B_R and B_L, which are created from the vacuum in the process of spectral flow under electric field $\mathbf{E} \parallel \mathbf{B}$.

results of both approaches coincide. Here we shall consider the infrared relativistic regime in the vicinity of the Fermi point to make the connection with the chiral anomaly in RQFT.

18.1.2 Chiral particle in magnetic field

Let us consider the chiral right-handed particle with electric charge q_R moving in magnetic and electric fields which are parallel to each other and are directed along the axis z. Let us first start with the effect of the magnetic field $\mathbf{B} = F_{12}\hat{\mathbf{z}}$. The Hamiltonian for the right-handed particle with electric charge q_R in the magnetic field is

$$\mathcal{H}_3 = c\sigma^3 p_3 + c\sigma^1 \left(p_1 - \frac{q_R}{2} F_{12} x^2 \right) + c\sigma^2 \left(p_2 + \frac{q_R}{2} F_{12} x^1 \right) . \qquad (18.1)$$

The motion of the particle in the (x, y) plane is quantized into the Landau levels shown in Fig. 18.2. The free motion is thus effectively reduced to 1D motion along the direction of magnetic field **B** with momentum $p_z (= p_3)$. Because of the chirality of the particles, the branch corresponding to the lowest ($n = 0$)

Landau level is asymmetric: the energy spectrum $E = cp_z$ of our right-handed particle crosses zero with positive slope (Fig. 18.2). Zero in this energy spectrum represents the Fermi surface in 1+1 spacetime which is described by the topological charge $N_1 = 1$ in eqn (8.3). The branches of the fermionic spectrum which cross zero energy are usually called the fermion zero modes. The number ν of fermion zero modes formed in the magnetic field equals the number of states at the Landau level. It is determined by the total magnetic flux $\Phi = \int dxdy F_{12}$ in terms of the elementary flux $\Phi_0 = 2\pi/|q_R|$:

$$\nu = \frac{\Phi}{\Phi_0} = \frac{|q_R F_{12}| L_1 L_2}{2\pi} , \qquad (18.2)$$

where L_1 and L_2 are the lengths of the system in the x and y directions. This is an example of the dimensional reduction of the Fermi point with topological charge N_3 in 3+1 spacetime (defect of co-dimension 3) to the Fermi surface with topological charge N_1 in 1+1 spacetime (defect of co-dimension 1) in the presence of the topologically non-trivial background, which in our case is provided by the magnetic flux.

18.1.3 Adler–Bell–Jackiw equation

Let us now apply an electric field **E** along axis z. According to the Newton law, $\dot{p}_3 = q_R E_3$, the electric field pushes the energy levels marked by the momentum p_3 from the negative side of the massless branch to the positive energy side. This is the spectral flow of levels generated by the electric field acting on fermion zero modes. As a result the particles filling the negative energy levels enter the positive energy world. The whole Dirac sea of fermions on the anomalous branch moves through the Fermi point transferring the electric charge q_R and the other fermionic charges carried by particles, say B_R, from the vacuum into the positive energy continuum of matter. The rate of production of chiral particles from the vacuum along one branch of fermion zero modes is $(L_3 \dot{p}_3)/2\pi$, where L_3 is the size of the system along the axis z. Accordingly the production of charge B per unit time is $\dot{B} = B_R(L_3\dot{p}_3)/2\pi = B_R q_R L_3/2\pi$. This must be multiplied by the number ν of fermion zero modes in eqn (18.2). Then dividing the result by the volume $V = L_1 L_2 L_3$ of the system one obtains the rate of production of fermionic charge B from the vacuum per unit time per unit volume:

$$\dot{B} = \frac{1}{4\pi^2} B_R q_R^2 \mathbf{E} \cdot \mathbf{B} . \qquad (18.3)$$

If we are interested in the production of electric charge we must insert $B_R = q_R$ to obtain $\dot{Q} = (1/4\pi^2) q_R^3 \mathbf{E} \cdot \mathbf{B}$.

The same spectral-flow mechanism of anomalous production of the charge B can be applied to the left-handed fermions. In the magnetic field the left-handed fermions give rise to the fermion zero modes with the spectrum $E = -cp_3$, i.e. they have negative slope. Now the force acting on the particles in the applied electric field has an opposite effect: it pushes the Dirac sea of left-handed particles down, annihilating the corresponding charge of the vacuum. If the left-handed particles have electric charge q_L and also the considered fermionic charge B_L, they contribute to the net production of the B-charge from the vacuum:

$$\dot{B} = \frac{1}{4\pi^2}\left(B_R q_R^2 - B_L q_L^2\right) \mathbf{E} \cdot \mathbf{B} \,. \tag{18.4}$$

It is non-zero if the mirror symmetry between the left and right worlds is not exact, i.e. if the fermionic charges of left and right particles are different. This is actually the equation for the anomalous production of fermionic charge, which has been derived by Adler (1969) and Bell and Jackiw (1969) for the relativistic systems.

It can be written in a more general form introducing the chirality C_a of fermionic species:

$$\dot{B} = \frac{1}{16\pi^2} F_{\mu\nu} F^{*\mu\nu} \sum_a C_a B_a q_a^2 \,. \tag{18.5}$$

Here q_a is the charge of the a-th fermion with respect to the gauge field $F^{\mu\nu}$; $F^{*\mu\nu} = (1/2)e^{\alpha\beta\mu\nu} F_{\alpha\beta}$ is the dual field strength.

Finally, if the chirality is not a good quantum number, as in ^{3}He-A far from the Fermi point, it can be written in terms of the **p**-space topological invariant in eqn (12.9):

$$\dot{B} = \frac{1}{16\pi^2} F^{\mu\nu} F^*_{\mu\nu} \mathbf{tr}\left(\mathbf{B}\mathbf{Q}^2\mathcal{N}\right) \,, \tag{18.6}$$

where **B** is the matrix of the fermionic charges B_a and **Q** is the matrix of the electric charges q_a.

In this form the Adler–Bell–Jackiw equation can be equally applied to the Standard Model and to ^{3}He-A. In the first case the corresponding gauge fields are hypercharge $U(1)$ and weak $SU(2)_L$ fields; in the second case the gauge field and fermionic charges must be expressed in terms of the ^{3}He-A observables. The effective 'magnetic' and 'electric' fields in ^{3}He-A are simulated by the space- and time-dependent $\hat{\mathbf{l}}$-texture. Since the effective gauge field is $\mathbf{A} = p_F \hat{\mathbf{l}}$, the effective magnetic and electric fields are $\mathbf{B} = p_F \vec{\nabla} \times \hat{\mathbf{l}}$ and $\mathbf{E} = p_F \partial_t \hat{\mathbf{l}}$.

Let us first apply eqn (18.6) to the Standard Model.

18.2 Anomalous non-conservation of baryonic charge

18.2.1 *Baryonic asymmetry of Universe*

In the Standard Model (Sec. 12.2) there are two additional, accidental global symmetries $U(1)_B$ and $U(1)_L$ whose classically conserved charges are the baryon number B and lepton number L. Each of the quarks has baryonic charge $B = 1/3$ and leptonic charge $L = 0$ with $B - L = 1/3$ in eqn (12.6). The leptons (neutrino and electron) have $B = 0$ and $L = 1$ with $B - L = -1$. The baryon number, as well as the lepton number, are not fundamental quantities, since they are not conserved in unified theories, such as G(224), where leptons and quarks are combined in the same multiplet in eqn (12.6). At low energy the matrix elements for the transformation of quarks to leptons become extremely small and the baryonic charge can be considered as a good quantum number with high precision.

The visible matter of our present Universe is highly baryon asymmetric: it consists of baryons and leptons, while the fraction of antibaryons and antileptons is negligibly small. The ratio of baryons to antibaryons now is a very big number. What is the origin of this number? This is one side of the puzzle.

Another side of the puzzle shows up when we consider what the baryonic asymmetry was in the early Universe and find that the Universe was actually too symmetric. To see this let us first assume that the Universe was baryon symmetric from the very beginning. Then at early times of the hot Universe ($t \ll 10^{-5}$ s), the thermally activated quark–antiquark pairs are as abundant as photons, i.e. $n_q + n_{\bar{q}} \sim n_\gamma$. As the Universe cools down, matter and antimatter annihilate each other until annihilation is frozen out. This occurs when the annihilation rate Γ becomes less than the expansion rate H of the Universe, so that nucleons and antinucleons become too rare to find one another. After annihilation is frozen out, one finds that only trace amounts of matter and antimatter remain, $n_q + n_{\bar{q}} \sim 10^{-18} n_\gamma$. This is much smaller than the observed nucleon to photon ratio $\eta = n_q/n_\gamma \sim 10^{-10}$. To prevent this so-called annihilation catastrophe the early Universe must be baryon asymmetric, but only with a very slight excess of baryons over antibaryons: $(n_q - n_{\bar{q}})/(n_q + n_{\bar{q}}) \sim \eta \sim 10^{-10}$. What is the origin of the small number η?

In modern theories which try to solve these two problems, it is assumed that originally the Universe was baryon symmetric. The excess of baryons could be produced in the process of evolution, if the three Sakharov criteria (1967b) are fulfilled: the Universe must be out of thermal equilibrium; the C (charge conjugation) and CP symmetries must be violated; and, of course, the baryon number is not conserved, otherwise the non-zero B cannot appear from the Universe which was initially baryon symmetric. One of the most popular mechanisms of the non-conservation of the baryonic charge is related to the axial anomaly (see the review papers by Turok (1992) and Trodden (1999)). In electroweak theory the baryonic charge can be generated from the vacuum due to spectral flow from negative energy levels in the vacuum to the positive energy levels of matter.

18.2.2 *Electroweak baryoproduction*

In the Standard Model there are two gauge fields whose 'electric' and 'magnetic' fields become a source for baryoproduction: the hypercharge field $U(1)_Y$ and the weak field $SU(2)_L$. Let us first consider the effect of the hypercharge field. According to eqn (18.6) the production rate of baryonic charge B in the presence of hyperelectric and hypermagnetic fields is

$$\dot{B}(Y) = \frac{1}{4\pi^2}\mathbf{tr}\left(BY^2\mathcal{N}\right)\mathbf{B}_Y\cdot\mathbf{E}_Y = \frac{N_g}{4\pi^2}(Y_{dR}^2 + Y_{uR}^2 - Y_{dL}^2 - Y_{uL}^2)\,\mathbf{B}_Y\cdot\mathbf{E}_Y, \quad (18.7)$$

where N_g is the number of families (generations) of fermions; Y_{dR}, Y_{uR}, Y_{dL} and Y_{uL} are hypercharges of right and left u and d quarks. Since the hypercharges of left and right fermions are different (see Fig. 12.1), one obtains the non-zero value of $\mathbf{tr}\left(\mathrm{BY}^2\mathcal{N}\right) = 1/2$, and thus a non-zero production of baryons by the hypercharge field

$$\dot{B}(Y) = \frac{N_g}{8\pi^2}\mathbf{B}_Y \cdot \mathbf{E}_Y. \tag{18.8}$$

The weak field also contributes to the production of the baryonic charge:

$$\dot{B}(T) = \frac{1}{4\pi^2}\mathbf{tr}\left(\mathrm{BT}_{3L}^2\mathcal{N}\right)\mathbf{B}_T^b \cdot \mathbf{E}_{bT} = -\frac{N_g}{8\pi^2}\mathbf{B}_T^b \cdot \mathbf{E}_{bT}. \tag{18.9}$$

Thus the total rate of baryon production in the Standard Model takes the form

$$\dot{B} = \frac{1}{4\pi^2}\left[\mathbf{tr}\left(\mathrm{BY}^2\mathcal{N}\right)\mathbf{B}_Y \cdot \mathbf{E}_Y + \mathbf{tr}\left(\mathrm{BT}_{3L}^2\mathcal{N}\right)\mathbf{B}_T^b \cdot \mathbf{E}_{bT}\right] \tag{18.10}$$

$$= \frac{N_g}{8\pi^2}\left(\mathbf{B}_Y \cdot \mathbf{E}_Y - \mathbf{B}_T^b \cdot \mathbf{E}_{bT}\right). \tag{18.11}$$

The same equation describes the production of the leptonic charge L: one has $\dot{L} = \dot{B}$ since $B - L$ is the charge related to the gauge group in GUT and thus is conserved due to anomaly cancellation. This means that production of one lepton is followed by production of three baryons.

The second term in eqn (18.10), which comes from non-Abelian $SU(2)_L$ field, shows that the nucleation of baryons occurs when the topological charge of the vacuum changes, say, by the sphaleron (Sec. 26.3.2) or due to de-linking of linked loops of the cosmic strings (Vachaspati and Field 1994; Garriga and Vachaspati 1995; Barriola 1995). This term is another example of the interplay between momentum space and real space topologies discussed in Chapter 11. It is the density of the topological charge in $\mathbf{r}$-space ($\propto \mathbf{B}_T^b \cdot \mathbf{E}_{bT}$) multiplied by the factor $\mathbf{tr}\left(\mathrm{BT}_{3L}^2\mathcal{N}\right)$, which is the topological charge in $\mathbf{p}$-space.

The first non-topological term in eqn (18.10) describes the exchange of the baryonic (and leptonic) charge between the hypermagnetic field and the fermionic degrees of freedom.

18.3 Analog of baryogenesis in ^{3}He-A

18.3.1 *Momentum exchange between superfluid vacuum and quasiparticle matter*

In ^{3}He-A the relevant fermionic charge B, which is important for the dynamics of superfluid liquid, is the linear momentum. There are three subsystems in superfluids which carry this charge: the superfluid vacuum, the texture (magnetic field) and the system of quasiparticles (fermionic matter). According to eqn (10.15) each of them contributes to the fermionic charges – the momentum of the liquid. Let us write down eqn (10.15) neglecting $\mathbf{P}^\mathrm{M}$ – the quasiparticle momentum transverse to $\hat{\mathbf{l}}$ which does not contribute to fermionic charge – since it is relatively small at low T:

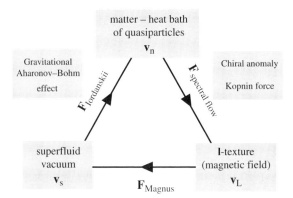

FIG. 18.3. Exchange of the fermionic charge in ^{3}He-A (linear momentum) between three subsystems moving with different velocities. The phenomenon of chiral anomaly regulates the exchange between $\hat{\mathbf{l}}$-texture and the normal component – the heat bath of quasiparticles. $\mathbf{v}_s$, $\mathbf{v}_n$ and $\mathbf{v}_L$ are velocities of the superfluid vacuum, normal component and texture (or vortex line, hence the index L) correspondingly.

$$\mathbf{P} = \mathbf{P}^{\text{F}} + \mathbf{P}\{\hat{\mathbf{l}}\} + mn\mathbf{v}_s \ , \ \mathbf{P}^{\text{F}} = \sum_{a,\mathbf{p}} \mathbf{p}^{(a)} f_a(\mathbf{p}) \ . \quad (18.12)$$

The first term $\mathbf{P}^{\text{F}}$ describes the fermionic charge carried by quasiparticles. Each quasiparticle of chirality C_a carries the fermionic charge $\mathbf{p}^{(a)} = -C_a p_F \hat{\mathbf{l}}$ (see eqn (10.25)). In addition the momentum $\mathbf{P}\{\hat{\mathbf{l}}\}$ is carried by the texture of the $\hat{\mathbf{l}}$-field which plays the role of magnetic field. Thus the second term, $\mathbf{P}\{\hat{\mathbf{l}}\}$, corresponds to the fermionic charge stored in the magnetic field. The superfluid vacuum moving with velocity $\mathbf{v}_s$ also carries momentum (the third term); this contribution is absent in the Standard Model: our physical vacuum is not a superfluid liquid.

One of the most important topics in superfluid dynamics is the momentum exchange between these three subsystems (Fig. 18.3). The axial anomaly describes the momentum exchange between two of them: the $\hat{\mathbf{l}}$-texture, which plays the role of the $U(1)$ gauge field, and the world of quasiparticles, the fermionic matter. This is what we discuss in this chapter.

The exchange between these two subsystems and the third one – the superfluid vacuum – is described by different physics which we discuss later. Of particular importance for superfluidity is the momentum exchange between the moving vacuum and quasiparticles, the third and the first terms in eqn (18.12), respectively. The force between the superfluid and normal components arising due to this momentum exchange is usually called the mutual friction force following Gorter and Mellink (1949), though the term friction is not very good since some or even the dominating part of this force is reversible and thus non-dissipative. In the early experiments in superfluid ^{4}He the momentum exchange between vacuum and matter was mediated by a chaotic, turbulent motion of the

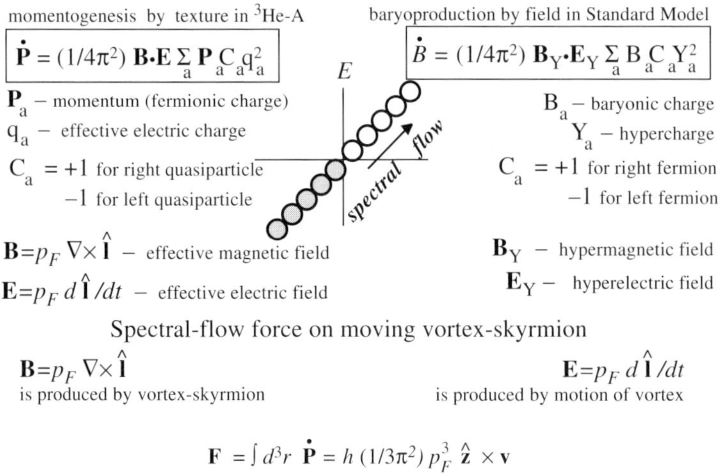

ANALOG OF BARYOGENESIS IN A-PHASE 243

momentogenesis by texture in ^{3}He-A	baryoproduction by field in Standard Model
$\dot{\mathbf{P}} = (1/4\pi^2)\, \mathbf{B}\cdot\mathbf{E}\, \sum_a P_a C_a q_a^2$	$\dot{\mathbf{B}} = (1/4\pi^2)\, \mathbf{B}_Y\cdot\mathbf{E}_Y \sum_a B_a C_a Y_a^2$
P_a – momentum (fermionic charge)	B_a – baryonic charge
q_a – effective electric charge	Y_a – hypercharge
$C_a = +1$ for right quasiparticle $\quad\;\; -1$ for left quasiparticle	$C_a = +1$ for right fermion $\quad\;\; -1$ for left fermion
$\mathbf{B} = p_F \nabla \times \hat{\mathbf{l}}$ – effective magnetic field	$\mathbf{B}_Y$ – hypermagnetic field
$\mathbf{E} = p_F\, d\hat{\mathbf{l}}/dt$ – effective electric field	$\mathbf{E}_Y$ – hyperelectric field

Spectral-flow force on moving vortex-skyrmion

$\mathbf{B} = p_F \nabla \times \hat{\mathbf{l}}$ $\qquad\qquad\qquad\qquad$ $\mathbf{E} = p_F\, d\hat{\mathbf{l}}/dt$
is produced by vortex-skyrmion $\qquad\qquad$ is produced by motion of vortex

$$\mathbf{F} = \int d^3r\, \dot{\mathbf{P}} = h\, (1/3\pi^2)\, p_F^3\, \hat{\mathbf{z}} \times \mathbf{v}$$

FIG. 18.4. Production of the fermionic charge in ^{3}He-A (linear momentum) and in the Standard Model (baryon number) described by the same Adler–Bell–Jackiw equation. Integration of the anomalous momentum production over the cross-section of the moving continuous vortex-skyrmion gives the loss of linear momentum and thus the additional force per unit length acting on the vortex due to spectral flow.

vortex tangle which produced the effective dissipative mutual friction between the normal and superfluid components.

Here we are interested in the momentum exchange between vacuum and matter mediated by the $\hat{\mathbf{l}}$-texture. The momentum exchange occurs in two steps: the momentum of the flowing vacuum (the third term) is transferred to the momentum carried by the texture (the second term), and then from the texture to the matter (the first term). The process $2 \to 1$ is due to the analog of chiral anomaly, while the process $3 \to 2$ corresponds to the lifting Magnus force acting on the vorticity of the $\hat{\mathbf{l}}$-texture from the moving vacuum.

There is also another specific momentum exchange $3 \to 1$ in Fig. 18.3 representing the analog of the gravitational Aharonov–Bohm effect, which will discussed in Chapter 31. It gives rise to the Iordanskii force.

In superfluids and superconductors with curl-free superfluid velocity $\mathbf{v}_s$, instead of the texture the singular quantized vortex serves as a mediator in the two-step process of momentum exchange. The modification of the chiral anomaly to the case of singular vortices will be discussed in Chapter 25.

18.3.2 Chiral anomaly in ^{3}He-A

Here we are interested in the process of the momentum transfer from the $\hat{\mathbf{l}}$-texture to quasiparticles. When a chiral quasiparticle crosses zero energy in its spectral flow it carries with it its linear momentum – the fermionic charge $\mathbf{p}^{(a)} = -C_a p_F \hat{\mathbf{l}}$.

We now apply the axial anomaly equation (18.5) to this process of transformation of the fermionic charge (momentum) carried by the magnetic field ($\hat{\mathbf{l}}$-texture) to the fermionic charge carried by chiral particles (normal component of the liquid) (Fig. 18.4). Substituting the relevant fermionic charge $\mathbf{p}^{(a)}$ into eqn (18.5) instead of B one obtains the rate of momentum production from the texture

$$\dot{\mathbf{P}}^{\mathrm{F}} = \frac{1}{4\pi^2}\mathrm{tr}\left(\mathbf{P}Q^2\mathcal{N}\right)\mathbf{B}\cdot\mathbf{E} = \frac{1}{4\pi^2}\mathbf{B}\cdot\mathbf{E}\sum_a \mathbf{p}^{(a)}C_a q_a^2 \ . \tag{18.13}$$

Here $\mathbf{B} = (p_F/\hbar)\nabla\times\hat{\mathbf{l}}$ and $\mathbf{E} = (p_F/\hbar)\partial_t\hat{\mathbf{l}}$ are effective 'magnetic' and 'electric' fields acting on fermions in ^{3}He-A; Q is the matrix of corresponding 'electric' charges in eqn (9.2): $q_a = -C_a$ (the 'electric' charge is opposite to the chirality of the ^{3}He-A quasiparticle, see eqn (9.4)). Using this translation to the ^{3}He-A language one obtains that the momentum production from the texture per unit time per unit volume is

$$\dot{\mathbf{P}}^{\mathrm{F}} = -\frac{p_F^3}{2\pi^2\hbar^2}\hat{\mathbf{l}}\left(\partial_t\hat{\mathbf{l}}\cdot(\nabla\times\hat{\mathbf{l}})\right) \ . \tag{18.14}$$

It is interesting to follow the history of this equation in ^{3}He-A. Volovik and Mineev (1981) considered the hydrodynamic equations for the superfluid vacuum in ^{3}He-A at $T = 0$ as the generalization of eqns (4.7) and (4.9) for the superfluid dynamics of ^{4}He which included the dynamics of the orbital momentum $\hat{\mathbf{l}}$. They found that according to these classical non-relativistic hydrodynamic equations the momentum of the superfluid vacuum $\mathbf{P}\{\hat{\mathbf{l}}\} + mn\mathbf{v}_{\mathrm{s}}$ is not conserved even at $T = 0$, when the quasiparticles are absent; and the production of the momentum is given by the right-hand side of eqn (18.14). They suggested that the momentum somehow escapes from the inhomogeneous vacuum to the world of quasiparticles. Later it was found by Combescot and Dombre (1986) that in the presence of the time-dependent $\hat{\mathbf{l}}$-texture quasiparticles are really nucleated, with the momentum production rate being described by the same eqn (18.14). Thus the total momentum of the system (superfluid vacuum and quasiparticles) has been proved to be conserved. In the same paper by Combescot and Dombre (1986) [95] it was first found that the quasiparticle states in ^{3}He-A in the presence of twisted texture of $\hat{\mathbf{l}}$ (i.e. the texture with $\nabla\times\hat{\mathbf{l}} \neq 0$) have a strong analogy with the eigenstates of a massless charged particle in a magnetic field. Then it became clear (Volovik 1986a) that equation (18.14) can be obtained from the axial anomaly equation in RQFT. Now we know why it happens: the spectral flow from the $\hat{\mathbf{l}}$-texture to the 'matter' occurs through the Fermi point and thus it can be described by the physics in the vicinity of the Fermi point, where the RQFT with chiral fermions necessarily arises and thus the anomalous flow of momentum can be described in terms of the Adler–Bell–Jackiw equation.

18.3.3 *Spectral-flow force acting on a vortex-skyrmion*

From the underlying microscopic theory we know that the total linear momentum of the liquid is conserved. Equation (18.14) thus implies that in the presence of a time-dependent texture the momentum is transferred from the texture (the distorted superfluid vacuum or magnetic field) to the heat bath of quasiparticles forming the normal component of the liquid (analog of matter). The rate of the momentum transfer gives an extra force acting on a moving $\hat{\mathbf{l}}$-texture. The typical continuous texture in ^{3}He-A is the doubly quantized vortex-skyrmion discussed in Sec. 16.2. The forces acting on moving vortex-skyrmions have been measured in experiments on rotating ^{3}He-A by Bevan *et al.* (1997*b*). So let us find the force acting on the skyrmion moving with respect to the normal component.

The stationary vortex has non-zero effective 'magnetic' field, $\mathbf{B} = (p_F \hbar) \nabla \times \hat{\mathbf{l}}$. If the vortex moves with velocity $\mathbf{v}_L$, then in the frame of the normal component the $\hat{\mathbf{l}}$-texture acquires the time dependence,

$$\hat{\mathbf{l}}(\mathbf{r}, t) = \hat{\mathbf{l}}(\mathbf{r} - (\mathbf{v}_L - \mathbf{v}_n)t) . \tag{18.15}$$

This time dependence induces the effective 'electric' field

$$\mathbf{E} = \frac{1}{\hbar} \partial_t \mathbf{A} = -\frac{p_F}{\hbar}((\mathbf{v}_L - \mathbf{v}_n) \cdot \nabla)\hat{\mathbf{l}} . \tag{18.16}$$

Since $\mathbf{B} \cdot \mathbf{E} \neq 0$, the motion of the vortex leads to the production of the quasiparticle momenta due to the spectral flow. Integrating eqn (18.14) over the cross-section of the simplest axisymmetric skyrmion with $n_1 = 2n_2 = 2$ in eqn (16.5), one obtains the momentum production per unit time per unit length and thus the force acting on unit length of the skyrmion line from the normal component:

$$\mathbf{F}_{\rm sf} = \int d^2\rho \frac{p_F^3}{2\pi^2 \hbar^2} \hat{\mathbf{l}} \left(((\mathbf{v}_n - \mathbf{v}_L) \cdot \nabla)\hat{\mathbf{l}} \cdot (\nabla \times \hat{\mathbf{l}})\right) = -\pi \hbar n_1 C_0 \hat{\mathbf{z}} \times (\mathbf{v}_L - \mathbf{v}_n), \tag{18.17}$$

where

$$C_0 = \frac{p_F^3}{3\pi^2 \hbar^3} . \tag{18.18}$$

The parameter C_0 has the following physical meaning: $(2\pi\hbar)^3 n_1 C_0$ is the volume inside the surface in the momentum space swept by Fermi points in the soft core of the vortex-skyrmion.

The above spectral-flow force in eqn (18.17) is transverse to the relative motion of the vortex with respect to the heat bath and thus is non-dissipative (reversible). In this derivation it was assumed that the quasiparticles and their momenta, created by the spectral flow from the inhomogeneous vacuum, are finally absorbed by the heat bath of the normal component. The retardation in the process of absorption and also the viscosity of the normal component lead also to a dissipative (friction) force between the vortex and the normal component:

$$\mathbf{F}_{\rm fr} = -\gamma(\mathbf{v}_L - \mathbf{v}_n) , \tag{18.19}$$

which will be discussed later for the case of singular vortices. Note that there is no momentum exchange between the vortex and the normal component if they move with the same velocity; according to Sec. 5.4 the condition that $\mathbf{v}_n = 0$ in the frame of a texture is one of the conditions of the global thermodynamic equilibrium, when the dissipation is absent.

18.3.4 Topological stability of spectral-flow force. Spectral-flow force from Novikov–Wess–Zumino action

The same result for the force in eqn (18.17) was obtained in a microscopic theory by Kopnin (1993). He used the so-called quasiclassical method (see Kopnin's book (2001)). This method is applicable at energies well above the first Planck scale, $E \gg \Delta_0^2/v_F p_F$, i.e. well outside the 'relativistic' domain, where RQFT is certainly not applicable. This reflects the fact that the spectral flow, the flow of levels along the anomalous branch of the energy spectrum, which governs the axial anomaly, does not depend on energy and can be calculated at any energy scale. In Kopnin's essentially non-relativistic calculations no notion of axial anomaly was invoked.

The spectral-flow force (18.17) does not depend on the details of the skyrmion structure as well. It can be derived not only for the axisymmetric skyrmion in eqn (16.5) but also for the general continuous vortex texture (Volovik 1992b); the only input is the topological charge of the vortex n_1. This force can be obtained directly from the topological Novikov–Wess–Zumino type of of action describing the anomaly (Volovik 1986b, 1993b). In the frame of the heat bath it has the same form as eqn (6.14) for ferromagnets in Sec. 6.1.5:

$$S_{\text{NWZ}} = -\frac{\hbar C_0}{2} \int d^3x \, dt \, d\tau \, \hat{\mathbf{l}} \cdot (\partial_t \hat{\mathbf{l}} \times \partial_\tau \hat{\mathbf{l}}) \,. \tag{18.20}$$

Here the unit vector $\hat{\mathbf{l}}$ along the orbital momentum of Cooper pairs substitutes the unit vector $\hat{\mathbf{m}}$ of spin magnetization in ferromagnets, while the role of the spin momentum $\mathbf{M}$ is played by the angular momentum

$$\mathbf{L}_{\text{anomalous}} = -\frac{\hbar}{2} C_0 \hat{\mathbf{l}} \,. \tag{18.21}$$

For the vortex 'center of mass' moving along the trajectory $\mathbf{r}_L(t,\tau)$ the vortex texture has the form $\hat{\mathbf{l}}(\mathbf{r},t,\tau) = \hat{\mathbf{l}}(\mathbf{r} - \mathbf{r}_L(t,\tau))$, and the action for the rectilinear vortex of length L becomes

$$S_{\text{NWZ}} = -\frac{C_0}{2} \int d^3x \, \hat{\mathbf{l}} \cdot \left(\frac{\partial \hat{\mathbf{l}}}{\partial x_L^i} \times \frac{\partial \hat{\mathbf{l}}}{\partial x_L^j} \right) \int dt \, d\tau \, \partial_t x_L^i \partial_\tau x_L^j \tag{18.22}$$

$$= -\pi n_1 C_0 L e_{ijk} \hat{z}^k \int dt \, d\tau \, \partial_t x_L^i \partial_\tau x_L^j = \frac{\pi}{2} n_1 C_0 L e_{ijk} \hat{z}^k \int dt \, v_L^i x_L^j \,. \tag{18.23}$$

Here we used eqn (16.8) for the mapping of the cross-section of the vortex to the sphere of unit vector $\hat{\mathbf{l}}$. Variation of the vortex-skyrmion action in eqn (18.23) over the vortex-skyrmion coordinate $\mathbf{x}_L(t)$ (with $\mathbf{v}_L = \partial_t \mathbf{x}_L$) gives the spectral-flow force acting on the vortex in eqn (18.17): $\mathbf{F}_{\text{sf}} = -\delta S_{\text{NWZ}}/\delta \mathbf{x}_L$.

18.3.5 Dynamics of Fermi points and vortices

The Novikov–Wess–Zumino action (18.23) is the product of two volumes in $\mathbf{p}$- and $\mathbf{r}$-space (Volovik 1993b):

$$S_{\text{NWZ}} = \pi\hbar \frac{V_{\mathbf{p}}V_{\mathbf{r}}}{(2\pi\hbar)^3} \ . \qquad (18.24)$$

The quantity $(2\pi\hbar)^3 n_1 C_0$ represents the volume within the surface in the $\mathbf{p}$-space spanned by Fermi points. The integral $(L/2)e_{ijk}\hat{z}^k \int dt \ v_L^i x_L^j$ represents the volume inside the surface swept by the vortex line in $\mathbf{r}$-space; it is written for the rectilinear vortex, but the expression in terms of the volume is valid for any shape of the vortex line (see Sec. 26.4.2 and eqn (26.12)). This demonstrates the close connection and similarity between Fermi points in $\mathbf{p}$-space and vortices in $\mathbf{r}$-space.

The action (18.24) in terms of the volume of the phase space spanned by Fermi points describes the general dynamics of the Fermi points of co-dimension 3, which is applicable even far outside the relativistic domain of the effective RQFT. This may show the route to the possible generalization of RQFT based on the physics of the Fermi points. Returning to the relativistic domain, one finds that in this low-energy limit, eqn (18.20) or (18.24) gives the action, whose variation represents the anomalous current in RQFT (see the book by Volovik 1992a). For, say, the hypermagnetic field this variation is

$$\delta S_{\text{NWZ}} = \frac{1}{2\pi^2} \text{tr}\left(\mathrm{Y}^3 \mathcal{N}\right) e^{\alpha\beta\mu\nu} \int d^3x \ dt \ A_\beta \partial_\mu A_\nu \delta A_\alpha \ . \qquad (18.25)$$

In the Standard Model the prefactor is zero due to anomaly cancellation (see eqn (12.17)).

18.3.6 Vortex texture as a mediator of momentum exchange. Magnus force

The spectral-flow Kopnin force $\mathbf{F}_{\text{sf}}$ is thus robust against any deformation of the $\hat{\mathbf{l}}$-texture which does not change its asymptote, i.e. the topological charge of the vortex – its winding number n_1. In this respect the spectral-flow force between the vortex texture and the bath of quasiparticles, which appears when the texture is moving with respect to the 'matter', resembles another force – the Magnus lifting force. The Magnus force describes the momentum exchange between the texture and the superfluid vacuum, the second and third terms in eqn (18.12), see Fig. 18.3. It acts on the vortex or vortex-skyrmion moving with respect to the superfluid vacuum:

$$\mathbf{F}_M = \pi\hbar n_1 n \hat{\mathbf{z}} \times (\mathbf{v}_L - \mathbf{v}_s(\infty)) \ , \qquad (18.26)$$

where again n is the particle density, the number density of ^{3}He atoms; $\mathbf{v}_s(\infty)$ is the uniform velocity of the superfluid vacuum far from the vortex. Here we marked by (∞) the external flow of the superfluid vacuum to distinguish it from the local circulating superflow around the vortex, but in future this mark will be omitted. In conventional notation used in canonical hydrodynamics, the Magnus force is

$$\mathbf{F}_M = \rho\kappa\hat{\mathbf{z}} \times (\mathbf{v}_L - \mathbf{v}_{\text{liquid}}) \ , \qquad (18.27)$$

where ρ is the mass density of the liquid and κ is the circulation of its velocity around the vortex. In our case $\rho = mn$ and $\kappa = n_1 \kappa_0$, with $\kappa_0 = \pi\hbar/m$ in superfluid ^{3}He and $\kappa_0 = 2\pi\hbar/m$ in superfluid ^{4}He.

Let us recall that the vortex texture (or quantized vortex in $U(1)$ superfluids) serves as mediator (intermediate object) for the momentum exchange between the superfluid vacuum moving with $\mathbf{v}_s$ and the fermionic heat bath of quasiparticles (normal component or 'matter') moving with $\mathbf{v}_n$. The momentum is transferred in two steps: first it is transferred from the vacuum to texture moving with velocity $\mathbf{v}_L$. This is described by the Magnus force acting on the vortex texture from the superfluid vacuum which depends on the relative velocity $\mathbf{v}_L - \mathbf{v}_s$. Then the momentum is transferred from the texture to the 'matter'. With a minus sign this is the spectral-flow force in eqn (18.17) which depends on the relative velocity $\mathbf{v}_L - \mathbf{v}_n$. In this respect the texture (or vortex) corresponds to the sphaleron (Sec. 26.3.2) or to the cosmic string in relativistic theories which also mediate the exchange of fermionic charges between the quantum vacuum and the matter.

If the other processes are ignored then in the steady state these two forces acting on the texture, from the vacuum and from the 'matter', must compensate each other: $\mathbf{F}_M + \mathbf{F}_{sf} = 0$. From this balance of the two forces one obtains that the vortex must move with the constant velocity determined by the velocities $\mathbf{v}_s$ and $\mathbf{v}_n$ of the vacuum and 'matter' respectively: $\mathbf{v}_L = (n\mathbf{v}_s - C_0\mathbf{v}_n)/(n - C_0)$. Note that in the Bose liquid, where the fermionic spectral flow is absent and thus $C_0 = 0$, this leads to the requirement that the vortex moves with the superfluid velocity.

However, this is valid only under special conditions. First, the dissipative friction must be taken into account. It comes in particular from the retardation of the spectral-flow process. The retardation also modifies the non-dissipative spectral-flow force as we shall see in the example of the ^{3}He-B vortex in Sec. 25.2. Second, the analogy with gravity shows that there is one more force of topological origin – the so-called Iordanskii (1964, 1966) force, see Fig. 18.3. It comes from the gravitational analog of the Aharonov–Bohm effect experienced by (quasi)particles moving in the presence of the spinning cosmic string and exists in the Bose liquid too (see Sec. 31.3.4).

18.4 Experimental check of Adler–Bell–Jackiw equation in ^{3}He-A

The spectral-flow force (the Kopnin force) acting on the vortex-skyrmion has been measured in experiments on vortex dynamics in ^{3}He-A by Bevan *et al.* (1997*a*,*b*). In such experiments a uniform array of vortices is produced by rotating the whole cryostat. In equilibrium the vortices and the normal component of the fluid (heat bath of quasiparticles) rotate together with the cryostat. An electrostatically driven vibrating diaphragm in Fig. 18.5 *left* produces an oscillating superflow, which via the Magnus force acting on vortices from the superfluid velocity field $\mathbf{v}_s$ generates the vortex motion. The normal component of the liquid remains clamped in the container frame due to the high viscosity of the system of quasiparticles in ^{3}He. Thus vortices move with respect to both the heat bath ('matter') and the superfluid vacuum. The vortex velocity $\mathbf{v}_L$ is determined by the overall balance of forces acting on the vortices. This includes the spectral-

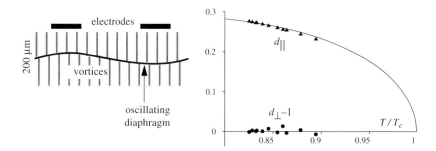

FIG. 18.5. Experimental verification of anomaly equation in ^{3}He-A. *Left*: A uniform array of vortices is produced by rotating the whole cryostat, and oscillatory superflow perpendicular to the rotation axis is produced by a vibrating diaphragm, while the normal fluid (thermal excitations) is clamped by viscosity, $\mathbf{v}_n = 0$. The velocity $\mathbf{v}_L$ of the vortex array is determined by the overall balance of forces acting on the vortices. *Right*: These vortices produce additional dissipation proportional to $d_\parallel$ and coupling between two orthogonal modes proportional to $1 - d_\perp$. (After Bevan *et al.* 1997a).

flow force $\mathbf{F}_{\rm sf}$ in eqn (18.17); the Magnus force $\mathbf{F}_{\rm M}$ in eqn (18.26); the friction force $\mathbf{F}_{\rm fr}$ in eqn (18.19); and the Iordanskii force in eqn (31.25):

$$\mathbf{F}_{\rm Iordanskii} = \pi n_1 \hbar n_n \hat{\mathbf{z}} \times (\mathbf{v}_s - \mathbf{v}_n) \ . \tag{18.28}$$

For the steady state motion of vortices the sum of all forces acting on the vortex must be zero:

$$\mathbf{F}_{\rm M} + \mathbf{F}_{\rm sf} + \mathbf{F}_{\rm Iordanskii} + \mathbf{F}_{\rm fr} = 0 \ , \tag{18.29}$$

From this force balance equation one has the following equation for $\mathbf{v}_{\rm L}$:

$$\hat{\mathbf{z}} \times (\mathbf{v}_{\rm L} - \mathbf{v}_{\rm s}) + d_\perp \hat{\mathbf{z}} \times (\mathbf{v}_n - \mathbf{v}_{\rm L}) + d_\parallel (\mathbf{v}_n - \mathbf{v}_{\rm L}) = 0 \ , \tag{18.30}$$

where

$$d_\perp = 1 - \frac{n - C_0}{n_{\rm s}(T)} \ , \quad d_\parallel = \frac{\gamma}{\pi n_1 n_{\rm s}(T)} \ . \tag{18.31}$$

Measurement of the damping of the diaphragm resonance and of the coupling between different eigenmodes of vibrations enables both dimensionless parameters, $d_\perp$ and $d_\parallel$ in eqn (18.31), to be deduced (Fig. 18.5 *right*). The most important for us is the parameter $d_\perp$, which gives information on the spectral-flow parameter C_0. The effect of the chiral anomaly is crucial for C_0: if there is no anomaly then $C_0 = 0$ and $d_\perp = n_n(T)/n_{\rm s}(T)$; if the anomaly is fully realized the parameter C_0 has its maximal value, $C_0 = p_F^3/3\pi^2\hbar^3$, which coincides with the particle density of liquid ^{3}He in the normal state, eqn (8.6). The difference between the particle density of liquid ^{3}He in the normal state C_0 and the particle density of liquid ^{3}He in superfluid ^{3}He-A state n at the same chemical potential μ is determined by the tiny effect of superfluid correlations on the particle density and is extremely small: $n - C_0 \sim n(\Delta_0/v_F p_F)^2 = n(c_\perp/c_\parallel)^2 \sim 10^{-6}n$; in this case one must have $d_\perp \approx 1$ for all practical temperatures, even including the region close to T_c, where the superfluid density $n_s(T) \sim n(1 - T^2/T_c^2)$ is small. ^{3}He-A experiments, made in the entire temperature range where ^{3}He-A is stable, gave precisely this value within experimental uncertainty, $|1 - d_\perp| < 0.005$ (Bevan et al. 1997b; see also Fig. 18.5).

This means that the chiral anomaly is fully realized in the dynamics of the $\hat{\mathbf{l}}$-texture and provides an experimental verification of the Adler–Bell–Jackiw axial anomaly equation (18.5), applied to ^{3}He-A. This supports the idea that baryonic charge (as well as leptonic charge) can be generated by electroweak gauge fields through the anomaly.

In the same experiments with the ^{3}He-B vortices the effect analogous to the axial anomaly is temperature dependent and one has the crossover from the regime of maximal spectral flow with $d_\perp \approx 1$ at high T to the regime of fully suppressed spectral flow with $d_\perp = n_n(T)/n_s(T)$ at low T (see Sec. 25.2.4 and Fig. 25.1). The reason for this is that eqn (18.5) for the axial anomaly and the corresponding equation (18.14) for the momentum production are valid only in the limit of continuous spectrum, i.e. when the distance ω_0 between the energy levels of fermions in the texture is much smaller than the inverse quasiparticle lifetime: $\omega_0 \tau \ll 1$. The spectral flow completely disappears in the opposite case $\omega_0 \tau \gg 1$, because the spectrum becomes effectively discrete. As a result, the force acting on a vortex texture differs by several orders of magnitude for the cases $\omega_0 \tau \ll 1$ and $\omega_0 \tau \gg 1$. The parameter $\omega_0 \tau$ is regulated by temperature. This will be discussed in detail in Chapter 25.

In the case of ^{3}He-A the vortices are continuous, the size of the soft core of the vortex is large and thus the distance ω_0 between the quasiparticle levels in the soft core is extremely small compared to $1/\tau$. This means that the spectral flow in ^{3}He-A vortices is maximally possible and the Adler–Bell–Jackiw anomaly equation is applicable there at all practical temperatures. This was confirmed experimentally.

Note in conclusion of this section that the spectral flow realized by the moving vortex can be considered as the exchange of fermionic charge between systems of different dimension: the 3+1 fermionic system outside the vortex core and the 1+1 fermions living in the vortex core (Volovik 1993b; Stone 1996; see also Chapter 25). In RQFT this corresponds to the Callan–Harvey (1985) process of anomaly cancellation in which also two systems with different dimension are involved.

19

ANOMALOUS CURRENTS

19.1 Helicity in parity-violating systems

Parity violation, the asymmetry between left and right, is one of the fundamental properties of the quantum vacuum of the Standard Model (see Sec. 12.2). This effect is strong at high energy on the order of the electroweak scale, but is almost imperceptible in low-energy condensed matter physics. At this scale the left and right particles are hybridized and only the left–right symmetric charges survive: electric charge Q and the charges of the color group $SU(3)_C$. Leggett's (1977a) suggestion to observe the macroscopic effect of parity violation using such a macroscopically coherent atomic system as superfluid ^{3}He-B is still very far from realization (Vollhardt and Wölfle 1990). On the other hand, an analog of parity violation exists in superfluid ^{3}He-A alongside related phenomena, such as the chiral anomaly which we discussed in the previous section and macroscopic chiral currents. So, if we cannot investigate the macroscopic parity-violating effects directly we can simulate analogous physics in ^{3}He-A.

Most of the macroscopic parity-violating phenomena are related to helicity: the energy of the system in which the parity is broken contains the helicity term $\lambda \mathbf{A} \cdot (\nabla \times \mathbf{A})$, where $\mathbf{A}$ is the relevant collective vector field. To have such terms the parity P must be violated together with all the combinations containing other discrete symmetries, such as CP, PT, CPT, PU$_2$ (where U$_2$ is the rotation by π), etc. Since they contain the first-order derivative of the order parameter, such terms sometimes cause instability of the vacuum toward the spatially inhomogeneous state, the so-called helical instability. In nematic liquid crystals, for example, the excess of the chiral molecules of one preferred chirality leads to the helicity term, $\lambda \hat{\mathbf{n}} \cdot (\nabla \times \hat{\mathbf{n}})$, for the nematic vector (director) field $\hat{\mathbf{n}}$. This leads to formation of the cholesteric structure, the helix.

The same phenomenon occurs in superfluid ^{3}He-A. The velocity $\mathbf{w} = \mathbf{v}_n - \mathbf{v}_s$ of the counterflow plays the role of the difference in chemical potentials between the left-handed and right-handed quasiparticles (Sec. 10.3.2). By creating the counterflow in the rotating vessel, one can generate an excess of quasiparticles of a given chirality and test experimentally the suggestion by Joyce and Shaposhnikov (1997) and Giovannini and Shaposhnikov (1998) that it exhibits helical instability. According to Joyce and Shaposhnikov, the system with an excess of right-handed electrons is unstable toward formation of the helical hypermagnetic field $\mathbf{B}_Y$. Below the electroweak transition, the formed field $\mathbf{B}_Y$ is transformed to the electromagnetic magnetic field $\mathbf{B}(\equiv \mathbf{B}_Q)$. Thus the helical instability can serve as a source of formation of primordial cosmological magnetic fields (see also the recent review paper on cosmic magnetic fields by Tornkvist (2000) and

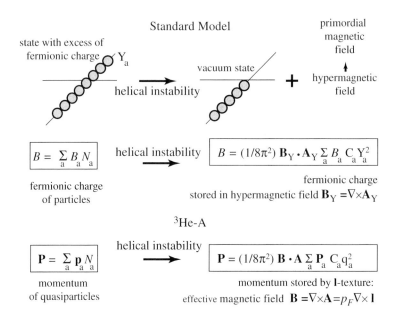

FIG. 19.1. Formation of magnetic field due to helical instability. The fermionic charge of right-handed particles minus that of left-handed ones is conserved at the classical level but not if quantum properties of the physical vacuum are taken into account. This charge can be transferred to the inhomogeneity of the vacuum via the axial anomaly in the process of helical instability. The inhomogeneity which absorbs the fermionic charge arises as a hypermagnetic field configuration in the Standard Model (*top*) and as the $\hat{\mathbf{l}}$-texture in ^{3}He-A, which is analogous to the magnetic field (*bottom*).

references therein). In Sec. 19.3 we show that the mechanism of formation of the hypermagnetic field in the relativistic plasma of right-handed electrons has a direct parallel with the formation of the helical $\hat{\mathbf{l}}$-texture in a rotating vessel. They are described by the same effective action, which contains the Chern–Simons helical term (Fig. 19.1).

19.2 Chern–Simons energy term

19.2.1 *Chern–Simons term in Standard Model*

Due to the axial anomaly, fermionic charge, say baryonic or leptonic, can be transferred to the 'inhomogeneity' of the vacuum. As a result the topologically non-trivial vacuum can acquire the fermionic charge. In particular, a monopole can acquire spin 1/2 due to fermion zero modes (Jackiw and Rebbi 1976*a*, 1984); in some theories of strong interactions protons and neutrons emerge as topological defects – skyrmions (Skyrme 1961) – whose baryonic charge and spin 1/2 are provided by the rearrangement of the fermionic vacuum in the presence of the defect (see the recent review by Gisiger and Paranjape 1998).

CHERN–SIMONS ENERGY TERM

In the Standard Model the typical inhomogeneity of the Bose fields, which absorbs the fermionic charge, is a helix of a magnetic field configuration, say, of hypermagnetic field. According to the axial anomaly equation (18.6), the fermionic charge density $\tilde{B}$ absorbed by the hypermagnetic field is

$$\tilde{B}\{\mathbf{A}_Y\} = \frac{1}{8\pi^2} \mathbf{A}_Y \cdot (\nabla \times \mathbf{A}_Y) \, \mathrm{tr}\left(\tilde{B} Y^2 \mathcal{N}\right) . \tag{19.1}$$

Let us recall that for the non-interacting relativistic fermions this equations reads

$$\tilde{B}\{\mathbf{A}_Y\} = \frac{1}{8\pi^2} \mathbf{A}_Y \cdot (\nabla \times \mathbf{A}_Y) \sum_a C_a \tilde{B}_a Y_a^2 , \tag{19.2}$$

where again a marks the fermionic species; $C_a = \pm 1$ is the chirality of the fermion; Y_a and $\tilde{B}_a$ are correspondingly the hypercharge of the a-th fermion and its relevant fermionic charge, whose absorption by the hyperfield is under discussion.

The fermionic charge which we are interested in here is $\tilde{B}_a = 3B_a + L_a$, where B_a and L_a are baryonic and leptonic numbers, i.e. $\tilde{B}_a = +1$ for quarks and leptons and $\tilde{B}_a = -1$ for antiquarks and antileptons. Both baryonic and leptonic numbers are extremely well conserved in our low-energy world at temperatures below the electroweak phase transition. But above the the electroweak transition separate conservations of B and L are violated by axial anomaly, while the combination $B - L$ is conserved (Sec. 12.2). For each quark and lepton in the Standard Model the charge $\tilde{B}_a = 1$; thus the number of chiral fermionic species in the Standard Model is $\sum_a \tilde{B}_a = 16 N_g$, where N_g is the number of fermionic families.

If fermionic species do not interact with each other then the number $N_a = \sum_\mathbf{p} \tilde{B}_a$ of the a-th fermionic particles is conserved separately, and one can introduce chemical potential μ_a for each species. Then the energy functional has the following term (compare with eqn (3.2)):

$$-\sum_a \mu_a N_a . \tag{19.3}$$

However, due to the chiral anomaly the charge $\tilde{B}_a$ can be distributed and redistributed between fermionic particles and the Bose field. As a result the total fermionic charge N_a has two contributions

$$N_a = \int d^3x \left(n_a^{\mathrm{fermion}} + \frac{1}{8\pi^2} C_a Y_a^2 \mathbf{A}_Y \cdot (\nabla \times \mathbf{A}_Y) \right) . \tag{19.4}$$

Here n_a is the number density of the a-th fermionic particles; the second term is the fermionic charge stored by the helix in the $U(1)_Y$ gauge field in eqn (19.2). This term gives the following contribution to the energy in eqn (19.3):

$$F_{\mathrm{CS}}\{\mathbf{A}_Y\} = -\sum_a \mu_a \int d^3x \, \tilde{B}_a\{\mathbf{A}_Y\} = -\frac{1}{8\pi^2} \sum_a C_a \mu_a Y_a^2 \int d^3x \, \mathbf{A}_Y \cdot (\nabla \times \mathbf{A}_Y) . \tag{19.5}$$

It describes the interaction of the helicity of hypermagnetic field with the chemical potentials μ_a of fermions and represents the Chern–Simons energy of the hypercharge $U(1)_Y$ field. It is non-zero because the parity is violated by non-zero values of chemical potentials μ_a. The corresponding Lagrangian is not gauge invariant, but the action is if μ_a are constant in space and time. The latter is natural since μ_a are Lagrange multipliers.

As an example let us consider the unification energy scale, where all the fermions have the same chemical potential μ, since they can transform to each other at this scale. From the generating function for the Standard Model in eqn (12.12) one has $\sum_a C_a Y_a^2 = \text{tr}\left(Y^2 \mathcal{N}\right) = 2N_g$ and thus the Chern–Simons energy of the hypercharge field at high energy becomes

$$F_{\rm CS}\{\mathbf{A}_Y\} = -\frac{\mu N_g}{4\pi^2} \int d^3x\, \mathbf{A}_Y \cdot (\nabla \times \mathbf{A}_Y) \ . \tag{19.6}$$

When $\mu = 0$ the CP and CPT symmetries of the Standard Model are restored, and the helicity term becomes forbidden.

19.2.2 Chern–Simons energy in ^{3}He-A

Let us now consider the ^{3}He-A counterpart of the Chern–Simons term. It arises in the $\hat{\mathbf{l}}$-texture in the presence of the homogeneous counterflow $\mathbf{w} = \mathbf{v}_{\rm n} - \mathbf{v}_{\rm s} = w\hat{\mathbf{z}}$ of the normal component with respect to the superfluid vacuum. As is clear, say, from eqn (10.32), the counterflow orients the $\hat{\mathbf{l}}$-vector along the counterflow, so that the equilibrium orientations of the $\hat{\mathbf{l}}$-field are $\hat{\mathbf{l}}_0 = \pm\hat{\mathbf{z}}$. Since $\hat{\mathbf{l}}$ is a unit vector, its variation $\delta\hat{\mathbf{l}} \perp \hat{\mathbf{l}}_0$. In the gauge field analogy, in which the effective vector potential is $\mathbf{A} = p_F \delta\hat{\mathbf{l}}$, this corresponds to the gauge choice $A_z = 0$.

The relevant fermionic charge in ^{3}He-A exhibiting the Abelian anomaly is the momentum of quasiparticles along $\hat{\mathbf{l}}_0$, i.e. $\mathbf{p}^{(a)} = -C_a p_F \hat{\mathbf{l}}_0$. According to eqn (19.2), where the fermionic charge $\tilde{B}_a$ is specified as $\mathbf{p}^{(a)}$ and the hypercharge is substituted by the charge $q_a = -C_a$ of quasiparticles, the helicity of the effective gauge field $\mathbf{A} = p_F \delta\hat{\mathbf{l}}$ carries the following linear momentum:

$$\mathbf{P}\{\mathbf{A}\} = \frac{1}{8\pi^2} \mathbf{A} \cdot (\nabla \times \mathbf{A}) \sum_a C_a \mathbf{p}^{(a)} q_a^2 = -\frac{p_F^3}{4\pi^2} \hat{\mathbf{l}}_0 \left(\delta\hat{\mathbf{l}} \cdot (\nabla \times \delta\hat{\mathbf{l}})\right) \ . \tag{19.7}$$

The total linear momentum density stored both in the heat bath of quasiparticles ('matter') and in the texture ('hyperfield') is thus

$$\mathbf{P} = \sum_{\mathbf{p},a} \mathbf{p}^{(a)} f_a(\mathbf{p}) + \mathbf{P}\{\delta\hat{\mathbf{l}}\} = \mathbf{P}^{\rm F} + \mathbf{P}\{\mathbf{A}\} \ . \tag{19.8}$$

This is in agreement with eqn (10.15) for the total current. Because of the Mermin–Ho relation between the superfluid velocity $\mathbf{v}_{\rm s}$ and the $\hat{\mathbf{l}}$-texture, the $\hat{\mathbf{l}}$-texture induces the $nm\mathbf{v}_{\rm s}$ contribution to the momentum which is also quadratic in $\delta\hat{\mathbf{l}}$. That is why, instead of the parameter C_0 in the anomalous current in eqn (10.11), one has $C_0 + n/2 \approx 3C_0/2$, and one obtains the factor $1/4\pi^2$ in eqn (19.7) instead of the expected factor $1/6\pi^2$.

The kinetic energy of the liquid, which is stored in the counterflow, is

$$-\mathbf{w} \cdot \mathbf{P} = -\mathbf{w} \cdot \sum_{\mathbf{p}} \mathbf{p} f(\mathbf{p}) - \mathbf{w} \cdot \mathbf{P}\{\mathbf{A}\} \approx -\mathbf{w} \cdot \mathbf{P}^{\mathrm{F}} - \mathbf{w} \cdot \mathbf{P}\{\mathbf{A}\} \,. \qquad (19.9)$$

The second term on the rhs of this equation is the analog of the Chern–Simons energy in eqn (19.5), which is now the energy stored in the $\hat{\mathbf{l}}$-field in the presence of the counterflow:

$$F_{\mathrm{CS}}\{\delta\hat{\mathbf{l}}\} = -\frac{1}{8\pi^2} \mathbf{A} \cdot (\nabla \times \mathbf{A}) \sum_a C_a \mu_a q_a^2 \equiv -\frac{p_F^3}{4\pi^2} (\hat{\mathbf{l}}_0 \cdot \mathbf{w}) \left(\delta\hat{\mathbf{l}} \cdot (\nabla \times \delta\hat{\mathbf{l}}) \right) \,. \qquad (19.10)$$

This term was earlier calculated in ^{3}He-A using the quasiclassical approach: it is the fourth term in eqn (7.210) of the book by Vollhardt and Wölfle (1990). The quasiclassical method is applicable in the energy range $\Delta_0 \gg T \gg \Delta_0^2/v_F p_F$, which is well above the first Planck scale $E_{\mathrm{Planck\ 1}} = \Delta_0^2/v_F p_F$ and thus outside the relativistic domain. In the present derivation we used RQFT well below $E_{\mathrm{Planck\ 1}}$, where the effect was discussed in terms of the Abelian anomaly. The results of the two approaches coincide indicating that the phenomenon of the anomaly is not restricted by the relativistic domain.

19.3 Helical instability and 'magnetogenesis' due to chiral fermions

19.3.1 *Relevant energy terms*

The Chern–Simons term in eqn (19.5) for the Standard Model and eqn (19.10) for ^{3}He-A is odd under spatial parity transformation, if the chemical potential and correspondingly the counterflow are fixed. Thus it can have a negative sign for properly chosen perturbations of the field or texture. This means that one can have an energy gain from the transformation of the fermionic charge from matter (quasiparticles) to the $U(1)$ gauge field ($\hat{\mathbf{l}}$-texture). This is the essence of the Joyce–Shaposhnikov (1997) scenario for the generation of primordial magnetic field starting from the homogeneous bath of chiral fermions. In ^{3}He-A language, the excess of the chiral fermions, i.e. the non-zero chemical potential μ_a for fermions, corresponds to non-zero counterflow: $\mu_a = -C_a p_F \mathbf{w} \cdot \hat{\mathbf{l}}$ (Sec. 10.3). Thus the Joyce–Shaposhnikov effect corresponds to the collapse of the flow of the normal component with respect to the superfluid vacuum toward the formation of $\hat{\mathbf{l}}$-texture – the analog of hypermagnetic field. The momentum carried by the flowing quasiparticles (the fermionic charge $\mathbf{P}^{\mathrm{F}}$) is transferred to the momentum $\mathbf{P}\{\delta\hat{\mathbf{l}}\}$ carried by the texture. Such a collapse of quasiparticle momentum toward $\hat{\mathbf{l}}$-texture has been investigated experimentally in the rotating cryostat (see Fig. 19.2 below) by Ruutu *et al.* (1996*b*, 1997) and was interpreted as an analog of magnetogenesis by Volovik (1998*b*).

Now let us write all the relevant energy terms in the Standard Model (thermal energy of fermions, Chern–Simons term and the energy of the hypermagnetic field) and their counterparts in ^{3}He-A:

$$W = W(T, \mu) + F_{\mathrm{CS}} + F_{\mathrm{hypermagn}} \,, \qquad (19.11)$$

$$W(T,\mu) = \frac{7\pi^2}{180}\sqrt{-g}T^4 \sum_a 1 - \frac{\sqrt{-g}}{12}T^2 \sum_a \mu_a^2 \,, \quad (19.12)$$

$$F_{\rm CS} = -\frac{1}{8\pi^2}\mathbf{A}_Y \cdot (\nabla \times \mathbf{A}_Y) \sum_a C_a \mu_a Y_a^2 \,, \quad (19.13)$$

$$F_{\rm hypermagn} = \frac{1}{96\pi^2}\ln\left(\frac{E_{\rm Planck}^2}{T^2}\right)\sqrt{-g}g^{ik}g^{mn}F_{imY}F_{knY}\sum_a Y_a^2 \,. \quad (19.14)$$

Equation (19.14) is the energy of the hypermagnetic field. In the logarithmically running coupling we left only the contribution of fermions to the polarization of the vacuum. This is what we need for application to ^{3}He-A, where only fermions are fundamental. Equation (19.12) is the thermodynamic energy of the gas of the relativistic quasiparticles at non-zero chemical potential, $|\mu_a| \ll T$. This energy is irrelevant in the Joyce–Shaposhnikov scenario, since it does not contain the field $\mathbf{A}_Y$. But in ^{3}He-A it is important because it gives rise to the mass of the hyperphoton.

19.3.2 *Mass of hyperphoton due to excess of chiral fermions (counterflow)*

In ^{3}He-A the chemical potential in eqn (19.12) depends on the $\hat{\mathbf{l}}$-vector: $\mu_a = -C_a p_F \mathbf{w} \cdot \hat{\mathbf{l}}$. The unit vector must be expanded up to second order in deviations:

$$\hat{\mathbf{l}} = \hat{\mathbf{l}}_0 + \delta\hat{\mathbf{l}} - (1/2)\hat{\mathbf{l}}_0(\delta\hat{\mathbf{l}})^2 \,. \quad (19.15)$$

Inserting this into eqn (19.12) and neglecting the terms which do not contain the $\delta\hat{\mathbf{l}}$-field, one obtains the energy whose translation to relativistic language represents the mass term for the $U(1)_Y$ gauge field (in ^{3}He-A $\mathbf{A}_Y = p_F \delta\hat{\mathbf{l}}$):

$$F_{\rm mass} = \frac{1}{4}mn_{\rm n\|}w^2(\delta\hat{\mathbf{l}}\cdot\delta\hat{\mathbf{l}}) \equiv \frac{1}{2}\sqrt{-g}g^{ik}A_{iY}A_{kY}M_{\rm hp}^2 \,, \quad (19.16)$$

$$M_{\rm hp}^2 = \frac{1}{6}\frac{T^2}{E_{\rm Planck\,2}^2}\sum_a Y_a^2 \mu_a^2 \,, \quad E_{\rm Planck\,2} = \Delta_0 \,. \quad (19.17)$$

In ^{3}He-A the mass $M_{\rm hp} \sim T\mu/E_{\rm Planck} = Tp_F w/\Delta_0$ of the 'hyperphoton' is physical and important for the dynamics of the $\hat{\mathbf{l}}$-vector. It determines the gap in the spectrum of orbital waves in eqn (9.24) – propagating oscillations of $\delta\hat{\mathbf{l}}$ (Leggett and Takagi 1978) – which play the role of electromagnetic waves. This hyperphoton mass appears due to the presence of the counterflow, which orients $\hat{\mathbf{l}}$ and thus provides the restoring force for oscillations of $\delta\hat{\mathbf{l}}$.

In principle, a similar mass can exist for the real hyperphoton. If the Standard Model is an effective theory, the local $U(1)_Y$ symmetry arises only in the low-energy corner and thus is approximate. It can be violated (not spontaneously but gradually) by the higher-order terms, which contain the Planck energy cut-off; in ^{3}He-A it is the second Planck scale $E_{\rm Planck\,2}$. Equation (19.17) suggests that the mass of the hyperphoton could arise if both the temperature T and the chemical potential μ_a are finite. This mass disappears in the limit of an infinite cut-off parameter or is negligibly small, if the cut-off is of Planck scale E_{Planck}. The ^{3}He-A example thus provides an illustration of how the non-renormalizable terms are suppressed by the small ratio of the energy to the fundamental energy scale of the theory (Weinberg 1999) and how the terms of order $(T/E_{\text{Planck}})^2$ appear in the effective quantum field theory (Jegerlehner 1998).

19.3.3 Helical instability condition

In ^{3}He-A eqns (19.11–19.14) give the following quadratic form of the energy in terms of the deviations $\delta\mathbf{l} \perp \hat{\mathbf{l}}_0 = \hat{\mathbf{z}}$ from the state with homogeneous counterflow along the axis $\hat{\mathbf{z}}$:

$$\frac{12\pi^2}{p_F^2 v_F} W\{\delta\hat{\mathbf{l}}\} = (\partial_z \delta\hat{\mathbf{l}})^2 \ln\frac{\Delta_0}{T} - 3m^* w \delta\hat{\mathbf{l}} \cdot (\hat{\mathbf{z}} \times \partial_z \delta\hat{\mathbf{l}}) + \pi^2 (m^* w)^2 \frac{T^2}{\Delta_0^2} (\delta\hat{\mathbf{l}})^2. \quad (19.18)$$

Since the Chern–Simons term depends only on the z-derivative, in the 'hypermagnetic' energy only the derivative along z are left. After the rescaling of the coordinates $\tilde{z} = z m^* w/\hbar$ one obtains

$$\tilde{W} = \frac{4W\{\delta\hat{\mathbf{l}}\}}{C_0 m^* w^2} = (\partial_{\tilde{z}} \delta\hat{\mathbf{l}})^2 \ln\frac{\Delta_0}{T} - 3\delta\hat{\mathbf{l}} \cdot (\hat{\mathbf{z}} \times \partial_{\tilde{z}} \delta\hat{\mathbf{l}}) + \pi^2 \frac{T^2}{\Delta_0^2} (\delta\hat{\mathbf{l}})^2 . \quad (19.19)$$

In Fourier components $\delta\hat{\mathbf{l}} = \sum_q \mathbf{a}_q e^{iq\tilde{z}}$ (with $\mathbf{a}_q \perp \hat{\mathbf{z}}$) this reads

$$\tilde{W} = \sum_q a^i_{-q} a^j_q \left[\left(q^2 \ln\frac{\Delta_0}{T} + \pi^2 \frac{T^2}{\Delta_0^2} \right) \delta_{ij} - 3iq e_{ij} \right] , \quad (19.20)$$

where $i = 1, 2$ and e_{ij} is the 2D antisymmetric tensor.

The quadratic form in eqn (19.20) becomes negative and thus the uniform counterflow becomes unstable toward the nucleation of the $\hat{\mathbf{l}}$-texture if

$$\frac{T^2}{\Delta_0^2} \ln\frac{\Delta_0}{T} < \frac{9}{4\pi^2} . \quad (19.21)$$

If this condition is fulfilled, the instability occurs for any value w of the counterflow.

In relativistic theories, where the temperature is always smaller than the Planck cut-off, the condition corresponding to eqn (19.21) is always fulfilled. Thus the excess of the fermionic charge is always unstable toward nucleation of the hypermagnetic field, if the fermions are massless, i.e. above the electroweak transition. In the scenario of magnetogenesis developed by Joyce and Shaposhnikov (1997) and Giovannini and Shaposhnikov (1998), this instability is responsible for the genesis of the hypermagnetic field well above the electroweak transition. The role of the subsequent electroweak transition is to transform this hypermagnetic field to the conventional (electromagnetic $U(1)_Q$) magnetic field due to the electroweak symmetry breaking.

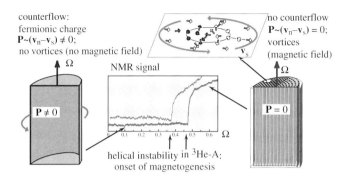

FIG. 19.2. Experimental 'magnetogenesis' in ^{3}He-A. *Left*: Initial vortex-free state in the rotating vessel contains a counterflow, $\mathbf{w} = \mathbf{v}_n - \mathbf{v}_s = \mathbf{\Omega} \times \mathbf{r} \neq 0$, since the average velocity of quasiparticles (the normal component) is $\mathbf{v}_n = \mathbf{\Omega} \times \mathbf{r}$, while the superfluid vacuum is at rest, $\mathbf{v}_s = 0$. The counterflow produces what would be a chemical potential in RQFT: $\mu_R = p_F(\hat{\mathbf{l}}_0 \cdot \mathbf{w})$ for right-handed particles, and $\mu_L = -\mu_R$, and quasiparticles have a net momentum $\mathbf{P} = mn_n \mathbf{w}$, analogous to the excess of the leptonic charge of right-handed electrons. *Middle*: When w reaches the critical value an abrupt jump of the intensity of the NMR satellite peak from zero signals the appearance of the $\hat{\mathbf{l}}$-texture, playing the part of the magnetic field (after Ruutu *et al.* 1996b, 1997). *Right*: The final result of the helical instability is an array of vortex-skyrmions (*top*). Vortices simulating the solid-body rotation of superfluid vacuum with $\langle \mathbf{v}_s \rangle = \mathbf{\Omega} \times \mathbf{r}$ reduce the counterflow. This means that the fermionic charge $\mathbf{P}^{\rm F}$ has been transformed into 'hypermagnetic' field.

19.3.4 *Mass of hyperphoton due to symmetry-violating interaction*

In ^{3}He-A the helical instability is suppressed by another mass of the 'hyperphoton', which comes from the symmetry-violating spin–orbit interaction $-g_D(\hat{\mathbf{l}} \cdot \hat{\mathbf{d}})^2$ in eqn (16.1). This gives an additional restoring force acting on $\hat{\mathbf{l}}$, and thus the additional mass to the gauge field. Using eqn (19.15) one obtains the following mass term:

$$-g_D(\hat{\mathbf{l}} \cdot \hat{\mathbf{d}})^2 = -g_D(\hat{\mathbf{l}}_0 \cdot \hat{\mathbf{d}})^2 + \frac{1}{2} g_D(\delta \hat{\mathbf{l}})^2 \equiv \text{constant} + \frac{1}{2}\sqrt{-g} M_{\rm hp}^2 g^{ik} A_{iY} A_{kY} ,\tag{19.22}$$

and the mass of the gauge boson induced by the symmetry-violating interaction is

$$M_{\rm hp} = \sqrt{\frac{g_D v_F}{p_F^2}} = E_D \sim \frac{\hbar v_F}{\xi_D} \sim 10^{-3} \Delta_0 . \tag{19.23}$$

This is much bigger than the Dirac mass $E_D^2 / E_{\text{Planck 2}}$ of a fermionic quasiparticle in the planar state induced by the same spin–orbit interaction, eqn (12.5).

As distinct from the mass of the 'gauge boson' in eqn (19.17), the mass in eqn (19.23) is independent of the counterflow (the chemical potential μ_a). As a result the helical instability occurs only if the counterflow exceeds the critical threshold $w_D \sim E_D/p_F$ determined by the spin–orbit mass of the hyperphoton.

19.3.5 Experimental 'magnetogenesis' in ^{3}He-A

This threshold has been observed experimentally by Ruutu et al. (1996b, 1997) (see Fig. 19.2). The initial state in the rotating ^{3}He-A is vortex-free. It contains a counterflow which simulates the fermionic charge stored in the heat bath of chiral quasiparticles. When the counterflow in the rotating vessel exceeds w_D, the intensive formation of the $\hat{\mathbf{l}}$-texture by helical instability is detected by NMR (Fig. 19.2 middle). This, according to our analogy, corresponds to the formation of the hypermagnetic field which stores the fermionic charge. There is no counterflow in the final state. Thus all the fermionic charge has been transferred from the fermions to the magnetic field.

The only difference from the Joyce–Shaposhnikov scenario is that the mass of the 'hyperphoton' provides the threshold for the helical instability. In principle, however, the similar threshold can appear in the Standard Model if there is a small non-renormalizable mass of the hyperphoton, $M_{\rm hp}$, which does not depend on the chemical potential. In this case the decay of the fermionic charge stops when the chemical potential of fermions becomes comparable to $M_{\rm hp}$. This leads to another scenario of baryonic asymmetry of the Universe. Suppose that the early Universe was leptonic asymmetric. Then the excess of the leptonic charge transforms to the hypermagnetic field until the mass of the hyperphoton prevents this process. After that the excess of leptonic (and thus baryonic) charge is no longer washed out. The observed baryonic asymmetry would be achieved if the initial mass of the hyperhoton at the electroweak temperature, $T \sim E_{\rm ew}$, were $M_{\rm hp} \sim 10^{-9} E_{\rm ew}$.

20
MACROSCOPIC PARITY-VIOLATING EFFECTS

20.1 Mixed axial–gravitational Chern–Simons term

20.1.1 Parity-violating current

The chiral anomaly phenomenon in RQFT can also be mapped to the angular momentum paradox in ^{3}He-A, which has possibly a common origin with the anomaly in the spin structure of hadrons as was suggested by Troshin and Tyurin (1997).

To relate the chiral anomaly and angular momentum paradox in ^{3}He-A let us consider the parity effects which occur for the system of chiral fermions under rotation. The macroscopic parity-violating effects in a rotating system with chiral fermions were first discussed by Vilenkin (1979, 1980a,b) The angular velocity of rotation $\boldsymbol{\Omega}$ defines the preferred direction of spin polarization, and right-handed fermions move in the direction of their spin. As a result, such fermions develop a current parallel to $\boldsymbol{\Omega}$. Assuming thermal equilibrium at temperature T and chemical potential of the fermions μ, one obtains the following correction to the particle distribution function due to interaction of the particle spin with rotation:

$$f = \left(1 + \exp\frac{cp - \mu L - (\hbar/2)\boldsymbol{\Omega}\cdot\boldsymbol{\sigma}}{T}\right)^{-1}. \tag{20.1}$$

Here L is the lepton number, which is $L = 1$ for the lepton and $L = -1$ for its antiparticle. For right-handed fermions the current is given by

$$\mathbf{j} = \frac{1}{2}c\,\text{tr}\sum_{\mathbf{p},L}\sigma L f = -\frac{\hbar c}{2}\text{tr}\,\sigma(\sigma\cdot\boldsymbol{\Omega})\sum_{\mathbf{p}}\partial_E f = \frac{1}{(\hbar c)^2}\left(\frac{T^2}{12} + \frac{\mu^2}{4\pi^2}\right)\boldsymbol{\Omega}. \tag{20.2}$$

The current $\mathbf{j}$ is a polar vector, while the angular velocity $\boldsymbol{\Omega}$ is an axial vector, and thus eqn (20.2) represents the macroscopic violation of the reflectional symmetry. Similarly, left-handed fermions develop a particle current antiparallel to $\boldsymbol{\Omega}$. If the number of left-handed and right-handed particles coincides, parity is restored and the odd current disappears.

If particles have charges q_a interacting with gauge field $\mathbf{A}$, then the particle current (20.2) is accompanied by the charge current, and the Lagrangian density acquires the term corresponding to the coupling of the current with the gauge field:

$$L = \frac{1}{\hbar^2 c^2}\boldsymbol{\Omega}\cdot\mathbf{A}\left(\frac{T^2}{12}\sum_a q_a C_a + \frac{1}{4\pi^2}\sum_a q_a C_a \mu_a^2\right). \tag{20.3}$$

20.1.2 Parity-violating action in terms of gravimagnetic field

The above equation contains the 'material parameter' – the speed of light c – and thus it cannot be applied to ^{3}He-A with anisotropic 'speed of light'. The necessary step is to represent eqn (20.3) in the covariant form applicable to any systems of the Fermi point universality class. This is achieved by expressing the rotation in terms of the metric field. Let us consider ^{3}He-A in the reference frame rotating with the container. In this frame all the fields including the effective metric are stationary in a global equilibrium. In this rotating frame the velocity of the normal component $\mathbf{v}_n = 0$, while the superfluid velocity in this frame is $\mathbf{v}_s = -\mathbf{\Omega} \times \mathbf{r}$. According to the relation (9.13) between the superfluid velocity and the effective metric one obtains the mixed components of the metric tensor: $g^{0i} = (\mathbf{\Omega} \times \mathbf{r})_i$. In general relativity the vector $\mathbf{G} \equiv g_{0i}$ represents the vector potential of the gravimagnetic field, and we obtain the familiar result that the rotation is equivalent to the gravimagnetic field.

Let us consider the axisymmetric situation when the orbital momentum axis $\hat{\mathbf{l}}$ is directed along $\mathbf{\Omega}$, and thus the superfluid velocity is perpendicular to $\hat{\mathbf{l}}$. Then the relevant 'speed of light' is $c_\perp$, and the effective gravimagnetic field is

$$\mathbf{B}_g = \nabla \times \mathbf{G} = 2\frac{\mathbf{\Omega}}{c_\perp^2} \; , \quad G_i \equiv g_{0i} = -\frac{v_{si}}{c_\perp^2}. \tag{20.4}$$

When the rotation is expressed in terms of the gravimagnetic field eqn (20.3) becomes

$$L = \frac{1}{\hbar^2} \mathbf{B}_g \cdot \mathbf{A} \left(\frac{T^2}{24} \sum_a q_a C_a + \frac{1}{8\pi^2} \sum_a q_a C_a \mu_a^2 \right) . \tag{20.5}$$

Now it does not explicitly contain the speeds of light, or any other material parameters of the system, such as the bare mass m of the ^{3}He atom or the renormalized mass m^*. Thus it is equally applicable to both systems: the Standard Model and ^{3}He-A. Equation (20.5) represents the mixed axial–gravitational Chern–Simons term, since instead of the conventional product $\mathbf{B} \cdot \mathbf{A}$ it contains the product of the magnetic and gravimagnetic fields (Volovik and Vilenkin 2000). Equation (20.5) is not Lorentz invariant, because the existence of a heat bath of fermions violates Lorentz invariance, since it provides a distinguished reference frame.

Finally let us write the mixed axial-gravitational Chern–Simons action in terms of the momentum space topological invariant:

$$S_{\text{mixed}} = \left(\frac{T^2}{24\hbar^2} \operatorname{tr}(\mathcal{QN}) + \frac{1}{8\pi^2 \hbar^2} \operatorname{tr}(\mathcal{QM}^2 \mathcal{N}) \right) \int d^4x\, \mathbf{B}_g \cdot \mathbf{A} \, , \tag{20.6}$$

where M is the matrix of the chemical potentials μ_a, and Q is the matrix of charges q_a interacting with the field $\mathbf{A}$. If all the fermions can transform to each other, the chemical potential becomes the same for all fermions, and the mixed term is determined by only one topological invariant $\operatorname{tr}(\mathcal{QN})$. In the Standard Model it is zero because of anomaly cancellation.

20.2 Orbital angular momentum in ^{3}He-A

Equation (20.5) is linear in the angular velocity $\mathbf{\Omega}$, and thus its variation over $\mathbf{\Omega}$ represents some angular momentum of the system which does not depend on the rotation velocity – the spontaneous angular momentum. In the Standard Model this angular momentum is proportional to the vector potential of the gauge field: $\mathbf{L}(T) - \mathbf{L}(T=0) = -\delta S_{\text{mixed}}/\delta\mathbf{\Omega} \propto \mathbf{A}$, which at first glance violates the gauge invariance. However, the total angular momentum, obtained by integration over the whole space, remains gauge invariant. Let us now proceed to the quantum liquid where what we found corresponds to the contributions to the spontaneous angular momentum of the liquid from thermal quasiparticles at non-zero temperature T and from the non-zero counterflow, which plays the role of the chemical potential. Let us consider first the temperature correction to the spontaneous angular momentum. Introducing the gauge field $\mathbf{A} = p_F\hat{\mathbf{l}}$ and the proper charges $q_a = -C_a$ of the chiral fermions in ^{3}He-A, one obtains

$$\mathbf{L}(T) - \mathbf{L}(T=0) = -\frac{\delta S_{\text{mixed}}}{\delta\mathbf{\Omega}} = -\frac{p_F T^2}{6\hbar^2 c_\perp^2}\hat{\mathbf{l}} \ . \tag{20.7}$$

Comparing this with eqns (10.29) and (10.42) one obtains that the prefactor can be expressed in terms of the longitudinal density of the normal component of the liquid

$$\mathbf{L}(T) - \mathbf{L}(T=0) = -\frac{\hbar}{2}n_{\text{n}\|}^0\hat{\mathbf{l}} = -\frac{\hbar}{2}\frac{m}{m^*}n_{\text{n}\|}\hat{\mathbf{l}} \ . \tag{20.8}$$

The total value of the angular momentum of ^{3}He-A has been the subject of a long-standing controversy (for a review see the books by Vollhardt and Wölfle (1990) and by Volovik (1992a). Different methods for calculating the angular momentum give results that differ by many orders of magnitude. The result is also sensitive to the boundary conditions, since the angular momentum in the liquid is not necessarily a local quantity, and to whether the state is strictly stationary or has a small but finite frequency. This is often referred to as the angular momentum paradox. The paradox is related to the axial anomaly induced by chiral fermions and is now reasonably well understood.

At $T=0$ the total angular momentum of the stationary homogeneous liquid with homogeneous $\hat{\mathbf{l}} = \hat{\mathbf{z}}$ can be found from the following consideration. According to eqn (7.53), $L_z - (1/2)N$ is the generator of the symmetry of the vacuum state in ^{3}He-A. Applying this to the vacuum state $(L_z - N/2)|vac\rangle = 0$, one obtains that the angular momentum of a homogeneous vacuum state is

$$\langle vac|L_z|vac\rangle = \frac{\hbar}{2}N \ , \tag{20.9}$$

where N is the number of particles in the vacuum. The physical meaning of the total angular momentum of ^{3}He-A is simple: each Cooper pair carries the angular momentum $\hbar$ in the direction of the quantization axis $\hat{\mathbf{l}}$ (let us recall that the Cooper pairs in ^{3}He-A have quantum numbers $L=1$, and $L_z=1$, where L_z is the projection of the orbital angular momentum onto the quantization axis $\hat{\mathbf{l}}$). This implies that the angular momentum density of the liquid at $T = 0$ is

$$\mathbf{L}(T=0) = \hat{\mathbf{l}}\,\frac{\hbar}{2}\,n\;, \qquad (20.10)$$

where n, as before, is the density of ^{3}He atoms. As distinct from the thermal correction, this vacuum term has no analog in effective field theory and can be calculated only within the microscopic (high-energy) physics. Adding the contribution to the momentum from fermionic quasiparticles in eqn (20.8), one may conclude that at non-zero temperature the spontaneous angular momentum is

$$\mathbf{L}(T) = \hat{\mathbf{l}}\,\frac{\hbar}{2}n^0_{\mathrm{s}\|}(T)\;,\;\;n^0_{\mathrm{s}\|}(T) = n - n^0_{\mathrm{n}\|}(T)\;. \qquad (20.11)$$

Such a value of the angular momentum density agrees with that obtained by Kita (1998) in a microscopic theory. This suggests that the non-renormalized value of the superfluid component density, $n^0_{\mathrm{s}\|}(T)/2$, is the effective number density of the 'superfluid' Cooper pairs which contribute to the angular momentum.

Equations (20.11) and (20.10) are, however, valid only for the static angular momentum. The dynamical angular momentum is much smaller, and is reduced by the value of the anomalous angular momentum in eqn (18.21) which comes from the spectral flow: $\mathbf{L}_{\mathrm{dyn}}(T=0) = \mathbf{L}(T=0) - \mathbf{L}_{\mathrm{anomalous}} = \frac{\hbar}{2}\hat{\mathbf{l}}\,(n - C_0)$. Let us recall that $(n - C_0)/n \sim c_\perp^2/c_\|^2 \sim 10^{-6}$, and thus the reduction is really substantial. The presence of the anomaly parameter C_0 in this almost complete cancellation of the dynamical angular momentum reflects the same crucial role of the axial anomaly as in the 'baryogenesis' by a moving texture discussed in Chapter 18. The dynamics of the $\hat{\mathbf{l}}$-vector is accompanied by the spectral flow which leads to the production of the angular momentum from the vacuum.

20.3 Odd current in ^{3}He-A

The parity-violating currents (20.2) could be induced in turbulent cosmic plasmas and could play a role in the origin of cosmic magnetic fields (Vilenkin and Leahy 1982). The corresponding ^{3}He-A effects are less dramatic but may in principle be observable.

Let us discuss the effect related to the second (temperature-independent) term in the mixed axial–gravitational Chern–Simons action in eqn (20.5). According to eqn (10.19) the counterpart of the chemical potentials μ_a of relativistic chiral fermions is the superfluid–normal counterflow velocity in ^{3}He-A. The relevant counterflow, which does not violate the symmetry and the thermodynamic equilibrium condition of the system, can be produced by superflow along the axis of the rotating container. Note that we approach the $T \to 0$ limit in such a way that the rotating reference frame is still active and determines the local equilibrium states. In the case of a rotating container this is always valid because of the interaction of the liquid with the container walls. For the relativistic counterpart we must assume that there is still a non-vanishing rotating thermal bath of fermionic excitations. This corresponds to the case when the condition $\omega\tau \ll 1$ remains valid, despite the divergence of the collision time τ at $T \to 0$. This is one of numerous subtle issues related to the anomaly, when the proper order of imposing limits is crucial.

We also assume that, in spite of rotation of the vessel, there are no vortices in the container. This is typical of superfluid ^{3}He-B, where the critical velocity for nucleation of vortices is comparable to the pair-breaking velocity $v_{\text{Landau}} = \Delta_0/p_F$ as was measured by Parts et al. (1995b; see Sec. 26.3.3). The critical velocity in ^{3}He-A, even in the geometry when the $\hat{\mathbf{l}}$-vector is not fixed, can reach 0.5 rad s^{-1} as was observed by Ruutu et al. (1996b, 1997). For the geometry with fixed $\hat{\mathbf{l}}$, it should be comparableto the critical velocity in ^{3}He-B. In addition, we assume that $\Omega r < c_\perp$ everywhere in the vessel, i.e. the counterflow velocity $\mathbf{v}_n - \mathbf{v}_s$ is smaller than the pair-breaking critical velocity $c_\perp = \Delta_0/p_F$ (the transverse 'speed of light'). This means that there is no region in the vessel where particles can have negative energy (ergoregion). The case when Ωr can exceed $c_\perp$ and effects caused by the ergoregion in rotating superfluids (Calogeracos and Volovik 1999a) will be discussed below in Chapter 31.

Although the mixed Chern–Simons terms have the same form in relativistic theories and in ^{3}He-A, their physical manifestations are not identical. In the relativistic case, the electric current of chiral fermions is obtained by variation with respect to $\mathbf{A}$, while in the ^{3}He-A case the observable effects are obtained by variation of the same term but with respect to ^{3}He-A observables. For example, the expression for the momentum carried by quasiparticles is obtained by variation of eqn (20.5) over the counterflow velocity $\mathbf{w}$. This leads to an extra fermionic charge carried by quasiparticles, which is odd in $\boldsymbol{\Omega}$:

$$\Delta \mathbf{P}^{\text{F}}(\boldsymbol{\Omega}) = -m \frac{p_F^3}{\pi^2 \hbar^2 c_\perp^2} \hat{\mathbf{l}} \, (\hat{\mathbf{l}} \cdot \mathbf{w}) \, (\hat{\mathbf{l}} \cdot \boldsymbol{\Omega}) \,. \tag{20.12}$$

From eqn (20.12) it follows that there is an $\boldsymbol{\Omega}$ odd contribution to the normal component density at $T \to 0$ in ^{3}He-A:

$$\Delta n_{\text{n}\|}(\boldsymbol{\Omega}) = \frac{\Delta \mathbf{P}^{\text{F}}(\boldsymbol{\Omega})}{m w_\|} = \frac{p_F^3}{\pi^2 \hbar^2} \frac{\hat{\mathbf{l}} \cdot \boldsymbol{\Omega}}{mc_\perp^2} \,. \tag{20.13}$$

The sensitivity of the normal density to the direction of rotation is the counterpart of the parity-violating effects in relativistic theories with chiral fermions. It should be noted though that, since $\hat{\mathbf{l}}$ is an axial vector, the right-hand sides of (20.12) and (20.13) transform, respectively, as a polar vector and a scalar, and thus (of course) there is no real parity violation in ^{3}He-A. However, a non-zero expectation value of the axial vector of the orbital angular momentum $\mathbf{L} = (\hbar/2) n_{\text{s}\|}^0(T) \hat{\mathbf{l}}$ does indicate a *spontaneously* broken reflectional symmetry, and an inner observer living in ^{3}He-A would consider this effect as parity violating.

The contribution (20.13) to the normal component density can have arbitrary sign depending on the sense of rotation with respect to $\hat{\mathbf{l}}$. This, however, does not violate the general rule that the overall normal component density must be positive: the rotation-dependent momentum $\Delta \mathbf{P}^{\mathrm{F}}(\mathbf{\Omega})$ was calculated as a correction to the rotation-independent current in eqn (10.33). This means that we used the condition $\hbar\Omega \ll mw^2 \ll mc_\perp^2$. Under this condition the overall normal density, given by the sum of (20.13) and (10.34), remains positive.

The 'parity' effect in eqn (20.13) is not very small. The rotational contribution to the normal component density normalized to the density of the ^{3}He atoms is $\Delta n_{\mathrm{n}\|}/n = 3\Omega/mc_\perp^2$, which is $\sim 10^{-4}$ for $\Omega \sim 3$ rad s^{-1}. This is within the resolution of the vibrating wire detectors in superfluid ^{3}He-A.

We finally mention a possible application of these results to the superconducting Sr$_2$RuO$_4$ if they really belong to the chiral superconductors as suggested by Rice (1998) and Ishida et al. (1998). An advantage of using superconductors is that the mass current $\Delta \mathbf{P}_{\mathrm{F}}$ in eqn (20.12) is accompanied by the electric current $(e/m)\Delta \mathbf{P}_{\mathrm{F}}$, and can be measured directly. An observation in Sr$_2$RuO$_4$ of the analog of the parity-violating effect that we discussed here (or of the other effects coming from the induced Chern–Simons terms (Goryo and Ishikawa 1999; Ivanov 2001) would be unquestionable evidence of the chirality of this superconductor.

In RQFT the Chern–Simons-type terms are usually discussed in relation to possible violation of Lorentz and CPT symmetries (Jackiw 2000; Perez-Victoria 2001). In our case it is the non-zero chemical potential of 'matter' which violates these symmetries. Consideration of the Chern–Simons terms in RQFT (Jackiw 2000; Perez-Victoria 2001) demonstrates that the factor in front of them is ambiguous within the effective theory. Quantum liquids provide an example of the finite high-energy system where the ambiguity in the relativistic corner is resolved by the underlying 'trans-Planckian' microscopic physics (Volovik 2001b).

21
QUANTIZATION OF PHYSICAL PARAMETERS

The dimensional reduction of the 3+1 system with Fermi points brings the anomaly to the (2+1)-dimensional systems with fully gapped fermionic spectrum. The most pronounced phenomena in these systems are related to the quantization of physical parameters, and to the fermionic charges of the topological objects – skyrmions. Here we consider both these effects. They are determined by the momentum space topological invariant $\tilde{N}_3$ in eqn (11.1). While its ancestor N_3 describes topological defects (singularities of the fermionic propagator) in the 4-momentum space, $\tilde{N}_3$ describes systems without momentum space defects and it characterizes the global topology of the fermionic propagator in the whole 3-momentum space (p_x, p_y, p_0). $\tilde{N}_3$ is thus responsible for the global properties of the fermionic vacuum, and it enters the linear response of the vacuum state to some special perturbations.

21.1 Spin and statistics of skyrmions in 2+1 systems
21.1.1 Chern–Simons term as Hopf invariant

Let us start with a thin film of ^{3}He-A. If the thickness a of the film is finite, the transverse motion of fermions – along the normal $\hat{\mathbf{z}}$ to the film – is quantized. As a result the fermionic propagator $\mathcal{G}$ not only is the matrix in the spin and Bogoliubov–Nambu spin spaces, but also acquires the indices of the transverse levels. This allows us to obtain different values of the invariant $\tilde{N}_3$ in eqn (11.1) by varying the thickness of the film. The Chern–Simons action, which is responsible for the spin and statistics of skyrmions in the $\hat{\mathbf{d}}$-field in ^{3}He-A film, is the following functional of $\hat{\mathbf{d}}$ (Volovik and Yakovenko 1989):

$$S_{\rm CS}\{\hat{\mathbf{d}}\} = \tilde{N}_3 \frac{\hbar}{64\pi} \int d^2x \, dt \, e^{\mu\nu\lambda} A_\mu F_{\nu\lambda} \; . \tag{21.1}$$

Here A_μ is the auxiliary gauge field whose field strength is expressed through the $\hat{\mathbf{d}}$-vector in the following way:

$$F_{\nu\lambda} = \partial_\nu A_\lambda - \partial_\lambda A_\nu = \hat{\mathbf{d}} \cdot \left(\partial_\nu \hat{\mathbf{d}} \times \partial_\lambda \hat{\mathbf{d}}\right) \; . \tag{21.2}$$

The field strength $F_{\nu\lambda}$ is related to the density of the topological invariant in the coordinate spacetime which describes the skyrmions. The topological charge of the $\hat{\mathbf{d}}$-skyrmion is (compare with eqn (16.12))

$$n_2\{\hat{\mathbf{d}}\} = \frac{1}{4\pi} \int d^2x F_{12} = \frac{1}{4\pi} \int dx \, dy \, \hat{\mathbf{d}} \cdot \left(\partial_x \hat{\mathbf{d}} \times \partial_y \hat{\mathbf{d}}\right) \; . \tag{21.3}$$

Let us recall the simplest anzats for a skyrmion with $n_2\{\hat{\mathbf{d}}\} = +1$:

$$\hat{\mathbf{d}} = \hat{\mathbf{z}} \cos \beta(\rho) + \hat{\boldsymbol{\rho}} \sin \beta(\rho) , \qquad (21.4)$$

where $\beta(0) = 0$ and $\beta(\infty) = \pi$.

The Chern–Simons action in eqn (21.1) is the product of the invariants in the momentum–frequency space (p_x, p_y, p_0) and in the coordinate spacetime (x, y, t):

$$S_{\text{CS}}\{\hat{\mathbf{d}}\} = \frac{\pi \hbar}{2} \tilde{N}_3 n_3\{\hat{\mathbf{d}}\} , \qquad (21.5)$$

where

$$n_3\{\hat{\mathbf{d}}(x,y,t)\} = \text{H}^{\text{Hopf}} = \frac{1}{32\pi^2} \int d^2x \, dt \, e^{\mu\nu\lambda} A_\mu F_{\nu\lambda} \qquad (21.6)$$

is the topological invariant – the Hopf invariant – which describes the mapping $S^3 \to S^2$. Here S^3 is the compactified 2+1 spacetime (we assume that at infinity the $\hat{\mathbf{d}}$-vector is constant, say $\hat{\mathbf{d}}(\infty) = \hat{\mathbf{z}}$, and thus the whole infinity is represented by a single point), while S^2 is the sphere of the unit vector $\hat{\mathbf{d}}$. This is the famous Hopf map $\pi_3(S^2) = Z$. The geometrical interpretation of the Hopf number is the linking number of two world lines $\hat{\mathbf{d}}(x,y,t) = \hat{\mathbf{d}}_1$ and $\hat{\mathbf{d}}(x,y,t) = \hat{\mathbf{d}}_2$, each corresponding to the constant value of $\hat{\mathbf{d}}$. In 3D space (x,y,z) the configurations described by the Hopf invariant $n_3\{\hat{\mathbf{l}}(x,y,z)\}$ (hopfions) have been investigated in ^{3}He-A experimentally by Ruutu et al. (1994) and theoretically by Makhlin and Misirpashaev (1995).

21.1.2 Quantum statistics of skyrmions

The quantum statistics of the $\hat{\mathbf{d}}$-skyrmions depends on how the wave function of the system with a single skyrmion behaves under adiabatic 2π rotation: $\Psi(2\pi) = \Psi(0) e^{i\theta}$ (Wilczek and Zee 1983). If θ equals an odd number of π, then the skyrmion must be a fermion; correspondingly if θ equals an even number of π, it is a boson. Since the phase of the wave function is $S/\hbar$, and only the Chern–Simons term in the action S is sensitive to the adiabatic 2π rotation, it is this term in eqn (21.5) which determines the quantum statistics of $\hat{\mathbf{d}}$-skyrmions. The process of the adiabatic 2π rotation of the $\hat{\mathbf{d}}$-skyrmion with $n_2\{\hat{\mathbf{d}}\} = +1$ in eqn (21.4) is given by

$$\hat{\mathbf{d}}(\rho,\phi,t) = \hat{\mathbf{z}} \cos \beta(\rho) + \sin \beta(\rho) \left(\hat{\boldsymbol{\rho}} \cos \alpha(t) + \hat{\boldsymbol{\phi}} \sin \alpha(t) \right) , \qquad (21.7)$$

where $\alpha(t)$ slowly changes from 0 to 2π. Substituting this time-dependent field into the action (21.5), one obtains that the action changes by the value $\Delta S_{\text{CS}} = \pi \hbar \tilde{N}_3/2$, i.e. the θ-factor $\Delta S/\hbar$ for the skyrmion is determined by the **p**-space topology of the vacuum:

$$\theta = \frac{\pi}{2} \tilde{N}_3 . \qquad (21.8)$$

Note that our system does not need to be relativistic, that is why eqn (21.9) is more general than the result based on the index of the Dirac operators in relativistic theories (Atiyah and Singer 1968, 1971; Atiyah et al. 1976, 1980).

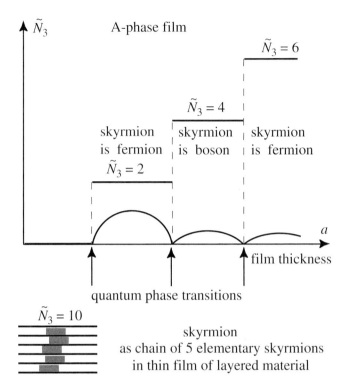

FIG. 21.1. Integer topological invariant $\tilde{N}_3$ as a function of the film thickness. Points where $\tilde{N}_3$ changes abruptly are quantum phase transitions at which the quantum statistics of $\hat{\mathbf{d}}$-skyrmions changes. Curves show the minimum of the quasiparticle energy spectrum. The spectrum becomes gapless at the quantum transition. *Bottom*: In layered materials, a skyrmion is a linear object consisting of elementary fermionic skyrmions. In a thin film consisting of 5 atomic layers with the total topological charge $\tilde{N}_3 = 10$, a skyrmion is a fermion.

Equation (21.9) means that the spin of the $\hat{\mathbf{d}}$-skyrmion with winding number $n_2\{\hat{\mathbf{d}}\} = +1$ is

$$s = \frac{\hbar \theta}{2\pi} = \frac{\hbar}{4} \tilde{N}_3 \,. \tag{21.9}$$

In ^{3}He-A films, the invariant $\tilde{N}_3$ is always even because of the spin degeneracy. That is why the spin of a skyrmion can be either integer or half an odd integer, i.e. a skyrmion is either the fermion or the boson depending on the thickness a of the film in Fig. 21.1. Roughly speaking, the invariant $\tilde{N}_3$ is proportional to the number $n_{\text{transverse}}$ of the occupied transverse levels (see the book by Volovik 1992a). In the BCS model of the weakly interacting Fermi gas one has $\tilde{N}_3 = 2n_{\text{transverse}}$. The number of the occupied levels $n_{\text{transverse}}$ is proportional to the thickness of the film. When the film grows the quantum transitions occur successively, at which the momentum space invariant $\tilde{N}_3$ and thus the quantum statistics of fermions abruptly change.

The change of the quantum statistics can be easily understood in an example of the layered systems in superconductors or semiconductors. If each layer of a thin film is characterized by an elementary topological charge $\tilde{N}_3 = 2$, the $\hat{\mathbf{d}}$-skyrmion in this film can be represented as a chain of elementary skyrmions, which are fermions (Fig. 21.1 *bottom*). Depending on the number of layers, the skyrmion as a whole contains an odd or even number of elementary fermionic skyrmions and thus is either the fermion or the boson.

As we discussed in Sec. 11.4, the quantum (Lifshitz) transitions between the states with different $\tilde{N}_3$ occur through the intermediate gapless regimes where $\tilde{N}_3$ is not well defined. Figure 21.1 shows that the quasiparticle energy spectrum becomes gapless at the transition. In principle, the intermediate gapless state between two plateaus can occupy a finite range of thicknesses.

21.2 Quantized response

21.2.1 *Quantization of Hall conductivity*

The topological invariants of the type in eqn (11.1) are also responsible for quantization of physical parameters of the systems, such as Hall conductivity (Ishikawa and Matsuyama 1986, 1987; Matsuyama 1987; Volovik 1988) and spin Hall conductivity (Volovik and Yakovenko 1989; Senthil *et al.* 1999; Read and Green 2000). Let us start with the (2+1)-dimensional semiconductor, whose valence band has the non-zero value of the topological invariant $\tilde{N}_3$, i.e. the integration in eqn (11.1) over the valence band gives $\tilde{N}_3 \neq 0$. Then there is the Chern–Simons term for the real electromagnetic field A_μ (Volovik 1988):

$$S_{\rm CS} = \tilde{N}_3 \frac{e^2}{16\pi\hbar} e^{\mu\nu\lambda} \int d^2x \, dt \, A_\mu F_{\nu\lambda} \,. \tag{21.10}$$

Variation of this term over A_x gives the current density along x, which is proportional to the transverse electric field:

$$j_x = \tilde{N}_3 \frac{e^2}{4\pi\hbar} E_y \,. \tag{21.11}$$

The electric current transverse to the applied electric field demonstrates that such a system exhibits the anomalous Hall effect, i.e. the Hall effect occurs without an external magnetic field. The Hall conductivity is quantized and this quantization is determined by the momentum space topological invariant:

$$\sigma_{xy} = \tilde{N}_3 \frac{e^2}{2h} \,. \tag{21.12}$$

The conditions for the quantized anomalous Hall effect are:

(i) The presence of the valence band separated by the gap from the conduction band. Only for the fully gapped system the quantization is exact.

(ii) This valence band must have the non-trivial momentum space topology, i.e. $\tilde{N}_3 \neq 0$. Let us recall that the non-zero value of $\tilde{N}_3$ means that the time reversal symmetry is broken (see Sec. 11.2.2). In other words, the system must have the ferromagnetic moment along the normal to the film.

(iii) Also what has been used in the derivation is that the gauge field can be introduced to the fermionic Lagrangian through the long derivative: $p_\mu \to p_\mu - eA_\mu$. This implies that gauge invariance is not violated, and this rules out (completely or partially) from consideration the systems exhibiting superconductivity (or superfluidity in electrically neutral systems). In superfluid/superconducting systems quantization is not exact and the spontaneous Hall current is not universal (Volovik 1988; Furusaki et al. 2001).

(iv) Finally, when the liquid state is considered, it is assumed that the Hall conductivity is measured in a static limit, i.e. the $\omega \to 0$ limit is taken first, and only after that the wave vector $q \to 0$. The $q \neq 0$ perturbations violate Galilean invariance, which prescribes that $\sigma_{xy} = 0$. This is another example of when the order of limits is crucial, as was discussed in the introduction to Chapter 18. For crystalline systems, Galilean invariance is violated by the crystal lattice.

21.2.2 Quantization of spin Hall conductivity

Let us consider again the semiconductor with $\tilde{N}_3 \neq 0$, and introduce an external magnetic field $\mathbf{H}(x,y,t)$, which interacts with electronic spins only. This interaction adds the Pauli term $\frac{1}{2}\gamma\boldsymbol{\sigma}\cdot\mathbf{H}$ to the Hamiltonian in the Theory of Everything, where γ is the gyromagnetic ratio for the particle spin (or for nuclear spin in liquid ^{3}He). As a result the Green function contains the long time derivative which contains spin:

$$-i\partial_t - \frac{1}{2}\gamma\sigma_i H^i \; , \qquad (21.13)$$

demonstrating that for the electrically neutral systems the magnetic field is equivalent to the A_0^i component of the external $SU(2)$ gauge field:

$$\mathbf{A}_0 = \gamma\mathbf{H} \; . \qquad (21.14)$$

The corresponding fermionic charge interacting with this non-Abelian gauge field is the particle spin $s = 1/2$. Let us now introduce the space components, $\mathbf{A}_1$ and $\mathbf{A}_2$, of the $SU(2)$ gauge field. These are auxiliary fields, which are useful for calculation of the spin current density. As a natural generalization of eqn (21.10) to the non-Abelian case, one obtains the following Chern–Simons action in terms of $\mathbf{A}_\mu$:

$$S_{\text{CS}} = \frac{\tilde{N}_3 s^2 \hbar}{16\pi} \int d^2x\, dt\, e^{\mu\nu\lambda} \left(\frac{1}{3}\mathbf{A}_\mu \cdot (\mathbf{A}_\nu \times \mathbf{A}_\lambda) + \mathbf{A}_\mu \cdot \mathbf{F}_{\nu\lambda} \right) , \qquad (21.15)$$

where

$$\mathbf{F}_{\mu\nu} = \partial_\mu \mathbf{A}_\nu - \partial_\nu \mathbf{A}_\mu - \mathbf{A}_\mu \times \mathbf{A}_\nu \qquad (21.16)$$

is the field strength of the non-Abelian $SU(2)$ gauge field.

The chiral spin current is obtained as the response of the action to auxiliary space components $\mathbf{A}_i$ of the $SU(2)$ gauge field in the limit $\mathbf{A}_i \to 0$ while the physical field $\mathbf{A}_0$ is retained intact:

$$\mathbf{j}_i = \left(\frac{\delta S_{\text{CS}}}{\delta \mathbf{A}_i}\right)_{\mathbf{A}_i=0} = \frac{\tilde{N}_3 s^2 \hbar}{4\pi} e_{ik} \mathbf{F}_{0k} = \frac{\gamma \tilde{N}_3}{16\pi} e_{ik} \frac{\partial \mathbf{H}}{\partial x_k} . \quad (21.17)$$

Equation (21.17) means that the spin conductance is quantized in terms of elementary quantum $\hbar/8\pi$:

$$\sigma_{xy}^s = \tilde{N}_3 \frac{\hbar}{16\pi} . \quad (21.18)$$

Here $\tilde{N}_3$ is an even number because of spin degeneracy.

21.2.3 Induced Chern–Simons action in other systems

In electrically neutral systems, such as ^{3}He-A film, there is also an analog of the Hall conductivity, if instead of electric current one considers the mass (or particle) current. However, in ^{3}He-A the $U(1)_N$ symmetry is spontaneously broken, and thus the condition (iii) in Sec. 21.2.1 is violated. This leads to extra terms in action, as a result quantization of the Hall conductivity is not exact (Volovik 1988; Furusaki et al. 2001). The same occurs for the spin Hall conductivity in ^{3}He-A films with one exception. The spin rotational symmetry $SO(3)_S$ is not completely broken in ^{3}He-A: there is still the symmetry under spin rotations about axis $\hat{\mathbf{d}}$. Thus, if $\hat{\mathbf{d}}$ is homogeneous, $\hat{\mathbf{d}} = \hat{\mathbf{z}}$, equation (21.18) is applicable for longitudinal spins, and the current of the longitudinal spin (oriented along $\hat{\mathbf{d}}$) is quantized:

$$j_i^z = \frac{\tilde{N}_3 \hbar}{16\pi} e_{ik} F_{0k}^z = \frac{\gamma \tilde{N}_3}{16\pi} e_{ik} \frac{\partial H^z}{\partial x^k} . \quad (21.19)$$

The same can also be applied to the triplet superconductors discussed in Sec. 11.2.1, where $\tilde{N}_3 = \pm 2$ per atomic layer.

For spin-singlet superconductors (i.e. for the superconductors where the spin of Cooper pairs is zero, $s = 0$), the $SO(3)_S$ group of spin rotations is not broken; thus eqn (21.18) is always applicable. In particular, for the spin-singlet $d_{x^2-y^2} \pm i d_{xy}$ superconductor, the topological invariant per spin projection is $\tilde{N}_3 = \pm 4$ per corresponding atomic layer (Sec. 11.2.3), and eqn (21.18) coincides with eqn (21) of Senthil et al. (1999).

We considered here the dimensional reduction of the invariant N_3 describing the Fermi points in 3+1 systems to the invariant $\tilde{N}_3$ describing fully gapped 2+1 fermionic systems. In Sec. 12.3.1 we discussed the Fermi points in 3+1 systems with zero topological invariant N_3 which are protected by symmetry. If the corresponding symmetry is parity P these systems are described by the symmetry-protected invariants $\mathbf{tr}(\text{P}\mathcal{N})$. The dimensional reduction of these invariants to the 2+1 systems, $\mathbf{tr}(\text{P}\tilde{\mathcal{N}})$, leads to quantization of other fermionic charges of skyrmions, in particular, of the electric charge of a skyrmion. The corresponding invariants and charges were discussed by Yakovenko (1989). The symmetry-protected invariants can be apllied if there are several Abelian gauge fields A_μ^I; the corresponding induced Chern–Simons term is

$$S_{\text{CS}} = \sum_{IJ} \mathbf{tr}\left(Q_I Q_J \tilde{\mathcal{N}}\right) \frac{1}{16\pi\hbar} e^{\mu\nu\lambda} \int d^2 x \, dt \, A_\mu^I F_{\nu\lambda}^J \,. \tag{21.20}$$

Here Q_I is the matrix of the charge I interacting with the gauge field A_μ^I. Such action has been used in the effective theory of the fractional quantum Hall effect (see e.g. Wen (2000) and references therein).

Part V

Fermions on topological objects
and brane world

22

EDGE STATES AND FERMION ZERO MODES ON SOLITON

The idea that our Universe lives on a brane embedded in higher-dimensional space (Rubakov and Shaposhnikov 1983; Akama 1982) is popular at the moment (see the review by Forste 2002). It is the further development of an old idea of extra compact dimensions introduced by Kaluza (1921) and Klein (1926). In the new approach the compactification occurs because the low-energy physics is concentrated within the brane; for example, in a flat 4D brane embedded in a 5D anti-de Sitter space with a negative cosmological constant (Randall and Sundrum 1999). Branes can be represented by topological defects, such as domain walls – membranes – (Rubakov and Shaposhnikov 1983) and cosmic strings (Abrikosov–Nielsen–Olesen vortices) (Akama 1983). It is supposed that we live inside the core of such a defect. Our 3+1 spacetime spans the extended coordinates of the brane, while the other (extra) dimensions are of the order of the core size. This new twist in the idea of extra dimensions is fashionable because by accommodatiing the core size one can bring the gravitational Planck energy scale close to the TeV range. That is why there is hope that the deviations from the Newton law can become observable at the distance of order 1 mm. At the moment the Newton law has been tested for distances > 0.2 mm by Hoyle *et al.* (2001).

The popular mechanism of why matter is localized on the brane is that the fermionic matter is represented by fermion zero modes, whose wave function is concentrated in the core region. Outside the core the fermions are massive and thus are frozen out at low T. Such an example of topologically induced Kaluza–Klein compactification of multi-dimensional space is provided by the condensed matter analogs of branes, namely domain walls and vortices. These topological defects contain fermion zero modes which can live only within the core of defects. These fermions form the 2+1 world within the domain wall and the 1+1 world in the core of the vortex. The modification of these condensed matter branes to higher dimensions is illuminating. The fermion zero modes in the 3+1 domain wall separating the 4+1 vacua of quantum liquid with different momentum space topology have Fermi points (see Sec. 22.2.4). Due to these Fermi points the chiral fermions, gauge and gravitational fields emerge in the same manner as was discussed in Sec. 8.2. As distinct from modern relativistic theories (Randall and Sundrum 1999), this scenario does not require the existence of 4+1 gravity in the bulk.

Another aspect of the interplay of fermion zero modes and topological defects is related to the drastic change of the fermionic vacuum in the presence of the topological defect. Because of this the defect itself in some cases acquires fermionic charge and half-integer spin; even the quantum statistics of the defect is reversed: it becomes a fermion (Jackiw and Rebbi 1976a, 1984), as we discussed in Sec. 21.1.2.

22.1 Index theorem for fermion zero modes on soliton

22.1.1 Chiral edge state – 1D Fermi surface

The **p**-space topology is also instrumental for investigating the response of the fermionic system to the non-trivial topological background in **r**-space. Here we consider the 2+1 system where the non-trivial topological background is provided by a domain wall (more specifically by a domain line, since a wall in 2D space is a linear object). We assume that each of the two vacua separated by the wall is fully gapped. Such vacua are described by the topological charges $\tilde{N}_3$, which are $\tilde{N}_3$(right) and $\tilde{N}_3$(left) for the vacua on the left and right sides of the wall respectively. Though the spectrum of quasiparticles is fully gapped outside the domain wall, it can be gapless inside this topological object; such mid-gap states in 1+1 systems have been analyzed by Su et al. (1979), see also the review paper by Heeger et al. (1988). Because of the gap in the bulk material all the low-temperature physics is determined by the gapless excitations living inside the topological object.

Inside the domain wall (the domain line) only the linear momentum $p_\parallel$, which is along the line, is a good quantum number. The fermion zero modes, by definition, are the branches of the quasiparticle spectrum $E_a(p_\parallel)$ that as functions of $p_\parallel$ cross the zero-energy level. Close to zero energy the spectrum of the a-th fermion zero mode is linear:

$$E_a(p_\parallel) = c_a(p_\parallel - p_a) \ . \tag{22.1}$$

The points p_a are zeros of co-dimension 1, and thus belong to the universality class of Fermi surfaces, described by the momentum space topological charge N_1 in eqn (8.3) for the Green function $\mathcal{G}^{-1} = ip_0 - E_a(p_\parallel)$. In our case the N_1 invariant coincides with the sign of the slope c_a of the fermionic spectrum: $N_{1a} = \text{sign } c_a$.

Let us introduce the algebraic number of the fermion zero modes in the core of the domain wall defined by the difference between the number of modes with positive and negative slopes:

$$\nu = \sum_a N_{1a} = \sum_a \text{sign } c_a \ . \tag{22.2}$$

We shall see that this total topological charge of fermion zero modes in the wall is determined by the difference of the topological charges $\tilde{N}_3$ of the vacua on the two sides of the interface:

$$\nu = \tilde{N}_3(\text{right}) - \tilde{N}_3(\text{left}) \ . \tag{22.3}$$

This illustrates the topology of dimensional reduction in momentum space: the momentum space topological invariant $\tilde{N}_3$ of the bulk 2+1 system gives rise to the 1+1 fermion zero modes described by the momentum space topological invariant N_1. This relation is similar to the Atiyah-Singer index theorem, which relates the number of fermion zero modes (actually the difference of the number of left-handed and right-handed modes) to the topological charge of the gauge field configuration (Atiyah and Singer 1968, 1971; Atiyah et al. 1976, 1980). In our consideration this is a topological property which does not depend on the precise form of the Hamiltonian, and it does not require relativistic invariance.

In systems exhibiting the quantum Hall effect, whose vacuum states are also described by the invariant $\tilde{N}_3$ (Ishikawa and Matsuyama 1986, 1987; Matsuyama 1987), such fermion zero modes are called chiral edge states (Halperin 1982; Wen 1990a; Stone 1990). For modes on the boundary of the system eqn (22.3) is also applicable, with $\tilde{N}_3 = 0$ on one side.

22.1.2 Fermi points in combined (p, r) space

As an illustration and also to derive the index theorem let us discuss the fermion zero modes in the domain wall between the vacua with $\tilde{N}_3 = \pm 1$: for example, the domain wall in a thin film of ^{3}He-A separating domains with $\hat{\mathbf{l}} = \hat{\mathbf{z}}$ and $\hat{\mathbf{l}} = -\hat{\mathbf{z}}$. The vacuum states in 2+1 systems with $\tilde{N}_3 = \pm 1$ have been discussed in Sec. 11.2.1. Let us introduce the coordinate y along the wall (line), the coordinate x normal to the line, and the operator $\mathcal{P}_x = -i\partial_x$ of the momentum transverse to the wall; the momentum p_y remains a good quantum number. Then the Hamiltonian for the fermions in the presence of the wall is given by the following modification of eqn (11.2):

$$\mathcal{H} = \check{\tau}^b g_b(\mathbf{p}) \ , \quad g_3 = \frac{\mathcal{P}_x^2 + p_y^2}{2m^*} - \mu \ , \quad g_1 = \frac{1}{2}\left(c(x)\mathcal{P}_x + \mathcal{P}_x c(x)\right) \ , \quad g_2 = c_0 p_y \ . \tag{22.4}$$

In ^{3}He-A film the parameter $c_0 = c_\perp$. The function $c(x)$ changes sign across the wall reflecting the change of orientation of $\hat{\mathbf{l}} = \pm\hat{\mathbf{z}}$ across the wall. As a result the topological invariant $\tilde{N}_3$ of the vacuum state also changes from $\tilde{N}_3(x = +\infty) = +1$ to $\tilde{N}_3(x = -\infty) = -1$ (Fig. 22.1 top left).

Since the topology of the spectrum of fermion zero modes does not depend on the details of the function $c(x)$, one may choose any function which reverses sign across the wall, $c(x = \mp\infty) = \mp c_0$: for example,

$$c(x) = c_0 \tanh\frac{x}{d} \ , \tag{22.5}$$

where d is the thickness of the domain wall. Later we shall use this example to discuss surfaces with infinite red shift: in this example the speed of light $c(x)$ becomes zero at the center of the wall (at $x = 0$).

Let us first consider the classical limit, where p_x and x are considered as independent coordinates:

$$\mathcal{H}_{\text{qc}} = \frac{p_x^2 + p_y^2 - p_F^2}{2m^*}\check{\tau}^3 + p_x c(x)\check{\tau}^1 + c_0 p_y \check{\tau}^2 \ . \tag{22.6}$$

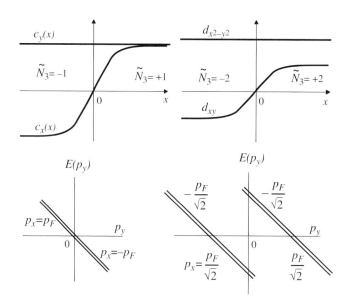

FIG. 22.1. Fermion zero modes on domain walls separating 2+1 vacua with different topological charge $\tilde{N}_3$. The domain wall in chiral p-wave superconductor separating vacua with $\tilde{N}_3 = 1$ and $\tilde{N}_3 = -1$ (*top left*) gives rise to two fermion zero modes (*bottom left*), or four if the spin degrees of freedom are taken into account. The domain wall in chiral d-wave superconductor separating vacua with $\tilde{N}_3 = 2$ and $\tilde{N}_3 = -2$ (*top right*) gives rise to four fermion zero modes (*bottom right*), or eight for two spin components. The asymmetry of the energy spectrum leads to the net momentum carried by occupied negative energy levels and thus to net mass and/or electric current in the y direction inside the domain wall discussed by Ho *et al.* (1984) for ^{3}He-A texture.

In the combined 3D space (x, p_x, p_y) the classical Hamiltonian contains two Fermi points, where $|\mathbf{g}| = 0$: at $(x_a, p_{xa}, p_{ya}) = (0, \pm p_F, 0)$. These two points are described by the non-zero topological invariant $N_3 = \pm 1$ in eqn (8.13) where now the integral is over the 2D surface surrounding the Fermi point at (x_a, p_{xa}, p_{ya}). This again illustrates the universality of the manifolds of zeros of co-dimension 3 – the Fermi points. The non-zero value of the invariant N_3 for the classical Hamiltonian gives rise to the fermion zero mode in the exact quantum mechanical spectrum.

22.1.3 *Spectral asymmetry index*

In the exact quantum mechanical problem in eqn (22.4) there is only one conserved quantum number, namely the momentum projection $p_y \equiv p_\parallel$ along the wall. We are interested in fermion zero modes and we want to know how many branches $E_a(p_y)$, if any, cross the zero-energy level. The algebraic sum ν of the branches of the gapless fermions is defined by the index of the spectral asymmetry $\nu(p_y)$ of the Bogoliubov operator $\mathcal{H}$ in eqn (22.4). This integer-valued index gives the difference between the numbers of positive and negative eigenvalues $E_a(p_y)$ of $\mathcal{H}$ at given momentum p_y; if the index $\nu(p_y)$ abruptly changes by unity at some p_{ya} this means that at this p_{ya} one of the energy levels $E_a(p_y)$ crosses zero energy. The spectral asymmetry index $\nu(p_y)$ can be expressed in terms of the Green function, $\mathcal{G}^{-1} = ip_0 - \mathcal{H}$:

$$\nu(p_y) = \mathbf{Tr} \int \frac{dp_0}{2\pi i} \mathcal{G} \partial_{p_0} \mathcal{G}^{-1} \qquad (22.7)$$

$$= \mathbf{Tr} \int \frac{dp_0}{2\pi} \mathcal{G} = -\mathbf{Tr} \int \frac{dp_0}{2\pi} \frac{\mathcal{H}}{p_0^2 + \mathcal{H}^2} = -\frac{1}{2} \sum_n \mathrm{sign} E_n(p_y) \ . \qquad (22.8)$$

Here **Tr** means the summation over all the states with given p_y. The algebraic sum of branches crossing zero as functions of p_y is

$$\nu = \nu(p_y = +\infty) - \nu(p_y = -\infty) \ . \qquad (22.9)$$

Note the apparent relation of $\nu(p_y)$ in eqn (22.7) to the momentum space invariant N_1 in eqn (8.3) which is responsible for the topological stability of the Fermi surface. Indeed, according to eqn (22.2), the index ν in eqn (22.9) is the number of Fermi surfaces in the 1D momentum space p_y.

22.1.4 Index theorem

Now let us relate the index ν in eqn (22.9) for the exact Hamiltonian $\mathcal{H}$ to the topological invariant N_3 of the Fermi points in the quasiclassical Hamiltonian $\mathcal{H}_{\mathrm{qc}}$. Since the core size d is of order of the coherence length $d \sim \xi = \hbar/m^* c_0$ and therefore is much larger than the wavelength p_F^{-1} of the excitations, one may use the gradient expansion for the Green function. The gradient expansion is a procedure in which the exact Green function $\mathcal{G}(p_0, p_y, -i\partial_x, x) = (ip_0 - \mathcal{H})^{-1}$ is expanded in terms of the Green function in the quasiclassical limit $\mathcal{G}_{\mathrm{qc}}(p_0, p_y, p_x, x) = (ip_0 - \mathcal{H}_{\mathrm{qc}})^{-1}$:

$$\mathcal{G}(p_0, p_y, -i\partial_x, x) = \mathcal{G}_{\mathrm{qc}}(p_0, p_y, p_x, x)$$
$$+ \frac{i}{2} \mathcal{G}_{\mathrm{qc}} (\partial_{p_x} \mathcal{G}_{\mathrm{qc}}^{-1} \mathcal{G} \partial_x \mathcal{G}_{\mathrm{qc}}^{-1} \mathcal{G}_{\mathrm{qc}} - \partial_x \mathcal{G}_{\mathrm{qc}}^{-1} \mathcal{G}_{\mathrm{qc}} \partial_{p_x} \mathcal{G}_{\mathrm{qc}}^{-1} \mathcal{G}_{\mathrm{qc}}) + \ldots \qquad (22.10)$$

Substituting this expansion into the spectral asymmetry index in eqn (22.7) one obtains

$$\nu(p_y) = \frac{1}{24\pi^2} e^{jkl} \mathrm{tr} \int dV \ \mathcal{G}_{\mathrm{qc}} \partial_j \mathcal{G}_{\mathrm{qc}}^{-1} \mathcal{G}_{\mathrm{qc}} \partial_k \mathcal{G}_{\mathrm{qc}}^{-1} \mathcal{G}_{\mathrm{qc}} \partial_l \mathcal{G}_{\mathrm{qc}}^{-1} \ . \qquad (22.11)$$

Here **tr** means trace only for matrix indices, and $dV = dp_0 dp_x dx$. This equation is well-defined only if the Green function has no singularities, which is fulfilled if the Fermi point in the quasiclassical Hamiltonian is not in the region of integration. For that it is necessary that $p_y \neq p_{ya}$.

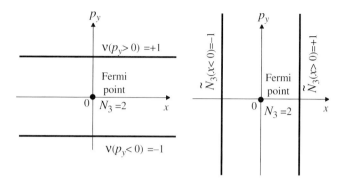

FIG. 22.2. The derivation of the index theorem for fermion zero modes on domain walls separating 2+1 vacua with different topological charge $\tilde{N}_3$. *Left*: The spectral asymmetry index $\nu(p_y)$ of the fermionic levels can be represented in terms of the 3D integral over (x, p_0, p_x) at given p_y (thick horizontal lines; axes p_0 and p_x are not shown). The difference of the integrals, $\nu(p_y > 0) - \nu(p_y < 0)$, represents the integral around the Fermi points in 4D space (x, p_y, p_0, p_x) which equals the topological charge N_3 of the Fermi points. *Right*: The same charge N_3 can be expressed in terms of the difference of two 3D integrals over (p_0, p_x, p_y) at given x (thick vertical lines; axes p_0 and p_x are not shown). Each of the integrals represents the topological charge $\tilde{N}_3$ of the 2+1 vacuum state of domains outside the wall. As a result one obtains the index theorem for the number of fermion zero modes in terms of the topological charges of two vacua: $\nu = \nu(p_y > 0) - \nu(p_y < 0) = \tilde{N}_3(x > 0) - \tilde{N}_3(x < 0)$.

Now the index theorem, eqn (22.3), which relates the number of fermion zero modes living within the domain wall to the topological charges of vacua on both sides of the interface comes from the following chain of equations

$$\nu = \nu(p_y = +\infty) - \nu(p_y = -\infty) \quad (22.12)$$

$$= \left(\int_{p_y=+\infty} dV - \int_{p_y=-\infty} dV \right) \left(\mathcal{G}_{qc}(p_0, x, p_x) \partial \mathcal{G}_{qc}^{-1}(p_0, x, p_x) \right)^3 \quad (22.13)$$

$$= N_3$$

$$= \left(\int_{x=+\infty} dV - \int_{x=-\infty} dV \right) \left(\mathcal{G}_{qc}(p_0, p_x, p_y) \partial \mathcal{G}_{qc}^{-1}(p_0, p_x, p_y) \right)^3 \quad (22.14)$$

$$= \tilde{N}_3(x = +\infty) - \tilde{N}_3(x = -\infty) . \quad (22.15)$$

In eqns (22.13) and (22.14) the abbreviation $\left(\mathcal{G}_{qc}\partial\mathcal{G}_{qc}^{-1}\right)^3$ stands for the 3-form in eqn (22.11). Integration regions in these two equations are different: in eqn (22.13) $dV = dp_0 dp_x dx$ (Fig. 22.2 *left*) and in eqn (22.14) $dV = dp_0 dp_x dp_y$ (Fig. 22.2 *right*), but both surround Fermi points in the 4D combined space (p_0, x, p_x, p_y), and thus the integrals are equal to the total charge N_3 of the

Fermi points. As a result the the spectral asymmetry index in eqn (22.12) is determined by the momentum space topology of the 2+1 vacua in the bulk, at $x = \pm\infty$, in eqn (22.15).

This proof can be applied to any kind of domain wall. We shall see in the following Chapter 23 that the same kind of index theorem exists for fermion zero modes living in the core of vortices, whose spectrum was first calculated by Caroli et al. (1964) in microscopic theory.

22.1.5 Spectrum of fermion zero modes

From the index theorem (22.3) it follows that within the interface separating vacua with opposite orientations of the $\hat{\bf l}$-vector in Sec. 22.1.2 (Fig. 22.1 *top left*) there must be $\nu = 2$ (or -2) gapless branches. Let us now calculate the spectrum of these two fermion zero modes explicitly, i.e. let us find the phenomenological parameters c_a and p_a in eqn (22.1). The parameters c_a represent the 'speeds of light' of the 1+1 low-energy fermions; their signs represent the 'helicity' of the fermions; the shift of the zero from the origin $p_a = e_a A_y$ introduces the vector potential A_y of the effective gauge field; and $e_a = \pm 1$ corresponds to the electric charge of the fermions. We shall use the quasiclassical approximation, which is valid if the size of the domain wall, $d \sim \xi$, is much larger than the characteristic wavelength $\lambda \sim p_F^{-1}$ of quasiparticles. The same approximation will be used for the calculation of the fermion zero modes in vortices in Chapter 23.

The lowest-energy states are concentrated in the vicinity of the Fermi points characterizing the classical energy spectrum in eqn (22.6), $p_{xa} = q_a p_F$, $p_{ya} = 0$, where $q_a = \pm 1$ is another fermionic charge. Near each of the two points one can expand the x-component of momentum

$$\mathcal{P} \approx q_a p_F - i\partial_x \,, \tag{22.16}$$

using the transformation of the wave function $\chi(x) \to \chi(x) \exp(iq_a p_F x)$. Leaving only the first-order term in $i\partial_x$ and taking into account that $m^* c_0 \ll p_F$, one obtains the Hamiltonian in the vicinity of each of the two points:

$$\mathcal{H}_a = -iq_a \frac{p_F}{m^*}\check\tau^3 \partial_x + q_a p_F c(x)\check\tau^1 + c_0 p_y \check\tau^2 \,. \tag{22.17}$$

To find the fermion zero modes let us consider first the case when p_y is just at its Fermi point value, i.e. at $p_y = 0$:

$$\mathcal{H}_a(p_y = 0, p_x = \pm p_F) = q_a p_F \left(-i \frac{1}{m^*}\check\tau^3 \partial_x + c(x)\check\tau^1 \right) \,. \tag{22.18}$$

This Hamiltonian is supersymmetric since (i) there is an operator anticommuting with $\mathcal{H}$, i.e. $\mathcal{H}\tau_2 = -\tau_2 \mathcal{H}$; and (ii) the potential $c(x)$ has a different sign at $x \to \pm\infty$. Thus at $p_y = 0$ each of the two Hamiltonians in eqn (22.18) must contain an eigenstate with exactly zero energy. These two eigenstates have the form

$$\chi_a(p_y=0,x) = e^{iq_a p_F x}\begin{pmatrix}1\\-i\end{pmatrix}\exp\left(-\int^x dx'\,c(x')\right).\tag{22.19}$$

These wave functions are normalizable just because $c(x)$ has a different sign at $x \to \pm\infty$.

Now we can consider the case of non-zero p_y. When p_y is small the third term in eqn (22.17) can be considered as a perturbation and its average over the wave function in eqn (22.19) gives the energy levels in terms of p_y. For both charges q_a one obtains

$$E_a(p_y) = -c_0 p_y\,\text{sign}(c(\infty) - c(-\infty)).\tag{22.20}$$

These are two anomalous branches of 1+1 fermions living in the domain wall. Their energy spectrum crosses zero energy as a function of the momentum $p_y \equiv p_\parallel$, in the considered case at $p_y = 0$ (Fig. 22.1 *bottom left*). For the given structure of the wall the energy spectrum appears to be doubly degenerate. Close to the crossing point these fermions are similar to relativistic chiral (left- or right-moving) fermions.

22.1.6 *Current inside the domain wall*

In general, the domain walls separating vacua with opposite orientations of the $\hat{\mathbf{l}}$-vector are current carrying (see e.g. the discussion in Volovik and Gor'kov 1985) This can be viewed in an example of ^{3}He-A where according to eqn (10.8) there is a mass (or particle) current proportional to $\nabla \times \hat{\mathbf{l}}$. This automatically leads to the mass current in the y direction inside the wall separating vacua with $\hat{\mathbf{l}} = +\hat{\mathbf{z}}$ and $\hat{\mathbf{l}} = -\hat{\mathbf{z}}$, since $\nabla \times \hat{\mathbf{l}} = -2\hat{\mathbf{y}}\delta(x)$. A particular contribution to this edge current is provided by fermion zero modes due to their spectral asymmetry as was discussed by Ho *et al.* (1984) for ^{3}He-A texture. The occupied negative energy levels in Fig. 22.1 *left bottom* are not symmetric with respect to the parity transformation $p_y \to -p_y$, that is why they carry momentum and thus the mass current in superfluids and the electric current in superconductors along the y direction.

The edge currents exist in most of the domain walls separating vacua with different $\tilde{N}_3$. The currents are forbidden by time reversal symmetry T and by parity. In our case T is violated by non-zero value of $\tilde{N}_3$, while the proper parity is violated for the general orientation of the domain wall with respect to crystal axes.

22.1.7 *Edge states in d-wave superconductor with broken T*

Let us consider another example illustrating the topological rule in eqn (22.3), namely fermion zero modes within the domain wall in a d-wave superconductor with broken time reversal symmetry T discussed in Sec. 11.2.3. The superconducting state with broken T is fully gapped, and thus the gapless fermions can live only within the brane. Since the wall separates domains with topological charges $\tilde{N}_3 = -2$ and $\tilde{N}_3 = +2$ per spin (Fig. 22.1 *top right*), from the index theorem in eqn (22.3) it follows that the index $\nu = 4$ (or -4), i.e. there must be four branches of fermion zero modes per spin per layer which cross zero energy as a function of $p_\parallel$.

Let us find these modes explicitly. The relevant Bogoliubov–Nambu Hamiltonian is given by eqn (11.7). Since the topological structure of fermion zero modes does not depend on the detailed structure of the order parameter within the wall, we shall choose the ansatz for the wall with $d_{xy}(x)$ changing sign, while $d_{x^2-y^2}$ is constant (Fig. 22.1 *top right*):

$$\mathcal{H} = \frac{\mathcal{P}^2 + p_y^2 - p_F^2}{2m^*}\check{\tau}^3 + \frac{p_y}{2}\{\mathcal{P}, d_{xy}(x)\}\check{\tau}^1 + (\mathcal{P}^2 - p_y^2)d_{x^2-y^2}\check{\tau}^2 \ , \ \mathcal{P} = -i\partial_x \ . \tag{22.21}$$

The (quasi)classical limit of this Hamiltonian,

$$\mathcal{H}_{qc} = \frac{p_x^2 + p_y^2 - p_F^2}{2m^*}\check{\tau}^3 + p_y p_x d_{xy}(x)\check{\tau}^1 + (p_x^2 - p_y^2)d_{x^2-y^2}\check{\tau}^2 \ , \tag{22.22}$$

has four Fermi points in (x, p_x, p_y) space situated at $x = 0$, $|p_{xa}| = |p_{ya}| = p_F/\sqrt{2}$. Using the transformation of the wave function $\chi(x) \to \chi(x)\exp(ip_{xa}x)$ we can expand the momentum $\mathcal{P}$ in the vicinity of these points, $\mathcal{P} = \pm p_F/\sqrt{2} - i\partial_x$, and eqn (22.21) acquires the following form:

$$\mathcal{H}_a = -i\frac{p_{xa}}{m^*}\check{\tau}^3\partial_x + p_{xa}p_{ya}d_{xy}(x)\check{\tau}^1 + c(p_y - p_{ya})\check{\tau}^2 \text{sign}(p_{ya}) \ , \ c = \sqrt{2}p_F d_{x^2-y^2} \ . \tag{22.23}$$

This is similar to eqn (22.18) and thus the same procedure can be applied as in Sec. 22.1.5 to find the energy spectrum. It gives four branches of the energy spectrum in Fig. 22.1 *bottom right*:

$$E_a(p_y) = c_a(p_y - p_{ya})\text{sign}(d_{xy}(\infty) - d_{xy}(-\infty)) \ , \ c_a = \sqrt{2}p_F d_{x^2-y^2} \ , \tag{22.24}$$

which cross zero as a function of $p_y \equiv p_\parallel$. As distinct from eqn (22.20) the crossing points are now split: $p_{ya} = e_a A_y$, where $e_a = \pm 1$ and $A_y = p_F/\sqrt{2}$ play the role of the electric charge and the vector potential of effective electromagnetic field respectively.

22.2 3+1 world of fermion zero modes

The dimensional reduction discussed in Sec. 22.1 can be generalized to higher dimensions. We shall start from a quantum liquid in 5+1 spacetime and obtain a 3+1 world with Fermi points and thus with all their attributes at low energy: chiral relativistic fermions, gauge fields and gravity. The dimensional reduction from 5+1 to 3+1 can be made in two steps, as in the case of the reduction from 3+1 to 1+1. First we consider a thin film in 5D space, which effectively reduces the space dimension to 4, and then consider the domain wall in this space. Fermions living within this wall form the effective 3+1 world. An alternative way from 5+1 to 3+1, which gives similar results, is to obtain the 3+1 world of fermion zero modes living in the core of a vortex. In relativistic theories the latter approach was developed by Akama (1983). However, in both cases it is not necessary to start with the relativistic theory in the original 5+1 spacetime. The only input is the momentum space topology of vacua on both sides of the interface, due to which RQFT emergently arises in the world of the fermion zero modes living within the brane.

22.2.1 Fermi points of co-dimension 5

In 5+1 spacetime a new topologically stable manifold of zeros in the quasi-particle energy spectrum can exist: zeros of co-dimension 5. Let us recall that co-dimension is the dimension of **p**-space minus the dimension of the manifold of zeros in the energy spectrum. In 5D momentum space the zeros of co-dimension 5 are points. The relativistic example of the propagator with topologically non-trivial zeros of co-dimension 5 is provided by $\mathcal{G}^{-1} = ip_0 - \mathcal{H}$, where the Hamiltonian in 5D space is $\mathcal{H} = \sum_{n=1}^{5} \Gamma^n p_n$, and Γ^{1-5} are 4×4 Dirac matrices satisfying the Clifford algebra $\{\Gamma^a, \Gamma^b\} = 2\delta^{ab}$. We can choose these matrices as $\Gamma^1 = \tau^3 \sigma^1$, $\Gamma^2 = \tau^3 \sigma^2$, $\Gamma^3 = \tau^3 \sigma^3$, $\Gamma^4 = \tau^1$, $\Gamma^5 = \tau^2$:

$$\mathcal{H}_5 = \tau^3 \sigma^i p_i + \tau^1 p_4 + \tau^2 p_5 \,, \quad i = 1, 2, 3, \quad \mathcal{H}^2 = E^2 = p^2 \,. \quad (22.25)$$

In this example the Fermi point of co-dimension 5 is at $\mathbf{p} = 0$. The topological stability of this point is provided by the integer-valued topological invariant

$$N_5 = C_5 \mathbf{tr} \int_\sigma dS^{\mu\nu\lambda\alpha\beta} \, \mathcal{G}\partial_{p_\mu} \mathcal{G}^{-1} \mathcal{G}\partial_{p_\nu} \mathcal{G}^{-1} \mathcal{G}\partial_{p_\lambda} \mathcal{G}^{-1} \mathcal{G}\partial_{p_\alpha} \mathcal{G}^{-1} \mathcal{G}\partial_{p_\beta} \mathcal{G}^{-1} \,. \quad (22.26)$$

The integral here is over the 5D surface in 6D space $p_\mu = (p_0, p_{1-5})$ around the point $(p_0 = 0, \mathbf{p} = 0)$; the prefactor C_5 is a normalization factor, so that $N_5 = +1$ for the particles obeying eqn (22.25). As in the case of 3+1 spacetime, if the vacuum has a Fermi point with $N_5 = +1$, then in the vicinity of this Fermi point, after rescaling and rotations, the Hamiltonian acquires the relativistic form of eqn (22.25). In other words, it is the topological invariant N_5 which gives rise to the low-energy chiral relativistic fermions in 5+1 spacetime.

22.2.2 Chiral 5+1 particle in magnetic field

Let us first consider the dimensional reduction produced by a magnetic field. We know that the motion of particles in the plane perpendicular to the magnetic field is quantized into the Landau levels. The free motion is thus effectively reduced to the rest dimensions, where the particles – fermion zero modes – are again chiral, but are described by the reduced topological invariant. Let us consider particles with charge q in the magnetic field $F_{45} = \partial A_5/\partial x^4 - \partial A_4/\partial x^5 = \text{constant}$. The Hamiltonian in this field becomes

$$\mathcal{H}_5 = \tau^3 \sigma^i p_i + \tau^1 \left(p_4 - \frac{q}{2} F_{45} x^5\right) + \tau^2 \left(p_5 + \frac{q}{2} F_{45} x^4\right) \,. \quad (22.27)$$

After Landau quantization one obtains the fermion zero mode – the massless branch on the first Landau level with the 'isospin' projection $\tau^3 = \text{sign } F_{45}$,

$$\mathcal{H}_3 = \sigma^i p_i \, \text{sign}\,(qF_{45}) \,, \quad i = 1, 2, 3 \,. \quad (22.28)$$

These modes represent chiral particles in $3 + 1$ spacetime, whose spectrum contains fermion zero modes of co-dimension 3. Their chirality depends on the direction of the magnetic field.

Thus the magnetic field reduces the Fermi point of co-dimension 5 in 5 + 1 momentum–energy space with topological charge N_5 to the Fermi point of co-dimension 5 with charge N_3 in 3 + 1 momentum–energy space. Simultaneously the dimension of the effective coordinate space in the low-energy corner, i.e. at energies $E^2 \ll |F_{45}|$, is reduced from 5 to 3. However, there is still the degeneracy of the fermion zero modes with respect to the position of orbits in the magnetic field. As a result the total number of fermion zero modes is determined by the number of flux quanta $\nu = \Phi/\Phi_0$, where $\Phi = \int dx_4 dx_5 F_{45}$.

22.2.3 Higher-dimensional anomaly

Adding magnetic field F_{12} in another two directions one obtains a further dimensional reduction and $1 + 1$ fermions in the lowest Landau level in this field. The energy of 1+1 fermions is $\mathcal{H}_1 = (\text{sign } F_{45})(\text{sign } F_{12})p_z$. The Fermi point at $p_z = 0$ is the 1D Fermi surface – the manifold of zeros of co-dimension 1. If now an electric field is introduced along the remaining space direction, $E_3 = F_{03}$, the chiral anomaly phenomenon arises – the chiral particles are produced from the vacuum due to spectral flow. The number of chiral particles produced per unit time and unit volume is proportional to the force qF_{03} acting on the 1+1 mode, and to the number of fermion zero modes $|qF_{12}L_1L_2/2\pi| \cdot |qF_{45}L_4L_5/2\pi|$, which comes from the degeneracy of the levels in magnetic fields F_{45} and F_{12} (see eqn (18.2)). As a result one obtains the (5+1)-dimensional version of the Adler–Bell–Jackiw equation for the particle production due to the chiral anomaly

$$\dot{n} = N_5 \frac{q^3}{(2\pi)^3} F_{03} F_{45} F_{12} = N_5 \frac{q^3}{720(2\pi)^3} e^{\alpha\beta\gamma\mu\nu\rho} F_{\alpha\beta} F_{\gamma\mu} F_{\nu\rho} , \qquad (22.29)$$

where N_5 is the integer-valued momentum space topological invariant for the Fermi point in 5+1 dimensions in eqn (22.26). Finally, in the general case of several fermionic species with electric charges q_a and some other fermionic charges B_a, the production of charge B per unit time per unit volume is given by

$$\dot{B} = \frac{1}{720(2\pi)^3} \text{tr} \left(B Q^3 \tilde{N}_5 \right) e^{\alpha\beta\gamma\mu\nu\rho} F_{\alpha\beta} F_{\gamma\mu} F_{\nu\rho} , \qquad (22.30)$$

where $\tilde{N}_5$ is the matrix of the topological invariant analogous to eqn (12.9) for 3+1 systems.

22.2.4 Quasiparticle world within domain wall in 4+1 film

Let us now discuss the alternative way of compactification. We start with a thin film of a quantum liquid in 5D space. If the motion normal to the film is quantized, there remains only the 4D momentum space along the film (p_1, p_2, p_3, p_4). The gapless quasiparticle spectrum becomes fully gapped in the film because of the transverse quantization. In the simplest case of one transverse level the Hamiltonian (22.25) becomes

$$\mathcal{H} = M\Gamma^5 + \sum_{n=1}^{4} \Gamma^i p_n = M\tau^2 + \tau^3 \sigma^i p_i + \tau^1 p_4 , \quad i = 1, 2, 3 . \qquad (22.31)$$

The vacuum of the fully gapped quantum liquid is characterized by non-trivial momentum space topology described by the invariant $\tilde{N}_5$ – an analog of $\tilde{N}_3$. It can be obtained by dimensional reduction from the invariant N_5 in eqn (22.26) of the 5+1 system (analog of N_3):

$$\tilde{N}_5 = C_5 \mathbf{tr} \int dS^{\mu\nu\lambda\alpha\beta} \, \mathcal{G}\partial_{p_\mu}\mathcal{G}^{-1}\mathcal{G}\partial_{p_\nu}\mathcal{G}^{-1}\mathcal{G}\partial_{p_\lambda}\mathcal{G}^{-1}\mathcal{G}\partial_{p_\alpha}\mathcal{G}^{-1}\mathcal{G}\partial_{p_\beta}\mathcal{G}^{-1} \,. \quad (22.32)$$

Here the integral is over the whole 4+1 momentum–frequency space.

Now we introduce the 3+1 domain wall (brane), which separates two domains, each with fully gapped fermions. Then everything can be obtained from the case of the quantum liquid in 2+1 spacetime just by adding the number 2 to all dimensions involved. In particular, the difference $\tilde{N}_5(\text{right}) - \tilde{N}_5(\text{left})$ of invariants on both sides of the brane (the analog of $\tilde{N}_3(\text{right}) - \tilde{N}_3(\text{left})$) gives rise to the 3+1 fermion zero modes within the domain wall. These fermion zero modes are described by the momentum space topological invariant N_3 (the analog of N_1) and thus have Fermi points – fermion zero modes of co-dimension 3.

In the same manner as in eqns (22.2) and (22.3) which relate the number of fermion zero modes to the topological invariants in bulk 2+1 domains, the total topological charge of Fermi points within the domain wall is expressed through the difference of the topological invariants in bulk 4+1 domains:

$$N_3 = \tilde{N}_5(\text{right}) - \tilde{N}_5(\text{left}) \,. \quad (22.33)$$

Close to the a-th Fermi point the fermion zero modes represent 3+1 chiral fermions, whose propagator has the general form expressed in terms of the tetrad field in eqn (8.18). The quantities $e_b^{\mu(a)}$ and $p_\mu^{(a)}$, which enter the fermionic spectrum, are dynamical variables. These are the low-energy collective bosonic modes which play the part of effective gravitational and gauge fields. The source of such emergent phenomena is the non-zero value of $\tilde{N}_5$ of vacuua in bulk. Thus, the brane separating the 4+1 vacua with different $\tilde{N}_5$ gives rise to the Fermi point universality class of quantum vacua, whose properties are dictated by momentum space topology. In principle, all the ingredients of the Standard Model can be obtained within the brane as emergent phenomena without postulating gravity and RQFT in the original 4+1 spacetime.

In a similar manner the gauge and gravity fields can arise as collective modes on the boundary of the 4+1 system exhibiting the quantum Hall effect as discussed by Zhang and Hu (2001). Both systems have similar topology. In the Zhang–Hu model the non-trivial topology is provided by the external field, while in our case it is provided by the non-zero topological charge $\tilde{N}_5$ of the vacua in the bulk. The difference is the same as in the 2+1 systems, where the QHE occurs in the external magnetic field, while the anomalous QHE is provided by the nontrivial topological invariant $\tilde{N}_3$ of the vacuum.

However, as we know, the non-trivial topology alone does not guarantee that the effective gravitational field will obey the Einstein equations: the proper (maybe discrete) symmetry and the proper relations between different Planck scales in the underlying fermionic system are required. The energy scale which provides the natural ultraviolet cut-off must be much smaller than the energy scale at which Lorentz invariance is violated. We hope that within this universality class one can obtain such the required hierarchy of Planck scales.

23

FERMION ZERO MODES ON VORTICES

23.1 Anomalous branch of chiral fermions

23.1.1 Minigap in energy spectrum

We shall start with the simplest axisymmetric vortex in a conventional isotropic superconductor, whose scalar order parameter asymptote is given by eqn (14.16). The spectrum of the low-energy bound states in the core of the vortex with winding number $n_1 = \pm 1$ was obtained in microscopic theory by Caroli et al. (1964). The quantum numbers which characterize the energy spectrum of fermionic excitations living on the line are the linear and angular momenta along the string, p_z and L_z. In the case of a vortex line the angular momentum must be modified, since the correct symmetry of the vacuum in the presence of the vortex is determined by the combination of rotation and global gauge transformation, whose generator Q is given by eqn (14.19): $Q = L_z - \frac{n_1}{2} N$, where for quasiparticles the generator $N = \check{\tau}^3$. Caroli et al. (1964) found that the energy spectrum of the bound states with lowest energy has the following form:

$$E(Q, p_z) = -n_1 \omega_0(p_z) Q .\qquad(23.1)$$

This spectrum is two-fold degenerate due to spin degrees of freedom; the generalized angular momentum Q of the fermions in the core was found to be a half-odd integer, $Q = n + 1/2$ (Fig. 23.1 *top right*). The level spacing $\omega_0(p_z)$ is small compared to the energy gap of the quasiparticles outside the core, $\omega_0(p_z) \sim \Delta_0^2/v_F p_F \ll \Delta_0$, but is nowhere zero. Its value at $p_z = 0$ is called the minigap, because $\omega_0(0)/2$ is the minimal energy of quasiparticles in the core, corresponding to $Q = \pm 1/2$.

Strictly speaking, the spectrum in the Caroli–de Gennes–Matricon equation (23.1) does not contain fermion zero modes. However, since the interlevel spacing is small, in many physical cases, say when temperature $T \gg \omega_0$, the discreteness of the quantum number Q can be ignored and in the quasiclassical approximation one can consider Q as continuous. Then from eqn (23.1) it follows that the spectrum as a function of continuous parameter Q contains an anomalous branch, which crosses zero energy at $Q = 0$ (Fig. 23.1 *top right*). The point $Q = 0$ represents the Fermi surface – the manifold of co-dimension 1 – in the 1D space of the angular momentum Q. The fermions in this 1D 'Fermi liquid' are chiral: positive energy fermions have a definite sign of the generalized angular momentum Q. These fermion zero modes exist only in the quasiclassical approximation, and we call them the pseudo-zero modes.

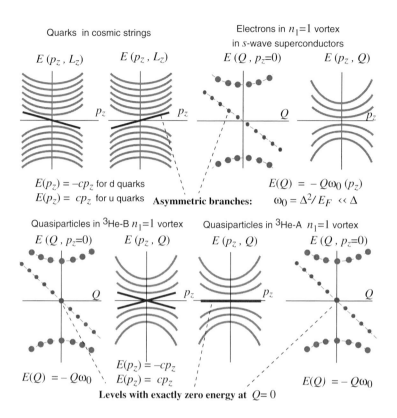

FIG. 23.1. Bound states of fermions on cosmic strings and vortices. In superconductors and Fermi superfluids the spectrum of bound states contains pseudo-zero modes – anomalous branches which as functions of the discrete quantum number Q cross the zero-energy level. The true fermion zero modes whose spectrum crosses zero energy as a function of p_z exist in the core of cosmic strings and ^{3}He-B vortices. The ^{3}He-A vortex with $n_1 = 1$ contains the branch with exactly zero energy for all p_z.

We shall see that for arbitrary winding number n_1, the number of fermion pseudo-zero modes, i.e. the number of branches crossing zero energy as a function of Q, equals $-2n_1$ (Volovik 1993a). This is similar to the index theorem for fermion zero modes in domain walls (Chapter 22) and in cosmic strings in RQFT in Fig. 15.1 (see Davis *et al.* (1997) and references therein). The main difference is that in strings and domain walls the spectrum of relativistic fermions crosses zero energy as a function of the linear momentum $p_\parallel (= p_z)$ (Fig. 23.1 *top left*), and the index theorem discriminates between left-moving and right-moving fermions. In condensed matter vortices the spectrum crosses zero energy as a function of the generalized angular momentum Q, and the index theorem discriminates between clockwise and counterclockwise rotating fermions: in quasiclassical approximation the quantity ω_0 in the Caroli–de Gennes–Matricon equation is the angular velocity.

23.1.2 Integer vs half-odd integer angular momentum of fermion zero modes

The topological properties of the spectrum $E(Q)$ and $E(p_z)$ of fermion zero modes in the vortex core and string core are universal and do not depend on the detailed structure of the core. Both modes belong to the universality class of Fermi surface – the manifold of co-dimension 1 either in momentum p_z-space and or in angular momentum Q-space.) If we proceed to non-s-wave superfluid or superconducting states, we find that the situation does not change so long as the quantum number Q is considered as continuous.

However, at a discrete level of description one finds the following observation: in some systems (such as $n_1 = 1$ vortex in s-wave superconductors) the quantum number Q is half-odd integer, while in others it is integer. In the latter vortices there is an anomalous fermion zero mode: at $Q = 0$ the quasiparticle energy in the Caroli–de Gennes–Matricon equation (23.1) is exactly zero for any p_z (Fig. 23.1 *bottom right*). Such anomalous, highly degenerate zero-energy bound states were first calculated in a microscopic theory by Kopnin and Salomaa (1991) for the $n_1 = \pm 1$ vortex in ^{3}He-A. These fermion zero modes are zeros of co-dimension 0 and thus are topologically unstable. That is why such a degeneracy can be lifted off if it is not protected by symmetry: for example, in ^{3}He-B vortices in Fig. 23.1 *bottom left* two anomalous branches corresponding to two spin projections split leaving the zero-energy state at $p_z = 0$ only. As a result one obtains the same situation as in cosmic strings in Fig. 23.1 *top left*, i.e. with the spectrum of fermion zero modes crossing zero as a function of p_z.

The difference between the two types of fermionic spectrum, with half-odd integer and integer Q, becomes important at low temperature $T < \omega_0$. We consider here representatives of these two types of vortices: the traditional $n_1 = \pm 1$ vortex in an s-wave superconductor (Fig. 23.1 *top right*) and the simplest form of the $n_1 = \pm 1$ vortex in ^{3}He-A with $\hat{\mathbf{l}}$ directed along the vortex axis (Fig. 23.1 *bottom right*). Their order parameters are

$$\Psi(\mathbf{r}) = \Delta_0(\rho) e^{in_1\phi} \; , \quad \oint d\mathbf{x} \cdot \mathbf{v}_s = n_1 \pi \hbar / m \; , \quad (23.2)$$

$$e_{\mu i} = \Delta_0(\rho) e^{in_1\phi} \hat{z}_\mu (\hat{x}_i + i\hat{y}_i) \; , \quad \oint d\mathbf{x} \cdot \mathbf{v}_s = n_1 \pi \hbar / m \; , \quad (23.3)$$

where z, ρ, ϕ are the coordinates of the cylindrical system with the z axis along the vortex line; and $\Delta_0(\rho)$ is the profile of the order parameter amplitude in the vortex core with $\Delta_0(\rho = 0) = 0$. Actually the core structure of the vortex is more complicated, and there are even some components which are non-zero at $\rho = 0$ (see the review by Salomaa and Volovik 1987). But since the structure of the spectrum of the fermion zero modes does not depend on such details as the profile of the order parameter in the core, we do not consider this complication.

23.1.3 Bogoliubov–Nambu Hamiltonian for fermions in the core

Let us consider first 2+1 superconductors and superfluids – thin films of ^{3}He or superconductors in layered systems. In such systems the motion along z is quantized, and thus there is no dependence of the energy spectrum on the momentum p_z. Also in both systems the energy spectrum is fully gapped when the vortex is absent, and the vacua of homogeneous 2+1 systems are characterized by the momentum space invariant $\tilde{N}_3$. It is given by eqn (11.1) in general, or by eqn (11.6) in the simplest case described by the 2×2 Hamiltonian. The invariant $\tilde{N}_3 = 0$ for the 2+1 superfluid/superconductor with s-wave pairing; $\tilde{N}_3 = 1$ per spin per transverse level for the ^{3}He-A film; $\tilde{N}_3 = 1$ per spin per crystal layer of Sr$_2$RuO$_4$ superconductor if it is really a chiral superconductor; and $\tilde{N}_3 = 2$ per spin per layer for d-wave high-temperature superconductors, if for some reason time reversal symmetry T is broken there and the order parameter component $d_{xy} \neq 0$.

In the 2D systems the vortex is a point defect. Since the topological consideration is robust to the deformation of the order parameter, we choose the simplest one – axisymmetric in real and momentum space. For each of the two spin components, the Bogoliubov–Nambu Hamiltonian for quasiparticles in the presence of a point vortex with winding number n_1 in the vacuum with topological charge $\tilde{N}_3$ can be written as

$$\mathcal{H} = \check{\tau}^b g_b(\mathbf{p}, \mathbf{r}) = \begin{pmatrix} M(p) & \Delta_0(\rho)e^{in_1\phi}\left(\frac{p_x+ip_y}{p_F}\right)^{\tilde{N}_3} \\ \Delta_0(\rho)e^{-in_1\phi}\left(\frac{p_x-ip_y}{p_F}\right)^{\tilde{N}_3} & -M(p) \end{pmatrix}. \quad (23.4)$$

Here, as before, $M(p) = (p_x^2 + p_y^2 - p_F^2)/2m^* \approx v_F(p - p_F)$.

Introducing the angle θ in momentum (p_x, p_y) space, eqn (23.4) can be rewritten in the form

$$\mathcal{H} = \begin{pmatrix} M(p) & \Delta_0(\rho)\left(\frac{p}{p_F}\right)^{\tilde{N}_3} e^{i(n_1\phi+\tilde{N}_3\theta)} \\ \Delta_0(\rho)\left(\frac{p}{p_F}\right)^{\tilde{N}_3} e^{-i(n_1\phi+\tilde{N}_3\theta)} & -M(p) \end{pmatrix}. \quad (23.5)$$

This form emphasizes the interplay between real space and momentum space topologies: non-trivial momentum space topology of the ground state (vacuum) enters the off-diagonal terms – the order parameter – in the same way as the topologically non-trivial background – the vortex. This is more pronounced if one takes into account that near the vortex axis $\Delta_0(\rho) \propto \Delta_0(\infty)(\rho/\xi)^{n_1}$. The difference between space and momentum dependence is in the diagonal elements $M(p)$. We shall discuss how the real space, momentum space and combined space topology determine and influence the fermion zero modes in the vortex core. We shall find that in the quasiclassical description the topology of fermion zero modes is completely determined by the real space topological charge n_1. However, the fine structure of the fermion zero modes, i.e. the two types of the fermion zero modes, with integer and half-odd integer Q, are determined by the combined real space and momentum space topology. These two classes have different parity (see below)

$$W = (-1)^{n_1+N_3} , \qquad (23.6)$$

which is constructed from the topological charges in real and momentum spaces (Volovik 1999b).

23.1.4 Fermi points of co-dimension 3 in vortex core

Since outside the core the fermionic spectrum is fully gapped, fermion zero modes are the only low-energy fermions in the discussed 2+1 systems. To understand the origin of fermion zero modes, we start with the classical description of the fermionic spectrum, in which the commutators between coordinates (x,y) and momenta (p_x, p_y) are neglected, and all four variables can be considered as independent coordinates of the combined 4D space (x, y, p_x, p_y). This is justified since the characteristic size ξ of the vortex core is much larger than the wavelength $\lambda = 2\pi/p_F$ of a quasiparticle: $\xi p_F \sim v_F p_F/\Delta_0 \gg 1$ and the quasiclassical approximation makes sense. The quasiparticle energy spectrum in this classical limit is given by $E^2(x, y, p_x, p_y) = |\mathbf{g}(\mathbf{r},\mathbf{p})|^2 = M^2(p) + \Delta_0^2(\rho)$, where it is taken into account that p is concentrated in the vicinity of p_F. This energy spectrum is zero when simultaneously $p_x^2 + p_y^2 = p_F^2$ (so that $M(p) = 0$) and $x = y = 0$ (so that $\Delta_0(\rho) = 0$) (Fig. 23.2 left). This is a line of zeros in the combined 4D space and thus a manifold of co-dimension 3. So we must find out what is the topological charge N_3 of this manifold.

It is easy to check that this charge is solely determined by the vortex winding number: $N_3 = n_1$ for each spin projection, irrespective of the value of the momentum space charge $\tilde{N}_3$ of the vacuum state. Let us surround an element of the line by the closed 2D surface σ (which is depicted as a closed contour in Fig. 23.2). Then the image $\tilde{\sigma}$ of this surface in the space of the Hamiltonian matrix is the sphere of unit vector $\hat{\mathbf{g}} = \mathbf{g}/|\mathbf{g}|$ in eqn (23.4). Thus σ encloses a topological defect, a $\hat{\mathbf{g}}$-hedgehog. The winding number of this hedgehog is $N_3 = n_1$. Thus each point on the circumference of zeros is described by the same topological invariant N_3, eqn (8.15), as the topologically non-trivial Fermi point describing the chiral particles. Figure 23.2 shows the case of a vortex with winding number $n_1 = 1$, for which hedgehogs have unit topological charge: $N_3 = 1$. Thus in the classical description the fermion zero modes on vortices are zeros of co-dimension 3 in the combined $(\mathbf{p}, \mathbf{r})$ space.

23.1.5 Andreev reflection and Fermi point in the core

To make the situation clearer let us remove one of the four 'coordinates', which is not relevant for the quasiclassical description of bound states. We shall exploit the fact that the characteristic size ξ of the vortex core is much larger than the wavelength of quasiparticle $\sim p_F^{-1}$. Due to this the classical trajectories of quasiparticles propagating through the core are almost straight lines. Such trajectories are characterized by the direction of the quasiparticle momentum $\mathbf{p}$ and the impact parameter b (Fig. 23.3 left). If the core is axisymmetric, the

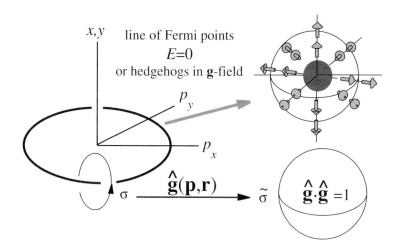

FIG. 23.2. In the core of the vortex with winding number n_1 the classical energy of a quasiparticle $E(x, y, p_x, p_y)$ is zero on a closed line in 4D space (x, y, p_x, p_y). Each point on this line is described by the topological invariant $N_3 = n_1$ for each spin projection. The 2D surface σ surrounding the element of the line is mapped to the 2D surface $\tilde{\sigma}$ in the space of the matrix $\mathcal{H}$ – the sphere of unit vector $\hat{\mathbf{g}} = \mathbf{g}/|\mathbf{g}|$. This is shown for the vortex with $n_1 = 1$, in which the singularity in the Hamiltonian represents the line of hedgehogs with $N_3 = 1$.

quasiparticle spectrum does not depend on the direction of the momentum, and we choose $\mathbf{p}$ along the y axis. Then the impact parameter b coincides with the coordinate x of the quasiparticle, and one has $M(p) \approx v_F(p_y - p_F)$. Thus p_x drops out of the Hamiltonian, and the classical spectrum $E^2(p_y, b, y) = v_F^2(p_y - p_F)^2 + \Delta_0^2(\rho)$ with $\rho = \sqrt{b^2 + y^2}$ is determined in the momentum–coordinate continuum (p_y, b, y).

During the motion of the quasiparticle in the vortex environment its energy is conserved. If the quasiparticle energy E is less than the maximum value of the gap, $\Delta_0^2(\infty)$, the quasiparticle moves back and forth along the trajectory in the potential well formed by the vortex core, being reflected at points $y = \pm y_0$, where $E = \Delta_0(\sqrt{b^2 + y_0^2})$ (Fig. 23.3 right). The main property of such a reflection is that the momentum of a quasiparticle remains the same, $p_y \approx p_F$, while its velocity $v_y = dE/dp_y = v_F^2(p_y - p_F)/E$ changes sign after reflection. In condensed matter physics this is called an Andreev reflection. The Caroli–de Gennes–Matricon bound states are obtained by quantization of this periodic motion of a quasiparticle (see Sec. 23.2 below).

One can check that the function $\mathbf{g}(b, y, p_y)$ has a hedgehog at the point $p_y = p_F$, $b = 0$, $y = 0$ with the topological charge $N_3 = n_1$. Thus we have a Fermi point of co-dimension 3 in a mixed space (b, y, p_y), which is the counterpart of a Fermi point in momentum space (p_x, p_y, p_z) describing a chiral particle.

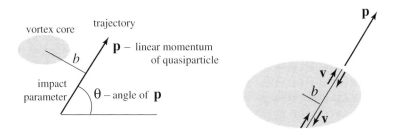

FIG. 23.3. Classical motion of quasiparticles in the (x,y) plane of the vortex core. *Left*: In the classical description the quasiparticle trajectories are straight lines. *Right*: Bound states of quasiparticles in the vortex core correspond to periodic motion along the straight lines with Andreev reflection from the walls of the potential well formed by the vortex core. After an Andreev reflection the momentum **p** of the quasiparticle did not change while its velocity **v** changed sign.

23.1.6 *From Fermi point to fermion zero mode*

Let us demonstrate how the quantization leads to dimensional reduction from the classical fermion zero modes of co-dimension 3 to the pseudo-zero modes of co-dimension 1 in Q space.

The first step to quantization is to take into account that the momentum p_y and the 'momentum' y do not commute. But this is just what happens with the chiral particles in a magnetic field, where the components of the generalized moment, $p_x - By$ and p_y, do not commute. Thus we arrive at the problem of a chiral particle in a magnetic field, with the effective magnetic field being parallel to the b axis. From this problem (see Secs 18.1.2 and 22.2.2) we know that the magnetic field realizes dimensional reduction: the Fermi point of co-dimension 3 described by the invariant N_3 gives rise to the Fermi point of co-dimension 1 described by the invariant N_1 representing 1D Fermi surfaces in p_z space. Since the role of the momentum p_z is played by the impact parameter b, the fermion zero mode of co-dimension 1 is the branch of the energy spectrum $E_a(b) = \kappa_a b$, which as a function of the coordinate b crosses zero.

The impact parameter b can be expressed in terms of the angular momentum, $b = L_z/p_F$. Since in the quasiclassical approximation the momentum L_z coincides with the quantum number Q, one obtains that close to the crossing point the spectrum of the a-th fermion zero mode must be $E_a(Q) = \omega_a Q$. Thus from the general topological arguments we obtained the form of the spectrum of fermion zero modes, which is the continuous limit of the Caroli–de Gennes–Matricon equation (23.1). In principle, the crossing point of the energy spectrum can be shifted from the origin, and the anomalous branch has the following general form:

$$E_a(Q) = \omega_a(Q - Q_a) \ . \tag{23.7}$$

The algebraic number of such branches is $N_1 = N_3 = n_1$ for each spin projection.

Thus the effective theory of fermion pseudo-zero modes is characterized by two sets of phenomenological parameters: slopes ω_a (effective speeds of light in the limit of continuous Q or minigaps in the case of discrete Q) and shifts of zeros Q_a (effective electromagnetic field). Let us now calculate the parameter ω_a for the simplest axisymmetric vortices with $n_1 = \pm 1$.

23.2 Fermion zero modes in quasiclassical description

23.2.1 Hamiltonian in terms of quasiclassical trajectories

Let us consider the quasiclassical quantization of the Andreev bound states whose classical motion is shown in Fig. 23.3. The low-energy trajectories through the vortex core are characterized by the direction θ of the trajectory (of the quasi-particle momentum $\mathbf{p}$) and the impact parameter b (Fig. 23.3). The magnitude of $\mathbf{p}$ is close to p_F, and thus for each θ the momentum $\mathbf{p}$ is close to the value

$$\mathbf{p} = p_F(\hat{\mathbf{x}} \cos\theta + \hat{\mathbf{y}} \sin\theta) \; ; \tag{23.8}$$

its group velocity velocity is close to $\mathbf{v}_F = \mathbf{p}/m^*$; and the Bogoliubov–Nambu Hamiltonian is

$$\mathcal{H} = \begin{pmatrix} M(p_x, p_y) & \Delta_0(\rho) e^{i(n_1\phi + \tilde{N}_3\theta)} \\ \Delta_0(\rho) e^{-i(n_1\phi + \tilde{N}_3\theta)} & -M(p_x, p_y) \end{pmatrix} . \tag{23.9}$$

Substituting $\chi \to e^{i\mathbf{p}\cdot\mathbf{r}}\chi$ and $\mathbf{p} \to \mathbf{p} - i\nabla$, and expanding in small ∇, one obtains the quasiclassical Hamiltonian for the fixed trajectory $(\mathbf{p}, b)$:

$$\mathcal{H}_{\mathbf{p},b} = -i\check{\tau}^3 \mathbf{v}_F \cdot \nabla + \Delta_0(\rho) \left(\check{\tau}^1 \cos(\tilde{N}_3\theta + n_1\phi) - \check{\tau}^2 \sin(\tilde{N}_3\theta + n_1\phi) \right) . \tag{23.10}$$

Since the coordinates ρ and ϕ are related by the equation $\rho \sin(\phi - \theta) = b$, the only argument is the coordinate along the trajectory $s = \rho \cos(\phi - \theta)$ and thus the Hamiltonian in eqn (23.10) has the form

$$\mathcal{H}_{\mathbf{p},b} = -iv_F \check{\tau}^3 \partial_s + \check{\tau}^1 \Delta_0(\rho) \cos\left(n_1\tilde{\phi} + (n_1 + \tilde{N}_3)\theta\right)$$
$$- \check{\tau}^2 \Delta_0(\rho) \sin\left(n_1\tilde{\phi} + (n_1 + \tilde{N}_3)\theta\right) , \tag{23.11}$$

where $\tilde{\phi} = \phi - \theta$ and ρ are expressed in terms of the coordinate s as

$$\tilde{\phi} = \phi - \theta \; , \quad \tan\tilde{\phi} = \frac{b}{s} \; , \quad \rho = \sqrt{b^2 + s^2} \; . \tag{23.12}$$

The dependence of the Hamiltonian on the direction θ of the trajectory can be removed by the following transformation:

$$\chi = e^{i(n_1 + \tilde{N}_3)\check{\tau}^3 \theta/2} \tilde{\chi} \; , \tag{23.13}$$

$$\tilde{\mathcal{H}}_{\mathbf{p},b} = e^{-i(n_1 + \tilde{N}_3)\check{\tau}^3 \theta/2} \mathcal{H} e^{i(n_1 + \tilde{N}_3)\check{\tau}^3 \theta/2}$$
$$= -iv_F \check{\tau}^3 \partial_s + \Delta_0(\sqrt{s^2 + b^2}) \left(\check{\tau}^1 \cos n_1\tilde{\phi} - \check{\tau}^2 \sin n_1\tilde{\phi} \right) . \tag{23.14}$$

The Hamiltonian in eqn (23.14) does not depend on the angle θ and on the topological charge $\tilde{N}_3$ and thus is the same for s-wave, p-wave and other pairing states. The dependence on $\tilde{N}_3$ enters only through the boundary condition for the wave function, which according to eqn (23.13) is

$$\tilde{\chi}(\theta + 2\pi) = (-1)^{n_1+\tilde{N}_3}\tilde{\chi}(\theta) \ . \tag{23.15}$$

With respect to this boundary condition, there are two classes of systems: with odd and even $n_1 + \tilde{N}_3$. The parity $W = (-1)^{n_1+\tilde{N}_3}$ in eqn (23.6) is thus instrumental for the fermionic spectrum in the vortex core.

23.2.2 Quasiclassical low-energy states on anomalous branch

In the quasiclassical approximation the quasiparticle state with the lowest energy corresponds to a trajectory which crosses the center of the vortex, i.e. with the impact parameter $b = 0$. Along this trajectory one has $\sin\tilde{\phi} = 0$ and $\cos\tilde{\phi} = \text{sign } s$. So eqn (23.14) becomes

$$\tilde{\mathcal{H}}_{b=0,\mathbf{p}} = -iv_F\tilde{\tau}^3\partial_s + \tilde{\tau}^1\Delta_0(|s|)\text{sign } s \ . \tag{23.16}$$

This Hamiltonian is supersymmetric since (i) there is an operator anticommuting with $\mathcal{H}$, i.e. $\mathcal{H}\tau_2 = -\tau_2\mathcal{H}$; and (ii) the potential $U(s) = \Delta_0(|s|)\text{sign } s$ has opposite signs at $s \to \pm\infty$. We have already discussed such a supersymmetric Hamiltonian for fermion zero modes in eqn (22.18). Supersymmetry dictates that the Hamiltonian contains an eigenstate with exactly zero energy. Let us write the corresponding eigenfunction including all the transformations made before:

$$\chi_{\theta,b=0}(s) = e^{ip_F s}e^{i(n_1+\tilde{N}_3)\tilde{\tau}^3\theta/2}\begin{pmatrix}1\\-i\end{pmatrix}\chi_0(s) \ , \tag{23.17}$$

$$\chi_0(s) = \exp\left(-\int^s ds'\frac{\Delta_0(|s'|)}{v_F}\text{sign } s'\right) \ . \tag{23.18}$$

Now we turn to the non-zero but small impact parameter $b \ll \xi$, when the third term in eqn (23.14) can be considered as a perturbation. Its average over the zero-order wave function in eqn (23.17) gives the energy levels in terms of b and thus in terms of the continuous angular momentum $Q \approx p_F b$:

$$E(Q,\theta) = -n_1 Q\omega_0 \ , \ \omega_0 = \frac{\int_0^\infty d\rho\frac{\Delta_0(\rho)}{p_F\rho}\exp\left(-\frac{2}{v_F}\int_0^\rho d\rho'\Delta_0(\rho')\right)}{\int_0^\infty d\rho\exp\left(-\frac{2}{v_F}\int_0^\rho d\rho'\Delta_0(\rho')\right)} \ . \tag{23.19}$$

Thus we obtain the parameter ω_0 in eqn (23.1) – the minigap – which originally was calculated using a microscopic theory by Caroli *et al.* (1964). From eqn (23.19) it follows that the minigap is of order $\omega_0 \sim \Delta_0/(p_F R)$ where R is the core radius. For singular vortices R is on the order of the coherence length $\xi = v_F/\Delta_0$ and the minigap is $\omega_0 \sim \Delta_0^2/(p_F v_F) \ll \Delta_0$. In a large temperature region $\Delta_0^2/(p_F v_F) \ll T \ll \Delta_0$ these bound fermionic states can be considered as fermion zero modes whose energy as a function of the continuous parameter Q crosses zero energy.

Below we shall extend this derivation to discrete values of Q, and also to the case of non-axisymmetric vortices, where the quasiclassical energy depends also on the direction of the trajectory (on the angle θ in Fig. 23.3 *left*). In the latter case the microscopic theory becomes extremely complicated, but the effective theory of the low-energy fermion zero modes works well. This effective theory of fermionic quasiparticles living in the vortex core is determined by the universality class again originating from the topology.

23.2.3 Quantum low-energy states and W-parity

In exact quantum mechanical problems the generalized angular momentum Q has discrete eigenvalues. Following Stone (1996) and Kopnin and Volovik (1997), to find the quantized energy levels we must take into account that the two degrees of freedom describing the low-energy fermions, the angle θ of the trajectory and the generalized angular momentum Q, are canonically conjugate variables. That is why the next step is the quantization of motion in the (θ, Q) plane. It can be obtained using the quasiclassical energy $E(\theta, Q)$ in eqn (23.19) as an effective Hamiltonian for fermion zero modes, with Q being an operator $Q = -i\partial_\theta$. For the axisymmetric vortex, this Hamiltonian does not depend on θ

$$\mathcal{H} = in_1\omega_0\partial_\theta \;, \qquad (23.20)$$

and has the eigenfunctions $e^{-iE\theta/n_1\omega_0}$ (we consider vortices with $n_1 = \pm 1$). The boundary condition for these functions, eqn (23.15), gives two different spectra of quantized energy levels, which depend on the W-parity in eqn (23.6):

$$E(Q) = -n_1 Q\omega_0 \;, \qquad (23.21)$$

$$Q = n \;, \quad W = +1 \;; \qquad (23.22)$$

$$Q = \left(n + \frac{1}{2}\right) \;, \quad W = -1 \;, \qquad (23.23)$$

where n is integer.

The phase $(n_1 + \tilde{N}_3)\tau_3\theta/2$ in eqn (23.13) plays the part of a Berry phase (see also March-Russel *et al.* 1992). It shows how the wave function of a quasiparticle changes when the trajectory in Fig. 23.3 is adiabatically rotated by angle θ. This Berry phase is instrumental for the Bohr–Sommerfeld quantization of fermion zero modes in the vortex core. It chooses between the only two possible quantizations consistent with the 'CPT-symmetry' of the energy levels: $E = n\omega_0$ and $E = (n + 1/2)\omega_0$. In both cases for each positive energy level E one can find another level with the energy $-E$. That is why the above quantization is applicable even to non-axisymmetric vortices, though the quantum number n is no longer the angular momentum, since the angular momentum of quasiparticles is not conserved in the non-axisymmetric environment.

23.2.4 Fermions on asymmetric vortices

Most vortices in condensed matter are not axisymmetric. In superconductors the rotational symmetry is violated by the crystal lattice. In superfluid ^{3}He-B the

$SO(2)$ rotational symmetry of the core is spontaneously broken in one of the two $n_1 = 1$ vortices (Sec. 14.2) as was demonstrated by Kondo et al. (1991). In superfluid ^{3}He-A the axisymmetry of vortices is violated due to A-phase anisotropy: far from the core there is a preferred orientation of the $\hat{\mathbf{l}}$-vector in the cross-sectional plane (Fig. 16.6).

Equation (23.7) describing the low-energy fermionic spectrum in the axisymmetric vortex in the limit of continuous Q can be generalized to the non-axisymmetric case (Kopnin and Volovik 1997, 1998a):

$$E_a(Q,\theta) = \omega_a(\theta)(Q - Q_a(\theta)) . \qquad (23.24)$$

For given θ the spectrum crosses zero energy at some $Q = Q_a(\theta)$. This form is provided by the topology of the vortex, which dictates the number of fermion zero modes. The energy levels can be obtained using the Bohr–Sommerfeld quantization of the adiabatic invariant $I = \oint Q(\theta, E) \, d\theta = 2\pi\hbar(n + \gamma)$, where $Q(\theta, E)$ is the solution of equation $\omega_a(\theta)(Q - Q_a(\theta)) = E$ (Kopnin and Volovik 1997, 1998a). The parameter $\gamma \sim 1$ (analog of the Maslov index) is not determined in this quasiclassical scheme, so we proceed further introducing the effective Hamiltonian for fermion zero modes.

The 'CPT'-symmetry of the Bogoliubov–Nambu Hamiltonian requires that if $E_a(Q,\theta)$ is the energy of the bound state fermion, then $-E_a(-Q,\theta + \pi)$ also corresponds to the energy of a quasiparticle in the core. Let us consider one pair of the conjugated branches related by the 'CPT'-symmetry. One can introduce the common effective gauge field $A_\theta(\theta, t)$ and the 'electric' charges $e_a = \pm 1$, so that $Q_a = e_a A_\theta$. Then the effective Hamiltonian for fermion zero modes becomes

$$\mathcal{H}_a = -\frac{n_1}{2} \left\{ \omega_a(\theta) \, , \, \left(-i\frac{\partial}{\partial \theta} - e_a A_\theta(\theta, t) \right) \right\} , \qquad (23.25)$$

where $\{\,,\,\}$ is the anticommutator. The Schrödinger equation for the fermions on the a-th branch is

$$\frac{i}{2}(\partial_\theta \omega_a)\Psi_a(\theta) + \omega_a(\theta)(i\partial_\theta + e_a A(\theta))\Psi_a(\theta) = n_1 E \Psi_a(\theta) . \qquad (23.26)$$

The normalized eigenfunctions are

$$\Psi_a(\theta) = \left\langle \frac{1}{\omega_a(\theta)} \right\rangle^{-1/2} \frac{1}{\sqrt{2\pi\omega_a(\theta)}} \exp\left(-i \int^\theta d\theta' \left(\frac{n_1 E}{\omega_a(\theta')} + e_a A(\theta') \right) \right) . \qquad (23.27)$$

Here the angular brackets mean the averaging over the angle θ.

For a single self-conjugated branch, according to the CPT-theorem one has $\int_0^{2\pi} d\theta A(\theta) = 0$. Then using the boundary conditions in eqn (23.15) one obtains the equidistant energy levels:

$$E_n = -n_1(n+\gamma)\omega_0 \, , \quad \omega_0 = \left\langle \frac{1}{\omega_a(\theta)} \right\rangle^{-1} \, , \quad \gamma = \frac{1-W}{4} , \qquad (23.28)$$

where n is integer, and W is the parity introduced in eqn (23.6). Thus in spite of the fact that the angular momentum Q is not a good quantum number, the spectrum of of fermion zero modes on non-axisymmetric vortices is the same as in eqns (23.21–23.23) for axisymmetric vortices, i.e. the spectrum is not disturbed by non-axisymmetric perturbations of the vortex core structure. The properties of the spectrum (chirality, equidistant levels and W-parity) are dictated by real and momentum space topology and are robust to perturbations. The non-axisymmetric perturbations lead only to renormalization of the minigap ω_0.

23.2.5 Majorana fermion with $E = 0$ on half-quantum vortex

In the 2D case, point vortices with parity $W = 1$ have fermion zero mode with exactly zero energy, $E_{n=0} = 0$. This mode is doubly degenerates due to spin, and this degeneracy can be lifted off by the spin–orbit coupling. The spin–orbit interaction splits the zero level in a symmetric way, i.e. without violation of the CPT-symmetry. This, however, does not happen for the half-quantum vortex, whose $E = 0$ state is not degenerate. Because of the CPT-symmetry, this level cannot be moved from its $E = 0$ position by any perturbation. Thus this exact fermion zero mode is robust to any perturbation which does not destroy the half-quantum vortex or the superconductivity. If there is another half-quantum vortex in the system, there are two zero levels, one per vortex. In this case the interaction between vortices and tunneling of quasiparticles between the two zero levels will split these levels.

Let us consider such a point-like Alice string in 2D ^{3}He-A. The order parameter outside the core is given by eqn (15.18):

$$e_{\mu i} = \Delta_0 \hat{d}_\mu (\hat{m}_i + i\hat{n}_i) = \Delta_0 \left(\hat{x}_\mu \cos \frac{\phi}{2} + \hat{y}_\mu \sin \frac{\phi}{2} \right) (\hat{x}_i + i\hat{y}_i) \, e^{i\phi/2}$$

$$= \frac{1}{2} (\hat{x}_i + i\hat{y}_i) \left((\hat{x}_\mu - i\hat{y}_\mu)e^{i\phi} + (\hat{x}_\mu + i\hat{y}_\mu) \right) \, . \quad (23.29)$$

For fermions with spin $s_z = -1/2$ the winding number of the order parameter is $n_1 = 1$, while for fermions with $s_z = +1/2$ the vacuum is vortex-free: its winding number is trivial, $n_1 = 0$. In this simplest realization, the Alice string is the $n_1 = 1$ vortex for a single spin population, and thus there is only one energy level with $E = 0$ in the Alice string.

There are many interesting properties related to this $E = 0$ level. Since the $E = 0$ level can be either filled or empty, there is a fractional entropy $(1/2)\ln 2$ per layer per vortex. The factor $(1/2)$ appears because in the pair-correlated superfluids/superconductors one must take into account that in the Bogoliubov–Nambu scheme we artificially doubled the number of fermions introducing both particles and holes. The quasiparticle excitation living at the $E = 0$ level coincides with its anti-quasiparticle, i.e. such a quasiparticle is a Majorana fermion (Read and Green 2000). Majorana fermions at the $E = 0$ level lead to non-Abelian quantum statistics of half-quantum vortices, as was found by Ivanov (2001, 2002): the interchange of two vortices becomes an identical operation (up to an overall phase) only on being repeated four times (see also Lo and Preskill

1993). It was suggested by Bravyi and Kitaev (2000) that this property of quasiparticles could be used for quantum computing.

Also the spin of the vortex in a chiral superconductor can be fractional, as well as the electric charge per layer per vortex (Goryo and Ishikawa 1999), but this is still not conclusive. The problem with fractional charge, spin and statistics related to topological defects in chiral superconductors is still open.

23.3 Interplay of p- and r-topologies in vortex core

23.3.1 *Fermions on a vortex line in 3D systems*

The above consideration can now be extended to 3D systems, where the energy levels of the vortex-core fermions depend on quantum number p_z, the linear momentum along the vortex line. The generalization of the quasiclassical spectrum in eqn (23.19) to excitations living in the vortex line is straightforward. The magnitude of the momentum of a quasiparticle along the trajectory is close to

$$\mathbf{p}_\perp = \sqrt{p_F^2 - p_z^2}(\hat{\mathbf{x}}\cos\theta + \hat{\mathbf{y}}\sin\theta) , \qquad (23.30)$$

and its the velocity is close to $|\mathbf{p}_\perp|/m^* = \sqrt{p_F^2 - p_z^2}/m^*$. One must substitute this velocity into eqn (23.19) instead of v_F, and also take into account that the modified angular momentum is determined by the transverse linear momentum: $Q \approx |\mathbf{p}_\perp|b$. Then for vortices in s-wave superconductors one obtains the following dependence of the effective angular velocity (minigap) on p_z:

$$\omega_0(p_z) = \frac{\int_0^\infty d\rho \frac{\Delta_0(\rho)}{|\mathbf{p}_\perp|\rho} \exp\left(-2\frac{m^*}{|\mathbf{p}_\perp|}\int_0^\rho d\rho' \Delta_0(\rho')\right)}{\int_0^\infty d\rho \exp\left(-2\frac{m^*}{|\mathbf{p}_\perp|}\int_0^\rho d\rho' \Delta_0(\rho')\right)} . \qquad (23.31)$$

The exact quantum mechanical spectrum of bound states in the n_1-vortex in s-wave superconductors is shown in Fig. 23.1 *top right*. For these vortices the W-parity in eqn (23.6) is $W = -1$, and the spectrum has no levels with exactly zero energy.

Figure 23.1 *bottom* shows the fermionic spectrum in $n_1 = 1$ vortices of the class characterized by the parity $W = 1$; this spectrum has levels with exactly zero energy. In ^{3}He-B, the $\sigma^i p_i$ interaction of spin degrees of freedom with momentum in eqn (7.48) leads to splitting of the doubly degenerate $E = 0$ levels (Misirpashaev and Volovik 1995). The splitting is linear in p_z (Fig. 23.1 *bottom left*) and thus one obtains the same topology of the energy spectrum of fermion zero modes as in cosmic strings in Fig. 23.1 *top left*. The most interesting situation occurs for the maximum symmetric vortices in ^{3}He-A with $n_1 = \pm 1$. The spin degeneracy is not lifted off, if the discrete symmetry is not violated, and the doubly degenerate $E = 0$ level exists for all p_z (Fig. 23.1 *bottom right*). This anomalous, highly degenerate branch of fermion zero modes of co-dimension 0 was first found in a microscopic theory by Kopnin and Salomaa (1991).

Fermion zero modes of co-dimension 1 in Fig. 23.1 originate from the nontrivial topology of a manifold of zeros of co-dimension 3 in the classical spectrum of quasiparticles in the core of a vortex. The same topology characterizes the homogeneous vacua in ^{3}He-A and Standard Model. This demonstrates again the emergency of Fermi points of co-dimension 3 characterized by the invariant N_3. Let us consider this topological equivalence in more detail.

23.3.2 Topological equivalence of vacua with Fermi points and with vortex

Let us consider a straight vortex along the z axis in such superfluids, where the states far from the vortex core are fully gapped. We start with the $n_1 = 1$ vortex in s-wave superconductors and the most symmetric $n_1 = 1$ vortex in ^{3}He-B. As follows from the Hamiltonian in eqn (23.4), the classical energy spectrum $E(\mathbf{r}, \mathbf{p})$ is zero when simultaneously $M(p) = 0$ and $\Delta_0(\rho) = 0$, which occurs when $p_x^2 + p_y^2 + p_z^2 = p_F^2$ and $\rho = \sqrt{x^2 + y^2} = 0$ (Fig. 23.4 *top*). The latter equations determine 3D manifold of zeros in 6D combined space (x, y, z, p_x, p_y, p_z): 1D line (the z axis) times 2D spherical surface $p_x^2 + p_y^2 + p_z^2 = p_F^2$. Thus the manifold of fermion zero modes in the classical spectrum has co-dimension 3. From the result of Sec. 23.1.4 it follows that this manifold is topologically non-trivial: each point on this hypersurface is described by the topological invariant $N_3 = \pm 1$ for each spin projection.

We can now compare this manifold with the manifold of zeros in the homogeneous vacuum of ^{3}He-A, where the quasiparticle energy spectrum in eqn (8.9) does not depend on $\mathbf{r}$ but becomes zero at two Fermi points in momentum space. In the combined 6D space (x, y, z, p_x, p_y, p_z) the fermion zero modes also form the 3D manifold: two Fermi points ($p_x = p_y = 0$, $p_z = +p_F$ and $p_x = p_y = 0$, $p_z = -p_F$) times the whole coordinate space. These fermion zero modes of co-dimension 3 are described by the π_3 topological invariant $N_3 = \mp 1$ for each spin projection.

Thus we obtain that the homogeneous vacuum of ^{3}He-A and the $n_1 = 1$ vortex in s-wave superconductors (or in ^{3}He-B) in Fig. 23.4 *top* contain manifolds of zeros of co-dimension 3. In the 6D space the orientation of these 3D manifolds is different, but they are described by the same topological invariant N_3. This implies that these two vacuum states (homogeneous and inhomogeneous) can be transformed into each other by continuous deformation of the 3D manifold (Volovik and Mineev 1982). Below we consider how this transformation occurs in the real ^{3}He-B vortex giving rise to a peculiar smooth core of this vortex with Fermi points in Fig. 23.4 *bottom*.

23.3.3 Smooth core of ^{3}He-B vortex

Fig. 23.4 shows two structures of the quasiclassical energy spectrum in the core of an axisymmetric vortex with winding number $n_1 = 1$ in ^{3}He-B. On the top it is the vortex with the maximum possible symmetry in eqn (14.20) (see also Fig. 14.5 *top left*). Its structure is similar to that of a vortex in s-wave superconductors. The amplitude of the B-phase order parameter $\Delta_B(\rho)$ becomes zero on the vortex axis, which means that superfluidity is completely destroyed and thus the $U(1)_N$ symmetry is restored. On the vortex axis one has therefore the normal Fermi liquid with a conventional Fermi surface.

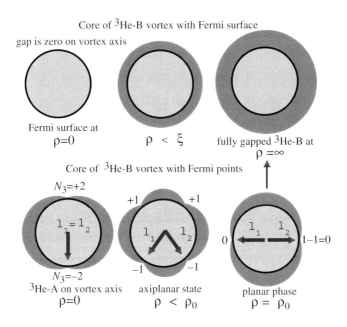

FIG. 23.4. *Top*: Singular core of a conventional vortex with $n_1 = 1$ in in s-wave superconductors and of the most symmetric vortex in ^{3}He-B. The order parameter is zero at $x = y = 0$, i.e. there is a normal state with Fermi surface on the vortex axis. The classical energy of quasiparticle $E(\mathbf{r}, \mathbf{p})$ is zero on a 3D manifold $p = p_F$, $x = y = 0$ in the combined 6D space (x, y, z, p_x, p_y, p_z) with topological invariant $N_3 = 2$. *Bottom*: The singularity on the vortex axis can be dissolved to form a smooth core. This happens in particular for the ^{3}He-B vortex with $n_1 = 1$ (Fig. 14.5 *top right*), where the state very close to ^{3}He-A appears on the vortex axis instead of the normal state, and vorticity becomes continuous in the core. Four Fermi points, each with $N_3 = \pm 1$, appear in the core region, as in the axiplanar state in Fig. 7.3. The directions to the nodes are marked by the unit vectors $\hat{\mathbf{l}}_1$ and $\hat{\mathbf{l}}_2$. At $\rho = \rho_0 \sim \xi$ the Fermi points with opposite N_3 annihilate each other, and at $\rho > \rho_0$ the system is fully gapped. The manifold of zeros is again the 3D manifold in the combined 6D space, which, however, is embedded in a different way. Within the smooth core the Fermi points with $N_3 = +1$ sweep a 2π solid angle each, while the Fermi points with $N_3 = -1$ sweep -2π each. This satisfies eqn (23.32) which connects the **p**-space and **r**-space topologies of a vortex.

However, it appears that such embedding of the 3D manifold of zeros in 6D space with a naked Fermi surface on the vortex axis is energetically unstable. The system prefers to protect the superfluidity everywhere including on the vortex axis, and in our case it is possible to reorient the 3D manifold of zeros in 6D space in such a way that there is no longer any naked Fermi surface, i.e. the $U(1)_N$ symmetry is broken everywhere in **r**-space. The singularity on the vortex axis with the naked Fermi surface becomes smoothly distributed throughout a finite region $\rho < \rho_0$ inside the vortex core in terms of Fermi points (Fig. 23.4 *bottom*).

On the axis (Fig. 23.4 *bottom left*), instead of the normal Fermi liquid with the Fermi surface, there are two Fermi points with $N_3 = \pm 2$ reflecting the existence of the A-phase in the vortex core (see Fig. 14.5 *top right*). Away from the vortex axis the Fermi points with $N_3 = \pm 2$ split into four Fermi ponts with unit topological charges, each with $N_3 = \pm 1$ (see Fig. 23.4 *bottom middle*, where the directions to the Fermi points are marked by the unit vectors $\hat{\mathbf{l}}_1$ and $\hat{\mathbf{l}}_2$). Locally the superfluid with four nodes is similar to the axiplanar phase in Fig. 7.3. Further from the axis the Fermi points with opposite charges attract each other and finally annihilate at some distance $\rho = \rho_0$ from the vortex axis (Fig. 23.4 *bottom right*). Since the point nodes exist at any point of **r**-space within the radius $\rho < \rho_0$, the manifold of zeros is again a 3D manifold in the combined 6D space, but it is embedded in a different way.

This again demonstrates the generic character of fermion zero modes of co-dimension 3. Moreover, in a given case the Fermi surface on the vortex axis appears to be energetically unstable toward the formation of the smooth distribution of Fermi points.

23.3.4 r-space topology of Fermi points in the vortex core

The interplay of real space and momentum space topologies also dictates the behavior of four Fermi points $\mathbf{p}^{(a)} = \pm p_F \hat{\mathbf{l}}_1, \pm p_F \hat{\mathbf{l}}_2$ as functions of the space coordinates (x, y). In the particular scenario of Fig. 23.4 *bottom*, the Fermi points with $N_3 = +1$ sweep 2π solid angle each, while the Fermi points with $N_3 = -1$ sweep -2π each. This suggests the general rule: if the a-th Fermi point with the momentum space topological charge N_{3a} sweeps the $4\pi \nu_a$ solid angle in the smooth or soft core of the vortex with winding number n_1, then one has the fundamental relation between three types of topological charges, n_1, N_{3a} and ν_a (Volovik and Mineev 1982):

$$n_1 = \frac{1}{2} \sum_a \nu_a N_{3a} \ . \tag{23.32}$$

Let us recall that n_1 is the π_1 topological charge of the **r**-space defect – the vortex – while N_{3a} is the π_3 topological charge of the **p**-space defect – the Fermi point. These **r**-space and **p**-space charges are connected via ν_a which is the π_2 topological charge characterizing the spatial dependence $\mathbf{p}^{(a)}(\mathbf{r})$ of the a-th Fermi point in the vortex core

$$\nu_a = \frac{1}{4\pi} \int dx \, dy \, \hat{\mathbf{p}}^{(a)} \cdot \left(\partial_x \hat{\mathbf{p}}^{(a)} \times \partial_y \hat{\mathbf{p}}^{(a)} \right) \ . \tag{23.33}$$

Because of the continuous distribution of Fermi points, the vorticity $\nabla \times \mathbf{v}_s$ within the smooth core is continuous, as in the case of vortex-skyrmions in ^{3}He-A.

The Mermin–Ho equation (9.17) relating continuous vorticity and the textures of Fermi points has the following general form:

$$\nabla \times \mathbf{v}_s = \frac{\hbar}{16m} e_{ijk} \sum_a N_{3a} \hat{p}_i^{(a)} \cdot \left(\nabla \hat{p}_j^{(a)} \times \nabla \hat{p}_k^{(a)} \right) \quad (23.34)$$

$$= \frac{\hbar}{8m} e_{ijk} \left(\hat{l}_{1i} \nabla \hat{l}_{1j} \times \nabla \hat{l}_{1k} + \hat{l}_{2i} \nabla \hat{l}_{2j} \times \nabla \hat{l}_{2k} \right) . \quad (23.35)$$

Equation (23.35) is applicable to the particular case when two $\hat{\mathbf{l}}$-vectors are involved, $\hat{\mathbf{l}}_1$ and $\hat{\mathbf{l}}_2$. In a pure ³He-A one has $\hat{\mathbf{l}}_1 = \hat{\mathbf{l}}_2 \equiv \hat{\mathbf{l}}$, and the original Mermin–Ho relation (9.17) is restored.

24
VORTEX MASS

24.1 Inertia of object moving in superfluid vacuum

24.1.1 Relativistic and non-relativistic mass

The mass (inertia) of an object is determined as the response of the momentum of the object to its velocity:

$$p_i = M_{ik} v^k . \tag{24.1}$$

If we are interested in the linear response, then the mass tensor is obtained from eqn (24.1) in the limit $\mathbf{v} \to 0$. Let us first consider a relativistic particle with the spectrum $E^2 = M^2 + g^{ik} p_i p_k$, where M is the rest energy. Since the velocity of the particle is $v^i = dE/dp_i$, its mass tensor is $M_{ik} = E g_{ik}$. In linear response theory one obtains the following relation between the mass tensor and the rest energy:

$$M_{ik}(\text{linear}) = M g_{ik} . \tag{24.2}$$

The same can be applied to the motion of an object in superfluids in the low-energy limit when it can be described by the 'relativistic' dynamics with an effective Lorentzian metric g_{ik} (acoustic metric in the case of ^{4}He and quasiparticle metric in superfluid ^{3}He-A).

Note that eqns $E^2 = M^2 + g^{ik} p_i p_k$ and (24.2) do not contain the speed of light c explicitly: the traditional Einstein relation $M = mc^2$ between the rest energy and the mass of the object (the mass–energy relation) is meaningless for the relativistic analog in anisotropic superfluids such as ^{3}He-A. What 'speed of light' enters the Einstein relation if this speed depends on the direction of propagation and is determined by two phenomenological parameters $c_\parallel$ and $c_\perp$? Of course, as we discussed above, for an observer living in the liquid the speed of light does not depend on the direction of propagation. Such a low-energy observer can safely divide the rest energy by his (or her) c^2, and obtain what he (or she) thinks is the mass of the object. But this mass has no physical meaning for the well-informed external observer living in either the trans-Planckian world or the Galilean world of the laboratory.

Here we discuss the inertia of an object moving in the quantum vacuum of the Galilean quantum liquid. If it is a foreign object like an atom of ^{3}He moving in the quantum vacuum of liquid ^{4}He, then in addition to its bare mass, the object acquires an extra mass since it involves some part of the superfluid vacuum into motion. If the object is an excitation of the vacuum, all its mass is provided by the liquid. A vortex moving with respect to the superfluid vacuum has no bare mass, since it is a topological excitation of the superfluid vacuum and thus does not exist outside of it. We find that there are several contributions to the effective mass of the vortex, which suggests different sources of inertia produced by distortions of the vacuum caused by the motion of the vortex. Till now, in considering the vortex dynamics we have ignored the vortex mass. The inertial term $M_{Lik}\partial_t v_L^i$ must be added to the balance of forces acting on the vortex in eqn (18.29). But this term contains the time derivative and thus at low frequencies of the vortex motion it can be neglected compared to the other forces, which depend on the vortex velocity $\mathbf{v}_L$. The vortex mass can show up at higher frequencies.

Here we estimate the vortex mass in superconductors and fermionic superfluids and relate it to peculiar phenomena in quantum field theory. But first we start with the 'relativistic' contribution, which dominates in superfluid ^{4}He.

24.1.2 *'Relativistic' mass of the vortex*

In superfluid ^{4}He the velocity field outside the core of a vortex and thus the main (logarithmic) part of the vortex energy are determined by the hydrodynamic equations. Because of the logarithm, whose infrared cut-off is supplied by the length of the vortex loop, the most relevant degrees of freedom have long wavelengths. Since the long-wavelength dynamics is governed by the acoustic metric (4.16) it is natural to expect that the hydrodynamic energy of a vortex (or soliton or other extended configuration of the vacuum fields) moving in a superfluid vacuum is connected with the hydrodynamic mass of the vortex by the 'relativistic' equation (24.2), where g_{ik} is the acoustic metric. If we are interested in the linear response, we must take the acoustic metric eqn (24.2) at zero superfluid velocity $\mathbf{v}_s = 0$ to obtain $g_{\mu\nu} = \mathrm{diag}(-1, c^{-2}, c^{-2}, c^{-2})$, where c is the speed of sound. Thus in accordance with eqn (24.2), the hydrodynamic mass of the vortex loop of length L at $T = 0$ must be

$$m_{L\ \mathrm{rel}} = \frac{E_{\mathrm{vortex}}}{c^2} = \frac{mn\kappa^2 L}{4\pi c^2} \ln \frac{L}{\xi} \ . \tag{24.3}$$

Here E_{vortex} is the 'rest energy' of the vortex – the energy of the vortex when it is at rest in the frame comoving with the superfluid vacuum (see e.g. eqn (14.7)); n is the particle density in the superfluid vacuum; κ is the circulation of superfluid velocity $\mathbf{v}_s$ around the vortex. The mass–energy relation for the vortex in eqn (24.3) is supported by a detailed analysis of the vortex motion in compressible superfluids by Duan and Leggett (1992), Duan (1994) and Wexler and Thouless (1996). According to eqn(24.3), the 'relativistic' mass of a vortex is $m_{L\ \mathrm{rel}} \sim mna_0^2 L \ln L/\xi$, where a_0 is the interatomic distance.

However, we shall see below that in Fermi superfluids this is not the whole story. The fermion zero modes attached to the vortex when it moves enhance the vortex mass. As a result, as was found by Kopnin (1978) the mass of a vortex in Fermi systems is on the order of the whole mass of the liquid concentrated in the vortex core

$$m_{\mathrm{LK}} \sim mn\xi^2 L \ . \tag{24.4}$$

Since the core size is on the order of the coherence length $\xi \gg a_0$, the Kopnin mass of a vortex exceeds its 'relativistic' mass by several orders of magnitude.

24.2 Fermion zero modes and vortex mass

According to the microscopic theory the Kopnin mass comes from the fermions trapped in the vortex core (Kopnin 1978, Kopnin and Salomaa 1991; van Otterlo et al. 1995; Kopnin and Vinokur 1998). This mass can also be derived using the effective theory for these fermion zero modes (Volovik 1997b). This effective theory is completely determined by the generic energy spectrum of fermion zero modes of co-dimension 1, eqn (23.7) for the axisymmetric vortex and eqn (23.24) if the axial symmetry is violated. The result does not depend on the microscopic details of the system.

24.2.1 *Effective theory of Kopnin mass*

If the axisymmetric vortex moves with velocity $\mathbf{v}_L$ with respect to the superfluid component, the fermionic energy spectrum in the stationary frame of the vortex texture is Doppler shifted: $\tilde{E} = E(Q, p_z) + \mathbf{p} \cdot (\mathbf{v}_s - \mathbf{v}_L)$. Due to this shift the vacuum – the continuum of the negative energy states – carries the fermionic charge, the momentum in eqn (10.35). The linear response of the momentum carried by the negative energy states of fermions to $\mathbf{v}_L - \mathbf{v}_s$ gives the Kopnin mass of the vortex:

$$\mathbf{P} = \sum_{Q,p_z} \mathbf{p}\Theta(-\tilde{E}) - \sum_{Q,p_z} \mathbf{p}\Theta(-E)$$

$$= \sum_{Q,p_z} \mathbf{p}(\mathbf{p} \cdot (\mathbf{v}_L - \mathbf{v}_s))\delta(E(Q,p_z)) = m_K(\mathbf{v}_L - \mathbf{v}_s) , \quad (24.5)$$

$$m_K = \frac{1}{2} \sum_{Q,p_z} \mathbf{p}_\perp^2 \delta(E) . \quad (24.6)$$

Using $\sum_{Q,p_z} = \int dQ dp_z dz / 2\pi\hbar$; $\mathbf{p}_\perp^2 = p_F^2 - p_z^2$; and the energy spectrum of fermion zero modes $E(Q, p_z) = \mp Q\omega_0(p_z)$ one obtains

$$m_K = L \int_{-p_F}^{p_F} \frac{dp_z}{4\pi\hbar} \frac{p_F^2 - p_z^2}{\omega_0(p_z)} . \quad (24.7)$$

For singular vortices, where $\omega_0 \sim \Delta_0^2/v_F p_F$, the estimation of the magnitude of the Kopnin mass gives eqn (24.4). This vortex mass is determined in essentially the same way as the normal density in the bulk system: $P_i^{\text{quasiparticles}} = mn_{nik}(v_n^k - v_s^k)$ in eqn (5.20). Here the role of the normal component is played by the fermion zero modes bound to the vortex, while the role of the normal component velocity $\mathbf{v}_n$ is played by the vortex velocity $\mathbf{v}_L$. We know that the normal component density can be non-zero even at $T = 0$. This occurs if the density of states (DOS) at zero energy, $N(0) = \sum \delta(E)$, is non-zero, which is typical for systems with a Fermi surface, i.e. for vacua with the fermion zero modes of co-dimension 1. An example of a finite DOS due to appearance of the Fermi surface has been discussed in Sec. 10.3.6. For fermion pseudo-zero modes of co-dimension 1 in the vortex core, the DOS is inversely proportional to the interlevel spacing, $N(0) \propto 1/\omega_0$, which is reflected in eqn (24.7) for the Kopnin mass.

Equation (24.7) is valid in the so-called clean limit case, when one can neglect the interaction of the fermion zero modes with impuritiesin superconductors or with the normal component in the bulk liquid outside the core. This Kopnin mass was obtained in the limit of low T. On the other hand, eqn (24.7) is valid only in the limit of continuous Q, i.e. the temperature must still be larger than the interlevel spacing. In the opposite limit $T \ll \omega_0$ the Kopnin mass disappears.

24.2.2 *Kopnin mass of smooth vortex: chiral fermions in magnetic field*

To illustrate the general character of eqn (24.7) let us calculate the Kopnin mass of a smooth vortex core, where one can use the classical energy spectrum of fermion zero modes. A smooth core is the best configuration to understand the origin of many effects related to the chiral fermions in the vortex core. Also the smooth core can be realized in many different situations. We have already seen in the example of ^{3}He-B vortices that the $1/r$-singularity of the superfluid velocity and the naked Fermi surface on a vortex axis can be removed by introducing Fermi points in the core region (Fig. 23.4 *bottom*). As a result the superfluid/superconducting state in the vortex core of any system can acquire the properties of the A-phase of superfluid ^{3}He, i.e. Fermi points of co-dimension 3 and continuous vorticity. As an example we can consider the continuous vortex-skyrmion in ^{3}He-A, but the result can be applicable to any vortex with a smooth core.

For a smooth or continuous vortex the non-zero DOS comes from the vicinity of the Fermi points, where the fermions are chiral and the $\hat{\mathbf{l}}$-textures play the role of an effective magnetic field. We know that in the presence of a magnetic field the chiral relativistic fermions have finite DOS at zeroth Landau level, eqn (18.2). To apply this equations to quasiparticles living in the vicinity of the Fermi points one must make the covariant generalization of the DOS by introducing the general metric tensor and then replace it by the effective ^{3}He-A metrics. The general form of the local DOS of the Weyl fermions in a magnetic field is

$$N(E=0,\mathbf{r}) = \frac{\sqrt{-g}}{2\pi^2 \hbar^2}\left(\frac{1}{2}g^{ij}g^{kl}F_{ik}F_{jl}\right)^{1/2}. \qquad (24.8)$$

In systems with Fermi points the effective metric tensor and effective gauge field are given by eqns (9.11–9.15). For the smooth core with radius $\rho_0 \gg \xi$ the contribution of the velocity field $\mathbf{v}_s$ can be neglected and one obtains the following local DOS at $E = 0$:

$$N(0,\mathbf{r}) = \frac{p_F}{2\pi^2 \hbar^2 c_\perp}|\hat{\mathbf{l}} \times (\nabla \times \hat{\mathbf{l}})|. \qquad (24.9)$$

This DOS can be inserted in eqn (10.36) to obtain the local density of the normal component which comes from the fermions trapped by the vortex at $T=0$:
$$mn_{nij}(\mathbf{r}, T \to 0) = \hat{l}_i \hat{l}_j p_F^2 N(0, \mathbf{r}) \ . \qquad (24.10)$$
For the axisymmetric vortex-skyrmion in eqn (16.5) one has
$$N(0, \rho) = \frac{p_F}{2\pi^2 \hbar^2 c_\perp} \sin\eta \, |\partial_\rho \eta|. \qquad (24.11)$$
The integral of the normal density tensor over the cross-section of the smooth core gives the mass trapped by the vortex
$$m_{\mathrm{K}} = \frac{1}{2}\int d^3x \, \hat{\mathbf{l}}_\perp^2 p_F^2 N(0,\rho) = L\frac{p_F^3}{2\pi\hbar^2 c_\perp}\int d\rho\, \rho\, \sin^3\eta(\rho)\, |\partial_\rho\eta| \ . \qquad (24.12)$$
This is the Kopnin mass expressed in terms of the distribution of the Fermi points – the $\hat{\mathbf{l}}$-field – in the coordinate space. It coincides with the momentum representation in eqn (24.7) after the interlevel distance $\omega_0(p_z)$ in the core of the vortex-skyrmion is expressed in terms of the $\hat{\mathbf{l}}$-texture (Volovik 1997b). The magnitude of the Kopnin mass of a smooth vortex in eqn (24.12) is $m_{\mathrm{K}} \sim mn\xi L\rho_0$, where ρ_0 is the size of the smooth core region (Kopnin 1995). For conventional singular vortices, ρ_0 must be substituted by the coherence length ξ and eqn (24.4) is restored.

In conclusion of this section, let us mention the relation to the RQFT. The local hydrodynamic energy of the normal component trapped by a smooth vortex is
$$F = \frac{m}{2} n_{nij}(\mathbf{r})(\mathbf{v}_{\mathrm{L}} - \mathbf{v}_{\mathrm{s}})_i (\mathbf{v}_{\mathrm{L}} - \mathbf{v}_{\mathrm{s}})_j \ . \qquad (24.13)$$
This can be rewritten in the following form, which is valid also for the chiral fermions in RQFT:
$$F = \frac{\mu_R^2 + \mu_L^2}{8\pi^2\hbar^2}\sqrt{-g}\sqrt{\frac{1}{2}g^{ij}g^{kl}F_{ik}F_{jl}} \ , \qquad (24.14)$$
where, as before in eqn (10.19), the chemical potentials of the left- and right-handed fermions in ^{3}He-A are expressed in terms of the counterflow: $\mu_R = -\mu_L = p_F(\hat{\mathbf{l}}\cdot\mathbf{w})$. Equation (24.14) represents the magnetic energy of the chiral particles with finite chemical potential in a strong magnetic field $B \gg \mu^2$ at $T \to 0$.

24.3 Associated hydrodynamic mass of a vortex

24.3.1 Associated mass of an object

There are other contributions to the vortex mass, which are related to the deformation of the vacuum fields due to the motion of the vortex. The most important of them is the associated (or induced) hydrodynamic mass, which is also proportional to the mass of the liquid in the volume of the core. An example of the associated mass is provided by an external body moving in an ideal liquid or in a superfluid – the mass of the liquid involved by the body in translational motion. This mass depends on the geometry of the body. For a moving cylindrical wire of radius R it is the mass of the liquid displaced by the wire:

$$m_{\text{L associated}} = m\pi R^2 L n \, . \tag{24.15}$$

This must be added to the bare mass of the cylinder to obtain the total inertial mass of the body. The associated mass arises because of the inhomogeneity of the density n of the liquid: $n(r > R) = n$ and $n(r < R) = 0$. When the wire moves with respect to the liquid, this produces the backflow around the wire and the liquid acquires a finite momentum proportional to the velocity.

In superfluids, this part of the superfluid vacuum which is involved in the motion together with the body can be considered as the normal density. But this normal density is produced not by the trapped quasiparticles, but by the inhomogeneity of the superflow around the object. Such an inhomogeneity is responsible for the normal density in porous materials and in aerogel, where some part of the superfluid is hydrodynamically trapped by the pores, and thus is removed from the overall superfluid motion. For vortices such an associated mass has been discussed by Baym and Chandler (1983) and Sonin et al. (1998). Here we estimate this contribution to the vortex mass for two situations.

First let us consider a vortex trapped by a cylindrical wire of radius $R \gg \xi$, such that the vortex core is represented by the wire. In this case eqn (24.15) gives the associated vortex mass which results from the backflow of the superfluid vacuum around the moving core. For such vortex a with the wire as the core this is the dominating mass of the vortex. The Kopnin mass, which can result from the normal excitations trapped near the surface of the wire, is considerably smaller, in particular because it represents a surface effect.

24.3.2 Associated mass of smooth-core vortex

As a second example, let us consider the associated mass of a vortex with a smooth core. This mass is also caused by the inhomogeneity of the liquid. In a given case it is the inhomogeneity of the superfluid density $n_s(\mathbf{r}) = n - n_n(\mathbf{r})$, where n_n is the non-zero local normal density in eqn (24.10) caused by fermion zero modes. Due to the profile of the local superfluid density the external flow is disturbed near the core according to the continuity equation. In the reference frame moving with the vortex, the texture is stationary and the trapped normal component is at rest: $\mathbf{v}_n = 0$. The continuity equation in this frame reads

$$\nabla \cdot (n_s \mathbf{v}_s) = 0 \, . \tag{24.16}$$

If the smooth core is large, $\rho_0 \gg \xi$, the DOS of fermion zero modes in eqn (24.9) is small, and thus the normal density produced by fermion zero modes in eqn (24.10) is small compared to the total density: $n_n \sim n\xi/\rho_0 \ll n$. As a result it can be considered as a perturbation. According to continuity equation, the disturbance $\delta \mathbf{v}_s = \nabla \alpha$ of the superflow (the backflow) caused by this perturbation is given by

$$n\nabla^2\alpha = -v_L^i \nabla_j n_{nij} , \qquad (24.17)$$

where we take into account that the asymptotic value of the superfluid velocity with respect to the vortex is with minus sign the velocity of the vortex, $\mathbf{v}_{s\infty} = -\mathbf{v}_L$.

In the simple approximation, when the normal component in eqn (24.10) trapped by the vortex is considered as isotropic, one obtains the following kinetic energy of the backflow which gives the associated vortex mass:

$$\frac{1}{2}mn \int d^3x\, (\nabla\alpha)^2 = \frac{1}{2}m_{L\ \text{associated}} v_L^2 , \quad m_{L\ \text{associated}} = mL \int d^2x \frac{n_n^2(r)}{2n} . \qquad (24.18)$$

Since $n_n \sim n\xi/\rho_0$ the associated mass is on the order of $mn\xi^2 L$ and does not depend on the core radius ρ_0: the large area ρ_0^2 of integration in eqn (24.18) is compensated by a small value of the normal component in the core with large ρ_0. That is why in the smooth core with $\rho_0 \gg \xi$ this mass is parametrically smaller than the Kopnin mass $\sim mn\xi\rho_0 L$ in eqn (24.12). But for conventional singular vortices $\rho_0 \sim \xi$ and these two contributions are of the same order of magnitude.

In conclusion, the behavior of the inertial mass of the vortex demonstrates that the effective relativistic mass–energy relation $m = E/c^2$, which is valid for vortices in Bose superfluids, is violated in Fermi superfluids due to the trans-Planckian physics of fermion zero modes and of the back reaction of the vacuum to the motion of the vortex.

25

SPECTRAL FLOW IN THE VORTEX CORE

25.1 Analog of Callan–Harvey mechanism of cancellation of anomalies

25.1.1 Analog of baryogenesis by cosmic strings

In Chapter 18 we discussed how massless chiral fermions – zeros of co-dimension 3 – influence the dynamics of a continuous vortex-skyrmion due to the effect of the chiral anomaly. Such a 'stringy texture' (as it is called in RQFT) serves as a mediator for the anomalous transfer of the fermionic charge – the linear momentum $\mathbf{P}$ – from the vacuum to the heat bath of fermions. This anomalous transfer is described by the Adler–Bell–Jackiw axial anomaly equation (18.5), where the effective $U(1)$ field acting on the Weyl fermions is produced by the $\hat{\mathbf{l}}$-fields. Now we shall discuss the same phenomenon of chiral anomaly when the mediator is a conventional singular vortex, while the corresponding massless fermions are fermion zero modes in the vortex core.

In ^{3}He-B and in conventional superconductors, fermions are massive outside the core, but they have a gapless (or almost gapless) spectrum of fermion (pseudo)zero modes of co-dimension 1 in eqn (23.1). These core fermions are chiral and they do actually the same job as chiral fermions in vortex-skyrmions. However, the spectral flow carried by the fermion zero modes in the singular core is not described by the Adler–Bell–Jackiw axial anomaly equation (18.5). Instead of the effective RQFT, we must use the effective theory of fermion zero modes of co-dimension 1 discussed in Chapter 23. However, we shall see that in the limit when the parameter Q of the fermionic spectrum can be considered as continuous, the result for the generation of momentum by a vortex with singular core exactly coincides with the axial anomaly result (18.17–18.18) for the generation of momentum by a moving vortex-skyrmion.

The process of momentum generation by vortex cores is similar to that of generation of baryonic charge by the cores of cosmic strings (Witten 1985; Vachaspati and Field 1994; Garriga and Vachaspati 1995; Barriola 1995; Starkman and Vachaspati 1996). Also the axial anomaly is instrumental for baryoproduction in the core of cosmic strings, and again the effect cannot be described by the the Adler–Bell–Jackiw equation, since the latter was derived using the energy spectrum of free fermions in the presence of homogeneous electric and magnetic fields. In cosmic strings these fields are no longer homogeneous, and the massless fermions exist only in the vortex core as fermion zero modes. Thus both the baryoproduction by cosmic strings and the momentogenesis by singular vortices must be studied using the effective theory of fermion zero modes of co-dimension

1. Let us see how the momentogenesis occurs in the core of a singular vortex.

25.1.2 Level flow in the core

Let us consider the spectral-flow force, which arises when a singular vortex moves with velocity $\mathbf{v}_L - \mathbf{v}_n$ with respect to the heat bath. In the heat-bath frame the order parameter (Higgs) field depends on the spacetime coordinates through the combination $\mathbf{r} - (\mathbf{v}_L - \mathbf{v}_n)t$ (compare with eqn (18.15) for a moving texture). We consider first the limit case when the fermionic quantum number Q can be treated as a continuous parameter and coincides with the angular momentum of a quasiparticle: $Q \approx L_z/\hbar$. Since $\mathbf{L} = \mathbf{r} \times \mathbf{p}$, in the heat-bath frame the parameter Q grows with time, and the energy of the fermion zero mode in the moving vortex with winding number n_1 becomes time dependent:

$$E(Q,\theta) = -n_1 \left(Q - \frac{1}{\hbar}\hat{\mathbf{z}} \cdot ((\mathbf{v}_L - \mathbf{v}_n) \times \mathbf{p}_\perp)t \right) \omega_0 . \tag{25.1}$$

The quantity

$$E_z = \frac{1}{\hbar}\hat{\mathbf{z}} \cdot ((\mathbf{v}_L - \mathbf{v}_n) \times \mathbf{p}_\perp) \tag{25.2}$$

acts on fermions localized in the core in the same way as an electric field E_z acts on chiral fermions on an anomalous branch in a magnetic field (Sec. 18.1.3) or on chiral fermion zero modes localized in a string in RQFT. The only difference is that under this 'electric' field the spectral flow in the vortex occurs along the Q axis ($\dot{Q} = E_z$) rather than along the direction p_z in strings where $\dot{p}_z = qE_z$. Since, according to the index theorem, for each quantum number Q there are $-2n_1$ quasiparticle levels, the fermionic levels cross zero energy at the rate

$$\dot{n} = -2n_1 \dot{Q} = -2n_1 E_z(\mathbf{p}_\perp) = -\frac{2}{\hbar} n_1 \hat{\mathbf{z}} \cdot ((\mathbf{v}_L - \mathbf{v}_n) \times \mathbf{p}_\perp) . \tag{25.3}$$

25.1.3 Momentum transfer by level flow

When the occupied level crosses zero, a quasiparticle on this level transfers its fermionic charges from the vacuum (from the negative energy states) along the anomalous branch into the heat bath ('matter'). For us the important fermionic charge is linear momentum. The rate at which the momentum $\mathbf{p}_\perp$ is transferred from the vortex to the heat bath due to spectral flow is obtained by integration over the remaining variable p_z (compare with eqn (24.5)):

$$\dot{\mathbf{P}} = \frac{1}{2}\sum_{p_z} \mathbf{p}_\perp \dot{n} = -\frac{n_1}{2\hbar}\hat{\mathbf{z}} \times (\mathbf{v}_L - \mathbf{v}_n) \sum_{p_z} \mathbf{p}_\perp^2 \tag{25.4}$$

$$= -\frac{n_1}{2\hbar}\hat{\mathbf{z}} \times (\mathbf{v}_L - \mathbf{v}_n) L \int_{-p_F}^{p_F} \frac{dp_z}{2\pi\hbar}(p_F^2 - p_z^2)$$

$$= -n_1 \frac{p_F^3}{3\pi\hbar^2} L\hat{\mathbf{z}} \times (\mathbf{v}_L - \mathbf{v}_n) . \tag{25.5}$$

The factor $1/2$ in eqn (25.4) is introduced to compensate the double counting of particles and holes in pair-correlated systems.

Thus the spectral-flow force acting on a vortex from the system of quasiparticles is (per unit length)

$$\mathbf{F}_{\text{sf}} = -\pi n_1 \hbar C_0 \hat{\mathbf{z}} \times (\mathbf{v}_{\text{L}} - \mathbf{v}_{\text{n}}). \tag{25.6}$$

This is in agreement with the result in eqn (18.17) obtained for the $n_1 = 2$ continuous vortex-skyrmion using the Adler–Bell–Jackiw equation. We recall that the parameter of the axial anomaly is $C_0 = p_F^3/3\pi^2\hbar^3 \approx n$.

The first derivation of the spectral-flow force acting on a singular vortex in conventional superconductors was made by Kopnin and Kravtsov (1976) who used the Gor'kov equations describing the fully microcopic BCS model. It was developed further by Kopnin and coauthors to other types of vortices: vortices in ^{3}He-B (Kopnin and Salomaa 1991); continuous vortex-skyrmions in ^{3}He-A (Kopnin 1993); non-axisymmetric vortices (Kopnin and Volovik 1997, 1998a); etc. That is why the spectral-flow force is also called the Kopnin force. Note that in all the cases the spectral-flow force can be obtained within the effective theory of fermion zero modes living in the vortex core and forming a special universality class of Fermi systems – Fermi surface in (Q, θ) space.

The process of transfer of a linear momentum from the superfluid vacuum to the normal motion of fermions within the core is a realization of the Callan–Harvey (1985) mechanism for anomaly cancellation. In the case of condensed matter vortices the anomalous non-conservation of linear momentum in the 1+1 world of the vortex core fermions and the anomalous non-conservation of momentum in the 3+1 world outside the vortex core compensate each other. As distinct from ^{3}He-A, where this process can be described in terms of Fermi points and Adler–Bell–Jackiw equation, the Callan–Harvey effect for singular vortices occurs in any Fermi superfluid. The anomalous fermionic Q branch, which mediates the momentum exchange, exists in any topologically non-trivial singular vortex: the chirality of these fermion zero modes and the anomaly are produced by the interplay of **r**-space and **p**-space topologies associated with the vacuum in the presence of a vortex. The effective theory of the Callan–Harvey effect does not depend on the detailed structure of the vortex core or even on the type of pairing, and is determined solely by the vortex winding number n_1 and the anomaly parameter C_0.

25.2 Restricted spectral flow in the vortex core

25.2.1 *Condition for free spectral flow*

In the above derivation it was implied that the discrete quantum number Q can be considered as continuous, otherwise the spectral flow along Q is impossible. To ensure the level flow, the interlevel distance ω_0 must be small compared to the width of the level $\hbar/\tau$, where τ is the lifetime of the fermions on the Q levels. The latter is determined by interaction of the fermion zero modes with the thermal fermions in the heat bath of the normal component (in superconductors the interaction of fermion zero modes with impurities usually dominates). Thus eqn (25.6) for the spectral-flow force is valid only in the limit of large scattering rate: $\omega_0\tau/\hbar \ll 1$. In the opposite limit $\omega_0\tau \gg 1$ the spectral flow is suppressed and the corresponding spectral-flow force becomes exponentially small. This also shows the limitation for exploring the macroscopic Adler–Bell–Jackiw anomaly equation in the electroweak model and in ^{3}He-A.

Since the spectral flow occurs through the zero energy, it is fully determined by the low-energy spectrum. Thus, to derive the spectral-flow force at arbitrary value of the parameter $\omega_0\tau$, one must incorporate the effective theory of fermion zero modes of co-dimension 1 in the vortex core. Such an effective theory is similar to the Landau theory of fermion zero modes in systems of the Fermi surface universality class – the Landau theory of Fermi liquid. Let us consider the kinetics of the low-energy quasiparticles on an anomalous branch in a vortex moving with respect to the heat bath.

25.2.2 Kinetic equation for fermion zero modes

We choose the frame of the moving vortex; in the steady–state regime the order parameter in this frame is stationary and the energy of quasiparticles is well determined. The effective Hamiltonian for quasiparticles in the moving vortex is given by

$$E_a(Q,\theta) = \omega_a(\theta)(Q - Q_a(\theta)) + (\mathbf{v}_\mathrm{s} - \mathbf{v}_\mathrm{L}) \cdot \mathbf{p}, \qquad (25.7)$$

where the last term comes from the Doppler shift and $\mathbf{v}_\mathrm{L}$ is the velocity of the vortex line. For simplicity we discuss the 2+1 case where $\mathbf{p} = (p_F\cos\theta, p_F\sin\theta)$, and the slope ω_a (minigap) does not depend on p_z.

According to the condition (ii) of Sec. 5.4 the global equilibrium takes place only if in the texture-comoving frame (i.e in the frame comoving with the vortex) the velocity of the normal component is zero, i.e. $\mathbf{v}_\mathrm{n} = \mathbf{v}_\mathrm{L}$. If the velocity of the normal component $\mathbf{v}_\mathrm{n} \neq \mathbf{v}_\mathrm{L}$, the motion of a vortex does not correspond to the true thermodynamic equilibrium and dissipation must take place, which at low T is determined by the kinetics of fermion zero modes. The effective theory of fermion zero modes in the vortex core is described in terms of canonically conjugated variables $\hbar Q$ and θ. The role of 'spatial' coordinate is played by the angle θ in momentum space, i.e. the effective space is circumference $U(1)$. This means that the chiral quasiparticles here are left or right moving along $U(1)$. The difference between the number of left- and right- moving fermionic species is $2n_1$ according to the index theorem. The kinetics of quasiparticles is determined by the Boltzmann equation for the distribution function $f_a(Q,\theta)$ (Stone 1996). For the simplest case of the $n_1 = \pm 1$ axisymmetric vortex with one anomalous branch, one has $\omega_a(\theta) = -n_1\omega_0$, $Q_a(\theta) = 0$, and the Boltzmann equation is

$$\hbar\partial_t f - n_1\omega_0\partial_\theta f - \partial_\theta((\mathbf{v}_\mathrm{s} - \mathbf{v}_\mathrm{L})\cdot\mathbf{p})\,\partial_Q f = -\hbar\frac{f(Q,\theta) - f_T(Q,\theta)}{\tau}. \qquad (25.8)$$

It is written in the so-called τ-approximation, where the collision term on the rhs is expressed in terms of the relaxation time τ due to collisions between fermion zero modes in the core and quasiparticles in the heat bath far from the core. The equilibrium distribution f_T corresponds to the global thermodynamic equilibrium state to which the system relaxes. This is the state where the vortex moves together with the heat bath, i.e. where $\mathbf{v}_L = \mathbf{v}_n$:

$$f_T(Q, \theta) = \left(1 + \exp\frac{-n_1\omega_0 Q + (\mathbf{v}_s - \mathbf{v}_n) \cdot \mathbf{p}}{T}\right)^{-1}. \quad (25.9)$$

When $\mathbf{v}_L \neq \mathbf{v}_n$ the equilibrium is violated, and the distribution function evolves according to the Boltzmann equation (25.8).

25.2.3 Solution of Boltzmann equation

Following Stone (1996) we introduce the new variable $l = Q - n_1(\mathbf{v}_s - \mathbf{v}_n) \cdot \mathbf{p}/\omega_0$ and obtain the equation for $f(l, \theta)$ which contains the velocity of a vortex only with respect to the heat bath:

$$\hbar \partial_t f - n_1 \omega_0 \partial_\theta f + \partial_\theta((\mathbf{v}_L - \mathbf{v}_n) \cdot \mathbf{p}) \, \partial_l f = -\hbar \frac{f(l,\theta) - f_T(l)}{\tau}. \quad (25.10)$$

Since we are interested in the momentum transfer from the vortex to the heat bath, we write the equation for the net momentum of quasiparticles

$$\mathbf{P} = \frac{1}{2}\int dl \int \frac{d\theta}{2\pi} f(l, \theta)\mathbf{p}, \quad (25.11)$$

which is

$$\dot{\mathbf{P}} - \frac{n_1}{\hbar}\omega_0 \hat{\mathbf{z}} \times \mathbf{P} + \pi \hbar C_0 \hat{\mathbf{z}} \times (\mathbf{v}_n - \mathbf{v}_L)(f_T(\Delta_0(T)) - f_T(-\Delta_0(T))) = -\frac{\mathbf{P}}{\tau}. \quad (25.12)$$

At the moment we consider only bound states below the gap $\Delta_0(T)$ and thus the integral $\int dl\partial_l n$ is limited by $\Delta_0(T)$. That is why the integral gives $f_T(\Delta_0(T)) - f_T(-\Delta_0(T)) = -\tanh(\Delta_0(T)/2T)$. The anomaly parameter C_0 which appears in eqn (25.12) is $C_0 = p_F^2/2\pi\hbar^2$ in the 2+1 case. As in the 3D case in eqn (18.18), $C_0 \approx n$ in the weak-coupling BCS systems.

In the steady state of the vortex motion one has $\dot{\mathbf{P}} = 0$, and the solution for the steady state momentum can be easily found (Stone 1996) As a result one obtains the following momentum transfer per unit time from the fermion zero modes to the normal component when the vortex moves with constant velocity with respect to the normal component:

$$\mathbf{F}_{\text{sf bound}} = \frac{\mathbf{P}}{\tau} = -\frac{\pi\hbar C_0}{1 + \omega_0^2 \tau^2} \tanh\frac{\Delta_0(T)}{2T}[(\mathbf{v}_L - \mathbf{v}_n)\omega_0\tau + n_1 \hat{\mathbf{z}} \times (\mathbf{v}_L - \mathbf{v}_n)]. \quad (25.13)$$

This is the contribution to the spectral-flow force due to bound states below $\Delta_0(T)$. This equation contains both the non-dissipative (the second term) and friction (the first term) forces. Now one must add the contribution of unbound states above the gap $\Delta_0(T)$. The spectral flow there is not suppressed, since the distance between the levels in the continuous spectrum is $\omega_0 = 0$. This gives the following spectral-flow contribution from the thermal tail of the continuous spectrum:

$$\mathbf{F}_{\text{sf unbound}} = -\pi\hbar n_1 C_0 \left(1 - \tanh\frac{\Delta_0(T)}{2T}\right) \hat{\mathbf{z}} \times (\mathbf{v}_L - \mathbf{v}_n) \,. \tag{25.14}$$

Finally the total spectral-flow force (Kopnin force) is the sum of two contributions, eqns (25.13–25.14). The non-dissipative part of the Kopnin force is

$$\mathbf{F}_{\text{sf}} = -\pi\hbar n_1 C_0 \left[1 - \frac{\omega_0^2 \tau^2}{1+\omega_0^2 \tau^2} \tanh\frac{\Delta_0(T)}{2T}\right] \hat{\mathbf{z}} \times (\mathbf{v}_L - \mathbf{v}_n) \,, \tag{25.15}$$

while the contribution of the spectral flow to the friction part of the Kopnin force is

$$\mathbf{F}_{\text{fr}} = -\pi\hbar C_0 \tanh\frac{\Delta_0(T)}{2T} \frac{\omega_0 \tau}{1+\omega_0^2 \tau^2} (\mathbf{v}_L - \mathbf{v}_n) \,. \tag{25.16}$$

This result coincides with that obtained by Kopnin in microscopic theory. In the limit $\omega_0\tau \to 0$, the friction force disappears, while the spectral-flow force reaches its maximum value, eqn (25.6), obtained for a continuous vortex-skyrmion using the Adler–Bell–Jackiw anomaly equation. In continuous vortices the interlevel spacing is very small, $\omega_0 \sim \hbar/(m\xi R_{\text{core}})$, and the spectral flow is not suppressed.

25.2.4 Measurement of Callan–Harvey effect in ^{3}He-B

Equations (25.15) and (25.16), with the anomaly parameter C_0 in eqn (18.18), can be applied to the dynamics of singular vortices in ^{3}He-B, where the minigap ω_0 is comparable to the inverse quasiparticle lifetime and the parameter $\omega_0\tau$ is regulated by temperature. Adding the missing Magnus and Iordanskii forces one obtains the following dimensionless parameters $d_\perp$ and $d_\parallel$ in eqn (18.30) for the balance of forces acting on a vortex:

$$d_\perp = \frac{C_0}{n_s}\left(1 - \frac{n}{n_s}\frac{\omega_0^2\tau^2}{1+\omega_0^2\tau^2}\tanh\frac{\Delta_0(T)}{2T}\right) - \frac{n_n}{n_s} \,, \tag{25.17}$$

$$d_\parallel = \frac{C_0}{n_s}\frac{\omega_0\tau}{1+\omega_0^2\tau^2}\tanh\frac{\Delta_0(T)}{2T} \,. \tag{25.18}$$

The regime of the fully developed axial anomaly occurs when $\omega_0\tau \ll 1$. This is realized close to T_c, since ω_0 vanishes at T_c. In this regime $d_\perp = (C_0 - n_n)/n_s \approx (n-n_n)/n_s = 1$. At lower T both ω_0 and τ increase, and finally at $T \to 0$ an opposite regime, $\omega_0\tau \gg 1$, is reached. In this limit the spectral flow becomes completely suppressed, the anomaly disappears and one obtains $d_\perp = -n_n/n_s \to 0$. This negative contribution to $d_\perp$ comes solely from the Iordanskii force.

Both extreme regimes and the crossover between them at $\omega_0\tau \sim 1$ have been observed in experiments with ^{3}He-B vortices by Bevan et al. (1995, 1997b, 1997a) and Hook et al. (1996). The friction force (Fig. 25.1 *top*) is maximal in the crossover region and disappears in the two extreme regimes, $\omega_0\tau \gg 1$ and

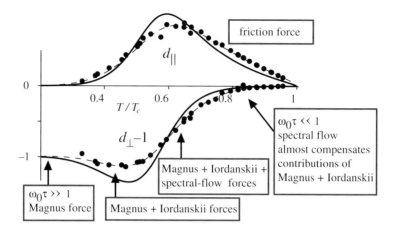

FIG. 25.1. Experimental momentogenesis by vortex-strings in ^{3}He-B: verification of the Callan–Harvey and gravitational Aharonov–Bohm effects. Solid lines are eqns (25.17) and (25.18). The spectral flow transfers the fermionic charge – the momentum – from the 1+1 fermions in the core to the 3+1 bulk superfluid. This is the analog of the Callan–Harvey effect in ^{3}He-B. The spectral flow is suppressed at low T but becomes maximal close to T_c. The negative value of $d_\perp$ at intermediate T demonstrates the Iordanskii force, which comes from the analog of the gravitational Aharonov–Bohm effect in Sec. 31.3. (After Bevan *et al.* 1997a.)

$\omega_0 \tau \ll 1$. In addition the experimental observation of the negative $d_\perp$ at low T (Fig. 25.1 *bottom*) verifies the existence of the Iordanskii force; thus the analog of the gravitational Aharonov–Bohm effect (Sec. 31.3) has also been measured in these experiments.

Part VI

Nucleation of quasiparticles and topological defects

26
LANDAU CRITICAL VELOCITY

The superfluid vacuum flows with respect to environment (the container walls) without friction until the relative velocity becomes so large that the Doppler-shifted energy of some excitations (quasiparticles or topological defects) becomes negative in the frame of the environment, and these excitations can be created from the vacuum. The threshold velocity v_{Landau} at which excitations of a given type acquire for the first time the negative energy is called the Landau velocity. This definition is applicable only for a single superfluid. When two or several superfluid components are involved the criterion will be the same but it cannot be expressed in terms of a single Landau velocity.

This does not mean that the supercritical flow of the vacuum (the flow with $v_s > v_{\text{Landau}}$) is not possible. If the excitations are macroscopic topological defects, the process of their creation requires overcoming a huge energy barrier. For example, in superfluid ^{3}He the Landau velocity for vortex nucleation can be exceeded by several orders of magnitude, and no vortices are created. In some cases the flow velocity can exceed the Landau velocity for nucleation of fermionic quasiparticles. Quasiparticles are created and fill the negative energy states. If it is possible to fill all the negative energy states without destroying the superfluid vacuum, then the dissipation stops and the supercritical superflow persists.

26.1 Landau critical velocity for quasiparticles

26.1.1 *Landau criterion*

Let us start with the Landau critical velocity for nucleation of quasiparticles. The energy of a quasiparticle in the reference frame of the environment (the container walls) is $\tilde{E}(\mathbf{p}) = E(\mathbf{p}) + \mathbf{p} \cdot \mathbf{v}_s$. It becomes negative for the first time when the the superfluid velocity with respect to container walls reaches the value

$$v_{\text{Landau}} = \min \frac{E(\mathbf{p})}{p_\|} . \tag{26.1}$$

Here $p_\|$ is the quasiparticle momentum along $\mathbf{v}_s$. It is important that at $T = 0$ there must be a preferred reference frame of the environment with respect to which the superfluid vacuum moves, otherwise quasiparticles do not know that they must be created. In superfluids depending on the physical situation such a reference frame can be provided by the container walls, by the external body moving in the liquid, by impurities forming the normal component, or by the texture. At $v_s > v_{\text{Landau}}$, nucleation of quasiparticles from the vacuum is allowed energetically, and they can be created due to interaction of the supercritically moving vacuum with the environment. The Landau critical velocity marks the onset of quantum friction. However, it is possible that the Landau velocity is exceeded at a place far from the boundaries, and nucleation of quasiparticles is suppressed. Such a situation will be discussed in Sec. 31.4 on quantum rotational friction.

In the anisotropic ^{3}He-A, the Landau critical velocity for nucleation of quasiparticles depends on the orientation of the flow with respect to the $\hat{\mathbf{l}}$-vector. From the quasiparticle spectrum (7.58) in the limit $c_\perp \ll v_F$ it follows that

$$v_{\text{Landau}}(\mathbf{v}_s \perp \hat{\mathbf{l}}) = c_\perp \ , \tag{26.2}$$

$$v_{\text{Landau}}(\mathbf{v}_s \cdot \hat{\mathbf{l}} \neq 0) = 0 \ . \tag{26.3}$$

26.1.2 Supercritical superflow in ^{3}He-A

For all orientations of the $\hat{\mathbf{l}}$-vector except for the transverse one, the Landau critical velocity is simply zero. Thirty years ago, before ^{3}He-A was discovered, it was thought that a non-zero value of the Landau velocity is the necessary condition for superfluidity. However, the ^{3}He-A example shows that this is not so. ^{3}He-A can still flow along the channel without friction. The reason for such supercritical suparflow is the following. In the process of quantum friction, the created fermionic quasiparticles fill the negative energy levels in the frame of the environment (container) thus transferring the momentum from the vacuum to the container. In ^{3}He-A, as we discussed in Sec. 10.3.5, it is possible to fill all the negative levels without completely destroying the superfluid vacuum. After all the negative levels are occupied the vacuum flows with the same velocity $\mathbf{v}_s$ without friction but with the reduced momentum: $m(n - n_{n\|})\mathbf{v}_s$. The reduction occurs due to the formed normal component with density $n_{n\|} \propto (\mathbf{v}_s \cdot \hat{\mathbf{l}})^2$ in eqn (10.34) which is trapped by the container walls. The only consequence of such a supercritical regime is the formation of the Fermi surface which gives the non-zero DOS and thus the finite density of the normal component at $T = 0$.

26.1.3 Landau velocity as quantum phase transition

For the transverse orientation of the $\hat{\mathbf{l}}$-vector, where $\mathbf{v}_s \cdot \hat{\mathbf{l}} = 0$, the situation is different: the Landau critical velocity coincides with the transverse 'speed of light' $c_\perp$ – the maximum attainable velocity of the low-energy 'relativistic' quasiparticles propagating in the directions perpendicular to $\hat{\mathbf{l}}$.

Can the flow velocity with respect to the walls exceed this value? This is not clear, since it depends on the behavior at the Planck scale. In the simplest model considered by Kopnin and Volovik (1998b) the interaction with the boundaries leads to the collapse of the supercritical superflow, when $v_s > c_\perp$.

But this is not the general rule, and in principle there can be two successive critical velocities (Fig. 26.1 left). The first one is v_{Landau}. Above this threshold the negative energy states appear and are occupied. This leads to the finite density of states at zero energy, and as a result this supercritical superflow has the finite normal density $n_n(T = 0)$ proportional to some power of $v_s - v_{\text{Landau}}$. At the

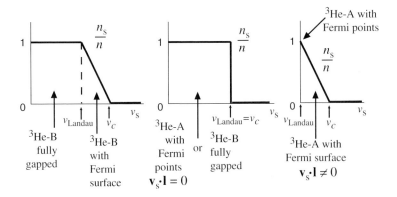

FIG. 26.1. *Left*: Two critical velocities of the superfluid vacuum with respect to a container. Above the Landau critical velocity v_{Landau} the vacuum is reconstructed due to created quasiparticles but remains superfluid (non-dissipative) with reduced but non-zero superfluid density $0 < n_s < n$. There is no dissipationless superfluid flow for velocities above the second critical velocity v_c. The Landau critical velocity v_{Landau} marks the quantum phase transition between vacua with different momentum space topology. *Middle*: ^{3}He-B at low pressure. The two critical velocities coincide. *Right*: ^{3}He-A with $\mathbf{v}_s \cdot \hat{\mathbf{l}} \neq 0$ and d-wave superconductor (Sec. 10.3.6). The Landau critical velocity is zero.

second threshold $v_c > v_{\text{Landau}}$ the superfluid density becomes zero, which means that superfluidity disappears: the flow becomes dissipative, non-stationary and finally turbulent.

The flow of ^{3}He-A, if it is not orthogonal to $\hat{\mathbf{l}}$, follows the scenario with $v_{\text{Landau}} = 0$ in Fig. 26.1 *right*. According to Nagai (1984), for the superflow along $\hat{\mathbf{l}}$, the critical velocity at which n_s becomes zero is $v_c = c_\perp \sqrt{e}$.

In ^{3}He-B both scenarios are possible depending on the range of microscopic parameters as was found by Vollhardt *et al.* (1980). At high external pressure, the scenario with two successive critical velocities, v_{Landau} and v_c, in (Fig. 26.1 *right*) takes place. At low external pressure, the two critical velocities coincide, $v_{\text{Landau}} = v_c$, and n_s drops from n to zero after the Landau critical velocity is reached (Fig. 26.1 *middle*).

In the region between the two critical velocities, $v_{\text{Landau}} < v_s < v_c$, the fermionic vacuum is well determined but it belongs to a different universality class – the class of vacua with a Fermi surface. The formation of a Fermi surface in ^{3}He-A when it flows with velocity $v_s > v_{\text{Landau}} = 0$ has been discussed in Sec. 10.3.6. In ^{3}He-B, the Fermi surface emerging in the supercritical regime can be found from the equation $\tilde{E}(\mathbf{p}) = 0$ with $\tilde{E}(\mathbf{p}) = E(p) + \mathbf{p} \cdot \mathbf{v}_s$ and $E^2(p) = v_F^2(p - p_F)^2 + \Delta_0^2$ in eqn (7.50).

In the region $v_s < v_{\text{Landau}} = \Delta_0/p_F$, the ^{3}He-B vacuum is topologically trivial with fully gapped spectrum. Thus in ^{3}He-B, v_{Landau} marks the quantum phase transition at $T = 0$ – the Lifshitz transition at which with the momentum space topology of fermion zero modes of the quantum vacuum changes. The finite density of states on the Fermi surface leads to a finite $n_n(T = 0)$, which thus plays the role of the order parameter in this quantum transition.

26.1.4 Landau velocity, ergoregion and horizon

Here we avoid discussing the vacuum stability at supercritical flow, simply by assuming that the region where the Landau velocity is exceeded is very far from the boundaries of the vessel. In this case the interaction with the boundaries can be neglected, the reference frame of the boundaries is lost and thus the boundaries no longer serve as the environment. Such a situation has a very close relation to the horizon problem in general relativity. Since the boundaries are effectively removed, the preferred reference frame is provided only by the inhomogeneity of the flow: for example, by the $\hat{\mathbf{l}}$-texture or by a spatial dependence of velocity field $\mathbf{v}_s(\mathbf{r})$. We know that in ^{3}He-A both of these fields simulate the effective gravity field. In the region where in the texture-comoving frame (i.e. in the frame where the texture is stationary) the superfluid velocity exceeds the Landau criterion, $v_s(\mathbf{r}) > v_{\text{Landau}}$, a quasiparticle can have negative energy. In general relativity such a region is called the ergoregion. The surface $v_s(\mathbf{r}) = v_{\text{Landau}}$ which bounds the ergoregion is called the ergosurface. We shall use the terms ergoregion and ergosurface in our non-relativistic physics of quantum liquids.

In the case when the Landau velocity coincides with the 'speed of light' c (as in eqn (26.2)), the equation for the ergosurface, $v_s(\mathbf{r}) = c$, coincides according to eqn (4.16) with the equation $g_{00}(\mathbf{r}) = 0$ for the 'acoustic' metric. This is just the conventional definition of the ergosurface in general relativity. When the velocity $\mathbf{v}_s(\mathbf{r})$ is normal to the ergosurface, such a surface is called a horizon. If $\mathbf{v}_s(\mathbf{r})$ is directed toward the region where $v_s(\mathbf{r}) > c$, this imitates the black-hole horizon (Unruh 1981, 1995). The low-energy 'relativistic' quasiparticles cannot escape from the region behind the horizon because their velocity c in the reference frame of the superfluid vacuum is less than the 'frame-dragging' velocity v_s. In the more general flow the horizon and the ergosurface are separated from each other, as in the case of a rotating black hole (Jacobson and Volovik 1998a; Visser 1998).

26.1.5 Landau velocity, ergoregion and horizon in case of superluminal dispersion

The definitions of the horizon and ergosurface must be modified when we leave the low-energy domain of relativistic physics and take into account the dispersion of the spectrum at higher energy. There are two possible cases: the initially relativistic spectrum $E(p)$ bends upward or downward at high energy. We shall mostly discuss the first case which is physically more attractive, i.e.

$$E(\mathbf{p}) = cp(1 + \gamma p^2 + \ldots) , \qquad (26.4)$$

with $\gamma > 0$. Such dispersion is realized for the fermionic quasiparticles in ^{3}He-A for the motion in the directions transverse to $\hat{\mathbf{l}}$. If one considers $p_z = p_F$, one obtains from eqn (7.58) the following dispersion of the spectrum: $E^2(\mathbf{p}_\perp) = c_\perp^2 p_\perp^2 + (p_\perp^2/2m^*)^2$. This gives $\gamma^{-1} = 8(m^*c_\perp)^2$, which shows that in this particular case the role of the Planck momentum, which scales the non-linear correction to the spectrum, is played by $p_{\text{Planck}} = m^*c_\perp$. The same expression is obtained for a weakly interacting Bose gas, where the quasiparticle spectrum in eqn (3.17) is $E^2(\mathbf{p}) = c^2 p^2 + (p^2/2m)^2$.

In this case of positive dispersion parameter $\gamma > 0$, the group velocity of massless quasiparticle is always 'superluminal':

$$v_G = dE/dp = c\left(1 + 3\gamma p^2\right) > c\,. \tag{26.5}$$

That is why there is no true horizon for quasiparticles: they are allowed to leave the black-hole region. It is, hence, a horizon only for quasiparticles living exclusively in the very low-energy corner $p \ll p_{\text{Planck}}$; they are not aware of the possibility of the 'superluminal' motion. Nevertheless, even the mere fact that there is a possibility of superluminal propagation at high energy is instrumental for an inner observer, who lives deep within the relativistic domain. Some physical results arising from the fact that $\gamma > 0$ do not depend on the value of γ (see Sec. 30.1.5).

The Landau critical velocity (26.1) for quasiparticles with spectrum (26.4) coincides with the 'speed of light', $v_{\text{Landau}} = c$. Thus the ergosurface is determined by the same equation $v_s(\mathbf{r}) = c$, as for fully relativistic quasiparticles.

26.1.6 Landau velocity, ergoregion and horizon in case of subluminal dispersion

In superfluid ^{4}He the negative dispersion of the quasiparticle spectrum is realized, $\gamma < 0$, with the group velocity $v_G = dE/dp < c$ (we ignore here a possible small upturn of the spectrum at low p in Fig. 6.1). In such superfluids the 'relativistic' ergosurface at $v_s(\mathbf{r}) = c$ does not coincide with the true ergosurface, at $v_s(\mathbf{r}) = v_{\text{Landau}}$, since the Landau velocity is determined by the roton part of the spectrum in Fig. 6.1: $v_{\text{Landau}} \approx E(p_0)/p_0$, where p_0 is the position of the roton minimum. v_{Landau} is about four times less than c. In the case of radial flow inward, the true ergosphere occurs at $v_s(r) = v_{\text{Landau}} < c$. There is also the inner surface $v_s(r) = c$, which marks the true horizon, since c is the maximum attainable speed for all quasiparticles (Fig. 32.3). This is in contrast to relativistically invariant systems, for which the ergosurface for a purely radial gravitational field of the chargless non-rotating black hole.

26.2 Analog of pair production in strong fields

Different values of the Landau velocity in ^{3}He-A for different orientations of the $\hat{\mathbf{l}}$-vector, eqn (26.2) and eqn (26.3), reflect the double role played by the superfluid velocity $\mathbf{v}_s$ in the effective low-energy theory of Fermi systems. According to eqns (9.11–9.15) $\mathbf{v}_s$ enters both the metric field and the potential of the effective electric field: $A_0 = p_F \hat{\mathbf{l}} \cdot \mathbf{v}_s$. The latter disappears only when the superflow is exactly orthogonal to $\hat{\mathbf{l}}$. For massless fermions any non-zero value of the potential A_0 leads to the formation of fermions from the vacuum, that is why $v_{\text{Landau}} = 0$ if $\hat{\mathbf{l}} \cdot \mathbf{v}_s \neq 0$.

An analog of the creation of massive fermions in strong fields is presented by ^{3}He-B, whose spectrum (7.49–7.50) is fully gapped. The Doppler-shifted energy spectrum in the container frame is

$$\tilde{E}(\mathbf{p}) = \pm\sqrt{M^2(\mathbf{p}) + c^2 p^2} + \mathbf{p} \cdot \mathbf{v}_s \approx \pm\sqrt{v_F^2 (p - p_F)^2 + \Delta_0^2} + p_F \hat{\mathbf{p}} \cdot \mathbf{v}_s \ . \quad (26.6)$$

This spectrum is non-relativistic; nevertheless the velocity term acts in a manner similar to the potential A_0 of an electric field, especially if the direction of the momentum $\hat{\mathbf{p}}$ is fixed. The quasiparticles are created when the Landau velocity $v_{\text{Landau}} \approx \Delta_0/p_F$ is reached somewhere in space. This corresponds to the case when the depth of the potential well created by the effective $A_0(\mathbf{r})$ for the massive fermions becomes comparable to their rest energy $M (= \Delta_0)$. A superfluid analog of the pair creation in a constant electric field – the Schwinger pair production – has been discussed by Schopohl and Volovik (1992).

26.2.1 Pair production in strong fields

Let us start with relativistic fermions. In nuclear physics a deep potential well can be obtained during the collision of two heavy bare nuclei with the total charge Z greater than the supercritical Z_c, at which the electron bound state enters the continuum spectrum in the 'valence' band with energy $E < -M$ (Gershtein and Zel'dovich 1969) [146]), and the production of electron–positron pairs from the vacuum becomes favourable energetically.

Let us recall the essential features of the Gershtein and Zel'dovich (1969) mechanism of pair production in strong fields [146] (see Calogeracos et al. (1996) for a detailed review). Consider an electron-attractive potential produced by a heavy bare nucleus, which has a vacant discrete level (Fig. 26.2(a)). The potential increases in strength when the second nucleus approaches. The level will cross $E = 0$ for some value of the potential, but there is nothing critical happening during the crossing. For some greater value the level crosses $E = -M$ and thus merges with the negative energy continuum – the valence band according to the terminology of solid-state physics (Fig. 26.2(b)). The vacant state is now occupied by an electron from the negative energy continuum, which means that the electron vacancy (positron) occupies a scattering state and escapes to infinity (Fig. 26.2(c)). When the second nucleus goes away, the potential becomes weak again (Fig. 26.2(d)) and the level returns to its original position, but the bound state is now filled by an electron. During the whole cycle the total electric charge is conserved, since both a positron and an electron are created, with the positron escaping to infinity and the electron filling the bound state.

26.2.2 Experimental pair production

The production of the electron–positron pairs has an analog in superfluids and superconductors, where it is called pair breaking, since it can be described as the

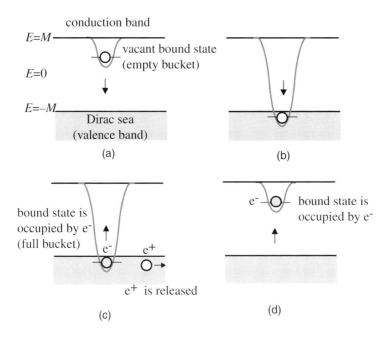

FIG. 26.2. Pumping of electron–positron pairs from the Dirac sea using draw-well.

breaking of a Cooper pair into two quasiparticles. The experiments, in which the mechanism of the pair production is similar to the Gershtein–Zel'dovich mechanism, have been conducted by Castelijns et al. (1986) and Carney et al. (1989). A cylindrical wire was vibrating in superfluid ^{3}He-B and the pair production was observed when the amplitude of the velocity of the wire exceeded some critical value. Since the Bogoliubov–Nambu fermions in ^{3}He-B are in many respects similar to Dirac electrons, we can map the quasiparticle radiation by a periodically driven wire in a supercritical regime to the particle production in an alternating electric field.

Let us discuss the rough model of how pair creation occurs in ^{3}He-B when the wire is oscillating with velocity $v = v_0 \cos(\omega t)$ (Lambert 1990; Calogeracos and Volovik 1999b). There are two features of the energy spectrum which are important: the continuous spectrum at $|E| > \Delta_0$ and bound states which appear near the surface of the wire, where the gap is reduced providing the potential well for quasiparticles in Fig. 26.3 and Fig. 26.4(a). Let $\Delta_0 - \epsilon$ be the energies of bound states in the range $\Delta_0 > \Delta_0 - \epsilon \geq \Delta_0 - \epsilon_0 \geq 0$. When the wire is oscillating, in the reference frame of the wire which can serve as an environmentthe the energy spectrum exhibits the Doppler shift. The velocity field around the wire is non-uniform: it equals the velocity of wire v at infinity and reaches the maximal value αv near the surface of the wire (for a perfect cylindrical wire $\alpha = 2$). That is why in the frame of the wire the continuous spectrum in the bulk and bound state

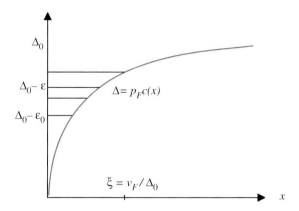

FIG. 26.3. Bound states at the surface of the superfluid. $\Delta_0 - \epsilon_0$ is the lowest bound state.

energies at the surface are Doppler shifted in a different way, with the maximal shift $\Delta_0 \to \Delta_0 \pm p_F|v|$ and $\Delta_0 - \epsilon \to \Delta_0 - \epsilon \pm \alpha p_F|v|$ in the bulk and at the surface correspondingly (Fig. 26.4(b)).

When the velocity of the wire increases one can reach the point where the unoccupied lowest bound state level with the energy $\Delta_0 - \epsilon_0 - \alpha p_F|v|$ merges with the negative energy continuum in the bulk at $-\Delta_0 + p_F|v|$ (Fig. 26.4(c)). Then the vacant state is now occupied by a quasiparticle from the negative energy continuum, while the quasihole escapes to infinity (Fig. 26.4(d)). After half a period the velocity of the wire becomes zero and the energy levels return to their original positions in Fig. 26.4(a). But a pair of quasiparticles have been created: one of them occupies the bound state at the surface, and the other is radiated away. The critical velocity, at which this mechanism of pair breaking occurs, is $v_0 = (2\Delta_0 - \epsilon_0)/(1+\alpha)p_F$. The measured pair-breaking critical velocity, at which the quasiparticle emission has been observed by Castelijns et al. (1986), Carney et al. (1989) and Fisher et al. (2001), appeared to be close to $v_0 = \Delta_0/3p_F$. This corresponds to a perfect cylindrical wire, where $\alpha = 2$, and a strongly suppressed gap at the surface, i.e. $\Delta_0 - \epsilon_0 = 0$.

26.3 Vortex formation

26.3.1 Landau criterion for vortices

Nucleation of vortices remains one of the most important problems in quantum liquids, with possible application to the formation of cosmic strings, magnetic monopoles and other topological defects in cosmology and quantum field theory.

From the energy consideration the formation of vortices becomes possible when the Landau criterion applied for the energy spectrum of vortices is reached. The energy spectrum of the vortex ring at $T = 0$ can be found from the equations for the energy and momentum of the ring in terms of its radius R:

$$\tilde{E} = E(R) + \mathbf{v}_s \cdot \mathbf{p}(R) , \qquad (26.7)$$

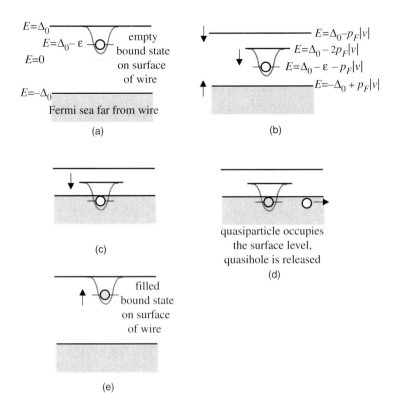

FIG. 26.4. The draw-well scenario of pair creation in ^{3}He-B in vibrating wire experiments.

$$E(R) = \frac{1}{2}mn\kappa^2 R \ln \frac{R}{R_{\text{core}}}, \qquad (26.8)$$

$$\mathbf{p}(R) = \pi mn\kappa R^2 \hat{\mathbf{p}}. \qquad (26.9)$$

Here $E(R)$ is the loop energy in the superfluid-comoving frame; $E(R)/2\pi R$ is the line tension; R_{core} is the core radius; $\tilde{E}$ is the loop energy in the reference frame of the boundaries of the container (the environment frame) in which the normal component velocity is zero in a global equilibrium, $\mathbf{v}_n = 0$; the linear momentum $\mathbf{p}(R)$ of the loop is directed perpendicular to the plane of the vortex ring.

From eqn (26.1) it follows that the Landau critical velocity for nucleation of vortex rings $v_{\text{Landau}} = \min[E(R)/p(R)] = 0$. For any non-zero velocity $\mathbf{v}_s$ of the flow of the superfluid vacuum with respect to the container, the energy $\tilde{E}$ of the vortex ring in the frame of the container becomes negative if the momentum (or the radius) of the vortex ring is large enough. A finite value of the Landau velocity is obtained if one takes into account the finite dimension of the container which provides the infrared cut-off for the Landau velocity: $v_{\text{Landau}} \sim \kappa/R_{\text{container}}$.

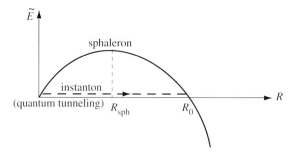

FIG. 26.5. Energy $\tilde{E} = E + \mathbf{p} \cdot \mathbf{v}_s$ of a vortex ring in the laboratory frame in the presence of superflow in eqn (26.7). Three states of the vortex loop are important. The state with zero radius $R = 0$ and energy $\tilde{E} = 0$ is the vortex-free vacuum state. The vortex ring with radius $R_{\rm sph}$ (eqn (26.10)) is the sphaleron. It is the saddle-point solution. The energy of the sphaleron determines the energy barrier and thus the thermal activation rate $\exp -(\tilde{E}_{\rm sph}/T)$. At low T thermal activation is substituted by quantum tunneling from the state with $R = 0$ to the vortex ring with radius $R = R_0$ given by eqn (26.19), whose energy is also zero, $\tilde{E} = 0$.

In reality the observed critical velocities for nucleation of singular vortices in ^{3}He-B are larger by several orders of magnitude (Parts *et al.* 1995*b*). This demonstrates that the energetical advantage for the nucleation of a vortex ring does not mean that vortex loops will be really nucleated. The vortex loop of large radius must be grown (together with the singularity in the core) from an initially smooth configuration. This requires the concentration of a large energy within a small region of the size of the core of the defect. Such a process involves an energy barrier (see Fig. 26.5), which must be overcome either by quantum tunneling or by thermal activation.

26.3.2 *Thermal activation. Sphaleron*

Thermal nucleation of vortices in the presence of counterflow was calculated by Iordanskii (1965).

In general, a thermally activated topological defect, which is the intermediate object between two vacuum states with different topological charges, is called a sphaleron (see the review paper by Turok 1992). It is the thermodynamically ustable, saddle-point stationary solution of, say, Ginzburg–Landau equation or other Euler–Lagrange equations. In superfluids, the sphaleron is represented by a metastable vortex loop, which, being the saddle-point solution, is stationary in the heat-bath frame. It is the critical vortex ring at the top of the barrier (Fig. 26.5). It represents the stationary solution of the equations in the presence of the counterflow $\mathbf{v}_n - \mathbf{v}_s$, and satisfies the thermal equilibrium conditions. Let us recall that in the global equilibrium the normal component velocity $\mathbf{v}_n$ must be zero in the frame of the container, while all the objects must be stationary in this frame. At the top of the barrier the group velocity of the vortex ring is zero, $d\tilde{E}/d\mathbf{p} = 0$, and thus the ring is at rest in the heat-bath frame of the normal component. However, this thermal equilibrium state is locally unstable: the ring as the saddle-point solution will either shrink or grow.

The radius and the energy of the sphaleron are

$$R_{\rm sph} = \frac{\kappa}{4\pi|\mathbf{v}_{\rm s} - \mathbf{v}_{\rm n}|} \ln \frac{R_{\rm sph}}{R_{\rm core}} \ , \quad \tilde{E}_{\rm sph} = \frac{1}{2} m n_{\rm s} \kappa^2 R_{\rm sph} \ln \frac{R_{\rm sph}}{R_{\rm core}} \ . \qquad (26.10)$$

The energy of the sphaleron determines the thermal activation rate, $e^{-\tilde{E}_{\rm sph}/T}$.

In a toroidal geometry the sphaleron is shown in Fig. 26.6(d). If the length of the toroidal channel is large enough, the intermediate saddle-point solution corresponds to a straight vortex line, which has the maximum length and thus the maximum energy among all intermediate vortices in Fig. 26.6. If the vortex has elementary circulation quantum, $\kappa = \kappa_0$, then the sphaleron is the intermediate object between the initial and final vacua with topological charges n_1 and $n_1 - 1$, which describe the circulation of the superflow along the channel.

In superfluid ^{3}He-B thermal nucleation is practically impossible, because of (i) the low temperature; and (ii) the huge barrier. The vortex ring is well determined only when its radius exceeds the core radius, which for the singular vortex is of order ξ, and for the continuous vortex-skyrmion is of order of dipole length ξ_D. As a result the maximal activation rate is about $\exp(-10^6)$ for a singular vortex and $\exp(-10^9)$ for a continuous one. This is too small in any units. Thermal activation or quantum tunneling can assist the nucleation only in the very closest vicinity of the instability threshold, where the energy barrier is highly suppressed. However, in this region external perturbations appear to be more effective.

26.3.3 *Hydrodynamic instability as mechanism of vortex formation*

Since thermal activation, and also quantum tunneling, are excluded from consideration in superfluid ^{3}He, the only remaining mechanism is local hydrodynamic instability of the laminar superfluid flow. This means that the threshold of the instability is determined not only by the Landau criterion, which marks the appearance of the negative energy states, but also by the requirement that the energy barrier between the initial vacuum state and the negative energy state disappears.

One can estimate the threshold of instability. According to eqn (26.10) the barrier for vortex nucleation in Fig. 26.5 disappears at such velocities where the radius of the sphaleron (the critical vortex ring) becomes comparable to the core size of the vortex:

$$v_c \sim \frac{\kappa}{2\pi R_{\rm core}} \ . \qquad (26.11)$$

At this velocity topology no longer supports the barrier between the vortex and vortex-free states. Let us stress that this is not the general rule for many reasons, in particular: (i) though at $R_{\rm sph} \sim R_{\rm core}$ the topology no longer provides

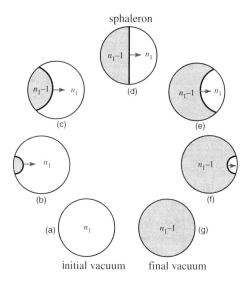

FIG. 26.6. (a–g) A vortex segment nucleated at the wall of the channel sweeps the cross-section of the channel and annihilates at the wall. If the vortex winding number is $n_1 = 1$, the topological charge of the vacuum – the circulation along the channel – changes in this process from $\kappa_0 n_1$ to $\kappa_0(n_1 - 1)$. The vortex serves as an intermediate object in the process of transition between vacuum (a) and vacuum (g), which have different topological charges. If the transition from (a) to (g) occurs via quantum tunneling, the process is called an instanton. If the transition occurs via thermal activation, the intermediate saddle-point stationary configuration (d) is called a sphaleron. Its energy determines the energy barrier between the two vacua.

the energy barrier for vortex nucleation, the non-topological barriers are still possible; (ii) the vortex formation can start earlier because of the roughness of the walls of container where the local superfluid velocity is enhanced.

Nevertheless the general trend is confirmed by experiments in three different liquids with substantially different sizes of the core (Fig. 26.7): in superfluid ^{4}He, $R_{\rm core} \sim a_0$, the interatomic space; in ^{3}He-B, $R_{\rm core} \sim \xi$; and in the case of continuous vortex-skyrmions in ^{3}He-A, $R_{\rm core} \sim \xi_D$. The large difference in critical velocities for nucleation of singular vortices in ^{3}He-B and vortex-skyrmions in ^{3}He-A allows for the creation of the interface between rotating ^{3}He-A and stationary ^{3}He-B in Fig. 17.9.

26.4 Nucleation by macroscopic quantum tunneling

26.4.1 *Instanton in collective coordinate description*

In superfluid ^{4}He the typical temperature is three orders of magnitude higher than in ^{3}He-B, while the minimal radius of the vortex loop is three orders of magnitude smaller. That is why the vortex formation by thermal activation and

NUCLEATION BY MACROSCOPIC QUANTUM TUNNELING

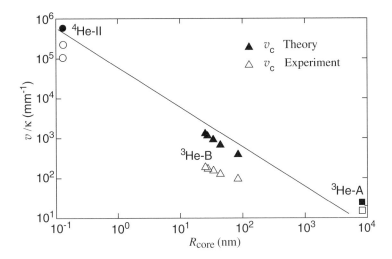

FIG. 26.7. Critical velocity of vortex formation in three different superfluids as a function of the size R_{core} of the core of nucleated vortices (after Parts et al. 1995b). Solid line is eqn (26.11).

even by quantum tunneling is possible. The parameters of high-temperature superconductors can also be favorable for this.

Macroscopic quantum nucleation of extended or topological objects is an interesting phenomenon not just in condensed matter: in cosmology the quantum nucleation of the Universe and black holes is considered; in RQFT the corresponding object is an instanton (Belavin et al. 1975). Quantum nucleation of vortices bears both features of instantons in RQFT. (i) It is the process of quantum tunneling through a barrier. (ii) It is the tunneling between vacuum states with different topology. In the toroidal geometry (Fig. 26.6), in the process of tunneling between the flow states with winding numbers n_1 and $n_1 - 1$, the vortex segment with winding number $n_1 = 1$ is nucleated at the wall of the container, sweeps the cross-section of the channel, and finally is annihilated at the wall.

The instanton is usually described by quantum field theory. But in many cases one can find the proper collective coordinates, and then the instanton is reduced to the process of quantum tunneling of a single effective particle instead of a field. Vortex nucleation in superfluid ^{4}He was calculated using both approaches, which gave very similar results: using the many-body wave function, which corresponds to quantum field theory (Sonin 1973); and in terms of collective coordinates for the vortex ring (Volovik 1972).

26.4.2 *Action for vortices and quantization of particle number in quantum vacuum*

The tunneling exponent is determined by the classical action for the vortex. This action is the same as in a classical perfect liquid:

$$S = \int dt \, \tilde{E}\{\mathbf{r}_L\} + \kappa\rho V_L\{\mathbf{r}_L\} \, . \tag{26.12}$$

$\tilde{E}$ in the first term is the energy of the vortex loop – the total hydrodynamic energy $\int d^3r (1/2)\rho \mathbf{v}_s^2$ of the superflow generated by the vortex. The energy depends on the position of the elements of the vortex line $\mathbf{r}_L(t,l)$, where l is the coordinate along the vortex line; $\rho = mn$ is the mass density of the liquid. In the infinite liquid the energy of the vortex loop is given by eqn (26.7).

The second term is topological, where $\kappa = n_1\kappa_0$, and V_L is the volume bounded by the area swept by the vortex loop between nucleation and annihilation (Rasetti and Regge 1975):

$$V_L\{\mathbf{r}_L\} = \frac{1}{3} \int dt \, dl \, \mathbf{r}_L \cdot (\partial_t \mathbf{r}_L \times \partial_l \mathbf{r}_L) \, . \tag{26.13}$$

The volume law for the topological part of vortex action follows from the general laws of vortex dynamics governed by the Magnus force; variation of the volume term leads to the classical Magnus force (18.27) acting on the vortex

$$\mathbf{F}_M = -\kappa\rho \frac{\delta V_L}{\delta \mathbf{r}_L} = \kappa\rho \partial_l \mathbf{r}_L \times \mathbf{v}_L \, , \tag{26.14}$$

where $\mathbf{v}_L = \partial_t \mathbf{r}_L$ is the velocity of the vortex line. This force is of topological origin and thus depends only on fundamental parameters of the system (see also eqn (18.24) for the similar volume law in the anomalous action: the axial anomaly is determined by the term in action which is proportional to the volume of the extended $(\mathbf{r}, \mathbf{p})$ space – the phase space). An inner observer living in the liquid who can measure the circulation around the vortex line and the force acting on the vortex would know such a fundamental quantity of the 'trans-Planckian' world as the mass density of the 'Planck' liquid.

The volume law for the topological action is a property of global vortices where the field $\mathbf{v}_s$ generated by a vortex line is not screened by the gauge field. For local cosmic strings and for fundamental strings the action is determined by the area swept by the string (Polyakov 1981). For vortices, the area law is obtained when the kinetic energy term $(1/2)M_L \mathbf{v}_L^2$ (where M_L is the mass of a vortex) is larger than the topological term. Vortex formation by quantum tunneling in this situation is similar to nucleation of electron–positron pairs in the presence of uniform electric field (Davis 1992).

For a quantized vortex in a quantum liquid, the topological volume term in action is related to quantization of the number of original bare particles comprising the vacuum, $N_{\text{vac}} = nV$, where V is the total volume of the system. This relation follows from the multi-valuedness of the topological action: ambiguity of the action with respect to the choice of the volume swept by the loop. For the closed loop nucleated from a point and then shrunk again to a point, it can be the volume V_L inside the surface swept by the loop, or the complementary volume $V_L - V$ outside the surface (we have written $V_L - V$ instead of $V - V_L$ since the sign of the volume is important). The multi-valuedness of the action has no consequences in classical physics, because the constant term is added to the action, when V_L is shifted to $V_L - V$. But in quantum mechanics this shift leads to an extra phase of the wave function of the system. Since the exponent $e^{iS/\hbar}$ should not depend on the choice of the volume, the difference between the two actions must be a multiple of $2\pi\hbar$:

$$S(V_L) - S(V_L - V) = m\kappa_0 n_1 N_{\text{vac}} = 2\pi\hbar p , \qquad (26.15)$$

where p is integer. In the same manner the Wess–Zumino action for ferromagnets in eqn (6.14) leads to the quantization of spin: $M = p\hbar/2$.

For ^{4}He, where $\kappa_0 = 2\pi\hbar/m$, eqn (26.15) suggests that for any winding number n_1 the quantity $N_{\text{vac}} n_1$ must be integer, and thus gives an integral value of N_{vac}. Equation (26.15) also gives the relation between the angular momentum of the liquid and N_{vac} in the presence of a rectilinear vortex with winding number n_1:

$$L_z = \frac{\hbar}{2}p = \hbar n_1 N_{\text{vac}} . \qquad (26.16)$$

For superfluid ^{3}He-B (or s-wave superfluids), where Cooper pairing takes place and thus the mass of the elementary boson – the Cooper pair – is twice the mass of the ^{3}He atom (or electron), the circulation quantum is $\kappa_0 = 2\pi\hbar/2m$. As a result $N_{\text{vac}} n_1 = 2p$, i.e. the same arguments prescribe an even number of atoms in the vacuum. This is justified, because with an odd number of atoms there is one extra atom that is not paired, and thus instead of the pure vacuum, the ground state represents the vacuum plus matter – a quasiparticle with energy Δ_0. The action in eqn (26.12) is applicable to the vacuum state only. Equation (26.16) relating angular momentum of the vortex to winding number n_1 and the number of particles in this system becomes $L_z = (\hbar/2) n_1 N_{\text{vac}}$ in agreement with eqn (14.19).

In general, if the mass of the elementary boson is km, where k is integer, the number of atoms in the vacuum must be a multiple of k, and the circulation quantum is $\kappa_0 = 2\pi\hbar/km$. The topological action for the vortex with winding number n_1 becomes

$$S_{\text{top}} = 2\pi\hbar \frac{n_1}{k} \mathcal{N} , \qquad (26.17)$$

where $\mathcal{N}$ is the number of atoms in the volume bounded by the area swept by the vortex loop between nucleation and annihilation.

26.4.3 Volume law for vortex instanton

In quantum tunneling the action along the instanton trajectory contains an imaginary part, which gives the nucleation rate $\Gamma \propto e^{-2\text{Im}S/\hbar}$. For nucleation of vortices at $T=0$ as well as for radiation of quasiparticles from the vacuum, we need two competing reference frames. One of them is provided by the superfluid vacuum. If only the homogeneous vacuum is present, its motion can be gauged away by Galilean transformation to obtain the liquid at rest. Thus we need another reference frame – the frame of the environment which violates Galilean invariance.

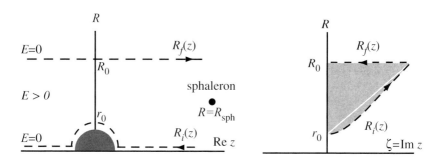

FIG. 26.8. *Left*: Two vortex trajectories with $\tilde{E} = 0$ around the pinning site with the shape of a hemisphere. $R_f(z)$ is the trajectory of the vortex ring with zero energy whose radius is close to R_0. $R_i(z)$ is the trajectory which corresponds to the vortex-free state. The vortex ring has zero radius outside the pinning site, while at the pinning site the radius follows the spherical shape of the pinning site. Such a trajectory can be obtained as the limit of trajectories with $\tilde{E} > 0$ when $\tilde{E} \to 0$. Two trajectories with $\tilde{E} = 0$ are separated by the region where the vortex has positive energy. The sphaleron saddle-point solution in this region is shown by the filled circle. *Right*: Two vortex trajectories with $\tilde{E} = 0$ meet each other on the imaginary axis $z = i\zeta$. The integral along the path gives the rate of the macroscopic quantum tunneling of the vortex from the initial trajectory $R_i(z)$ through the energy barrier to the final trajectory $R_f(z)$. The tunneling exponent is proportional to the number of atoms inside the surface swept by the vortex ring.

It can be provided by an external body, by texture of the order parameter, by the normal component or by boundaries of the vessel. Let us consider the quantum nucleation of a vortex line in the simplest geometry of a smooth wall with one pinning site on the wall providing the preferred reference frame (Volovik 1972). Choosing the pinning site in the form of a hemisphere of radius r_0 (Fig. 26.8 *left*), one can eliminate the plane boundary by reflection in the plane, and the problem becomes equivalent to the superfluid vacuum moving with respect to the static spherical body with the velocity at infinity $\mathbf{v}_s(\infty) = v_0\hat{\mathbf{z}}$. The superfluid motion is stationary, i.e. time independent in the frame of the body.

We shall use the collective coordinate description of the tunneling in which all the degrees of quantum field theory are reduced to the slowest ones corresponding to the bosonic zero mode on a moving vortex loop. Because of the spherical symmetry of the impurity the nucleated vortex loop will be in the form of a vortex ring whose plane is oriented perpendicular to the flow. The zero mode is described by two collective coordinates, the radius of the loop R and the position z of the plane of the vortex ring. According to the topological volume term in the action in eqns (26.12) and (26.13), these variables can be made canonically conjugate, since the momentum of the vortex ring, directed along $\hat{\mathbf{z}}$, is determined by its radius, $p_z = \pi \rho n_1 \kappa_0 R^2$ (see eqn (26.9)). The topological term in the action in such a simple geometry is the adiabatic invariant

$$S_{\text{top}}(V_L) = \int dz \, p_z \, . \tag{26.18}$$

There are two important trajectories, the initial and final, $z_i(R)$ and $z_f(R)$ (Fig. 26.8 *left*). Both have zero energy with respect to the energy of the moving liquid. The trajectories are separated by the energy barrier and can be connected only if the coordinate z is extended to the complex plane (Fig. 26.8 *right*). The trajectory $z_i(R)$ asymptotically corresponds to the vortex-free vacuum state, while $z_f(R)$ corresponds to the vortex moving away from the pinning site. Far from the pinning center the energy of the vortex loop of radius R (Fig. 26.8 *left*) is given by eqn (26.7).

The vortex nucleated from the vacuum state must have zero energy, $\tilde{E} = 0$. This determines the radius of the nucleated vortex:

$$R_0 = \frac{|\kappa|}{2\pi v_0} \ln \frac{R_0}{a_0} \, , \tag{26.19}$$

where a_0 is interatomic space which determines the core size in ^{4}He. Thus far from the pinning center the trajectory in the final state is $R_f(z) = R_0$ and does not depend on z.

The initial state trajectory corresponding to the vacuum state can be approximated by the vortex loop moving very close to the surface of the spherical body, in the layer of atomic size a_0. Its trajectory is thus $z^2 + R^2 = r_0^2$, or $R_i(z) = \sqrt{r_0^2 - z^2}$. On the imaginary axis $z = i\zeta$, the radius of the vortex on such a trajectory can reach the size R_0 of the moving vortex, and thus the trajectory $R_i(z)$ crosses the trajectory $R_f(z) = R_0$ in the complex plane (Fig. 26.8 *right*). The tunneling exponent is produced by the change of the action along the path on imaginary axis $z = i\zeta$. If $R_0 \gg r_0$, in the main part of the trajectory one has $R_i(z) = i\zeta$ and the tunneling exponent is given by the topological action:

$$2\text{Im}S = 2\text{Im} \int dz(p_{zf}(z) - p_{zi}(z)) = 4\pi\kappa_0 n_1 \rho \int_0^{R_0} d\zeta \, (R_0^2 - \zeta^2) \tag{26.20}$$

$$= \frac{8\pi^2}{3} \hbar n_1 n R_0^3 = 2\pi \hbar n_1 N_0 \, . \tag{26.21}$$

Here we introduced the effective number of atoms N_0 involved in quantum nucleation of the vortex loop. In the considered limit $v_0 \ll \kappa_0 n_1/r_0$ this number does not depend on the size r_0 of the impurity: $N_0 = n(4\pi/3)R_0^3$. It is the number of atoms within the solid sphere whose radius coincides with the radius R_0 of the nucleated vortex loop. This reflects the volume law of the topological term in the action for vortex dynamics.

The volume law for the tunneling exponent was also found by Sonin (1973), who calculated the tunneling rate $\Gamma = e^{-2\text{Im}S}$ as the overlapping integral of two many-body wave functions, $\Gamma = |\langle \Psi_f | \Psi_i \rangle|^2$, where the initial state represents the vortex-free vacuum, and the final state is the vacuum with the vortex. The effective action ImS was then minimized with respect to the velocity field around the vortex. The extremal trajectory corresponds to the formation of the intermediate state of the vortex line with the deformed velocity field around the vortex loop. The resulting effective particle number N_0 is logarithmically reduced compared to eqn (26.21) obtained in the collective coordinate description, which assumed the equilibrium velocity field:

$$2\text{Im}S/\hbar = 2\pi n_1 N_0 \ , \quad N_0 = \frac{27}{\pi \ln \frac{R_0}{a}} n R_0^3 \ . \tag{26.22}$$

The linear dependence of the tunneling exponent on the number N_0 of particles, effectively participating in the tunneling, was also found in other systems. For example, it was found by Lifshitz and Kagan (1972) and Iordanskii and Finkelstein (1972) for the quantum nucleation of the true vacuum from the false one in quantum solids.

Experiments on vortex nucleation in superfluid ^{4}He and their possible interpretation in terms of quantum tunneling are discussed in the review paper by Avenel *et al.* (1993). The problem of vortex tunneling was revived due to experiments on vortex creep in superconductors (see the review paper by Blatter *et al.* 1994). For the vortex tunneling in superconductors, and also in fermionic superfluids such as ^{3}He-A and ^{3}He-B, the situation is more complicated because of the fermion zero modes in the vortex core discussed in Part V, which lead to the spectral flow. Nucleation of vortices is supplemented by the nucleation of fermionic charges, as happens in the instanton process in RQFT. Also the entropy of the fermion zero modes is important: it increases the probability of quantum nucleation of vortices in the same manner as discussed by Hawking *et al.* (1995) for quantum nucleation of black holes.

27

VORTEX FORMATION BY KELVIN–HELMHOLTZ INSTABILITY

In the presence of the flexible surface the critical velocity of the hydrodynamic instability becomes substantially lower. The free surface of the superfluid liquid or the interface between two superfluids provides the new soft mode which becomes unstable in the presence of the counterflow. This is the Kelvin–Helmholtz (KH) type of instability (Helmholtz 1868; Kelvin 1910). It gives the reasonably well-understood example of the vortex formation in superfluids. We shall modify the result obtained for classical liquids to our case of superfluid liquids. In Chapter 32.3 we shall see that there is a close relation between the KH instability in superfluids and the physics of black holes on the brane between two quantum vacua.

27.1 Kelvin–Helmholtz instability in classical and quantum liquids

27.1.1 Classical Kelvin–Helmholtz instability

Kelvin–Helmholtz instability belongs to a broad class of interfacial instabilities in liquids, gases, plasma, etc. (see the review paper by Birkhoff 1962). It refers to the dynamic instability of the interface of the discontinuous flow, and may be defined as the instability of the vortex sheet. Many natural phenomena have been attributed to this instability. The most familiar ones are the generation by the wind of waves in water, whose Helmholtz instability was first analyzed by Lord Kelvin (1910), and the flapping of sails and flags analyzed by Lord Rayleigh (1899).

Many of the leading ideas in the theory of instability were originally inspired by considerations about inviscid flows. The corrugation instability of the interface between two ideal liquids sliding along each other was investigated by Lord Kelvin (1910). The critical relative velocity $|v_1 - v_2|$ for the onset of instability toward generation of surface waves (the capillary–gravity waves) is given by

$$\frac{1}{2}\frac{\rho_1\rho_2}{\rho_1+\rho_2}(v_1-v_2)^2 = \sqrt{\sigma F} \ . \qquad (27.1)$$

Here σ is the surface tension of the interface between two liquids; ρ_1 and ρ_2 are their mass densities; and F is related to the external field stabilizing the position of the interface: in the case of two liquids it is the gravitational field

$$F = g(\rho_1 - \rho_2) \ . \qquad (27.2)$$

The surface mode which is excited first has the wave number corresponding to the inverse 'capillary length'

$$k_0 = \sqrt{F/\sigma} \, , \qquad (27.3)$$

and frequency

$$\omega_0 = k_0 \frac{\rho_1 v_1 + \rho_2 v_2}{\rho_1 + \rho_2} \, . \qquad (27.4)$$

From eqn (27.4) it follows that the excited surface mode propagates along the interface with the phase velocity $v_{\text{phase}} = (\rho_1 v_1 + \rho_2 v_2)/(\rho_1 + \rho_2)$.

However, among the ordinary liquids one cannot find an ideal one. That is why in ordinary liquids and gases it is not easy to correlate theory with experiment. In particular, this is because one cannot properly prepare the initial state – the plane vortex sheet is never in equilibrium in a viscous fluid; it is not the solution of the hydrodynamic equations. This is why it is not so apparent whether one can properly discuss its 'instability'.

Superfluids are the only proper ideal objects where these ideas can be implemented without reservations, and where the criterion of instability does not contain viscosity. Recently the first experiment has been performed with two sliding superfluids, where the non-dissipative initial state was well determined, and the well-defined threshold was reported by Blaauwgeers et al. (2002). The initial state is the non-dissipative vortex sheet separating two sliding superfluids in Fig. 17.9 *bottom*. One of the superfluids performs the solid-body-like rotation together with the vessel, while in the other one the superfluid component is in the so-called Landau state, i.e. it is vortex free and thus is stationary in the inertial frame. The threshold of the Kelvin–Helmholtz type of instability has been marked by the formation of vortices in the vortex-free stationary superfluid: this initially stationary superfluid starts to spin up due to the neighboring rotating superfluid.

27.1.2 Kelvin–Helmholtz instabilities in superfluids at low T

The extension of the consideration of classical KH instability to superfluids adds some new physics, which finally leads to the possibility of simulation of the black-hole event horizon and ergoregion (Chapter 32.3). First of all, it is now the two-fluid hydrodynamics with superfluid and normal components which must be incorporated. Let us first consider the limit case of low T, where the fraction of the normal component is negligibly small, and thus the complication of the two-fluid hydrodynamics is avoided. In this case one may guess that the classical result (27.1) obtained for the ideal inviscid liquids is applicable to superfluids too, and the only difference is that in the experiments by Blaauwgeers et al. (2002) the role of gravity in eqn (27.2) is played by the applied gradient of magnetic field H, which stabilizes the position of the interface between ^{3}He-A and ^{3}He-B (see Fig. 27.1):

$$F = \frac{1}{2} \nabla \left((\chi_A - \chi_B) H^2 \right) \, . \qquad (27.5)$$

Here χ_A and χ_B are magnetic susceptibilities of the A- and B-phases respectively.

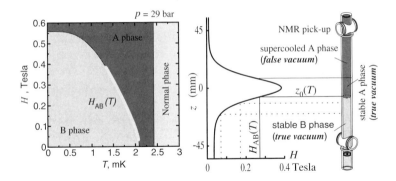

FIG. 27.1. Regulation of the 'gravity' field F in experiments by Blaauwgeers et al. (2002). The role of gravity is played by the gradient of magnetic field in eqn (27.5). The position z_0 of the interface is determined by the equation $H(z_0) = H_{AB}(T)$, where $H_{AB}(T)$ marks the first-order phase transition line between A- and B-phases in the (H, T) plane (*left*). When the temperature T is varied (or the magnitude of the field regulated by the valve current I) the position $z_0(T)$ of the interface is shifted changing the field gradient dH/dz at the position of the interface. The plane $z = z_1$, where also $H(z_0) = H_{AB}(T)$, separates the region where the ^{3}He-A vacuum is true ($z < z_1$) and false ($z > z_1$). The false (metastable) state appears to be extremely stable, and is destroyed only by ionizing radiation (see Sec. 28.1.4).

However, this is not the whole story. The criterion of KH instability in eqn (27.1) depends only on the relative velocity of the sliding liquids. However, there always exists a preferred reference frame of environment. It is the frame of the container, or, if the container walls are far away, the frame where the inhomogeneity of magnetic field H is stationary. Due to interaction of the interface with the environment, the instability can start earlier. The energy of the excitations of the surface, ripplons (quanta of the capillary–gravity waves), becomes negative in the reference frame of environment before the onset of the classical KH instability. The new criterion, which corresponds to appearance of the ergoregion, will depend on the velocities of superfluid vacua with respect to the preferred reference frame of environment. Only if this interaction is neglected will the KH criterion (27.1) be restored. The latter corresponds to the appearance of the metric singularity in general relativity, as will be discussed in Chapter 32.3.

Let us consider these two criteria. We repeat the same derivation as in the case of classical KH instability, assuming the same boundary conditions, but with one important modification: in the process of the dynamics of the interface one must add the friction force arising when the interface is moving with respect to the preferred reference frame of the environment – the frame of container walls, which coincides with the frame of the stable position of the interface. The instability criterion should not depend on the form and magnitude of the friction force; the only requirement is that it is non-zero and thus the interaction with the preferred reference frame of the environment is established in the equations of two-fluid dynamics. That is why we choose the simplest form

$$F_{\text{friction}} = -\Gamma \left(\partial_t \zeta - v_{nz} \right) , \tag{27.6}$$

where $\zeta(x,y,t)$ is perturbation of the position of the interface in the container frame. It is assumed that the z axis is along the normal to the interface (see Fig. 27.1), and the normal component velocity is fixed by the container walls, $v_{nz} = 0$. The friction force in eqn (27.6) is Galilean invariant if the whole system – AB-interface and container – are considered. For the interface alone, the Galilean invariance is violated if $\Gamma \neq 0$. This reflects the interaction with the environment which provides the preferred reference frame. This symmetry violation is the main reason for the essential modification of the KH instability in superfluids.

The parameter Γ in the friction force has been calculated for the case where the interaction between the interface and container is transferred by the remnant normal component. In this case the friction occurs due to Andreev scattering of ballistic quasiparticles by the interface (Yip and Leggett 1986; Kopnin 1987; Leggett and Yip 1990; see Sec. 29.3 and eqn (29.18)).

The relevant perturbation is

$$\zeta(x,t) = a \sin(kx - \omega t) , \tag{27.7}$$

where the x axis is along the local direction of the velocities. In the rotating container the velocities $\mathbf{v}_{s1}$ and $\mathbf{v}_{s2}$ of both superfluids are along the wall of the container, and thus are parallel to each other. Because of the friction, the spectrum of these surface perturbations (ripplons) is modified compared to the classical result in the following way:

$$\rho_1 \left(\frac{\omega}{k} - v_1 \right)^2 + \rho_2 \left(\frac{\omega}{k} - v_2 \right)^2 = \frac{F + k^2 \sigma}{k} - i\Gamma \frac{\omega}{k} . \tag{27.8}$$

or

$$\frac{\omega}{k} = \frac{\rho_1 v_1 + \rho_2 v_2}{\rho_1 + \rho_2} \pm \frac{1}{\sqrt{\rho_1 + \rho_2}} \sqrt{\frac{F + k^2 \sigma}{k} - i\Gamma \frac{\omega}{k} - \frac{\rho_1 \rho_2}{\rho_1 + \rho_2}(v_1 - v_2)^2} . \tag{27.9}$$

If $\Gamma = 0$ the instability occurs when the classical threshold value in eqn (27.1) is reached. At this KH threshold the spectrum of ripplons with $k = k_0$ in eqn (27.3) acquires the imaginary part, Im $\omega(k) \neq 0$. This imaginary part has both signs, and thus above the classical threshold perturbations grow exponentially in time. This is the conventional KH instability.

The frame-fixing parameter Γ changes the situation completely. Because of the friction, the ripplon spectrum always has the imaginary part Im $\omega(k) \neq 0$ reflecting the attenuation of the surface waves. At some velocities the imaginary part Im $\omega(k)$ crosses zero and the attenuation transforms to amplification causing the instability. This occurs first for ripplons with the same value of the wave

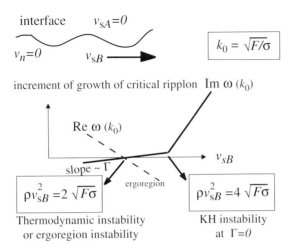

FIG. 27.2. Sketch of imaginary and real parts of frequency of critical ripplon (with $k = k_0$) at the interface between ^{3}He-A and ^{3}He-B under the conditions $v_{sA} = v_{nA} = v_{nB} = 0$, and $T \to 0$, i.e. $v_1 = 0$, $v_2 = v_{sB} \neq 0$; and $\rho_1 = \rho_2 = \rho$. The imaginary part crosses zero, and the attenuation of ripplons transforms to the amplification, just at the same moment when the real part of the ripplon frequency crosses zero. The region where Re $\omega < 0$, i.e. where the ripplon has negative energy, is called ergoregion. The slope of the imaginary part is proportional to the friction parameter Γ. If Γ is strictly zero, and thus the connection with the frame of the environment is lost, the surface instability starts to develop when the classical KH criterion in eqn (27.1) is reached.

vector, as in eqn (27.3) (see Fig. 27.2 for the case when $v_1 = v_{sA} = 0$ and $v_2 = v_{sB} \neq 0$). The onset of instability is given by

$$\frac{1}{2}\rho_1 v_1^2 + \frac{1}{2}\rho_2 v_2^2 = \sqrt{\sigma F} \:. \tag{27.10}$$

The group velocity of the critical ripplon is $v_{\text{group}} = d\omega/dk = 0$, i.e. the critical ripplon is stationary in the reference frame of the container. The frequency of the critical ripplon is $\omega = 0$, i.e. both the real and imaginary parts of the spectrum cross zero at the threshold. The negative value of the real part above the threshold means the appearance of the ergoregion in the frame of the environment, that is why this instability is the instability in the ergoregion. We call it the ergoregion instability.

The ergoregion instability is important for physics of rotating black holes (see Kang 1997 and references therein). As distinct from the Zel'dovich–Starobinsky mechanism of the black-hole ergoregion instability which will be discussed in Sec. 31.4.1, the instability of the ergoregion in the brane world of the AB-interface is caused by the interaction with the bulk environment. The relativistic version of this mechanism will be discussed in Sec. 32.3.

The criterion for the ergoregion instability in eqn (27.10) does not depend on the relative velocities of superfluids, but is determined by the velocities of each of the two superfluids with respect to the environment (to the container or to the remnant normal component). According to this criterion the instability will occur even in the following cases. (1) Two liquids have equal densities, $\rho_1 = \rho_2$, and move with the same velocity, $v_1 = v_2$. This situation is very similar to the phenomenon of a flag flapping in the wind, discussed by Rayleigh in terms of the KH instability – the instability of the passive deformable membrane between two distinct parallel streams having the same density and the same velocity (see the latest experiments by Zhang et al. 2000). [518]) In our case the role of the flag is played by the interface, while the flagpole which pins the flag serves as the reference frame of the environment which violates the Galilean invariance. (2) The superfluids are on the same side of the surface, i.e. there is a free surface of a superfluid which contains two or more interpenetrating superfluid components, say neutron and proton components in neutron stars. (3) There is only a single superfluid with a free surface. This situation which corresponds to $\rho_2 = 0$ was discussed by Korshunov (1991; 2002).

Note that eqn (27.10) does not depend on the frame-fixing parameter Γ and thus is valid for any non-zero Γ even in the limit $\Gamma \to 0$. But eqn (27.10) does not coincide with the classical equation (27.1) obtained when Γ is exactly zero. Such a difference between the limit and exact cases is known in many areas of physics, where the gapless bosonic or fermionic modes are involved. Even in classical hydrodynamics the normal mode of inviscid theory may not be the limit of a normal mode of viscous theory (Lin and Benney 1962). Below we discuss how the crossover between the two criteria, (27.10) and (27.1), occurs when $\Gamma \to 0$.

27.1.3 Ergoregion instability and Landau criterion

Let us first compare both results, (27.10) and (27.1), with the Landau criterion in (26.1). The energy $E(p)$ (or $\omega(k)$) in this Landau criterion is the quasiparticle energy (or the mode frequency) in the superfluid-comoving frame. In our case there are two moving superfluids, which is why there is no unique superfluid-comoving frame. The latter appears either when instead of the interface one considers the surface of a single liquid (i.e. if $\rho_2 = 0$), or if both supefluids move with the same velocity $v_1 = v_2$. In these particular cases the Landau criterion in its simplest form must work. Using the well-known spectrum of capillary–gravity waves on the interface between two stationary liquids

$$\frac{\omega^2(k)}{k^2} = \frac{1}{\rho_1 + \rho_2} \frac{F + k^2 \sigma}{k} , \tag{27.11}$$

one obtains the following Landau critical velocity:

$$v_{\text{Landau}}^2 = \min \frac{\omega^2(k)}{k^2} = \frac{2}{\rho_1 + \rho_2} \sqrt{F\sigma} . \tag{27.12}$$

The Landau criterion coincides with eqn (27.10) for the ergoregion instability if $v_1 = v_2$, or if $\rho_2 = 0$ (for the case when $\rho_2 = 0$ see Andreev and Kompaneetz

1972). But this does not coincide with the canonical KH result (27.1): there is no instability at $v_1 = v_2$ in the canonical KH formalism.

Let us now consider the general case when $v_1 \neq v_2$, and the Landau criterion in the form of equation (27.12), i.e. for a single superfluid velocity, is no longer applicable. For example, in the case of $v_1 = 0$, $v_2 = v$, $\rho_1 = \rho_2 = \rho$ one obtains:

$$v_{\text{naive Landau}}^2 = \frac{1}{\rho}\sqrt{F\sigma} \ , \quad v_{\text{ergoregion}}^2 = \frac{2}{\rho}\sqrt{F\sigma} \ , \quad v_{\text{KH}}^2 = \frac{4}{\rho}\sqrt{F\sigma} \ . \quad (27.13)$$

Such a criterion agrees neither with the canonical KH criterion for $\Gamma = 0$, nor with the ergoregion criterion for $\Gamma \neq 0$. This demonsrates that in our case of the two sliding superfluids the Landau criterion must be used in its more fundamental formulation given in the beginning of Chapter 26 : the instability occurs when in the frame of the environment the frequency of the surface mode becomes zero for the first time: $\omega(k; v_1, v_2) = 0$. This corresponds to the appearance of the ergoregion and coincides with the criterion in eqn (27.10). As distinct from the Landau criterion in the form of (26.1) valid for a single superfluid velocity, where it is enough to know the quasiparticle spectrum in the superfluid-comoving frame, in the case of two or several superfluid velocities one must calculate the quasiparticle spectrum as a function of all the velocities in the frame of the environment.

27.1.4 Crossover from ergoregion instability to Kelvin–Helmholtz instability

The difference in the result for the onset of the interface instability in the two regimes – with $\Gamma = 0$ and with $\Gamma \neq 0$ – disappears in the case when two superfluids move in such a way that in the reference frame of the environment the combination $\rho_1 v_1 + \rho_2 v_2 = 0$. In this arrangement, according to eqn (27.4), the frequency of the critical ripplon created by classical KH instability is zero in the container frame. Thus at this special condition the two criteria, original KH instability (27.1) and ergoregion instability (27.10), must coincide; and they really do.

If $\rho_1 v_1 + \rho_2 v_2 \neq 0$, the crossover between the two regimes occurs by varying the observation time. Let us consider this in an example of the experimental set-up (Blaauwgeers et al. 2002) with the vortex-free B-phase and the vortex-full A-phase in the rotating vessel. In the container frame one has $\mathbf{v}_1 = \mathbf{v}_{sA} = 0$, $\mathbf{v}_2 = \mathbf{v}_{sB} = -\mathbf{\Omega} \times \mathbf{r}$; the densities of the two liquids, ^{3}He-A and ^{3}He-B, are the same with high accuracy: $\rho_A = \rho_B = \rho = mn$. If $\Gamma \neq 0$ the instability occurs at the boundary of the vessel, where the velocity of the ^{3}He-B is maximal, and when this maximal velocity reaches the value $v_{\text{ergoregion}}$ in eqn (27.13). This velocity is smaller by $\sqrt{2}$ than that given by the classical KH equation: $v_{\text{ergoregion}} = v_{\text{KH}}/\sqrt{2}$.

From eqn (27.8) it follows that slightly above the threshold $v_c = v_{\text{ergoregion}}$ the increment of the exponential growth of the critical ripplon is

$$\text{Im } \omega(k_0) = \frac{\Gamma k_0}{2\rho}\left(\frac{v_{sB}}{v_c} - 1\right) \ , \ \text{at } v_{sB} - v_c \ll v_c \ . \quad (27.14)$$

In the limit of vanishing frame-fixing parameter, $\Gamma \to 0$, the increment becomes small and the ergoregion instability of the interface has no time to develop if the observation time is short enough. The interface becomes unstable only at a higher velocity of rotation when the classical threshold of KH instability, v_{KH} in eqn (27.1), is reached (see Fig. 27.2). Thus, experimental results in the limit $\Gamma \to 0$ would depend on the observation time – the time one waits for the interface to be coupled to the environment and for the instability to develop. For sufficiently short time one will measure the classical KH criterion (27.1). For example, in experiments with constant acceleration of rotation $\dot{\Omega}$, the classical KH regime is achieved if the acceleration is fast enough: $\dot{\Omega} \gg \lambda \Gamma$, where $\lambda = v_{KH}^3/\sigma R$, and R is the radius of the vessel. For slow acceleration, $\dot{\Omega} \ll \lambda \Gamma$, the ergoregion instability criterion (27.10) will be observed.

27.2 Interface instability in two-fluid hydrodynamics

27.2.1 *Thermodynamic instability*

Let us now consider the case of non-zero T, where each of the two liquids contain superfluid and normal components. In this case the analysis requires the 2×2-fluid hydrodynamics. This appears to be a rather complicated problem, taking into account that in some cases the additional degrees of freedom related to the interface itself must also be added. The two-fluid hydrodynamics has been used by Korshunov (1991; 2002) to investigate of the instability of the free surface of superfluid ^{4}He triggered by the relative motion of the normal component of the liquid with respect to the superfluid one. We avoid all these complications assuming that the viscosity of the normal components of both liquids is high, as actually happens in superfluid ^{3}He. In this high-viscosity limit we can ignore the dynamics of the normal components, which are clamped by the container walls. Then the problem is reduced to a problem of the thermodynamic instability of the superflow in the presence of the interface.

We start with the following initial non-dissipative state corresponding to the thermal equilibrium in the presence of the interface and superflows. In thermal equilibrium the normal component must be at rest in the container frame, $\mathbf{v}_{n1} = \mathbf{v}_{n2} = 0$, while the superfluids can move along the interface with velocities $\mathbf{v}_{s1}$ and $\mathbf{v}_{s2}$ (here the velocities are in the frame of the container). The onset of instability can be found from free energy consideration: when the free energy of static perturbations of the interface becomes negative in the frame of the environment (the container frame) the initial state becomes thermodynamically unstable. The free energy functional for the perturbations of the interface in the reference frame of the container contains the potential energy in the 'gravity' field, surface energy due to surface tension, and kinetic energy of velocity perturbations $\tilde{\mathbf{v}}_{s1} = \nabla \Phi_1$ and $\tilde{\mathbf{v}}_{s2} = \nabla \Phi_2$ caused by deformations of the interface:

$$\mathcal{F}\{\zeta\} = \frac{1}{2} \int dx \left(F\zeta^2 + \sigma(\partial_x \zeta)^2 + \int_{-\infty}^{\zeta} dz \rho_{s1ik} \tilde{v}_{s1}^i \tilde{v}_{s1}^k + \int_{\zeta}^{\infty} dz \rho_{s2ik} \tilde{v}_{s2}^i \tilde{v}_{s2}^k \right) .$$
(27.15)

For generality we discuss anisotropic superfluids, whose superfluid densities $\rho_s = mn_s$ are tensors. The velocity perturbation fields $\tilde{\mathbf{v}}_{sk} = \nabla \Phi_k$, obeying the continuity equation $\partial_i(\rho_s^{ik}\tilde{v}_{sk}) = 0$, have the following form:

$$\Phi_1(x, z<0) = A_1 e^{k_1 z} \cos kx \,, \quad \Phi_2(x, z>0) = A_2 e^{-k_2 z} \cos kx \,, \quad (27.16)$$

$$\rho_{s1z} k_1^2 = \rho_{s1x} k^2 \,, \quad \rho_{s2z} k_2^2 = \rho_{s2x} k^2 \,. \quad (27.17)$$

The connection between the deformation of the interface, $\zeta(x) = a \sin kx$, and the velocity perturbations follows from the boundary conditions. Because of the large viscosity of the normal component it is clamped by the boundaries of the vessel. Then from the requirement that the mass and the heat currents are conserved across the interface, one obtains that the superfluid velocity in the direction normal to the wall must be zero: $\mathbf{v}_{s1} \cdot \mathbf{n} = \mathbf{v}_{s2} \cdot \mathbf{n} = 0$. This gives the following boundary conditions for perturbations:

$$\partial_z \Phi_1 = v_{s1} \partial_x \zeta \,, \quad \partial_z \Phi_2 = v_{s2} \partial_x \zeta \,. \quad (27.18)$$

Substituting this in the free-energy functional (27.15), one obtains the quadratic form of the free energy of the surface modes

$$\mathcal{F}\{\zeta\} = \frac{1}{2} \sum_k |\zeta_k|^2 \left(F + k^2 \sigma - k \left(\sqrt{\rho_{sx1}\rho_{sz1}} v_{s1}^2 + \sqrt{\rho_{sx2}\rho_{sz2}} v_{s2}^2 \right) \right) \,. \quad (27.19)$$

This energy becomes negative for the first time for the critical ripplon with $k_0 = (F/\sigma)^{1/2}$ when

$$\frac{1}{2} \left(\sqrt{\rho_{sx1}\rho_{sz1}} v_{s1}^2 + \sqrt{\rho_{sx2}\rho_{sz2}} v_{s2}^2 \right) = \sqrt{\sigma F} \,. \quad (27.20)$$

At $T = 0$, when the normal components of the liquids disappear and one has $\rho_{sx1} = \rho_{sz1} = \rho_1$ and $\rho_{sx2} = \rho_{sz2} = \rho_2$, this transforms to the criterion (27.10) for the ergoregion instability. Equation (27.20) reproduces the experimental data obtained by Blaauwgeers et al. (2002) without any fitting parameters (see Fig. 27.3).

27.2.2 Non-linear stage of instability

In experiments by Blaauwgeers et al. (2002) the onset of the interface ergoregion instability is marked by the appearance of the vortex lines in ^{3}He-B which are monitored in NMR measurements. Vortices appear at the non-linear stage of the instability. The precise mechanism of the vortex formation is not yet known. One may guess that the A-phase vorticity concentrated in the vortex layer at the interface (Fig. 17.9) bottom) is pushed by the Magnus force toward the vortex-free B-phase region as was suggested by Krusius et al. (1994) to interpret their experiments on vortex penetration through the moving AB-interface. When the potential well for vortices is formed by the corrugation of the interface (see Fig. 27.4), the interfacial vortices are pushed there further deepening the potential well until it forms the droplet of the A-phase filled by vorticity. The vortex-full

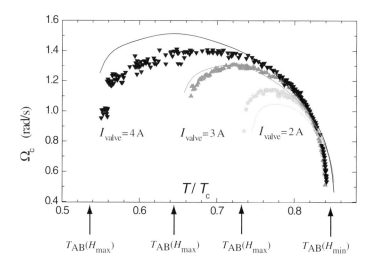

FIG. 27.3. Critical velocity of KH type of instability of the AB-interface as a function of T in experiments by Blaauwgeers et al. (2002). When the temperature changes the equilibrium position of the interface also changes, together with the critical velocity determined by the field gradient at the interface playing the role of gravity. The critical velocity of the instability tends to zero when the position of the interface approaches minimal or maximal values of the magnetic field in the cell, where $dH/dz \rightarrow 0$ (see Fig. 27.1). Temperatures at which $dH/dz \rightarrow 0$ are marked by arrows. Three sets of data correspond to three different values of the maximum of the applied magnetic field, which is regulated by applying different currents I_{valve}. Solid curves are from the ergoregion instability criterion in eqn (27.20) at the conditions $v_{s1} \equiv v_{sB} = \Omega R$, $v_{s2} \equiv v_{sA} = v_{nA} = v_{nB} = 0$. No fitting parameters were used.

droplet propagates to the bulk B-phase where such a multiply-quantized vortex relaxes to the singly-quantized vortex lines.

Under the conditions of the experiment, penetration of vortices into the B-phase decreases the counterflow there below the instability threshold, and the vortex formation is stopped. That is why one may expect that the vortex-full droplet is nucleated during the development of the instability from a single seed. The size of the seed is about one-half of the wavelength $\lambda_0 = 2\pi/k_0$ of the perturbation. The number of created vortices is found from the circulation of superfluid velocity carried by a piece of the vortex sheet of size $\lambda_0/2$, which is determined by the jump of superfluid velocity across the sheet: $\kappa = |\mathbf{v}_{sB} - \mathbf{v}_{sA}|\lambda_0/2$. Dividing this by the circulation quantum κ_0 of the created B-phase vortices, one obtains the number of vortices produced as the result of growth of one segment of the perturbation:

INTERFACE INSTABILITY IN TWO-FLUID HYDRODYNAMICS

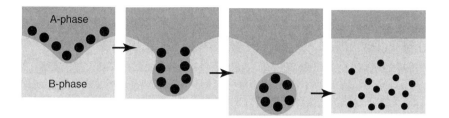

FIG. 27.4. Possible scenario of vortex formation by the surface instability.

$$N = \frac{\kappa}{\kappa_0} \sim \frac{v_c \lambda_0}{2\kappa_0} . \qquad (27.21)$$

It is about 10 vortices per event under the conditions of the experiment. This is in good agreement with the measured number of vortices created per event (Blaauwgeers et al. 2002), which confirms the droplet mechanism of vortex penetration.

Probably, the experiments on surface instability in superfluids will allow the solution of the similar problem of the non-linear stage of instability in ordinary liquids (see e.g. Kuznetsov and Lushnikov (1995)).

The vortex formation by surface instability is a rather generic phenomenon. It occurs in laser-manipulated Bose gases (Madison et al. 2001; Sinha and Castin 2001). It can be applied to different kinds of interfaces, in particular to the boundary between the normal and superfluid liquids whose shear flow instability has been discussed by Aranson et al. (2001), see Sec. 28.2.4. Such an interface naturally appears, for example, as a boundary of a gaseous Bose–Einstein condensate, or at the rapid phase transition into the superfluid state as will be discussed in the next chapter.

The instability of the free surface of superfluid under the relative flow of the normal and superfluid components of the same liquid has been recently reexamined by Korshunov (2002), who also obtained two criteria of instability depending on the interaction with the environment. In his case, the frame-fixing parameter which regulates the interaction with the environment is the viscosity η of the normal component of the liquid. For $\eta \neq 0$ the critical counterflow for the onset of surface instability is η-independent:

$$\frac{1}{2}\rho_s (v_s - v_n)^2 = \sqrt{\sigma F} . \qquad (27.22)$$

It corresponds to eqn (27.20), but as expected it does not coincide with the result obtained for exactly zero viscosity. The same eqn (27.22) was obtained by Kagan (1986) and Uwaha and Nozieres (1986) for the threshold of excitation of crystallization waves at the solid–liquid interface by the liquid flow

$$\frac{1}{2}\rho_s (v_s - v_{\text{solid}})^2 = \sqrt{\sigma F} . \qquad (27.23)$$

The environment reference frame here is provided by the crystal lattice.

One can argue that the formation of the singular-core vortices on the rough sample boundary (Parts *et al.* 1995*b*) is also an example of surface instability. When the Landau criterion for quasiparticles is reached at the sharpest surface spike, a bubble of normal liquid is created around this spike. The bubble's interface with respect to the surrounding superflow, which moves at a high relative velocity, then undergoes the surface instability creating vortices (Fig. 28.6).

28

VORTEX FORMATION IN IONIZING RADIATION

28.1 Vortices and phase transitions

Let us now return to vortex nucleation in a single superfluid liquid ^{3}He-B. The threshold v_c of the hydrodynamic instability of the flow in ^{3}He-B in Fig. 26.7, at which vortices are nucleated, is several orders of magnitude larger than the Landau criterion for vortex nucleation. Because of a huge energy barrier for vortex nucleation in superfluid ^{3}He, the energetically metastable superflow with $v_{\rm s} \gg v_{\rm Landau}$ will persist on a geological time scale if $v_{\rm s} < v_c$. However, Ruutu et al. (1996a, 1998) observed that ionizing radiation helps vortices to overcome the barrier. They found that under neutron irradiation vortices are formed below v_c, in the velocity region $v_{\rm neutron} < v_{\rm s} < v_c$. The new experimental threshold $v_{\rm neutron}$ roughly corresponds to the Landau velocity for formation of vortices in the confined region of the size R_b of the fireball formed by a neutron (see Fig. 28.1), $v_{\rm neutron} \sim \kappa_0/\pi R_b$.

According to current belief, the vortex formation observed in the subcritical regime $v < v_c$ under ionizing radiation occurs via the Kibble–Zurek (KZ) mechanism, which was originally developed to describe the phase transitions in the early Universe. In this scenario a network of cosmic strings is formed during a rapid non-equilibrium second-order phase transition, owing to thermal fluctuations.

The formation of topological defects in non-equilibrium phase transitions is a very generic phenomenon. It is sometimes called the phase ordering, which reflects the process of the establishment of the homogeneous order parameter state when the defects generated by the transition are decaying. That vortices are necessarily produced in non-equilibrium phase transition can be viewed from the general consideration of the superfluid phase transition in terms of proliferation of vortices (Onsager 1949; Feynman 1955; see Fig. 28.2). The description of the broken-symmetry phase transition in terms of topological defects is especially useful for such transitions where in the low-T state the order parameter cannot be introduced. This happens for example in 2+1 systems where the transition – the Berezinskii–Kosterlitz–Thouless transition – is exclusively described in terms of the unbinding of vortex pairs and formation of the vortex plasma.

28.1.1 Vortices in equilibrium phase transitions

Qualitatively the phase transition with broken $U(1)$ symmetry can be described as the topological transition. Let us consider the algebraic sum of the topological charges $n_1(C)$ of all vortices which cross the surface σ stretched on a big loop C as a function of the length of the loop L.

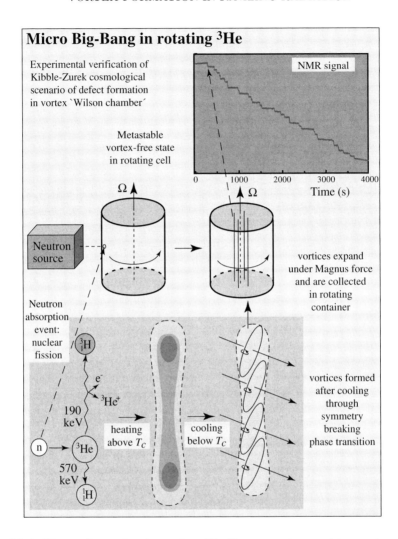

FIG. 28.1. Vortex formation in a micro Big-Bang event caused by neutron irradiation in experiments by Ruutu et al. (1996a).

The disordered state is characterized by an infinite cluster of vortex lines – 'a connected tangle throughout the liquid' according to Onsager (1949). The number of positive and negative charges is equal on average, and thus their algebraic sum is determined by statistical noise. Below the superfluid transition there are finite-size vortex loops instead of the infinite cluster, and thus the statistical properties of the topological noise become different. In the case of 2D systems the statistical noise of the topological charges above and below the phase transition has been discussed by Kosterlitz and Thouless (1978).

Above the phase transition the isolated point vortices form the plasma state,

Equilibrium states

Disordered phase:
infinite cluster of defects

Ordered phase:
defects form closed loops

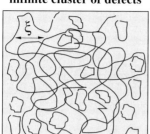
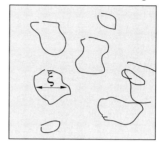

Nonequilibrium phase transition:
infinite cluster cannot disappear immediately due to topology,
it survives in the ordered phase

intervortex distance ξ_V gradually increases

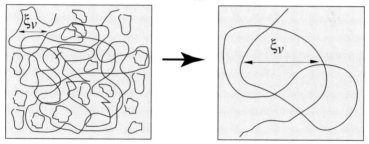

FIG. 28.2. Infinite vortex cluster in disordered state is topologically different from the array of vortex loops in the ordered state. That is why after the quench of the disordered state the infinite cluster persists in the superfluid state.

so that the amplitude of the noise is proportional to the square-root of the total number of point vortices. Thus the algebraic number of vortices in the disordered phase is proportional the square root of area of the surface σ: $|n_1(L)| \sim \sqrt{S/\xi^2} \sim L/\xi$ when $L/\xi \to \infty$. In the ordered superfluid state, only pairs of vortices are present. In this case the statistical noise $|n_1(L)|$ is produced by the number of pairs cut by the loop C. As a result the topological charge is proportional to the square root of the perimeter of the loop, $|n_1(L)| \sim (L/\xi)^{1/2}$.

Such a difference between the superfluid (ordered) and non-superfluid (disordered) states can be expressed in terms of the loop function (see e.g. Toulouse 1979):

$$g(L) = \left\langle e^{i(2\pi/\kappa_0) \oint_C \mathbf{v}_s \cdot d\mathbf{r}} \right\rangle , \qquad (28.1)$$

where κ_0 is the circulation quantum. In the superfluid state of 2D system the loop function decays as $g(L) \sim e^{-L/\xi}$, while in the normal state the exponent contains an area of the loop: $g(L) = e^{-L^2/\xi^2}$. This can be mapped to the quantum phase transitions in gauge theories, if $\mathbf{v}_s$ is substituted by the gauge field; the state with the area law is called the confinement phase, because the charges are confined there (see the book by Polyakov 1987).

The equilibrium phase transition between the area law and the perimeter law occurs at the temperature at which the coherence length $\xi(T)$ in eqn (10.5) becomes infinite, $\xi(T_c) = \infty$.

28.1.2 Vortices in non-equilibrium phase transitions

In the non-equilibrium rapid cooling through the transition, the coherence length cannot follow its thermodynamic equilibrium value determined by temperature because of the slow-down of processes near the transition point. Thus ξ does not cross infinity, and as a result the infinite cluster of the disordered state inevitably persists below T_c. This is observed as formation of $U(1)$ vortices after a rapid phase transition into the state with broken $U(1)$ symmetry.

The qualitative theory of defect formation after quench has been put forward by Zurek (1985). His theory determines the moment – called Zurek time – at which the vortices or strings belonging to the infinite cluster become well defined in the low-T state, i.e. when they become resolved from the background thermal fluctuations in the ordered phase. The density of vortices in this cluster, which starting from this point can be experimentally investigated, is called the initial vortex density. It is determined by the interplay of the cooling time τ_Q and the internal characteristic relaxation time $\tau(T)$ of the system, which diverges at $T \to T_c$ (the so-called critical slow-down). At infinite cooling time an equilibrium situation is restored, and no vortices of infinite length will be observed in the superfluid state. The cooling rate shows how fast the temperature changes when the phase transition temperature is crossed: $1/\tau_Q = \partial_t T/T_c$. If $t = 0$ is the moment when T_c is crossed one has

$$\frac{T(t)}{T_c} = 1 - \frac{t}{\tau_Q}. \tag{28.2}$$

Far above transition the intervortex distance ξ_v in the infinite cluster is determined by the coherence length: $\xi_v = \xi(T) = \xi_0|1-T/T_c|^{-1/2}$ if the Ginzburg–Landau expansion is applicable. When $T \to T_c$ the intervortex distance $\xi_v(t)$ increases together with the thermodynamic coherence length. The relaxation time of the system also increases; if the time-dependent Ginzburg–Landau theory (see below) is applicable, one has

$$\tau(t) = \tau_0 \left|1 - \frac{T(t)}{T_c}\right|^{-1} = \frac{\tau_0 \tau_Q}{t}. \tag{28.3}$$

At some moment $t = -t_Z$ the relaxation time approaches the characteristic time of variation of temperature. After that the system parameters can no longer

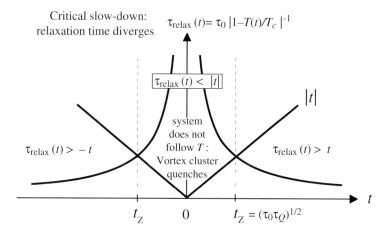

FIG. 28.3. Non-equilibrium region where the system does not follow the temperature change and the infinite vortex cluster is quenched.

follow their equilibrium thermodynamic values (Fig. 28.3). This is the Zurek time t_Z, which is found from the equation $t_Z = \tau(t_Z)$:

$$t_Z = \sqrt{\tau_0 \tau_Q} \ . \tag{28.4}$$

Thus in the vicinity of the phase transition, at $|t| < t_Z$, the infinite vortex cluster is quenched with the intervortex distance being equal to $\xi_v = \xi(t_Z)$. This vortex spaghetti persists into the ordered state, and finally becomes observable when the core size $\xi(T)$ becomes smaller than ξ_v, i.e. at $t > t_Z$. Thus $\xi_v(t_Z)$

$$\xi_v = \xi(t_Z) \sim \xi_0 \left(\frac{\tau_Q}{\tau_0}\right)^{1/4} \tag{28.5}$$

determines the initial density of vortices in the ordered state: $n_v \sim \xi_v^{-2} = \xi_0^{-2}\sqrt{\tau_0/\tau_Q}$.

At $t > t_Z$ the system represents the superfluid state contaminated by the vortex spaghetti gradually relaxing to the thermodynamic equilibrium superfluid state. The topological noise – an extra topological charge accumulated in the infinite vortex cluster – escapes to the boundaries, the cluster is rarefied, and finally disappears. However, it is possible that one or several vortex lines remain pinned between the opposite boundaries forming the remnant vorticity. If the cosmological phase transition occurred in the expanding (and thus non-equilibrium) Universe, the 'remnant vorticity' could also arise: the 'infinite' cosmic string can stretch from horizon to horizon (see the reviews by Hindmarsh and Kibble (1995) and Vilenkin and Shellard (1994)).

Here we discussed the formation of topological defects of a global $U(1)$ group. On the formation of topological defects in gauge field theories see the review paper by Rajantie (2002). This review contains the most complete list of references to papers dealing with the KZ scenario.

28.1.3 Vortex formation by neutron radiation

The details of experiments by Ruutu *et al.* (1996*a*) in superfluid ^{3}He-B are shown in Fig. 28.1. A cylindrical container is rotating with angular velocity below the hydrodynamic instability threshold $\Omega_c = v_c/R$, where R is the radius of the container. Thus the superfluid component is in the vortex-free Landau state, i.e. it is at rest with respect to the inertial frame. Such a state stores a huge amount of kinetic energy in the frame of the environment which is now the frame of the rotating container. In a container with typical radius $R = 2.5$ mm and height $L = 7$ mm, rotating with angular velocity $\Omega = 3$ rad s^{-1}, the kinetic energy of the counterflow is $(1/2) \int dV \, \rho_s (v_s - v_n)^2 \sim 10$ GeV. This energy cannot be released, since as we know the intrinsic half-period of the decay of the superflow due to vortex formation is essentially larger than, say the proton lifetime.

Ionizing radiation assists in releasing this energy by producing vortex loops. Experiments with irradiated superfluid ^{3}He were started in 1992 in Stanford, where it was found that the irradiation assists the transition of supercooled ^{3}He-A to ^{3}He-B (see the discussion by Leggett 1992). Later Bradley *et al.* (1995) demonstrated that the neutron irradiation of ^{3}He-B produces a shower of quasiparticles. Then Ruutu *et al.* (1996*a*) observed that in a rotating vessel, neutrons produce vortices. In experiments with stationary ^{3}He-B, Bäuerle *et al.* (1996) accurately measured the energy deposited from a single neutron to quasiparticles in the low-T region and found an energy deficit indicating that in addition to quasiparticles vortices are also formed.

The decay products from the neutron absorption reaction $n + {}^3_2\text{He} = p + {}^3_1\text{H} + 764$ keV generate ionization tracks. The details of this process are not well known in liquid ^{3}He, but it causes heating which drives the temperature in a small volume of ~ 100 μm size above the superfluid transition (Fig. 28.1). Inside the hot spot the $U(1)$ symmetry is restored. Subsequently the heated bubble cools back below T_c into the broken-symmetry state with a thermal relaxation time of order $1\,\mu$s. This process forms the necessary conditions for the KZ mechanism within the cooling bubble. But as in each real experiment there are several complications when compared to the assumptions made when the theory was derived.

Most of these complications are related to the finite size of the fireball. The temperature distribution within the cooling bubble is non-uniform; the mean free path of quasiparticles is comparable to the characteristic size of the bubble; the phase of the order parameter is fixed outside the bubble; there is an external bias – the counterflow; etc. All this raises concerns whether the original KZ scenario is responsible for the vortex formation observed in neutron experiments (Leggett 2002). There are, however, some experimental observations which are in favor of the KZ mechanism. (i) As was found by Ruutu *et al.* (1996*a*) the statistics of the number of vortices created per event is in qualitative and even in quantitative agreement with the KZ theory; (ii) further measurements by Eltsov *et al.* (2000) demonstrated that the spin–mass vortex (the combination of a conventional (mass) vortex and spin vortex, see Sec. 14.1.4) is also formed and is directly observed in the neutron irradiation experiment. This strengthens the importance of the KZ mechanism, which is applicable to the nucleation of different topological defects, and places further constraints on the interplay between it and other competing effects, which must be included as modifications of the original scenario.

28.1.4 Baked Alaska vs KZ scenario

Unfortunately, the complete theory of the vortex formation in ^{3}He-B after nuclear reaction, as well as the theory of the formation of ^{3}He-B bubbles in supercooled ^{3}He-A by ionizing radiation, must include several very different energy scales, and thus it cannot be described in terms of the effective theory only. If the mean free path is long and increases with decreasing energy, a 'Baked Alaska' effect takes place, as has been described by Leggett (1992). A thin shell of the radiated high-energy particles expands with the Fermi velocity v_F, leaving behind a region at reduced T. In this region, which is isolated from the outside world by a warmer shell, a new phase can be formed. This scenario provides a possible explanation of the formation of the B-phase in the supercooled A-phase.

Such a Baked Alaska mechanism for generation of the bubble with the false vacuum has also been discussed in high-energy physics, where it describes the result of a hadron–hadron collision. In the relativistic case the thin shell of energetic particles expands with the speed of light. In the region separated from the exterior vacuum by the hot shell a false vacuum with a chiral condensate can be formed (Bjorken 1997; Amelino-Camelia *et al.* 1997).

However, there can be other scenarios of the process of the formation of the B-phase in the supercooled A-phase. In particular, an extension of the KZ mechanism to the formation of the domain walls – interfaces between A- and B-phases – has been suggested (Volovik 1996). It was numerically simulated by Bunkov and Timofeevskaya (1998). A- and B-phases represent local minima of almost equal depth, but are separated from each other by a large energy barrier. The interface between them can be considered in the same manner as a topological defect, and can be nucleated in thermal quench together with vortices. This also has an analogy in cosmology (see the book by Linde 1990). The unification symmetry group at high energy ($SU(5)$, $SO(10)$ or $G(224)$) can be broken in different ways: into the phase $G(213) = SU(3) \times SU(2) \times U(1)$, which is our world, and into $G(14) = SU(4) \times U(1)$, which corresponds to a false vacuum with higher energy. In some models both phases represent local minima of almost equal depth, but are separated from each other by a large energy barrier in the same manner as the A- and B-phases of ^{3}He. The non-equilibrium (cosmological) phase transition can create a network of AB interfaces ($G(213)$–$G(14)$ interfaces) which evolves either to the true or false vacuum.

In the case of the vortex formation the competing mechanism to the KZ scenario is the instability of the propagating front of the second-order phase transition (Aranson *et al.* 1999, 2001, 2002), which is an analog of the surface instability discussed in Chapter 27.

Due to all these complications, at the moment one can discuss only some fragments of the processes following the 'Big Bang' caused by neutron reaction. Here we shall concentrate on two such fragments: modification of the KZ scenario to the cases when (i) there is a temperature gradient; and (ii) in addition the counterflow is present. This can be considered using the effective theory – the time-dependent Ginzburg–Landau theory. Though the applicability of this theory to the real situation is under question, the results reveal some generic behavior, which will most probably survive beyond the effective theory.

28.2 Vortex formation at normal–superfluid interface

28.2.1 *Propagating front of second-order transition*

In the presence of the temperature gradient the phase transition does not occur simultaneously in the whole space. Instead the transition propagates as a phase front between the normal and superfluid phases. For a rough understanding of the modification of the KZ scenario of vortex formation to this situation let us consider the time-dependent Ginzburg–Landau (TDGL) equation for the one-component order parameter $\Psi = \Delta/\Delta_0$:

$$\tau_0 \frac{\partial \Psi}{\partial t} = \left(1 - \frac{T(\mathbf{r},t)}{T_c}\right)\Psi - \Psi|\Psi|^2 + \xi_0^2 \nabla^2 \Psi \ . \tag{28.6}$$

Here, as before, $\tau_0 \sim \hbar/\Delta_0$ and ξ_0 are correspondingly the relaxation time of the order parameter and the coherence length far from T_c.

If the quench occurs homogeneously in the whole space $\mathbf{r}$, the temperature depends only on one parameter, the quench time τ_Q:

$$T(t) \approx \left(1 - \frac{t}{\tau_Q}\right)T_c \ . \tag{28.7}$$

In the presence of a temperature gradient, say along x, a new parameter appears:

$$T(x - ut) \approx \left(1 - \frac{t - x/u}{\tau_Q}\right)T_c \ . \tag{28.8}$$

Here u is the velocity of the temperature front which is related to the temperature gradient

$$\nabla_x T = \frac{T_c}{u\tau_Q} \ . \tag{28.9}$$

The limit case $u = \infty$ corresponds to the homogeneous phase transition, where the KZ scenario of vortex formation is applicable.

In the opposite limit case $u \to 0$ at $\tau_Q u = \text{constant}$, the order parameter is almost in equilibrium and follows the transition temperature front:

$$|\Psi(x,t)|^2 = \left(1 - \frac{T(x-ut)}{T_c}\right), \quad T < T_c. \tag{28.10}$$

In this case the phase coherence is preserved behind the transition front and thus no defect formation is possible. It is clear that there exists some characteristic critical velocity u_c of the propagating temperature front, which separates two regimes: at $u \geq u_c$ the vortices are formed, while at $u \leq u_c$ the defect formation either is strongly suppressed (Kibble and Volovik 1997; Kopnin and Thuneberg 1999) or completely stops (Dziarmaga et al. 1999).

28.2.2 Instability region in rapidly moving interface

Let us consider how vortices are formed in the limit of large velocity of the front. As was found by Kopnin and Thuneberg (1999), if $u \gg u_c$ the phase transition front – the interface between the normal and superfluid liquids – cannot follow the temperature front: it lags behind (see Fig. 28.4). In the space between these two boundaries the temperature is already below the phase transition temperature, $T < T_c$, but the phase transition has not yet happened, and the order parameter is still not formed, $\Psi = 0$.

Let us estimate the width of the region of the supercooled normal phase. In steady laminar motion the order parameter depends on $x - ut$. Introducing a dimensionless variable $\tilde{x}$ and a dimensionless parameter a,

$$\tilde{x} = (x - ut)(u\tau_Q \xi_0^2)^{-1/3}, \quad a = \left(\frac{u\tau_0}{\xi_0}\right)^{4/3}\left(\frac{\tau_Q}{\tau_0}\right)^{1/3}, \tag{28.11}$$

one obtains the linearized TDGL equation in the following form:

$$\frac{d^2\Psi}{d\tilde{x}^2} + a\frac{d\Psi}{d\tilde{x}} - \tilde{x}\Psi = 0, \tag{28.12}$$

or

$$\Psi(\tilde{x}) = \text{constant} \cdot e^{-a\tilde{x}/2}\chi(\tilde{x}), \quad \frac{d^2\chi}{d\tilde{x}^2} - \left(\tilde{x} + \frac{a^2}{4}\right)\chi = 0. \tag{28.13}$$

This means that Ψ is an Airy function, $\chi(\tilde{x} - \tilde{x}_0)$, centered at $\tilde{x} = \tilde{x}_0 = -a^2/4$ and attenuated by the exponential factor $e^{-a\tilde{x}/2}$.

At large velocity of the front $a \gg 1$, it follows from eqn (28.13) that $\Psi(\tilde{x})$ quickly vanishes as $\tilde{x}$ increases above $-a^2/4$. Thus there is a supercooled region $-a^2/4 < \tilde{x} < 0$, where $T < T_c$, but the order parameter is not yet formed: the solution is essentially $\Psi = 0$. The lag between the order parameter and temperature fronts is $\tilde{x}_0 = a^2/4$ or in conventional units

$$x_0(u) = \frac{1}{4}\frac{u^3 \tau_Q \tau_0^2}{\xi_0^2}. \tag{28.14}$$

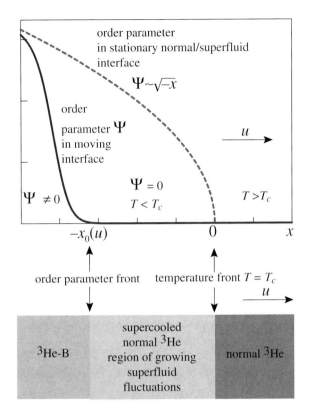

FIG. 28.4. The superfluid order parameter in the rapidly moving temperature front of the second-order transition in the frame of the front according to Kopnin and Thuneberg (1999). The phase transition front lags behind the temperature front forming the metastable region where $T < T_c$ but the mean-field order parameter is not developed. This region represents the supercooled metastable normal ^{3}He. Fluctuations growing in this region produce vortices according to the KZ scenario.

28.2.3 Vortex formation behind the propagating front

The existence of the supercooled normal phase in a wide region in Fig. 28.4 was obtained as the solution of the TDGL equation, which does not take into account thermal fluctuations of the order parameter. In reality such a supercooled region is unstable toward the formation of bubbles of the superfluid phase with $\Psi \neq 0$, which, however, cannot occur if there are no seeds provided by external or thermal noise. If the region is big enough the growth occurs independently in different regions of the space, which is the source of the vortex formation according to the KZ mechanism. At a given point of space **r** the development of the instability from the seed can be found from the linearized TDGL equation, since during the initial growth of the order parameter $\Psi = |\Psi|e^{i\Phi}$ the cubic term can be neglected:

$$\tau_0 \frac{\partial \Psi}{\partial t} = \frac{t}{\tau_Q} \Psi . \tag{28.15}$$

This gives an exponentially growing order parameter, which starts from some seed Ψ_{fluc}, caused by fluctuations:

$$\Psi(\mathbf{r}, t) = \Psi_{\text{fluc}}(\mathbf{r}) \exp \frac{t^2}{2\tau_Q \tau_0} . \tag{28.16}$$

Because of the exponential growth, even if the seed is small, the modulus of the order parameter reaches its equilibrium value $|\Psi_{\text{eq}}| = \sqrt{1 - T/T_c}$ after the Zurek time t_Z:

$$t_Z = \sqrt{\tau_Q \tau_0} . \tag{28.17}$$

This occurs independently in different regions of space and thus the phases of the order parameter in each bubble are not correlated. The spatial correlation between the phases becomes important at distances ξ_v where the gradient term in eqn (28.6) becomes comparable to the other terms at $t = t_Z$. Equating the gradient term $\xi_0^2 \nabla^2 \Psi \sim (\xi_0^2/\xi_v^2) \Psi$ to, say, the term $\tau_0 \partial \Psi/\partial t|_{t_Z} = \sqrt{\tau_0/\tau_Q} \Psi$, one obtains the characteristic Zurek length scale:

$$\xi_v = \xi_0 \left(\tau_Q/\tau_0 \right)^{1/4} . \tag{28.18}$$

At this scale the bubbles with different phases Φ of the order parameter touch each other forming vortices if the phases do not match properly. This determines the initial distance between the defects in homogeneous quench.

We can estimate the lower limit of the characteristic value of the fluctuations $\Psi_{\text{fluc}} = \Delta_{\text{fluc}}/\Delta_0$, which serve as a seed for the vortex formation. If there is no other source of fluctuations, caused, say, by external noise, the initial seed is provided by thermal fluctuations of the order parameter in the volume ξ_v^3. The energy of such fluctuation is $\xi_v^3 \Delta_{\text{fluc}}^2 N_F(0)/E_F$, where E_F is the Fermi energy and $N_F(0)$ the fermionic density of states in the normal Fermi liquid. Equating this energy to the temperature $T \approx T_c$ one obtains the magnitude of the thermal fluctuations of the order parameter

$$\frac{|\Psi_{\text{fluc}}|}{|\Psi_{\text{eq}}|} \sim \left(\frac{\tau_0}{\tau_Q} \right)^{1/8} \frac{T_c}{E_F} . \tag{28.19}$$

Since the fluctuations are initially rather small their growth time exceeds the Zurek time by the factor $\sqrt{\ln |\Psi_{\text{eq}}|/|\Psi_{\text{fluc}}|}$.

Now we are able to estimate the threshold u_c above which the vortex formation becomes possible. The defects are formed when the time of growth of fluctuations, $\sim t_Z = \sqrt{\tau_Q \tau_0}$, is shorter than the time $t_{\text{sw}} = x_0(u)/u$ required for the transition front to travel through the instability region in Fig. 28.4 whose size is $x_0(u)$ in eqn (28.14). Thus the equation $t_Z = x_0(u_c)/u_c$ gives an estimate for the critical value u_c of the velocity of the temperature front, at which the laminar propagation becomes unstable:

$$u_c \sim \frac{\xi_0}{\tau_0} \left(\frac{\tau_0}{\tau_Q}\right)^{1/4}. \qquad (28.20)$$

This agrees with the estimate $u_c = \xi_v/t_Z$ by Kibble and Volovik (1997).

In the case of the fireball formed by a neutron the velocity of the temperature front is $u \sim R_b/\tau_Q$, which makes $u \sim 10$ m s^{-1}. The critical velocity u_c we can estimate to possess the same order of magnitude value. This estimate suggests that the thermal gradient should be sufficiently steep in the neutron bubble such that defect formation can be expected.

The further fate of the vortex tangle formed under the KZ mechanism is the phase ordering process: the intervortex distance continuously increases until it reaches the critical size $R_c \sim \kappa_0/2\pi v_s$ in eqn (26.10), when the vortex loops are expanded by the counterflow and leave the bubble. The number of nucleated vortices per bubble is thus $n_1 \sim (R_b/R_c)^3 \sim (v_s/v_{\text{neutron}})^3$, where $v_{\text{neutron}} \sim \kappa_0/2\pi R_b$ is the critical velocity for formation of vortices by neutron radiation, and we consider the limit of large velocities, $v_s \gg v_{\text{neutron}}$. Both the cubic law

$$n_1 \sim \left(\frac{v_s}{v_{\text{neutron}}}\right)^3 - 1 \qquad (28.21)$$

and the dependence of the critical velocity on the bubble size are in agreement with experiments by Ruutu *et al.* (1996*a*).

28.2.4 *Instability of normal–superfluid interface*

Due to external counterflow used in neutron experiments in rotating vessel, another source of the vortex nucleation becomes important – the surface instability of the propagating front (Aranson *et al.* 1999, 2001, 2002). This is a variant of KH instability discussed in Chapter 27. Now it is the interface between normal and superfluid liquids, which is destabilized by the shear flow. The corresponding shear flow is caused by the counterflow: the normal fluid is at rest with the rotating vessel due to its viscosity, while the superfluid is in the Landau vortex-free state and thus is moving in the rotating frame. Such a version of the KH instability has been calculated analytically using TDGL theory, while the vortex formation at the non-linear stage of the development of this instability has been simulated in 3D numerical calculations by Aranson *et al.* (1999, 2001, 2002).

Let us consider a simple scenario of how such KH instability can occur under the conditions of the experiment, which is consistent with the numerical simulations made by Aranson *et al.* (1999, 2001, 2002). The process of development of this shear flow instability can be roughly split into two stages (see Fig. 28.5). At the first stage the heated region of the normal liquid surrounded by the superflow undergoes a superfluid transition. The transition should occur in the state with the lowest energy, which corresponds to the superfluid at rest, i.e. with $\mathbf{v}_s = 0$. Thus the superfluid–superfluid interface appears, which separates the

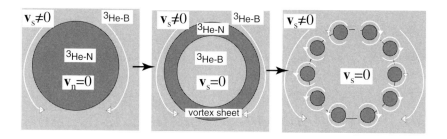

FIG. 28.5. KH instability scenario of vortex formation in a 'Big-Bang' event. *Left*: The normal liquid formed in the hot spot is at rest with the vessel. *Middle*: It transforms to the superfluid state with $\mathbf{v}_s = 0$. Thus we have a superfluid region with $\mathbf{v}_s = 0$, which is separated from the moving bulk superfluid by the vortex sheet. *Right*: Due to the shear-flow instability the vortex sheet decays to vortex lines.

state with superflow (outside) from the state without superflow (inside). Such a superfluid–superfluid interface with tangential discontinuity of the superfluid velocity represents a vortex sheet by definition. The vortex sheet in ^{3}He-B is not supported by topology. It experiences the shear-flow instability and breaks up into a chain of vortex lines. The development of this instability represents the second stage of the process in Fig. 28.5.

28.2.5 Interplay of KH and KZ mechanisms

In numerical simulations made by Aranson *et al.* (1999, 2001, 2002), in addition to vortices formed by the shear-flow instability at the propagating front, the formation of vortices by the KZ mechanism discussed in Sec. 28.2.2 has also been observed. It occurs during shrinking of the interior region with normal fluid in Fig. 28.5 *left*. The interplay of the KZ mechanism and KH instability depends on the parameters of the system. In numerical simulations by Aranson *et al.* (1999, 2001, 2002) the chain of vortices formed in shear-flow instability screen the external superflow so tightly that the vortices formed in the interior region due to the KZ mechanism do not have a superflow bias and decay.

What the interplay of the bulk and surface effects is in experiments can be deduced from the experiments. In the share-flow instability scenario the chain of vortices is formed at the interface. If the counterflow is large the number of quantized vortices in this chain is $n_1 \approx \kappa/\kappa_0$, where $\kappa = \pi v_s R_b$ is the circulation in the piece of the vortex sheet of length πR_b, and κ_0 is the elementary circulation quantum of singular vortex in ^{3}He-B. This gives the number of vortices nucleated in one event due to KH instability of the propagating front. The vortex formation starts at threshold velocity v_c, at which $\pi v_c R_b/\kappa_0$ reaches unity. Thus the number of nucleated vortices can be extrapolated as $n_1 \approx v_s/v_c - 1$, which is what is seen in numerical simulations. This linear law is, however, in disagreement with the experiments demonstrating a cubic law in eqn (28.21) which is the characteristic of the KZ mechanism occurring in the interior of the hot spot. This shows that

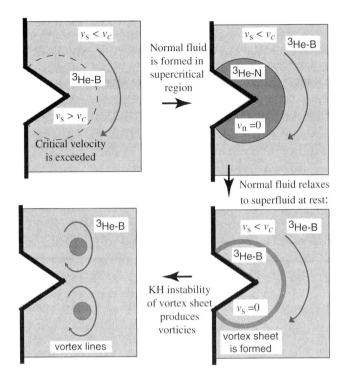

FIG. 28.6. Shear-flow instability scenario of vortex formation near a proturberance at the container wall. The normal liquid formed in the region where the Landau criterion is exceeded is at rest with respect to the vessel boundary, $\mathbf{v}_n = 0$. The normal liquid is unstable and transforms to the superfluid state with $\mathbf{v}_s = 0$ separated from the moving bulk superfluid by a vortex layer. The latter is unstable toward formation of vortex lines.

the bulk effects can be more important.

28.2.6 *KH instability as generic mechanism of vortex nucleation*

Each of these mechanisms of vortex formation, by KH shear-flow instability and due to the KZ fluctuation mechanism, can be either derived analytically for a simple geometry, or understood qualitatively with a simple physical picture in mind. They do not depend much on the geometry and parameters of the TDGL equation. Probably both mechanisms hold even if TDGL theory cannot be applied. This suggests that both of them are generic and fundamental mechanisms of vortex formation in superfluids.

One may guess that most (if not all) of the observed events of formation of singular vortices in superfluids are related to one of the two mechanisms, KH and KZ, or to their combination. For example, let us discuss the possible scenario of conventional vortex formation, i.e. without ionizing radiation (Fig. 28.6). When the superfluid velocity exceeds the Landau criterion or the critical velocity v_c

of the instability of superflow, the state with the normal phase can be formed as an intermediate state. Then the discussed hydrodynamic instability of the normal/superfluid interface in the presence of the tangential flow will result in the creation of vortices.

Part VII

Vacuum energy and vacuum in non-trivial gravitational background

29

CASIMIR EFFECT AND VACUUM ENERGY

29.1 Analog of standard Casimir effect in condensed matter

There are several macroscopic phenomena which can be directly related to the properties of the physical quantum vacuum. The Casimir effect is probably the most accessible effect of the quantum vacuum. Casimir predicted in 1948 that there must be an attractive force between two parallel conducting plates placed in the vacuum, the force being induced by the vacuum fluctuations of the electromagnetic field. The Casimir force modified by the imperfections of conducting plates has been measured in several experiments using different geometries, see the review paper by Lambrecht and Reynaud (2002).

The calculation of the vacuum pressure is based on the regularization schemes, which allows the effect of the low-energy modes of the vacuum to be separated from the huge diverging contribution of the high-energy degrees of freedom. There are different regularization schemes: Riemann's zeta-function regularization; introduction of the exponential cut-off; dimensional regularization, etc. People are happy when different regularization schemes give the same results. But this is not always so (see e.g. Esposito *et al.* 1999, 2000; Ravndal 2000; Falomir *et al.* 2001), and in particular the divergences occurring for spherical geometry in odd spatial dimension are not cancelled (Milton 2000; Cognola *et al.* 2001). This raises some criticism about the regularization methods (Hagen 2000, 2001) or even some doubts concerning the existence and the magnitude of the Casimir effect.

The same type of Casimir effect arises in condensed matter: for example, in quantum liquids. The advantage of the quantum liquid is that the structure of the quantum vacuum is known at least in principle. That is why one can calculate everything starting from the first principles of microscopic theory – the Theory of Everything in eqn (3.2). One can calculate the vacuum energy under different external conditions, without invoking any cut-off or regularization scheme. Then one can compare the results with what can be obtained within the effective theory which deals only with the low-energy phenomena. The latter requires the regularization scheme in order to cancel the ultraviolet divergency, and thus one can judge whether and which of the regularization schemes are physically relevant for a given physical situation.

In the analog of the Casimir effect in condensed matter we can use the analogies discussed above. Let us summarize them. The ground state of the quantum liquid corresponds to the vacuum of RQFT. The low-energy bosonic and fermionic quasiparticles in the quantum liquid correspond to matter. The low-energy modes with linear spectrum $E = cp$ can be described by the relativistic-type effective theory. The speed c of sound or of other collective bosonic or fermionic modes (spin waves, gapless Bogoliubov fermions, etc.) plays the role of the speed of light. This 'speed of light' is the 'fundamental constant' for the effective theory. It enters the corresponding effective theory as a phenomenological parameter, though in principle it can be calculated from the more fundamental microscopic physics playing the role of trans-Planckian physics. The effective theory is valid only at low energy which is much smaller than the Planck energy cut-off E_{Planck}. In quantum liquids the analog of E_{Planck} is determined either by the bare mass m of the atom of the liquid, $E_{\text{Planck}} \sim mc^2$, or by the Debye temperature, $E_{\text{Planck}} \sim \hbar c / a_0$, where a_0 is the interatomic distance which plays the role of the Planck length. In liquid ^{4}He the two Planck scales have the same order of magnitude reflecting the stability of the liquid in the absence of the environment.

The typical massless modes in quantum Bose liquids are sound waves. The acoustic field is described by the effective theory corresponding to the massless scalar field. The walls of the container provide the boundary conditions for the sound wave modes; usually these are the Neumann boundary conditions: the mass current through the wall is zero, i.e. $\hat{s} \cdot \nabla \Phi = 0$, where $\hat{s}$ is the normal to the surface. Because of the quantum hydrodynamic fluctuations the Casimir force must occur between two parallel plates immersed in the quantum liquid. Within the effective theory the Casimir force is given by the same equation as the Casimir force acting between the conducting walls due to quantum electromagnetic fluctuations. The only modifications are: (i) the speed of light must be substituted by the spin of sound; (ii) the factor $1/2$ must be added, since we have the scalar field of the longitudinal sound wave instead of two polarizations of transverse electromagnetic waves. If a is the distance between the plates and A is their area, then the a-dependent contribution to the ground state energy of the quantum liquid at $T = 0$ which follows from the effective theory must be

$$E_C = E_{\text{vac restricted}} - E_{\text{vac free}} = \frac{1}{2} \sum_{\nu \text{ restricted}} E_\nu - \frac{1}{2} \sum_{\nu \text{ free}} E_\nu \quad (29.1)$$

$$= \frac{\hbar c}{2} V \left(\sum_n \int \frac{d^2 k_\perp}{(2\pi)^2} \sqrt{k_\perp^2 + \frac{\pi^2 n^2}{a^2}} - \int \frac{d^3 k}{(2\pi)^3} k \right) \quad (29.2)$$

$$= -\frac{\hbar c \pi^2 A}{1440 a^3} . \quad (29.3)$$

Here ν are the quantum numbers of phonon modes: $\nu = (\mathbf{k}_\perp, n)$ in the restricted vacuum, and $\nu = \mathbf{k}$ in the free vacuum; $V = Aa$ is the volume of the space between the plates. Equation (29.3) corresponds to the additional negative pressure in the quantum liquid at $T = 0$ in the region between the plates

$$P_C = -\frac{1}{A} \partial_a E_C = -\frac{\hbar c \pi^2}{480 a^4} = \frac{3 E_C}{A} . \quad (29.4)$$

This pressure is induced by the boundaries.

Such microscopic quantities of the quantum liquid as the mass of the atom m or the interatomic space $a_0 \equiv L_{\text{Planck}}$ do not explicitly enter eqns (29.3) and (29.4): the analog of the standard Casimir force is completely determined by the 'fundamental' parameter c of the effective scalar field theory and by the geometry of the system. Of course, as we already know, the total vacuum energy of the quantum liquid cannot be described in terms of the zero-point energy of the phonon field

$$E_{\text{vac}} \neq E_{\text{zero point}} = \frac{1}{2} \sum_\nu E_\nu \,. \qquad (29.5)$$

The value of the total vacuum energy, E_{vac}, as well as of the total pressure in the liquid, cannot be determined by the effective theory, and the microscopic 'trans-Planckian' physics must be evoked for that. The latter shows that even the sign of the vacuum energy can be different from that given by the zero-point energy of phonons (see Sec. 3.3.1). Nevertheless, it appears that the phonon modes are the relevant modes for the calculation of the small corrections to the vacuum energy and pressure. The huge ultraviolet divergence of each of the two terms in eqn (29.2) is simply canceled out by using the proper regularization scheme. The main contribution to the vacuum energy difference comes from the large wavelength of order $a \gg a_0 \equiv L_{\text{Planck}}$, where the effective theory does work. As a result the vacuum energy difference can be expressed in terms of the zero-point energy of the phonon field in eqn (29.1).

There are several contributions to the vacuum energy imposed by the restricted geometry. They comprise different fractions $(L_{\text{Planck}}/a)^n$ of the main (bulk) energy of the non-perturbed vacuum, which is of order $E_{\text{vac}} = \epsilon_{\text{vac}} V \sim V E_{\text{Planck}}^4/(\hbar^3 c^3)$. (i) The lowest-order correction, i.e. with $n = 1$, comes from the surface energy or surface tension of the liquid, and is of order $E_{\text{surface}} \sim A E_{\text{Planck}}^3/(\hbar^2 c^2) \sim E_{\text{vac}}(L_{\text{Planck}}/a)$. The effect of the surface tension will be discussed later in Sec. 29.4.3. (ii) If the boundary conditions for the order parameter field gives rise to texture, the textural contribution to the vacuum energy is $E_{\text{texture}} \sim E_{\text{vac}}(L_{\text{Planck}}/a)^2$. Finally, (iii) the standard Casimir energy in eqn (29.4) is of order $E_C \sim E_{\text{vac}}(L_{\text{Planck}}/a)^4$. Zero-, first- and second-order contributions substantially depend on the microscopic Planck-scale physics, while the Planck scale completely drops out from the standard Casimir effect.

Since the surface energy does not depend on the distance a between the walls, it does not produce the force on the wall. Thus, in the absence of textures, only the standard Casimir pressure with $n = 4$ can be measured by an inner observer, who has no information on the trans-Planckian physics of the quantum vacuum. However, in Sec. 29.5 we shall discuss the possibility of the mesoscopic Casimir effect with $n = 3$.

The Casimir type of forces in condensed matter comes from the change of the vacuum pressure provided by the boundary conditions imposed by the restricted geometry. The similar phenomenon occurs at non-zero temperature and even in classical systems: the boundary conditions modify the spectrum of thermal fluctuations (see the review paper by Kardar and Golestanian 1999).

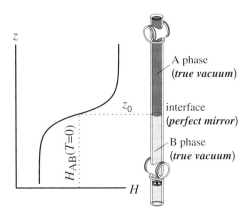

FIG. 29.1. Interface between two vacua stabilized by the vacuum pressure induced by an external magnetic field. The interface is at $z = z_0$, where $H(z_0) = H_{AB}$. It separates the true vacuum of ^{3}He-A from the true vacuum of ^{3}He-B.

29.2 Interface between two different vacua

29.2.1 Interface between vacua with different broken symmetry

The interface between ^{3}He-A and ^{3}He-B, which we considered in Chapter 27 and Sec. 17.3, appears to be useful for the consideration of the vacuum energy and Casimir effect in quantum liquids, Fig. 29.1. Such an interface is interesting because it separates not only the vacuum states with different broken symmetry, but also between vacua of different universality classes (Fig. 29.2).

The vacua in ^{3}He-A and ^{3}He-B have different broken symmetries, neither of which is the subgroup of the other (Sec. 7.4). Thus the phase transition between the two superfluids is of first order. The interface between the two vacua is stable and is stationary if the two phases have the same vacuum energy $\tilde\epsilon$ (or the same free energy if $T \neq 0$). At $T = 0$ the difference between the vacuum energies of ^{3}He-A and ^{3}He-B is regulated by the magnetic field. In the presence of an external magnetic field H the vacuum energy becomes $\epsilon_A(H) = \epsilon_A(H=0) - (1/2)\chi_A H^2$ and $\epsilon_B(H) = \epsilon_B(H=0) - (1/2)\chi_B H^2$ for A- and B-phases respectively. Due to the different spin structure of Cooper pairs in ^{3}He-A and ^{3}He-B, the spin susceptibilities of the two liquids are different. For the proper orientation of the $\hat{\mathbf{d}}$-vector of the A-phase with respect to the direction of the magnetic field, one has $\chi_A > \chi_B$, and at some value H_{AB} of the magnetic field the energies of the two vacua become equal. This allows us to stabilize the interface between the two vacua in the applied gradient of magnetic field even at $T = 0$. The position of the interface is given by the equation $H(\mathbf{r}) = H_{AB}$.

If the liquids are isolated from the environment, the pressure in the true equilibrium vacuum state (at $T = 0$) must be zero and the chemical potential μ must be constant throughout the system. Outside the interface, where the A and B vacua are well determined, one has

INTERFACE BETWEEN TWO DIFFERENT VACUA

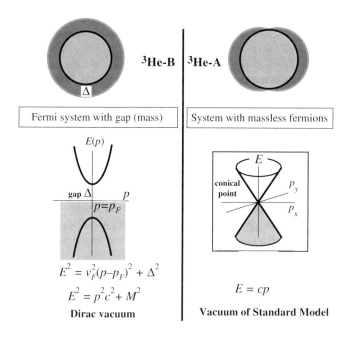

FIG. 29.2. Interface between two vacua of different universality classes. The interface between ^{3}He-A and ^{3}He-B corresponds to the interface between the vacuum of the Standard Model with massless chiral fermions and the vacuum of Dirac fermions with Planck masses. Since the low-energy (quasi)particles cannot penetrate from the right vacuum into the left one, the interface represents a perfect mirror.

$$0 = P = \mu n_B - \epsilon_B(H=0) + \frac{1}{2}\chi_B H^2(z) = \mu n_A - \epsilon_A(H=0) + \frac{1}{2}\chi_A H^2(z) \,. \quad (29.6)$$

Equation (29.6) determines the position z_0 of the interface in Fig. 29.1 at $T = 0$ and $P = 0$. By reducing the magnitude H of the magnetic field one shifts the interface upward.

Note that the vacuum energy density $\tilde{\epsilon} = -P$ is zero in both vacua outside the interface. The initial difference in the energies (and pressures) of the two liquids is compensated by the energy (and pressure) induced by the interaction with an external magnetic field. As a result the total vacuum energy of the system comes from the interface only. This is the surface energy – the surface tension of the interface multiplied by the area A of the interface. The surface energy of the interface does not contribute to the pressure if the interface is planar, i.e. if it is not curved. The same occurs for the topological domain wall separating two degenerate vacua.

29.2.2 Why the cosmological phase transition does not perturb the zero value of the the cosmological constant

Applying this consideration to the vacuum in RQFT one can suggest that the first-order phase transition between two vacuum states should not change the vacuum energy and thus the cosmological constant. The cosmological constant remains zero after the phase transition, when complete equilibrium is reached again. In the presence of the interface between the two vacua, the system acquires the vacuum energy, which is the energy of the interface. This implies that, while the homogeneous vacuum is weightless, the domain wall separating two vacua is gravitating, as well as other topological defects, strings and monopoles.

We arrived at this conclusion by considering the first-order phase transition between the two vacua in quantum liquids. However, the same is valid for the second-order phase transition when symmetry breaking occurs; for example, the phase transition from the normal to the superfluid state of a quantum liquid. At first glance our conjecture, that such a phase transition does not change the zero value of the vacuum energy and of the cosmological constant, seems paradoxical. We know for sure that the broken-symmetry phase transition occurs when the symmetric vacuum becomes the saddle point of the energy functional. Thus it is energetically advantageous to develop the order parameter, and the energy of the vacuum must decrease. This is certainly true: the energy difference between the broken-symmetry superfluid state and the symmetric normal states in eqn (7.28) is negative, and this is the reason for the phase transition to the superfluid state to occur.

There is, however, no paradox. The energy difference in eqn (7.28) is considered at a fixed chemical potential in the normal state, $\mu = \mu_{\rm normal}$. We know that the chemical potential of the system and the particle density n are adjusted to nullify the relevant vacuum energy: $\tilde{\epsilon} = 0$. This means that, after the phase transition to the superfluid state occurs at $T = 0$ and $P = 0$, the chemical potential will change from $\mu_{\rm normal}$ to $\mu_{\rm super}$ to satisfy the equilibrium condition in a new (superfluid) vacuum. Thus the second-order broken-symmetry phase transition also does not violate the zero condition for the vacuum energy in equilibrium. The wrong unstable vacuum and the true broken-symmetry vacuum at $P = 0$ and $T = 0$ have different values of the chemical potential, $\mu_{\rm super} \neq \mu_{\rm normal}$. But both vacua have zero vacuum energy $\tilde{\epsilon} = 0$.

In superfluid ^{3}He (and in superconductors) the difference between chemical potentials in superfluid (superconducting) and normal states has the following order of magnitude: $|\mu_{\rm super} - \mu_{\rm normal}| \sim \Delta_0^2/v_F p_F$. This difference is important for the physics of Abrikosov vortices in superconductors: the core of a vortex acquires electric charge (Khomskii and Freimuth 1995; Blatter et al. 1996). The electric charge per unit length of a vortex with core size ξ can be on the order of $e(dn/d\mu)\xi^2|\mu_{\rm super} - \mu_{\rm normal}|$, but actually it must be less due to the screening effects in superconductors. A discussion of both the theory of the vortex charge and experiments where the vortex charge has been measured using the NMR technique can be found in the review paper by Matsuda and Kumagai (2002).

Probably the same effect could lead to the non-zero electric or other fermionic charge accumulated by the cosmic string due to the asymmetry of the vacuum.

29.2.3 Interface as perfectly reflecting mirror

At $T \neq 0$ the equilibrium condition for the interface and its dynamics are influenced by the fermionic quasiparticles (Bogoliubov excitations). In the ^{3}He-A vacuum, fermionic quasiparticles are chiral and massless, while in the ^{3}He-B vacuum they are fully gapped. At temperatures T well below the temperature T_c of the superfluid transition and thus much smaller than the gap Δ_0, the gapped ^{3}He-B fermions are frozen out, and the only thermal fermions are present on the ^{3}He-A side of the interface.

Close to the gap nodes, the energy spectrum of the gapless ^{3}He-A fermions is relativistic. These low-energy massless fermions cannot propagate through the AB-interface to the ^{3}He-B side, where the minimum energy of fermions is on the order of Planck scale, see Fig. 29.2, and thus they completely scatter from the interface. Actually such scattering corresponds to Andreev (1964) reflection (see below in Sec. 29.3.1). However, from the point of view of the inner observer, the reflection is conventional, and for the 'relativistic' world of the ^{3}He-A quasiparticles the AB-interface represents a perfectly reflecting mirror. By moving the interface one can simulate the dynamic Casimir effects – the response of the quantum vacuum to the motion of the mirror (see e.g. Maia Neto and Reynaud 1993; Law 1994; Kardar and Golestanian 1999). By moving the interface with 'superluminal' velocity one can investigate the problems related to the quantum vacuum in the presence of the ergoregion and event horizon.

On the other hand, the relativistic invariance emerging at low energy simplifies the calculation of forces acting on a moving interface in the limit of low T. This is instrumental, for example, for studying the peculiarities of Kelvin–Helmholtz instability in superfluids (Chapter 27).

29.2.4 Interplay between vacuum pressure and pressure of matter

Considering the interface at non-zero but low temperatures we can discuss the problem of vacuum energy in the presence of matter. At low T, the matter appears only on the A-phase side of the interface – the gas of relativistic quasiparticles which form the normal component of the liquid. This relativistic gas adds positive pressure of the matter, eqn (10.22), to the pressure of the A-phase vacuum on the rhs of eqn (29.6). The total pressure must remain zero: $P_{\rm vac} + P_{\rm matter} = 0$. Thus the vacuum energy and the vacuum pressure of ^{3}He-A become non-zero, while those of ^{3}He-B are zero since there is no matter in ^{3}He-B:

$$\tilde{\epsilon}_{\rm vac\ A} = -P_{\rm vac\ A} = P_{\rm matter\ A} = \frac{1}{3}\epsilon_{\rm matter\ A} = \frac{7\pi^2 N_F}{360}\sqrt{-g}T^4 , \quad (29.7)$$

$$\tilde{\epsilon}_{\rm vac\ B} = -P_{\rm vac\ B} = 0 . \quad (29.8)$$

In this arrangement one obtains the following relation between the vacuum energy and the energy of matter on the A-phase side of the Universe: $\tilde{\epsilon}_{\rm vac\ A} =$

$(1/3)\epsilon_{\text{matter A}}$. In this example, where the only source of perturbation of the vacuum state is the matter, the order of magnitude coincidence between the vacuum energy density (the cosmological constant) and the energy density of matter naturally arises and does not look puzzling.

The interplay between the 'vacuum' pressure and the pressure of the 'matter' (quasiparticles) has been observed in experiments with slow (adiabatic) motion of the AB-interface at low T by Bartkowiak (*et al.* 1999) in a geometry similar to that in Fig. 29.1. In these experiments the relativistic character of the low-energy fermionic quasiparticles in ^{3}He-A has also been verified. In the adiabatic process the entropy is conserved. Since the entropy is mostly concentrated in the relativistic gas of ^{3}He-A quasiparticles, the total number of thermal quasiparticles in ^{3}He-A must be conserved under adiabatic motion of the interface. When one decreases the magnetic field H, the interface moves upward decreasing the volume of the ^{3}He-A Universe. As a result the entropy density and the quasiparticle density increase, and, since both of them are proportional to T^3, the temperature rises. The released thermal energy – the latent heat of the phase transition – has been measured and the T^4-dependence of the thermal energy of the relativistic gas has been observed.

In the reversed process, when the volume of the A-phase increases, the temperature drops. This can be used as a cooling process at low temperature $T \ll \Delta_0$.

29.2.5 Interface between vacua with different speeds of light

Similar situations could occur if both vacua have massless excitations, but their 'speeds of light' are different: for example, at the interface between Bose condensates with different density n and speed of sound c on the left and right sides of the interface. The simplest example is

$$g_L^{00} = -\frac{1}{c_L^2} \, , \, g_R^{00} = -\frac{1}{c_R^2} \, , \, g_L^{ij} = g_R^{ij} = \delta^{ij} \, . \tag{29.9}$$

This effective spacetime is flat everywhere except for the interface itself where the curvature is non-zero:

$$R_{0101} = c\partial_z^2 c = \frac{1}{2}(c_R^2 - c_L^2)\delta'(z) - (c_R - c_L)^2 \delta^2(z) \, . \tag{29.10}$$

The difference in pressure of the matter on two sides of the interface (see eqn (5.31))

$$\Delta P = \frac{\pi^2}{90\hbar^3}T^4 \left(\frac{\sqrt{-g_R}}{g_{00R}^2} - \frac{\sqrt{-g_L}}{g_{00L}^2} \right) = \frac{\pi^2}{90\hbar^3}T^4 \left(\frac{1}{c_R^3} - \frac{1}{c_L^3} \right) \tag{29.11}$$

will be compensated by the difference in the vacuum pressures $P_{\text{vac L}} - P_{\text{vac R}} = \tilde\epsilon_{\text{vac R}} - \tilde\epsilon_{\text{vac L}}$. The situation similar to the AB-interface corresponds to the case when, say, $c_L \gg c_R$, so that there are practically no quasiparticles on the lhs of the interface, even if the temperature is finite.

29.3 Force on moving interface

29.3.1 Andreev reflection at the interface

Let us now consider the AB-interface moving with constant velocity $\mathbf{v}_{AB}$. If $\mathbf{v}_{AB} = \mathbf{v}_n \neq \mathbf{v}_s$, the interface is stationary in the heat-bath frame and thus remains in equilibrium with the heat bath of quasiparticles. The dissipation and thus the friction force are absent. The counterflow $\mathbf{w} = \mathbf{v}_n - \mathbf{v}_s$ modifies the pressure of the matter on the A-phase side. If the interface is moving with respect to the normal component, $\mathbf{v}_{AB} \neq \mathbf{v}_n$, global equilibrium is violated leading to the friction force experienced by the mirror moving with respect to the heat bath.

The motion of the AB-interface in the ballistic regime for quasiparticles has been considered by Yip and Leggett (1986), Kopnin (1987), Leggett and Yip (1990) and Palmeri (1990). In this regime the force on the interface comes from Andreev reflection at the interface of the ballistically moving, thermally distributed fermionic quasiparticles.

Let us recall the difference between the conventional and Andreev reflection in superfluid ^{3}He-A. For a low-energy quasiparticle its momentum is concentrated in the vicinity of one of the two Fermi points, $\mathbf{p}^{(a)} = -C_a p_F \hat{\mathbf{l}}$, where $C_a = \pm 1$ plays the role of chirality of a quasiparticle. In the conventional reflection the quasiparticle momentum is reversed, which means that after reflection the quasiparticle acquires an opposite chirality C_a. The momentum transfer in this process is $\Delta \mathbf{p} = \pm 2 p_F \hat{\mathbf{l}}$, whose magnitude $2p_F$ is well above the 'Planck' momentum $m^* c_\perp$. The probability of such a process is exponentially small unless the scattering center has an atomic (Planck) size $a_0 \equiv L_{\rm Planck} \sim \hbar/p_F$. The thickness of the interface is on the order of the coherence length which is much larger than the wavelength of a quasiparticle: $\xi = \hbar v_F/\Delta_0 = (c_\|/c_\perp)\hbar/p_F \sim 10^3 \hbar/p_F$. As a result the conventional scattering from the AB-interface is suppressed by the huge factor $\exp(-p_F \xi/\hbar) \sim \exp(-c_\|/c_\perp)$. This means that, though the non-conservation of chirality is possible due to the trans-Planckian physics, it is exponentially suppressed. In RQFT such non-conservation of chirality can occur in lattice models, where the distance in momentum space between Fermi points of opposite chiralities is also of order of the Planck momentum (Nielsen and Ninomiya 1981).

In Andreev reflection, the momentum $\mathbf{p}$ of a quasiparticle remains in the vicinity of the same Fermi point, i.e. the chirality of a quasiparticle does not change. Instead, the deviation of the momentum from the Fermi point changes sign, $\mathbf{p} - \mathbf{p}^{(a)} \to -(\mathbf{p} - \mathbf{p}^{(a)})$, and the velocity of the quasiparticle is reversed. For an external observer Andreev reflection corresponds to the transformation of a particle to a hole, while for an inner observer this is the conventional reflection. In this process the momentum change can be arbitrarily small, which is why there is no exponential suppression for Andreev reflection, and it is the dominating mechanism of scattering from the interface.

Let us consider this scattering in the texture-comoving frame – the reference frame of the moving interface, where the order parameter (and thus the metric) is time independent and the energy of quasiparticles is well-defined. We assume that in this frame the superfluid and normal velocities are both along the normal to the interface: $\mathbf{v}_s = \hat{\mathbf{z}} v_s$ and $\mathbf{v}_n = \hat{\mathbf{z}} v_n$. Since Andreev reflection does not change the position $\mathbf{p}^{(a)}$ of the Fermi point, we can count the momentum from this position. In addition we can remove the anisotropy of the speed of light by rescaling, to obtain $E(p) = cp$. Due to the boundary condition for the $\hat{\mathbf{l}}$-vector at the interface, $\hat{\mathbf{l}} \cdot \hat{\mathbf{z}} = 0$, one has $\mathbf{p}^{(a)} \cdot \mathbf{v}_s = 0$. As a result the Doppler-shifted spectrum of quasiparticles becomes $\tilde{E} = cp + \mathbf{p} \cdot \mathbf{v}_s$.

In the ballistic regime, the force acting on the interface from the gas of these massless relativistic quasiparticles living in the half-space $z > 0$ is

$$F_z = \sum_{\mathbf{p}} \Delta p_z v_{Gz} f_T(\mathbf{p}) \,. \tag{29.12}$$

Here $\mathbf{v}_G$ is the group velocity of incident particles:

$$v_{Gz} = \frac{d\tilde{E}}{dp_z} = c\cos\theta + v_s \,, \tag{29.13}$$

Δp_z is the momentum transfer after reflection

$$\Delta p_z = 2p \frac{\cos\theta + v_s/c}{1 - v_s^2/c^2} \tag{29.14}$$

and θ is the angle between the momentum $\mathbf{p}$ of incident quasiparticles and the normal to the interface $\hat{\mathbf{z}}$. Far from the interface, at the distance of the order of the mean free path, quasiparticles are in a local thermal equilibrium with equilibrium distribution function $f_T(\mathbf{p}) = 1/(1 + e^{(E(p) - \mathbf{p} \cdot \mathbf{w})/T})$, which does not depend on the reference frame since it is determined by the Galilean invariant counterflow velocity $\mathbf{w} = \mathbf{v}_n - \mathbf{v}_s$ (compare with eqn (25.9) for the case of the moving vortex).

29.3.2 *Force acting on moving mirror from thermal relativistic fermions*

It follows from eqns (29.12–29.14) that the force per unit area acting from the gas of relativistic fermions on the reflecting interface is

$$\frac{F_z(v_s, w)}{A} = -\hbar c \frac{7\pi^2}{60} \frac{T^4}{(\hbar c)^4} \alpha(v_s, w) \,, \tag{29.15}$$

$$\alpha(v_s, w) = \frac{1}{1 - v_s^2} \int_{-1}^{-v_s} d\mu \frac{(\mu + v_s)^2}{(1 - \mu w)^4} = \frac{(1 - v_s)^2}{3(1 + w)^3 (1 + v_s)(1 + v_s w)} \,. \tag{29.16}$$

Equation (29.16) is valid in the range $-c < v_s < c$, where v_s is the superfluid velocity with respect to the interface and $c = 1$. The force (29.15) disappears at $v_s \to c$, because the quasiparticles cannot reach the interface if it moves away from them with the 'speed of light'. The force diverges at $v_s \to -c$, when all quasiparticles become trapped by the interface, so that the interface resembles the black-hole horizon.

If the normal component is at rest in the interface frame, i.e. at $v_n = v_{AB} = 0$, the system is in a global thermal equilibrium (see Sec. 5.4) with no dissipation. In this case eqn (29.15) gives a conventional pressure of 'matter' acting on the interface from the gas of (quasi)particles:

$$\frac{F_z(v_s, w = -v_s)}{A} = \Omega = \hbar c \frac{7\pi^2}{180} \frac{T^4}{(\hbar c)^4 \left(1 - \frac{v_s^2}{c^2}\right)^2} = \frac{7\pi^2}{180\hbar^3} \sqrt{-g} T_{\text{eff}}^4 \, . \quad (29.17)$$

Here again $T_{\text{eff}} = T/\sqrt{-g_{00}} = T/\sqrt{1 - v_s^2/c^2}$ is the temperature measured by an inner observer, see eqn (5.29), while the real thermodynamic temperature T measured by the external observer plays the role of the Tolman temperature in general relativity.

Now let us consider a small deviation from equilibrium, i.e. $v_n \neq v_{AB} = 0$ but small. Then the friction force appears, which is proportional to the velocity of the interface with respect to the normal component $v_{AB} - v_n$:

$$\frac{F_{\text{friction}}}{A} = -\Gamma(v_{AB} - v_n) \, , \quad \Gamma = \frac{3}{c}\Omega \, . \quad (29.18)$$

Equation (29.18) can be extrapolated to the zero-temperature case, when the temperature T must be substituted by frequency $\hbar\omega$ of the motion of the wall. This leads to a friction force which is proportional to the fourth-order time derivative of the wall velocity, $F_{\text{friction}} \sim (\hbar A/c^4) d^4 v_{AB}/dt^4$, in agreement with the result obtained by Davis and Fulling (1976) for a mirror moving in the vacuum. This is somewhat counterintuitive, since according to this equation the energy dissipation is proportional to $\dot{E} \sim v_{AB} F_{\text{friction}} \sim (\hbar A/c^4)(d^2 v_{AB}/dt^2)^2$ and is absent for the constant acceleration dv_{AB}/dt of the interface in the liquid. The reason is that we consider here the linear response. The non-linear friction can lead to the energy dissipation containing the acceleration: $\dot{E} \sim (\hbar A/c^6)(dv_{AB}/dt)^4$. The motion with constant acceleration contains a lot of interesting physics which can be checked using the analogous effects in superfluids. In particular this is related to the Unruh (1976) effect: the motion of a body in the vacuum with constant acceleration dv_{AB}/dt leads to the thermal radiation of quasiparticles with the Unruh temperature $T_U = \hbar |dv_{AB}/dt|/2\pi c$, and thus to the energy dissipation $\dot{E} \sim (AT_U^4/\hbar^3 c^2) \sim (\hbar A/c^6)(dv_{AB}/dt)^4$.

29.3.3 Force acting on moving AB-interface

Now let us apply the results obtained to the ^{3}He-A, which has an anisotropic 'speed of light' and also contains the vector potential $\mathbf{A} = p_F \hat{\mathbf{l}}$. Typically $\hat{\mathbf{l}}$ is parallel to the AB-interface, which is dictated by boundary conditions. In such a geometry the effective gauge field is irrelevant: the constant vector potential $\mathbf{A} = p_F \hat{\mathbf{l}}_0$ can be gauged away by shifting the momentum. The scalar potential $A_0 = \mathbf{A} \cdot \mathbf{v}_s$, which is obtained from the Doppler shift, is zero since $\hat{\mathbf{l}} \perp \mathbf{v}_s$ in the considered geometry. In the same way the effective chemical potential $\mu_a = -C_a p_F (\hat{\mathbf{l}} \cdot \mathbf{w})$ is also zero in this geometry. Thus if $\mathbf{p}$ is counted from $e\mathbf{A}$

the situation becomes the same as that discussed in theprevious section, and one can apply eqns (29.17) and (29.18) modified by the anisotropy of the 'speed of light' in ^{3}He-A. In the limit of small relative velocity $v_{AB} - v_n$ eqns (29.17) and (29.18) give

$$\frac{F_z}{A} = -\frac{7\pi^2}{180}\sqrt{-g}T_{\text{eff}}^4\left(1 + 3\frac{v_{AB} - v_n}{c_\perp}\right) \;, \quad T_{\text{eff}} = \frac{T}{\sqrt{-g_{00}}} \;. \tag{29.19}$$

The first term represents the pressure of the matter, where now $\sqrt{-g} = 1/c_\perp^2 c_\parallel$, and $g_{00} = 1 - (\mathbf{v}_s - \mathbf{v}_{AB})^2/c_\perp^2$ (see eqns (9.11–9.13)).

The friction coefficient $\Gamma \sim T^4/c_\perp^3 c_\parallel$ obtained here is valid in the relativistic regime, i.e. below the first Planck scale $T \ll m^* c_\perp^2 = \Delta_0^2/v_F p_F$. In the non-relativistic quasiclassical regime above the first Planck scale, in the region $\Delta_0 \gg T \gg m^* c_\perp^2$, the friction coefficient has been obtained by Kopnin (1987). For the same geometry of $\hat{\mathbf{l}}$ parallel to the interface it is $\Gamma \sim T^3 m^*/c_\perp c_\parallel$. These two results match each other at the temperature of order of the first Planck scale, $T \sim m^* c_\perp^2$.

29.4 Vacuum energy and cosmological constant

Now, with all our experience with quantum liquids, where some kind of gravity arises in the low-energy corner, what can be said about the following issues?

29.4.1 *Why is the cosmological constant so small?*

For quantum liquids the vacuum energy was considered in Sec. 3.3.4 for Bose superfluid ^{4}He and in Sec. 7.3.4 for Fermi superfluid ^{3}He-A. In both cases, if the liquid is freely suspended and the surface effects are neglected, its vacuum energy density is exactly zero:

$$\rho_\Lambda \equiv \tilde{\epsilon} = \frac{1}{V}\langle \text{vac}|\mathcal{H} - \mu\mathcal{N}|\text{vac}\rangle_{\text{equilibrium}} = 0 \;. \tag{29.20}$$

This result is universal, i.e. it does not depend on details of the interaction of atoms in the liquid and on their quantum statistics. This also means that it does not depend on the structure of the effective theory, which arises in the low-energy corner.

On the contrary, an inner observer who is familiar with the effective theory only and is not aware of a very simple thermodynamic identity, which follows from the microscopic physics, will compute the vacuum energy density whose magnitude is determined by the Planck energy scale, and whose sign depends on the fermionic and bosonic content.

In superfluid ^{4}He the effective theory contains phonons as elementary bosonic particles and no fermions. The vacuum energy computed by an inner observer is represented by the zero-point energy of phonons:

$$\rho_\Lambda \sqrt{-g} = \frac{1}{2V}\sum_{\text{phonons}} cp \sim \frac{1}{c^3}E_{\text{Debye}}^4 = \sqrt{-g}\, E_{\text{Planck}}^4 \;. \tag{29.21}$$

Here c is the speed of sound; the 'Planck' energy cut-off is determined by the Debye temperature $E_{\text{Planck}} \equiv E_{\text{Debye}} = \hbar c/a_0$ with a_0 being the interatomic space, which plays the role of the Planck length, $a_0 \equiv L_{\text{Planck}}$; g is the determinant of the acoustic metric: $\sqrt{-g} = 1/c^3$.

The vacuum energy density of the ^{3}He-A liquid estimated by an inner observer living there comes from the Dirac vacuum of quasiparticles:

$$\rho_\Lambda \sqrt{-g} = -\frac{1}{V} \sum_{\text{quasiparticles}} \sqrt{g^{ik} p_i p_k} \sim -\sqrt{-g}\, E_{\text{Planck 2}}^4 \,. \qquad (29.22)$$

Here $\sqrt{-g} = 1/c_\| c_\perp^2$, and the second 'Planck' energy cut-off $E_{\text{Planck 2}} \sim \hbar c_\perp/a_0$.

Equations (29.21) and (29.22) are particular cases of the estimate (2.9) of the vacuum energy within the effective RQFT. Both inner observers will be surprised to know that their computations are extremely far from reality, just as we are surprised that our estimates of the vacuum energy are in huge disagreement with cosmological experiments.

Disadvantages of calculations of the vacuum energy within the effective field theory are: (i) the result depends on the cut-off procedure; (ii) the result depends on the choice of the zero from which the energy is counted: a shift of the zero level leads to a shift in the vacuum energy. These drawbacks disappear in exact microscopic theory, i.e. in the Theory of Everything in eqn (3.2): (i) eqn (29.20) does not depend on cut-off(s) of the effective theory(ies); (ii) the energy in eqn (29.20) does not depend on the choice of zero-energy level: the overall shift of the energy in $\mathcal{H}$ is exactly compensated by the shift of the chemical potential μ; and finally the most important (iii) in both quantum liquids the vacuum energy density is exactly zero without any fine tuning. This fundamental result does not depend on microscopic details of quantum liquids, and thus we can hope that it is applicable also to the 'cosmic fluid' – the quantum vacuum.

The exact theory demonstrates that not just the low-energy degrees of freedom of the effective theory (phonons or Bogoliubov quasiparticles) must be taken into account, but all degrees of freedom of the quantum liquid, including 'Planckian' and 'trans-Planckian' domains. According to simple thermodynamic arguments, the latter completely compensate the contribution (29.21) or (29.22) from the low-energy domain.

In exact theory, nullification of the vacuum energy occurs for the liquid-like states only, which can exist as isolated systems. For the gas-like states the chemical potential is positive, $\mu > 0$, and thus these states cannot exist without an external pressure. That is why, one can expect that the solution of the cosmological constant problem can be provided by the mere assumption that the vacuum of RQFT is the liquid-like rather than the gas-like state. However, this is not necessary. As we have seen in the example of the Bose gas and also in the example of Fermi gas in Sec. 7.3.5, the gas-like state suggests its own solution of the cosmological constant problem. The gas-like state can exist only under external pressure, and thus its vacuum energy is non-zero. However, this non-zero energy is not gravitating if the vacuum is in equilibrium.

Thus in both cases, it is the vacuum stability that guarantees the solution of the main cosmological constant problem. Being guided by these examples, in which the trans-Planckian physics is completely known, one may extend the general principle of the vacuum stability to the physical vacuum too. Then the problem of the cosmological constant disappears together with the cosmological constant in equilibrium. Note that this principle of the vacuum stability is applicable even to such vacuum states which represent the locally unstable saddle point: what is needed is the stationarity of the vacuum energy with respect to small linear perturbations.

As was mentioned by Bjorken (2001b) 'The original (cosmological constant) problem does not go away, but is restated in terms of why the collective modes of such a quantum liquid (quantum vacuum) so faithfully respect gauge invariance and Lorentz covariance, as well as general covariance for the emergent gravitons'. The universal properties of the fermionic quantum liquids with Fermi points in momentum space probably show the route to the solution of the restated problem.

Actually, what we need from condensed matter, where the trans-Planckian physics is known, is to extract some general principles which do not depend on details of the substance, but which cannot be obtained within the effective theory. Superfluid ^{3}He, with its highly distorted gravity and caricature quantum field theory, is still a confined vacuum; that is why one can hope that general conclusions based on its properties can be extended to a more complicated 'many-body' system such as the vacuum of RQFT. The principles of a non-gravitating vacuum, and of zero vacuum energy, together with the Fermi point universality could be of this kind.

29.4.2 *Why is the cosmological constant of order of the present mass of the Universe?*

Another problem related to vacuum energy is the cosmic coincidence problem. According to the Einstein equations the cosmological constant ρ_Λ (the vacuum energy density) must be constant in time, while the energy density ρ_M of (dark) matter must decrease as the Universe expands. Why are these two quantities of the same order of magnitude just at the present time, as is indicated by recent astronomical observations (Perlmutter *et al.* 1999; Riess *et al.* 2000)? At the moment it is believed that $\rho_\Lambda \sim 2$–$3\rho_M$, where ρ_M is mostly concentrated in the invisible dark matter, while the density of ordinary known matter, such as baryonic matter, is relatively small being actually within the noise.

In a quantum liquid there is a natural mechanism for the complete cancellation of the vacuum energy. But such a cancellation works only under perfect conditions. If the quantum liquid is perturbed, the vacuum pressure and energy will respond and must be proportional to perturbations of the vacuum state. The perturbations, which disturb zero value of the vacuum pressure, can be: the non-zero energy density of the matter in the Universe; the non-zero temperature of background radiation; space curvature; time dependence caused by expansion of the Universe; some long-wavelength fields – quintessence, etc., and their self-

VACUUM ENERGY AND COSMOLOGICAL CONSTANT

consistent combinations. This indicates that the cosmological constant is not a constant at all but is a dynamical quantity arising as a response to perturbations of the vacuum. At the moment all the deviations from the perfect vacuum state are extremely small compared to the Planck energy scale. This is the reason why the cosmological constant must be extremely small.

Thus from the condensed matter point of view it is natural that $\rho_\Lambda \sim \rho_M$. These two quantities must even be of the same order of magnitude if the perturbations of the vacuum caused by the matter are dominating. In Section 29.2.4 we considered how this happens if one completely ignores all other perturbations, including the gravitational field. In that example the quantum vacuum in ^{3}He-A responds to the 'matter' formed by massless relativistic quasiparticles at $T \neq 0$ (see eqn (29.7)). The ultra-relativistic matter obeys the equation of state

$$P_M = \frac{1}{3}\rho_M = \frac{7\pi^2 N_F}{360\hbar^3}T^4 \ . \tag{29.23}$$

For an isolated liquid the partial pressure of matter P_M must be compensated by the negative vacuum pressure P_Λ to support the zero value of the external pressure,

$$P_{\text{total}} = P_\Lambda + P_M = 0 \ . \tag{29.24}$$

As a result one obtains the following relation between the energy densities of the vacuum and ultrarelativistic matter:

$$\rho_\Lambda = -P_\Lambda = P_M = \frac{1}{3}\rho_M \ . \tag{29.25}$$

The puzzle transforms to the technical problem of how to explain the observed relation $\rho_\Lambda \sim 2\text{--}3\rho_M$. Most probably this ratio depends on the details of the trans-Planckian physics: the Einstein equations must be modified to allow the cosmological constant to vary in time, and this is where the trans-Planckian physics may intervene.

29.4.3 Vacuum energy from Casimir effect

Another example of the induced non-zero vacuum energy density is provided by the boundaries of the system. Let us consider a finite droplet of liquid ^{3}He or liquid ^{4}He with radius R. The stability of the droplet against decay into isolated ^{3}He or ^{4}He atoms is provided by the negative values of the chemical potential: $\mu_3 < 0$ and $\mu_4 < 0$. If this droplet is freely suspended the surface tension leads to non-zero vacuum pressure P_Λ, which at $T = 0$ must compensate the pressure caused by the surface tension due to the curvature of the surface. For a spherical droplet one has

$$P_{\text{total}} = P_\Lambda + P_\sigma = 0 \ , \ \sqrt{-g}P_\sigma = -\frac{2\sigma}{R} \ , \tag{29.26}$$

where σ is the surface tension. As a result one obtains the negative vacuum energy density:

$$\sqrt{-g}\rho_\Lambda = -\sqrt{-g}P_\Lambda = -\frac{2\sigma}{R} \sim -\frac{E_{\text{Debye}}^3}{\hbar^2 c^2 R} \equiv -\sqrt{-g}E_{\text{Planck}}^3 \frac{\hbar c}{R} \ . \tag{29.27}$$

This is analogous to the Casimir effect, in which the boundaries of the system produce a non-zero vacuum pressure. The strong cubic dependence of the vacuum pressure on the 'Planck' energy $E_{\text{Planck}} \equiv E_{\text{Debye}}$ reflects the trans-Planckian origin of the surface tension $\sigma \sim E_{\text{Debye}}/a_0^2$: it is the energy (per unit area) related to the distortion of atoms in the surface layer of atomic size a_0. The term of order E_{Planck}^3/R in the Casimir energy has been considered by Ravndal (2000). Such vacuum energy, with R being the size of the cosmological horizon, has been connected by Bjorken (2001a) to the energy of the Higgs condensate in the electroweak phase transition.

The partial pressure P_σ induced by the surface tension can serve as an analog of the quintessence in cosmology (on quintessence see e.g. Caldwell *et al.* 1998). The equation of state for the surface tension is

$$P_\sigma = -(2/3)\rho_\sigma \ , \tag{29.28}$$

where $\sqrt{-g}\rho_\sigma$ is the surface energy $4\pi R^2 \sigma$ divided by the volume of the droplet. In cosmology the quintessence with the same equation of state is represented by a wall wrapped around the Universe or by a tangled network of cosmic domain walls (Turner and White 1997). In our case the quintessence is also related to the wall – the boundary of the droplet.

29.4.4 *Vacuum energy induced by texture*

The non-zero vacuum energy density, with a weaker dependence on E_{Planck}, is induced by the inhomogeneity of the vacuum. Let us discuss how the vacuum energy density in an isolated quantum liquid responds to the texture. We choose the simplest example which can be discussed in terms of the interplay of partial pressures of the vacuum and texture. It is the soliton in ^{3}He-A, which has the same topology as the soliton in Fig. 16.2, but with a non-zero twist of the $\hat{\mathbf{l}}$-field: $\hat{\mathbf{l}} \cdot (\nabla \times \hat{\mathbf{l}}) \neq 0$. As was shown in Sec. 10.5.3 such a texture carries the non-zero Riemann curvature in the effective gravity of ^{3}He-A. This will allow us to relate this to the problem of vacuum energy induced by the Riemann curvature in general relativity.

Within the twist soliton the field of the unit vector $\hat{\mathbf{l}}$ is given by $\hat{\mathbf{l}}(z) = \hat{\mathbf{x}}\cos\phi(z) + \hat{\mathbf{y}}\sin\phi(z)$, where ϕ is the angle between $\hat{\mathbf{l}}$ and $\hat{\mathbf{d}}$ which changes from $\phi(-\infty) = 0$ to $\phi(+\infty) = \pi$. The profile of ϕ is determined by the interplay of the spin–orbit energy (16.1) and the gradient energy of the field ϕ which can be found from eqn (10.9). Since the spin–orbital interaction is the part of the vacuum energy which depends on ϕ we shall write the energy densities in the following form:

$$\rho = \rho_\Lambda + \rho_{\text{grad}} \ , \tag{29.29}$$

$$\sqrt{-g}\rho_{\text{grad}} = K_t(\hat{\mathbf{l}} \cdot (\nabla \times \hat{\mathbf{l}}))^2 = K_t(\partial_z \phi)^2 \ , \tag{29.30}$$

VACUUM ENERGY AND COSMOLOGICAL CONSTANT

$$\sqrt{-g}\rho_\Lambda = \sqrt{-g}\rho_\Lambda(\phi = 0) + g_D \sin^2\phi \ . \tag{29.31}$$

The solitonic solution of the sine–Gordon equation for ϕ is $\tan(\phi/2) = e^{z/\xi_D}$, where ξ_D is the dipole length: $\xi_D^2 = K_t/g_D$. From this solution it follows that the vacuum and gradient energy densities have the following profile:

$$\sqrt{-g}\left(\rho_\Lambda(z) - \rho_\Lambda(\phi=0)\right) = \sqrt{-g}\rho_{\mathrm{grad}}(z) = \frac{K_t}{\xi_D^2 \cosh^2(z/\xi_D)} \ . \tag{29.32}$$

Let us consider the 1D case, which allows us to use the space-dependent partial pressures. We have two subsystems: the vacuum with the equation of state $P_\Lambda = -\rho_\Lambda$, and the texture with the equation of state $P_{\mathrm{grad}} = \rho_{\mathrm{grad}}$. In cosmology the latter equation of state describes the so-called stiff matter. In equilibrium, in the absence of the environment the total pressure is zero, and thus the positive pressure of the texture (stiff matter) must be compensated by the negative pressure of the vacuum:

$$P_{\mathrm{total}} = P_\Lambda(z) + P_{\mathrm{grad}}(z) = 0 \ . \tag{29.33}$$

Using the equations of state one obtains another relation between the vacuum and the gradient energy densities:

$$\rho_\Lambda(z) = -P_\Lambda(z) = P_{\mathrm{grad}}(z) = \rho_{\mathrm{grad}}(z) \ . \tag{29.34}$$

Comparing this equation with eqn (29.32) one finds that

$$\rho_\Lambda(\phi = 0) = 0 \ . \tag{29.35}$$

This means that even in the presence of a soliton the main vacuum energy density – the energy density of the bulk liquid far from the soliton – is exactly zero if the liquid is in equilibrium at $T = 0$. Within the soliton the vacuum is not homogeneous, and the vacuum energy density induced by such a perturbation of the vacuum equals the energy density of the perturbation: $\rho_\Lambda(z) = \rho_{\mathrm{grad}}(z)$.

The induced vacuum energy density in eqn (29.32) is inversely proportional to the square of the size of the region where the field is concentrated:

$$\rho_\Lambda(R) \sim E_{\mathrm{Planck}}^2 \left(\frac{\hbar c}{R}\right)^2 \ . \tag{29.36}$$

Similar behavior for the vacuum energy density in the interior region of the Schwarzschild black hole, with R being the Schwarzschild radius, was discussed by Chapline et al. (2001,2002) and Mazur and Mottola (2001). In the case of the soliton the size of the perturbed region is $R \sim \xi_D$.

In cosmology, the vacuum energy density obeying eqn (29.36) with R proportional to the Robertson–Walker scale factor has been suggested by Chapline (1999), and with R proportional to the size of the cosmological horizon has been suggested by Bjorken (2001a).

29.4.5 Vacuum energy due to Riemann curvature and Einstein Universe

In Sec. 10.5.3 we found an equivalence between the gradient energy of twisted texture in ^{3}He-A and the Riemann curvature of effective space with the time independent metric

$$-\frac{1}{16\pi G}\int d^3r\sqrt{-g}\mathcal{R} \equiv K_t\int d^3r(\hat{\mathbf{l}}\cdot(\nabla\times\hat{\mathbf{l}}))^2 \ . \tag{29.37}$$

Thus using the example considered in the above Sec. 29.4.4, one may guess that the space curvature perturbs the vacuum and induces the vacuum energy density $\rho_\Lambda = -\frac{1}{16\pi G}\mathcal{R}$. Of course, this is true only as an order of magnitude estimate since we extended the (1+1)-dimensional consideration of the texture to the 3+1 gravity.

Let us consider an example from general relativity, where one can exactly find the response of the vacuum energy to the space curvature and to the energy density of matter. The necessary condition for the applicability of the above considerations to the vacuum in general relativity is the stationarity of the system (vacuum + matter). As in condensed matter systems, the stationary state does not necessarily mean the local minimum of some energy functional, but can be the locally unstable saddle point as well. This example is provided by the static closed Universe with positive curvature obtained by Einstein (1917) in his work where for the first time he introduced the cosmological term. The most important property of this solution is that the cosmological constant is not fixed but is self-consistently found from the Einstein equations. In other words, the vacuum energy density in the Einstein Universe is adjusted to the perturbations caused by matter and curvature, in the same manner as we observed in quantum liquids.

Let us consider how this happens using only the equilibrium conditions. In the static Universe two equilibrium conditions must be fulfilled:

$$P_{\text{total}} = P_M + P_\Lambda + P_\mathcal{R} = 0 \ , \ \rho_{\text{total}} = \rho_M + \rho_\Lambda + \rho_\mathcal{R} = 0 \ . \tag{29.38}$$

Here $\rho_\mathcal{R}$ is the energy density stored in the spatial curvature and $P_\mathcal{R}$ is the partial pressure of the spatial curvature:

$$\rho_\mathcal{R} = -\frac{\mathcal{R}}{16\pi G} = -\frac{3k}{8\pi GR^2} \ , \ P_\mathcal{R} = -\frac{1}{3}\rho_\mathcal{R} \ . \tag{29.39}$$

Here R is the cosmic scale factor in the Friedmann–Robertson–Walker metric,

$$ds^2 = -dt^2 + R^2\left(\frac{dr^2}{1-kr^2} + r^2d\theta^2 + r^2\sin^2\theta d\phi^2\right) \ ; \tag{29.40}$$

the parameter $k = (-1, 0, +1)$ for an open, flat or closed Universe respectively. The equation of state on the rhs of eqn (29.39) follows from the equation $P_\mathcal{R} = -d(\rho_\mathcal{R} R^3)/d(R^3)$.

The first equation in (29.38) is the requirement that for the 'isolated' Universe the external pressure must be zero. The second equation in (29.38) reflects the gravitational equilibrium, which requires that the total mass density must be zero (actually the 'gravineutrality' corresponds to the combination of two equations in (29.38), $\rho_{\text{total}}+3P_{\text{total}} = 0$, since $\rho+3P$ serves as a source of the gravitational field in the Newtonian limit). The gravineutrality is analogous to the electroneutrality condition in condensed matter.

We must solve eqns (29.38) together with the equations of state $P_a = w_a \rho_a$, where $w = -1$ for the vacuum, $w = -1/3$ for the space curvature, $w = 0$ for the cold matter and $w = 1/3$ for the radiation field. The simplest solution of these equations is the flat Universe without matter. The vacuum energy density in such a Universe is zero.

The solution with matter depends on equation of state for matter. For the cold Universe with $P_M = 0$, eqns (29.38) give the following value of the vacuum energy density:

$$\rho_\Lambda = \frac{1}{2}\rho_M = -\frac{1}{3}\rho_\mathcal{R} = \frac{k}{8\pi G R^2} \, . \tag{29.41}$$

For the hot Universe with the equation of state for the radiation matter $P_M = (1/3)\rho_M$, one obtains

$$\rho_\Lambda = \rho_M = -\frac{1}{2}\rho_\mathcal{R} = \frac{3k}{16\pi G R^2} \, . \tag{29.42}$$

Since the energy of matter is positive, the static Universe is possible only for positive curvature, $k = +1$, i.e. for the closed Universe.

This is a unique example of the stationary state, in which the vacuum energy on the order of the energy of matter and curvature is obtained within the effective theory of general relativity. It is quite probable that the static states of the Universe are completely within the responsibility of the effective theory and are determined by eqns (29.38), which do not depend on the details of the trans-Planckian physics.

Unfortunately (or maybe fortunately for us) the above static solution is unstable, and our Universe is non-stationary. For the non-stationary case we cannot find the relation between the vacuum energy and the energies of other ingredients being within the effective theory. The reason is that ρ_Λ must be adjusted to perturbations and thus must change in time, which is forbidden within the Einstein equations due to Bianchi identities. That is why the effective theory must be modified to include the equation of motion for ρ_Λ. Such a modification is not universal and depends on the details of the Planckian physics. Thus the only what we can say at the moment is that the cosmological constant tracks the development of the Universe.

29.4.6 Why is the Universe flat?

The connection to the Planckian physics can also give some hint on how to solve the flatness problem. At the present time the Universe is almost flat, which means that the energy density of the Universe is close to the critical density ρ_c. This implies that the early Universe was extremely flat: since $(\rho - \rho_c)/\rho_c =$

$3k/(8\pi G\rho_c R^3) \propto R$, in the early Universe where R is small one has $|\rho-\rho_c|/\rho_c \ll 1$. At $t = 1$ s after the Big Bang this was about 10^{-16}.

What is the reason for such fine tuning? The answer can be provided by the inflationary scenario in which the curvature term is exponentially suppressed, since the exponential inflation of the Universe simply irons out curved space to make it extraordinarily flat. The analogy with quantum liquids suggests another solution of the flatness problem.

According to the 'cosmological principle' the Universe must be homogeneous and isotropic. This is strongly confirmed by the observed isotropy of cosmic background radiation. Within general relativity the Robertson–Walker metric describes the spatially homogeneous and isotropic distribution of matter and thus satisfies the cosmological principle. The metric field is not homogeneous, but in general relativity the property of the homogeneity must be determined in a covariant way, i.e. it should not depend on coordinate transformation. In the Robertson–Walker Universe the covariant quantity – the curvature – is constant, and thus this Universe is homogeneous.

However, if general relativity is an effective theory, the invariance under the coordinate transformations exists at low energy only. At higher energy, the contravariant metric field $g^{\mu\nu}$ itself becomes the physical quantity: an external observer belonging to the trans-Planckian world can distinguish between different metrics even if they are equivalent for the inner low-energy observer. According to eqn (29.40) the metric field is not homogeneous unless $k = 0$. If $k \neq 0$ the 'Planck' observer views the Robertson–Walker metric as space dependent. Moreover, the r^2-dependence of the contravariant metric element $g^{rr} \propto 1 - kr^2$ implies a huge deformation of the 'Planck liquid', which is strongly prohibited. That is why, according to the trans-Planckian physics, the Universe must be flat at all times, i.e. $k = 0$ and $\rho_\Lambda + \rho_M = \rho_c$. This means that the cosmological principle of homogeneity of the Universe can well be an emergent phenomenon reflecting the Planckian physics.

29.4.7 What is the energy of false vacuum?

It is commonly believed that the vacuum of the Universe underwent one or several broken-symmetry phase transitions. Each of the transitions is accompanied by a substantial change in the vacuum energy. Moreover, there can be false vacua separated from the true vacuum by a large energy barrier. Why is the true vacuum so distinguished from the others that it has exactly zero energy, while the energies of all other false vacua are enormously large? Where does the principal difference between vacua come from?

The quantum liquid answer to this question is paradoxical (see Sec. 29.2.2): in the absence of external forces all the vacua including the false ones have zero energy density and thus zero cosmological constant. There is no paradox, however, because the positive energy difference between the false and the true vacuum is obtained at fixed chemical potential μ. Let us suppose that the liquid is in the false vacuum state. Then its vacuum energy density $\tilde{\epsilon} = \epsilon - \mu n \equiv \sqrt{-g}\rho_\Lambda = 0$. After the transition from the false vacuum to the true one has occurred there is an energy release. However, when the equilibrium state of the true vacuum is reached, the chemical potential μ will be automatically adjusted to cancel the energy density of the new vacuum. Thus in an isolated system the vacuum energy density is zero both below and above the phase transition.

This means that the energies of all condensates in RQFT (gluon and quark condensates in QCD, Higgs field in electroweak theory, etc.) do not violate the zero value of the cosmological constant in equilibrium.

29.4.8 Discussion: why is vacuum not gravitating?

We found that vacuum energy density is exactly zero, if the following conditions are fulfilled: there are (i) no external forces acting on the liquid; (ii) no quasiparticles (matter) in the liquid; (iii) no curvature and inhomogeneity; (iv) no boundaries which give rise to the Casimir effect; and (v) no time dependence and non-equilibrium processes. Each of these five factors perturbs the vacuum state and induces a non-zero value of the vacuum energy density of order of the energy density of the perturbation. Applying this to the vacuum in the Universe, one may expect that in each epoch the vacuum energy density is of order of the matter density of the Universe, and/or of its curvature, and/or of the energy density of the smooth component – the quintessence. At the present moment all the perturbations are extremely small compared to the Planck scales, which is why we have an extremely small cosmological constant. In other words, the cosmological constant is small because the Universe is old. Small perturbations, such as the expansion of the Universe and the energy density of matter, represent small ripples on the surface of the great pacific ocean of the Dirac vacuum. They do not disturb the depth of the ocean which is extremely close to the complete quietness now. And actually it was quiet even during the violent processes in early Universe, whose energy scales were still much smaller than the Planck scale.

However, the actual problem for cosmology is not why the vacuum energy is zero (or very small), but why the vacuum is not (or almost not) gravitating. These two problems are not necessarily related since in the effective theory the equivalence principle is not the fundamental physical law, and thus does not necessarily hold when applied to the vacuum energy. The condensed matter analogy gives us examples of how the effective gravity appears as an emergent phenomenon in the low-energy corner. In these examples the gravity is not fundamental: it is one of the low-energy collective modes of the quantum vacuum. This dynamical mode provides the effective metric for the low-energy quasiparticles serving as an analog of matter. The gravity does not exist on the microscopic (trans-Planckian) level and emerges only in the low-energy limit simultaneously with the relativity, with relativistic matter and with the interaction between gravity and matter.

The vacuum state of the quantum liquid is the outcome of the microscopic interactions of the underlying ^{4}He or ^{3}He atoms. These atoms, which live in the 'trans-Planckian' world and form the vacuum state there, do not experience the 'gravitational' attraction experienced by the low-energy quasiparticles, since the effective gravity simply does not exist at the microscopic scale (of course, we ignore the real gravitational attraction of bare atoms, which is extremely small in quantum liquids). That is why the vacuum energy cannot interact with gravity, and cannot serve as a source of the effective gravity field: the vacuum is not gravitating.

On the other hand, the long-wavelength perturbations of the vacuum are within the sphere of influence of the low-energy effective theory, and such perturbations can be the source of the effective gravitational field. Deviations of the vacuum from its equilibrium state, described by the effective theory, are gravitating.

29.5 Mesoscopic Casimir force

In this section we introduce the type of the Casimir effect, whose source is missing in the effective theory. In quantum liquids it is related to the discrete quantity – the particle number N, i.e. the number of bare ^{3}He or ^{4}He atoms of the liquid. The particle number is the quantity which is missing by an inner observer, but is instrumental for the Planckian physics. The bare atoms are responsible for the construction of the vacuum state. The conservation law for particle number leads to the relevant vacuum energy density $\tilde{\epsilon} = \epsilon - \mu n$, which is invariant under the shift of the energy, and gives us the mechanism for cancellation of the vacuum energy density in equilibrium.

Now we turn to the problem of whether the discreteness of N in the vacuum can be probed. For that we consider the type of the Casimir effect, which is determined not by the finite size of the system, but by the finite-N effect, which in condensed matter is referred to as the mesoscopic effect. If in the Casimir-type effects we naively replace the finite volume $V = a^3$ of the system by the finite $N = nV \sim a^3/a_0^3$, we obtain the following dependence on N of different contributions to the vacuum energy:

$$E_{\text{bulk vacuum}} \sim E_{\text{Planck}} N , \qquad (29.43)$$
$$E_{\text{surface}} \sim E_{\text{Planck}} N^{2/3} , \qquad (29.44)$$
$$E_{\text{texture}} \sim E_{\text{Planck}} N^{1/3} , \qquad (29.45)$$
$$E_{\text{mesoscopic}} \sim E_{\text{Planck}} N^0 , \qquad (29.46)$$
$$E_{\text{Casimir}} \sim E_{\text{Planck}} N^{-1/3} . \qquad (29.47)$$

We assumed here that all three dimensions of the system are of the same order; $E_{\text{Planck}} = \hbar c/a_0$, where a_0 is the interatomic space; and c is the analog of the speed of light. We also included the missing contribution from the mesoscopic effect, eqn (29.46), which the effective theory is not able to predict, since it is the property of the microscopic high-energy degrees of freedom. This mesoscopic effect cannot be described by the effective theory dealing with the continuous medium, even if the theory includes the real boundary conditions with the frequency dependence of dielectric permeability.

The mesoscopic effect has an oscillatory behavior which is characteristic of mesoscopic phenomena and under certain conditions it can be more pronounced than the standard Casimir effect.

29.5.1 Vacuum energy from 'Theory of Everything'

Let us consider this finite-N mesoscopic effect using the quantum 'liquid' where the microscopic Theory of Everything is extremely simple, and the mesoscopic Casimir forces can be calculated exactly without invoking any regularization procedure. This is the 1D gas of non-interacting fermions with a single spin component and with the spectrum $E(p) = p^2/2m$. The 3+1 case of the ideal Fermi gas in slab geometry can be found in the paper by Bulgac and Magierski (2001) and in references therein.

At $T = 0$ fermions occupy all the energy levels below the positive chemical potential μ, i.e. with $E(p) - \mu < 0$, forming the zero-dimensional Fermi surfaces at $p_z = \pm p_F$ (zeros of co-dimension 1 with topological charges $N_1 = \pm 1$). In the infinite system, the vacuum energy density expressed in terms of the particle density $n = p_F/\pi\hbar$ is

$$\epsilon(n) = \int_{-p_F}^{p_F} \frac{dp_z}{2\pi\hbar} \frac{p^2}{2m} = \frac{p_F^3}{6\pi\hbar m} = \frac{\pi^2 \hbar^2}{6m} n^3 \ . \tag{29.48}$$

The equation of state comes from the thermodynamic identity relating the pressure P and the energy (see eqn (3.27)):

$$P = -\tilde{\epsilon} = \mu n - \epsilon = n\frac{d\epsilon}{dn} - \epsilon = 2\epsilon \ . \tag{29.49}$$

The speed of sound is $c = p_F/m$, which coincides with the slope of the relativistic spectrum of quasiparticles in the vicinity of the Fermi surface: $E(p) \approx \pm c(p \pm p_F)$. The Planck energy scale, which marks the violation of 'Lorentz invariance', is played by $E_{\text{Planck}} = cp_F$.

In a 1D cavity of size a, if particles cannot penetrate through the boundaries their energy spectrum becomes discrete:

$$E_k = \frac{\hbar^2 \pi^2}{2ma^2} k^2 \ , \tag{29.50}$$

where k is integer. The total energy of N fermions in the cavity is

$$E(N,a) = \sum_{k=1}^{N} E_k = \frac{\hbar^2 \pi^2}{12ma^2} N(N+1)(2N+1) \ . \tag{29.51}$$

29.5.2 Leakage of vacuum through the wall

In the traditional Casimir effect the quantum vacuum is the same on both sides of the conducting plate, otherwise the force on the palte comes from the bulk effect, i.e. from the infinite difference between the vacuum energy densities across the

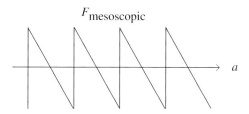

FIG. 29.3. Mesoscopic Casimir effect. *Top*: Two 1D slabs with Fermi gas separated by a penetrable membrane. *Bottom*: The force acting on the membrane as a function of its position. Jumps occur due to the discrete number of fermions in the slabs.

plate. To simulate this situation we consider the force acting on the membrane between two slabs of sizes a and b with the same Fermi gas (Fig.29.3 *top*). We assume the same boundary conditions at all three walls, but we allow the particles to transfer through the membrane between the slabs to provide the same bulk pressure in the two slabs avoiding the bulk effect. The connection can be done due to, say, a very small holes (tunnel junctions) in the membrane. This does not violate the boundary conditions and does not disturb the particle energy levels, but still allows the particle exchange between the two vacua.

In the traditional Casimir effect, the force between the conducting plates arises because the electromagnetic fluctuations of the vacuum in the slab are modified due to boundary conditions imposed on the electric and magnetic fields. In reality the reflection is not perfect, and these boundary conditions are applicable only in the low-frequency limit. The plates are transparent for the high-frequency electromagnetic modes, as well as for the other degrees of freedom of real vacuum (fermionic and bosonic), that can easily penetrate through the conducting wall. That is why the high-frequency degrees of freedom, which produce the divergent terms in the vacuum energy, do not contribute to the force. Thus the imperfect reflection produces the natural regularization scheme.

The dispersion of dielectric permeability weakens the real Casimir force. Only in the so-called retarded limit $a \gg c/\omega_0$, where ω_0 is the characteristic frequency at which the dispersion becomes important, does the Casimir force acquire the universal behavior which depends only on geometry but does not depend on how easily the high-energy vacuum leaks through the conducting walls. (Let us recall that in both the retarded limit $a \gg c/\omega_0$ and the unretarded limit $a \ll c/\omega_0$, the Casimir force comes from the zero-point fluctuations. But in the retarded limit these are the zero-point fluctuations of the electromagnetic vacuum, which are important, and they give rise to the Casimir force. In the unretarded limit the main contribution comes from the zero-point fluctuations of the dipole moments of atoms, and the Casimir force is transformed to the van der Waals force.)

According to this consideration, our example in which fermions forming the quantum vacuum are almost totally reflected from the membrane and the penetration of the quantum vacuum through the membrane is suppressed, must be extremely favorable for the universal Casimir effect. Nevertheless, we shall show that even in this favorable case the mesoscopic finite-N effects deform the Casimir effect: the contribution of the diverging terms to the Casimir effect becomes dominating. They produce highly oscillating vacuum pressure in condensed matter (Fig.29.3 *bottom*). The amplitude of the mesoscopic fluctuations of the vacuum pressure in this limit exceeds by a factor $p_{\text{Planck}} a/\hbar$ the value of the standard Casimir pressure.

29.5.3 Mesoscopic Casimir force in 1D Fermi gas

The total vacuum energy of fermions in the two slabs in Fig.29.3 *top* is

$$E(N_a, N_b, a, b) = \frac{\hbar^2 \pi^2}{12m} \left(\frac{N_a(N_a+1)(2N_a+1)}{a^2} + \frac{N_b(N_b+1)(2N_b+1)}{b^2} \right), \tag{29.52}$$

where

$$N_a + N_b = N . \tag{29.53}$$

Since particles can transfer between the slabs, the global vacuum state in this geometry is obtained by minimization over the discrete particle number N_a at fixed total number N of particles in the vacuum. If the mesoscopic $1/N_a$ corrections are ignored, one obtains $N_a/a = N_b/b = n$, and the pressures acting on the wall from the two vacua compensate each other.

However, N_a and N_b are integer valued, and this leads to mesoscopic fluctuations of the Casimir force. Within a certain range of parameter a there is a global minimum characterized by integers (N_a, N_b). In the neighboring intervals of parameters a, one has either $(N_a + 1, N_b - 1)$ or $(N_a - 1, N_b + 1)$. The force acting on the wall in the state (N_a, N_b) is obtained by variation of $E(N_a, N_b, a, b)$ over a at fixed N_a and N_b:

$$F(N_a, N_b, a, b) = -\frac{dE(N_a, N_b, a, b)}{da} + \frac{dE(N_a, N_b, a, b)}{db} . \tag{29.54}$$

When a increases then at some critical value of a, where $E(N_a, N_b, a, b) = E(N_a + 1, N_b - 1, a, b)$, one particle must cross the wall from the right to the left. At this critical value the force acting on the wall changes abruptly (we do not discuss here the interesting physics arising just at the critical values of a, where the degeneracy occurs between the states (N_a, N_b) and $(N_a + 1, N_b - 1)$; at these positions of the membrane the particle numbers N_a and N_b are undetermined and are actually fractional due to the quantum tunneling between the slabs

(Andreev 1998)). Using the spectrum in eqn (29.52) one obtains for the jump of the Casimir force

$$F(N_a \pm 1, N_b \mp 1) - F(N_a, N_b) \approx \pm \frac{\hbar^2 \pi^2}{m}\left(\frac{N_a^2}{a^3} + \frac{N_b^2}{b^3}\right) \approx \pm \frac{\hbar^2 \pi^2 n^2}{m}\left(\frac{1}{a} + \frac{1}{b}\right). \tag{29.55}$$

In the limit $a \ll b$ the amplitude of the mesoscopic Casimir force and thus the difference in the vacuum pressure on two sides of the wall is

$$|\Delta F_{\text{meso}}| = 2\frac{E_{\text{Planck}}}{a}, \tag{29.56}$$

where $E_{\text{Planck}} = c p_F$. It is a factor $1/N_a$ smaller than the vacuum energy density in eqn (29.48). On the other hand it is a factor $p_F a/\hbar \equiv p_{\text{Planck}} a/\hbar$ larger than the traditional Casimir pressure, which in the 1D case is $P_C \sim \hbar c/a^2$.

The divergent term which linearly depends on the Planck energy cut-off E_{Planck} as in eqn (29.56) has been revealed in many different calculations which used the effective theory (see e.g. Cognola *et al.* 2001), and attempts have been made to invent a regularization scheme which would cancel this divergent contribution.

29.5.4 *Mesoscopic Casimir forces in a general condensed matter system*

Equation (29.56) for the amplitude of the mesoscopic fluctuations of the vacuum pressure can be immediately generalized for 3D space. The mesoscopic random pressure comes from the discrete nature of the underlying quantum liquid, which represents the quantum vacuum. When the volume V of the vessel changes continuously, the equilibrium number N of particles changes in a stepwise manner. This results in abrupt changes of pressure at some critical values of the volume:

$$P_{\text{meso}} \sim P(N \pm 1) - P(N) = \pm\frac{dP}{dN} = \pm\frac{mc^2}{V} \equiv \pm\frac{E_{\text{Planck}}}{V}, \tag{29.57}$$

where c is the speed of sound. Here the microscopic quantity – the mass m of the atom – explicitly enters, since the mesoscopic pressure is determined by microscopic trans-Planckian physics.

For the pair-correlated systems, such as Fermi superfluids with a finite gap in the energy spectrum, the amplitude must be two times larger. This is because the jumps in pressure occur when two particles (the Cooper pair) tunnel through the junction, $\Delta N = \pm 2$. The transition with $\Delta N = \pm 1$ requires breaking the Cooper pair and costs energy equal to the gap.

29.5.5 *Discussion*

Equation (29.57) is applicable, for example, for a spherical shell of volume $V = (4\pi/3)a^3$ immersed in the quantum liquid. Let us compare this finite-N mesoscopic effect with vacuum pressure in the traditional Casimir effect obtained within the effective theories for the same spherical shell geometry. In the case of the original Casimir effect the effective theory is quantum electrodynamics.

In superfluid ^{4}He this is the low-frequency quantum hydrodynamics, which is equivalent to the relativistic scalar field theory. In other superfluids, in addition to phonons other low-energy modes are possible, namely the massless fermions and the sound-like collective modes such as spin waves. The massless modes with linear ('relativistic') spectrum in quantum liquids play the role of the relativistic massless scalar fields and chiral fermions. The boundary conditions for the scalar fields are typically the Neumann boundary conditions, corresponding to the (almost) vanishing mass or spin current through the wall (let us recall that there must be some leakage through the shell to provide the equal bulk pressure on both sides of the shell).

If we believe in the traditional regularization schemes which cancel out the ultraviolet divergence, then from the effective scalar field theory one must obtain the Casimir pressure $P_C = -dE_C/dV = K\hbar c/8\pi a^4$, where $K = -0.4439$ for the Neumann boundary conditions, and $K = 0.005639$ for the Dirichlet boundary conditions (Cognola et al. 2001). The traditional Casimir pressure is completely determined by the effective low-energy theory, and does not depend on the microscopic structure of the liquid: only the 'speed of light' c enters this force. The same mesoscopic pressure due to massless bosons is valid for the pair-correlated fermionic superfluids, if the fermionic quasiparticles are gapped and their contribution to the Casimir pressure is exponentially small compared to the contribution of the bosonic collective modes.

However, at least in our case, the result obtained within the effective theory is not correct: the real Casimir pressure in eqn (29.57) is produced by the finite-N mesoscopic effect. It essentially depends on the Planck cut-off parameter, i.e. it cannot be determined by the effective theory, it is much bigger, by the factor $E_{\text{Planck}}a/\hbar c$, than the traditional Casimir pressure, and it is highly oscillating. The regularization of these oscillations by, say, averaging over many measurements, by noise, or due to quantum or thermal fluctuations of the shell, etc., depends on the concrete physical conditions of the experiment.

This shows that in some cases the Casimir vacuum pressure is not within the responsibility of the effective theory, and the microscopic (trans-Planckian) physics must be evoked. If two systems have the same low-energy behavior and are described by the same effective theory, this does not mean that they necessarily experience the same Casimir effect. The result depends on many factors, such as the discrete nature of the quantum vacuum, the ability of the vacuum to penetrate through the boundaries, dispersion relation at high frequency, etc. It is not excluded that even the traditional Casimir effect which comes from the vacuum fluctuations of the electromagnetic field is renormalized by the high-energy degrees of freedom

Of course, the extreme limit of the almost impenetrable wall, which we considered, is not applicable to the standard (electromagnetic) Casimir effect, where the overwhelming part of the fermionic and bosonic vacua easily penetrates the conducting walls, and where the mesoscopic fluctuations must be small. This difference is fundamental. In the Casimir effect measured in the physical vacuum of our Universe, the boundaries or walls are made of the excitations of the vacuum. On the contrary, in the Casimir effect discussed for the vacuum of the quantum liquid, the wall of the container is made of a substance which is foreign to (i.e. external to) the vacuum of the quantum liquid. That is why the particles of the vacuum cannot penetrate through the walls of the container. In our world, we cannot create such a hard wall using our low-energy quasiparticle material. That is why the vacuum easily penetrates through the wall, and the mesoscopic effect discussed above must be small. But is it negligibly small? In any case the condensed matter example demonstrates that the cut-off problem is not a mathematical, but a physical one, and the physics dictates the proper regularization scheme or the proper choice of the cut-off parameters.

30

TOPOLOGICAL DEFECTS AS SOURCE OF NON-TRIVIAL METRIC

Topological defects in ^{3}He-A represent the topologically stable configurations of the order parameter. Since some components of the order parameter serve as the metric field of effective gravity, one can use the defects as the source of the non-trivial effective metric. In this chapter we consider two such defects in ^{3}He-A, the domain wall and disclination line. In general relativity they correspond respectively to planar and linear singularities in the field of vierbein. In the domain wall (Sec. 30.1) one or several vectors of the dreibein change sign across the wall. At such a wall in 3D space (or at the 3D hypersurface in 3+1 space) the vierbein is degenerate, so that the determinant of the contravariant metric $g^{\mu\nu}$ becomes zero on the surface. In disclination, the vierbein field and the metric are degenerate on the line (the axis of a disclination), and the dreibein rotates by 2π around this defect line. This metric has conical singularity (Sec. 30.2), which is similar to the gravitational field of the local cosmic string. The degenerate metric corresponding to the field of a spinning cosmic string, which is reproduced by the vortex line, will be considered in Chapter 31. In general relativity, there can be vierbein defects of even lower dimension: the point defects in 3+1 spacetime (Hanson and Regge 1978; d'Auria and Regge 1982). We do not discuss analogs of such instanton defects here.

30.1 Surface of infinite red shift

30.1.1 Walls with degenerate metric

In general relativity, two types of walls with the degenerate metric were considered: with degenerate contravariant metric $g^{\mu\nu}$ and with degenerate covariant metric $g_{\mu\nu}$. Both types of walls could be generic. The case of degenerate $g_{\mu\nu}$ was discussed in details by Bengtsson (1991) and Bengtsson and Jacobson (1998). According to Horowitz (1991), for a dense set of coordinate transformations the generic situation is the 3D hypersurface where the covariant metric $g_{\mu\nu}$ has rank 3.

The physical origin of walls with the degenerate contravariant metric $g^{\mu\nu}$ has been discussed by Starobinsky (1999). Such a wall can arise after inflation, if the field which produces inflation – the inflaton field – has a Z_2 degenerate vacuum. This is the domain wall separating domains with two different vacua of the inflaton field. The metric $g^{\mu\nu}$ can everywhere satisfy the Einstein equations in vacuum, but at the considered surfaces the metric $g^{\mu\nu}$ cannot be diagonalized to the Minkowski metric $g^{\mu\nu} = \text{diag}(-1, 1, 1, 1)$. Instead, on such a surface the diagonal metric becomes degenerate, $g^{\mu\nu} = \mathrm{diag}(-1,0,1,1)$. This metric cannot be inverted, and thus $g_{\mu\nu}$ has a singularity on the domain wall. In principle, the domains can have different spacetime topology, as was emphasized by Starobinsky (1999).

The degeneracy of the contravariant metric implies that the 'speed of light' propagating in some direction becomes zero. We consider here a condensed matter example when this direction is normal to the wall, and thus the 'relativistic' quasiparticles cannot communicate across the wall. Let us recall that in the effective gravity in liquid ^{4}He, the emergent physical gravitational fields arise as the contravariant metric $g^{\mu\nu}$. In ^{3}He-A, the emergent physical gravitational field is the field of the tetrads – the square root of the contravariant metric $g^{\mu\nu}$ – which appear in the propagator for fermions in the vicinity of a Fermi point in eqn (8.18). In both cases $g^{\mu\nu}$ is the property of the quasiparticle energy spectrum at low energy. That is why the fact that this matrix $g^{\mu\nu}$ cannot be inverted at the wall is not crucial for the quantum liquid. But it is crucial for an inner observer living in the low-energy relativistic world, since the surface of the degenerate metric (or the surface of the degenerate vierbein – the vierbein wall) drastically changes the geometry of the relativistic world, so that some parts of the physical space are not accessible for an inner observer.

The vierbein wall can be simulated by domain walls in superfluids and superconductors, at which some of the three 'speeds of light' cross zero. Such walls can exist in ^{3}He-B (Salomaa and Volovik 1988; Volovik 1990b); in chiral p-wave superconductors (Matsumoto and Sigrist 1999; Sigrist and Agterberg 1999); in d-wave superconductors (Volovik 1997a); and in thin ^{3}He-A films (Jacobson and Volovik 1998b; Volovik 1999c; Jacobson and Koike 2002). We consider the latter domain wall. The structure of the order parameter within this domain wall was considered in Sec. 22.1.2. When this vierbein wall moves, it splits into a black-hole/white-hole pair, which experiences the quantum friction force due to Hawking radiation discussed in Sec. 32.2.5 (Jacobson and Volovik 1998b; Jacobson and Koike 2002; Sec. 32.1.1).

30.1.2 *Vierbein wall in ^{3}He-A film*

The vierbein wall we are interested in separates in superfluid ^{3}He-A film domains with opposite orientations of the unit vector $\hat{\mathbf{l}}$ of the orbital momentum of Cooper pairs: $\hat{\mathbf{l}} = \pm\hat{\mathbf{z}}$ (Sec. 22.1.2). Here $\hat{\mathbf{z}}$ is along the normal to the film, and the coordinate x is across the domain wall, which is actually the domain line along the y axis. Let us start with the wall at rest with respect to both the heat bath and superfluid vacuum. Since the wall is topologically stable, stationary with respect to the heat bath, and static (i.e. there is no superflow across the wall), such a wall does not experience any dissipation.

In the 2+1 spacetime of a thin film, the Bogoliubov–Nambu Hamiltonian for fermionic quasiparticles in the wall background is given by eqn (22.4):

$$\mathcal{H} = \frac{p_x^2 + p_y^2 - p_F^2}{2m^*}\check{\tau}^3 + \mathbf{e}_1(x)\cdot\mathbf{p}\;\check{\tau}^1 + \mathbf{e}_2(x)\cdot\mathbf{p}\check{\tau}^2\;. \tag{30.1}$$

When $|\mathbf{p}| \to 0$ this Hamiltonian describes the massive Dirac fermions in the field of the zweibein $\mathbf{e}_1$ and $\mathbf{e}_2$. We assume the following order parameter texture (the field of zweibein) within the wall:

$$\mathbf{e}_1(x) = \hat{\mathbf{x}} c(x) \ , \quad \mathbf{e}_2 = \hat{\mathbf{y}} c_\perp \ . \tag{30.2}$$

Here the speed of 'light' propagating along the y axis is constant, while the speed of 'light' propagating along the x axis changes sign across the wall, for example, as in eqn (22.5):

$$c(x) = c_\perp \tanh \frac{x}{d} \ . \tag{30.3}$$

The exact solution for the order parameter within the wall (Salomaa and Volovik 1989) is slightly different from this ansatz, but this does not change the topology of the wall and therefore is not important for the discussed phenomena. At $x = 0$ the dreibein is degenerate: the vector product $\mathbf{e}_1 \times \mathbf{e}_2 = 0$, and thus the unit vector $\hat{\mathbf{l}} = \mathbf{e}_1 \times \mathbf{e}_2 / |\mathbf{e}_1 \times \mathbf{e}_2|$ is not determined.

Since the momentum projection p_y is the conserved quantity, we come to a pure 1+1 problem. Now we shall manipulate p_y and other parameters of the system in order to obtain the relativistic 1+1 physics in the low-energy limit, which we want to simulate. For that we assume that (i) $p_F \ll m^* c_\perp$; (ii) $p_y = \pm p_F$; and (iii) the thickness d of the domain wall is larger than the 'Planck' length scale: $d \gg \hbar / m^* c_\perp$. Then, rotating the Bogoliubov spin and ignoring the non-commutativity of the p_x^2 term and $c(x)$, one obtains the following Hamiltonian for the 1+1 particle:

$$\mathcal{H} = M(\mathcal{P}_x)\check{\tau}^3 + \frac{1}{2}(c(x)\mathcal{P}_x + \mathcal{P}_x c(x))\check{\tau}^1 \ , \tag{30.4}$$

$$M^2(\mathcal{P}_x) = \frac{\mathcal{P}_x^4}{4m^2} + c_\perp^2 p_F^2 \ , \tag{30.5}$$

where, as before, the momentum operator $\mathcal{P}_x = -i\hbar \partial_x$. If the non-linear term ($\mathcal{P}_x^4$ term in eqn (30.5)) is completely ignored, one obtains the 1+1 massive Dirac fermions

$$\mathcal{H} = M\check{\tau}^3 + \frac{1}{2}(c(x)\mathcal{P}_x + \mathcal{P}_x c(x))\check{\tau}^1 \ , \tag{30.6}$$

$$M^2 = M^2(\mathcal{P}_x = 0) = c_\perp^2 p_F^2 \ . \tag{30.7}$$

30.1.3 Surface of infinite red shift

The classical energy spectrum of the low-energy quasiparticles obeying eqn (30.6)

$$E^2 - c^2(x) p_x^2 = M^2 \ , \tag{30.8}$$

gives rise to the effective contravariant metric

$$g^{00} = -1 \ , \quad g^{xx} = c^2(x) \ . \tag{30.9}$$

The inverse (covariant) metric $g_{\mu\nu}$ describes the effective spacetime in which the relativistic quasiparticles propagate. The corresponding line element is

$$ds^2 = -dt^2 + \frac{1}{c^2(x)} dx^2 \ . \tag{30.10}$$

The 'vierbein wall' at $x = 0$, where the 'speed of light' becomes zero, represents the surface of the infinite red shift for relativistic fermions. Let us consider a relativistic quasiparticle with the rest energy M moving from this surface to infinity. Since the metric is stationary the energy of a quasiparticle is conserved, $E =$ constant, and its momentum is coordinate dependent:

$$p(x) = \frac{\sqrt{E^2 - M^2}}{c(x)} \ . \tag{30.11}$$

The wavelength of a quasiparticle is infinitely small at the surface $x = 0$, that is why the distant observer finds that quasiparticles emitted from the vicinity of this surface are highly red-shifted.

Though the metric element g_{xx} is infinite at $x = 0$, the Riemann curvature is everywhere zero. Thus eqn (30.10) represents a *flat* effective spacetime for any function $c(x)$. In general relativity this means that the 'coordinate singularity' at $x = 0$, where $g_{xx} = \infty$, can be removed by the coordinate transformation. Indeed, if the inner observer who lives in the $x > 0$ domain introduces a new coordinate $\xi = \int dx/c(x)$, then the line element takes the standard flat form

$$ds^2 = -dt^2 + d\xi^2 \ , \quad -\infty < \xi < \infty \ . \tag{30.12}$$

The same will be the result of the transformation made by the inner observer living within the left domain ($x < 0$). For each of the two inner observers their half-space is a complete space: they are not aware of existence of their partner in the neighboring sister Universe.

However, real physical spacetime, as viewed by the Planck-scale external observer, contains both sister Universes. This is an example of the situation when the effective spacetime, which is complete from the point of view of the low-energy observer, appears to be only a part of the more fundamental underlying spacetime. For the external observer, there is no general covariance at the fundamental level, and thus the general coordinate transformation is not the symmetry operation. Moreover, the coordinate transformation $\xi = \int dx/c(x)$ is not physical because it moves the wall to infinity and thus completely removes one of the two domains from physical space. This demonstrates that if general relativity is an effective theory, then not all coordinate transformations are innocent: some of them must be forbidden on physical grounds. But such grounds cannot be derived within the effective theory, and thus cannot be noticed by an inner observer, unless he (or she) is able to probe the small corrections to the relativistic physics coming from the Planckian scale.

In our case it is the superluminal dispersion of the spectrum which restores the correct geometry of the system. Since the 'speed of light' $c(x)$ becomes zero at the wall, the two flat spacetimes are disconnected in the relativistic approximation. But this approximation breaks down near $x = 0$, where the momentum in eqn (30.11) becomes large and the non-linearity in the energy spectrum becomes important. The superluminal non-linearity allows the two inner observers to communicate across the surface of the infinite red shift.

There are other examples of surfaces of the infinite red shift in condensed matter, at which the speed of 'light' for some collective modes becomes zero. These are the surface of the vanishing speed of sound discussed by Chapline et al. (2001, 2002), and of the vanishing speed of second sound in superfluids discussed by Mohazzab (2000). The practical realization of such a surface with an infinite red shift can be achieved at the second-order phase boundary between two superfluids, ^{3}He-A and ^{3}He-A$_1$, which can be stabilized by the applied gradient of magnetic field. At this phase boundary the speed of spin waves becomes zero. The advantage of this arrangement is that the interface can exist even at $T = 0$, which means that the quantum effects in the vacuum in the presence of such a surface can be simulated. In gravity, an example of the infinite-redshift surface is also provided by the extremal black hole, whose condensed matter analog will be discussed in Sec. 32.1.4 (Fig. 32.2).

30.1.4 Fermions across static vierbein wall

Considering the classical dynamics of the low-energy quasiparticles, we found that within the relativistic domain communication between the two sister Universes is forbidden. Does this conclusion persist if quantum mechanics is applied? As we shall see in the example of the event horizon, quantum mechanics does change the physics of the black hole: the Hawking (1974) radiation from the black hole appears. Is the physics changing in our situation? The answer is 'Yes': quantum mechanics opens the route to communication across the wall. Of course, as in case of similar subtle problems, there is an uncertainty related to the analytic continuation of the quasiparticle wave function across the coordinate singularity.

We shall see that in a given problem this uncertainty is resolved within the relativistic domain merely by the assumption that there is a superluminal communication at high energy, while the details of the non-linear dispersion are not important. This will be confirmed by direct consideration of the superluminal dispersion in Sec. 30.1.5.

There are two ways to treat the problem in the relativistic theory. In one approach one makes first the coordinate transformation $\xi = \int dx/c(x)$ in one of the domains, say, at $x > 0$, where the line element becomes eqn (30.12). The standard solution for the wave function of the Dirac particle with the rest energy M propagating in the flat 1+1 spacetime of this domain is

$$\chi(\xi) = \frac{A}{\sqrt{2}} \exp(i\xi\epsilon) \begin{pmatrix} Q \\ Q^{-1} \end{pmatrix} + \frac{B}{\sqrt{2}} \exp(-i\xi\epsilon) \begin{pmatrix} Q \\ -Q^{-1} \end{pmatrix},$$

$$\epsilon = \sqrt{E^2 - M^2}, \quad Q = \left(\frac{E+M}{E-M}\right)^{1/4}. \quad (30.13)$$

Here A and B are arbitrary constants. In this approach it makes no sense to discuss any connection to the other domain, which simply does not exist in this representation.

In the second approach we do not make the coordinate transformation and work with both domains. The wave function for the Hamiltonian (30.6) at $x > 0$ follows from the solution in eqn (30.13) after restoring the old coordinates:

$$\chi(x > 0) = \frac{A}{\sqrt{2c(x)}} \exp\left(i\xi(x)\epsilon\right) \begin{pmatrix} Q \\ Q^{-1} \end{pmatrix}$$

$$+ \frac{B}{\sqrt{2c(x)}} \exp\left(-i\xi(x)\epsilon\right) \begin{pmatrix} Q \\ -Q^{-1} \end{pmatrix}, \quad \xi(x) = \int^x \frac{dx}{c(x)}. \quad (30.14)$$

A similar solution exists at $x < 0$. We can now connect the solutions for the right and left half-spaces using (i) the analytic continuation across the point $x = 0$; and (ii) the conservation of the quasiparticle current across the interface. The quasiparticle current, e.g. at $x > 0$, is

$$j = c(x)\chi^\dagger \check{\tau}^1 \chi = |A|^2 - |B|^2. \quad (30.15)$$

The analytic continuation depends on the choice of the contour around the point $x = 0$ in the complex x plane. If the point $x = 0$ is shifted to the lower part of the complex plane one obtains the following solution in the left domain:

$$\chi(x < 0) = \frac{-iA}{\sqrt{2|c(x)|}} \exp\left(-\frac{\epsilon}{2T_H}\right) \exp\left(i\xi(x)\epsilon\right) \begin{pmatrix} Q \\ Q^{-1} \end{pmatrix}$$

$$+ \frac{-iB}{\sqrt{2|c(x)|}} \exp\left(\frac{\epsilon}{2T_H}\right) \exp\left(-i\xi(x)\epsilon\right) \begin{pmatrix} Q \\ -Q^{-1} \end{pmatrix}, \quad (30.16)$$

where T_H is

$$T_H = \frac{\hbar}{2\pi} \left.\frac{dc}{dx}\right|_{x=0}. \quad (30.17)$$

As we shall see later in eqn (32.9) the quantity T_H has the meaning of Hawking temperature, though there is no radiation when the wall is at rest. The Hawking temperature marks the crossover from the classical to quantum regime. If the quasiparticle energy $\epsilon \gg T_H$, one has $\partial\lambda/\partial x = 4\pi^2 T_H/\epsilon \ll 1$, where $\lambda = 2\pi\hbar/|p_x| = 2\pi\hbar c/\epsilon$ is the de Broglie wavelength of the quasiparticle. In this energy range one can use the quasiclassical approximation for the wave function.

The conservation of the quasiparticle current (30.15) across the point $x = 0$ gives the connection between coefficients A and B:

$$|A|^2 - |B|^2 = |B|^2 \exp\left(\frac{\epsilon}{T_H}\right) - |A|^2 \exp\left(-\frac{\epsilon}{T_H}\right). \quad (30.18)$$

If $\epsilon \gg T_H$ one has $|A| = |B|\exp\left(\frac{\epsilon}{2T_H}\right) \gg |B|$, and the wave function across the wall becomes

$$\chi(x > 0) = \frac{1}{\sqrt{2|c(x)|}} \exp\left(i\xi(x)\epsilon\right) \begin{pmatrix} Q \\ Q^{-1} \end{pmatrix}, \quad (30.19)$$

$$\chi(x<0) = \frac{-i}{\sqrt{2|c(x)|}} \exp\left(-i\xi(x)\epsilon\right) \begin{pmatrix} Q \\ -Q^{-1} \end{pmatrix}. \qquad (30.20)$$

Though at the classical level the two sister Universes on both sides of the singularity are not connected, there is a quantum mechanical interaction between them. The wave functions across the wall are connected by the relation $\chi(-x) = i\check{\tau}^3 \chi^*(x)$.

30.1.5 Communication across the wall via non-linear superluminal dispersion

In the above derivation we relied upon the analytic continuation and on the conservation of the quasiparticle current across the wall. Let us justify this using the non-linear correction in eqn (30.5), which was ignored before. We shall work in the quasiclassical regime, which holds if $\epsilon \gg T_H$. In a purely classical limit one has the following non-linear spectrum with superluminal dispersion at high energy:

$$E^2 = M^2 + c^2(x)p_x^2 + \frac{p_x^4}{4(m^*)^2}. \qquad (30.21)$$

This spectrum determines two classical trajectories

$$p_x(x) = \pm\sqrt{2m^*\left(\sqrt{\epsilon^2 + (m^*)^2 c^4(x)} - m^* c^2(x)\right)}. \qquad (30.22)$$

Both of them have no singularity at $x = 0$: the two trajectories continuously cross the domain wall in opposite directions, while the Bogoliubov spin continuously changes its direction to the opposite one. Far from the wall, the quasiclassical wave function corresponding to the trajectory of a quasiparticle propagating from the left to the right transforms to eqn (30.20) obtained within the relativistic domain. This wave function describes the propagation through the wall without reflection: in the quasiclassical limit the reflection is exponentially suppressed.

The second trajectory, with minus sign in eqn (30.22), describing the propagation from the right to the left, gives rise to the quantum mechanical solution obtained by the analytic continuation when the point $x = 0$ is shifted to the upper part of the complex plane.

This confirms that there is a quantum mechanical coherence between the two flat worlds, which do not interact classically across the vierbein wall. The coherence is established by non-linear correction of the spectrum: $E^2(p) = M^2 + c^2 p^2 + \gamma p^4$. In our consideration the non-linear dispersion parameter γ was chosen as positive, which allows superluminal propagation across the wall at high momenta p. The result, however, does not depend on the magnitude of γ, i.e. the only relevant input of the 'Planck' physics is the mere possibility of superluminal communication between the worlds across the wall which justifies the analytic continuation. Thus quantum mechanics may contain more information on the sister Universe than classical physics. This is similar to the problem of the black-hole horizon.

In principle, the two worlds across the vierbein wall can be smoothly connected by a path which does not cross the wall, say, in a Möbius strip geometry

(Silagadze 2001; Foot and Silagadze 2001). In ^{3}He-A film with Möbius geometry, the domains with opposite orientations of $\hat{\mathbf{l}}$ can be smoothly connected.

Finally, let us consider the limit of the small rest energy, $M \to 0$, when quasiparticles become chiral with the spin directed along or opposite to the momentum p_x. The spin structure of the wave function in a semiclassical approximation is given by

$$\chi(x) = \exp\left(i\check{\tau}^2 \alpha/2\right) \chi(+\infty) , \quad \tan \alpha = \frac{p_x}{2m^* c(x)}. \tag{30.23}$$

Since α changes by π across the wall, the spin of the chiral quasiparticle rotates by π. This means that the right-handed quasiparticle transforms to the left-handed one when it crosses the vierbein wall.

30.2 Conical space and antigravitating string

Now let us turn to linear defects in the vierbein field at which the metric is degenerate. The analog of such a vierbein line in ^{3}He-A is represented by the radial disclination in eqn (15.13) (Fig. 30.1). Around this defect, one of the vectors of the dreibein in eqn (7.16), say $\mathbf{e}_2$, remains constant, $\mathbf{e}_2 = c_\perp \hat{\mathbf{z}}$, while the other two are rotating by 2π:

$$\hat{\mathbf{l}}(\mathbf{r}) = \hat{\rho} , \quad \mathbf{e}_1 = c_\perp \hat{\phi} . \tag{30.24}$$

The vector $\hat{\mathbf{l}}$ is radial, whence the name radial disclination. The interval corresponding to the effective metric in eqn (7.18) outside the core of disclination is

$$ds^2 = -dt^2 + \frac{1}{c_\perp^2} dz^2 + \frac{1}{c_\parallel^2} \left(d\rho^2 + \frac{c_\parallel^2}{c_\perp^2} \rho^2 d\phi^2 \right) . \tag{30.25}$$

Rescaling the radial and axial coordinates, $\rho = c_\parallel R$ and $z = c_\perp Z$, one obtains

$$ds^2 = -dt^2 + dZ^2 + dR^2 + a^2 R^2 d\phi^2 , \quad a^2 = \frac{c_\parallel^2}{c_\perp^2} \gg 1 . \tag{30.26}$$

In relativistic theories this is the conical metric. Such a metric, but with $a < 1$, arises outside the local cosmic strings with the singular energy density concentrated in the string core

$$T_0^0 = \frac{1-a}{4Ga} \delta_2(\mathbf{R}) , \tag{30.27}$$

where G is the gravitational constant. The space outside the string core is flat, but the proper length $2\pi R a$ of the circumference of radius R around the axis is smaller than $2\pi R$, if $a < 1$. A similar angle deficit takes place in the conical geometry, hence the name. Though the space is effectively flat outside the defect, the conical singularity gives rise to the curvature concentrated on the disclination axis, at $R = 0$, see Sokolov and Starobinsky (1977) and Banados et al. (1992):

$$\mathcal{R}_{R\phi}^{R\phi} = 2\pi \frac{a-1}{a} \delta_2(\mathbf{R}) , \quad \delta_2(\mathbf{R}) = \delta(X)\delta(Y) . \tag{30.28}$$

This curvature and the mass density of the string in eqn (30.27) are coupled via the Einstein equations.

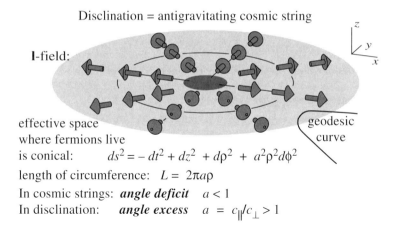

FIG. 30.1. The radial disclination in ^{3}He-A is equivalent to a cosmic string with an excess angle. Since all the geodesic curves are repelling from the string, the dysgyration serves as an example of the antigravitating string.

For the radial disclination in ^{3}He-A, one has $a^2 > 1$, i.e. there is an excess of the angle, or the 'negative angle deficit'. Moreover, we have $a = c_\parallel/c_\perp \gg 1$, which would correspond to a cosmic string with a large negative mass on the order of Planck scale, i.e. the string in ^{3}He-A is antigravitating – the quasiparticle trajectories are repelled from the string (Fig. 30.1).

31

VACUUM UNDER ROTATION AND SPINNING STRINGS

Here we discuss the properties of the quantum vacuum in superfluids in the presence of the analog of a gravimagnetic field. Such an effective field arises either under rotation which is equivalent to the constant in the space gravimagnetic field, or in the presence of conventional $U(1)$-vortices, where the effective gravimagnetic field is concentrated in the core of the vortex – the gravitational analog of the Aharonov–Bohm tube with magnetic flux.

31.1 Sagnac effect using superfluids

31.1.1 *Sagnac effect*

Let us first consider the real (not effective) gravimagnetic field, which is equivalent to the metric in the rotating frame. As we know (see Sec. 20.1.1), the rotation with angular velocity $\mathbf{\Omega}$ is equivalent to the following constant gravimagnetic field:

$$\mathbf{B}_g = \nabla \times \mathbf{G} = 2\frac{\mathbf{\Omega}}{c^2} \ , \quad G_i \equiv g_{0i} \ . \tag{31.1}$$

Here c is a real speed of light in the real vacuum, and g_{0i} are non-diagonal elements of the metric of the Minkowski spacetime in the rotating frame:

$$ds^2 = -\left(1 - \frac{\Omega^2 \rho^2}{c^2}\right) dt^2 + 2\frac{\Omega}{c^2} \rho^2 d\phi dt + \frac{1}{c^2}\left(dz^2 + d\rho^2 + \rho^2 d\phi^2\right)$$

$$= -dt^2 + \frac{1}{c^2}\left(dz^2 + d\rho^2 + \rho^2(d\phi + \Omega dt)^2\right) \ . \tag{31.2}$$

The field $\mathbf{B}_g$ is called gravimagnetic because it interacts with particles in a similar way as the conventional $U(1)$ magnetic field; the role of the charge e of a particle is played by its energy E. It is illustrated by the Coriolis force acting on a particle in the rotating frame:

$$\mathbf{F}_{\text{Coriolis}} = E\mathbf{v} \times \mathbf{B}_g = 2\frac{E}{c^2}\mathbf{v} \times \mathbf{\Omega} \ , \tag{31.3}$$

where $\mathbf{v}$ is the velocity of the particle. The corresponding term in the action, which gives rise to the Coriolis force, can be written as $E \oint \mathbf{G} \cdot d\mathbf{r}$, which is equivalent to the action $e \oint \mathbf{A} \cdot d\mathbf{r}$ for electrically charged particles in an electromagnetic field.

The Sagnac (1914) effect is the appearance of the phase difference between the two beams of particles, counterclockwise (ccw) and clockwise (cw), as measured by a rotating detector (see e.g. Ryder and Mashhoon 2001). Using the

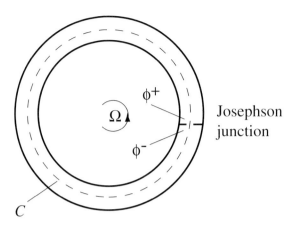

FIG. 31.1. Schematic illustration of the Sagnac effect in superfluids.

correspondence between the gravimagnetic and magnetic fields and eqn (31.1), one obtains a phase difference between the ccw and cw beams measured by a detector rotating along the path $C = d\sigma$:

$$\phi^+ - \phi^- = \Delta\phi_{\text{Sagnac}} = \omega \oint_C \mathbf{G} \cdot d\mathbf{r} = \omega \int_\sigma d\mathbf{S} \cdot \mathbf{B}_g = \frac{2\omega}{c^2} \int_\sigma d\mathbf{S} \cdot \mathbf{\Omega} \ . \quad (31.4)$$

Here $\omega = E/\hbar$ is either the de Broglie frequency of matter waves or the frequency of photons.

31.1.2 Superfluid gyroscope under rotation. Macroscopic coherent Sagnac effect

One can use the important property of superfluids – the phase coherence – to amplify and observe the effect of rotation (Cerdonio and Vitale 1984, Varoquaux et al. 1992, Packard and Vitale 1992). In the phase coherent systems, equation (31.4) gives the shift of the macroscopic phase of the order parameter. This has been probed using the Josephson effect which allowed the rotation of the Earth to be measured by superfluid interferometry in superfluid ^{4}He and ^{3}He (Avenel and Varoquaux 1996; Schwab et al. 1997; Avenel et al. 1997; Packard 1998; Avenel et al. 1998; Mukharsky et al. 2000). Let us recall that in the Bose superfluid ^{4}He, the coherent phase entering eqn (31.4) is just the phase of the order parameter, $\phi_{\text{coh}} = \Phi$. For the Fermi superfluids, where the Cooper pair consists of two atoms, it is one-half of the phase of the order parameter, $\phi_{\text{coh}} = \Phi/2$.

In the experiment, the superfluid is confined in the channel in the form of a closed loop C containing the weak link – the Josephson junction (Fig. 31.1). This channel is rotating with constant angular velocity; for example, it is rotating together with the Earth. In the rotating frame, channel walls and textures are stationary; that is why it is easier to consider the effect in this frame. The superfluid velocity in the rotating frame is $\mathbf{v}_s = (\hbar/m)\nabla\phi_{\text{coh}} - \mathbf{\Omega} \times \mathbf{r}$. If one neglects the small current through the Josephson junction and if the channel is thin enough, the superfluid velocity in the loop is almost zero in the rotating frame, $\mathbf{v}_s = (\hbar/m)\nabla\phi_{\text{coh}} - \mathbf{\Omega} \times \mathbf{r} \approx 0$, while in the inertial frame $\mathbf{v}_s = (\hbar/m)\nabla\phi_{\text{coh}} \approx \mathbf{\Omega} \times \mathbf{r}$, i.e. the motion of the superfluid in the channel mimics the rotational motion to a good precision. Of course, due to the curl-free condition for $\mathbf{v}_s$ in the inertial frame, this is not an exact equation, but in a given almost 1D geometry of thin channel the rotational motion is almost indistinguishable from the translational one. As a result, the phase shift around the loop is

$$\Delta\phi_{\text{coh}} = \oint_C \nabla\phi_{\text{coh}} \cdot d\mathbf{r} = \frac{m}{\hbar}\int_C d\mathbf{r}\cdot(\mathbf{\Omega}\times\mathbf{r}) = \frac{2\omega}{c^2}\int_\sigma d\mathbf{S}\cdot\mathbf{\Omega}\,,\ \omega = \frac{mc^2}{\hbar}\,. \quad (31.5)$$

This coincides with eqn (31.4) for the Sagnac effect with one important improvement: the phase is now coherent and thus the phase difference can be easily measured using the Josephson effect. A few comments must be made: (i) c here is a real speed of light (not effective), and thus it is the real Sagnac effect, while the superfluid serves as a tool for the measurements; (ii) if all relativistic effects are taken into account, the rest energy mc^2 of the atom must be substituted by the relativistic chemical potential μ, which takes into account the energy of the interaction of the atom with the other atoms of the liquid, i.e. $\omega = \mu/\hbar$; (iii) the same equation (31.5) is applicable to the phase difference around any closed loop in superfluids which does not contain vortices.

The phase difference across the Josephson junction, given by the Sagnac equation (31.5), has been measured in superfluids. Observation of the Sagnac effect in superfluids is an example of the phenomenon of amplification caused by the macroscopic superfluid coherence discussed by Leggett (1977, 1998).

31.2 Vortex, spinning string and Lense–Thirring effect

31.2.1 Vortex as vierbein defect

Now let us turn to the effective gravimagnetic field emerging in superfluids. The gravitational analog of the gravimagnetic tube is represented by quantized vortices. A quantized vortex in ^{3}He-A (Fig. 31.2 *top*) is another example of linear topological defects in the vierbein field at which the metric is degenerate. We shall use the simplest ansatz for the structure of a singular vortex with winding number n_1 given by eqn (23.3), which can be written in terms of the zweibein vectors $\mathbf{e}_1$ and $\mathbf{e}_2$ in eqn (7.16). In the case of a vortex with winding number $n_1 = 1$, these vectors are rotating by 2π around the vortex axis and are zero on the vortex axis:

$$\hat{\mathbf{l}} = \hat{\mathbf{z}}\,,\ \mathbf{e}_1 = c(\rho)\hat{\boldsymbol{\rho}}\,,\ \mathbf{e}_2 = c(\rho)\hat{\boldsymbol{\phi}}\,,\ c(0) = 0\,,\ c(\infty) = c_\perp\,. \quad (31.6)$$

The superfluid velocity circulating around the axis is, according to eqn (9.16),

$$\mathbf{v}_s = \frac{\hbar}{2m\rho}\hat{\boldsymbol{\phi}}\,,\ \oint d\mathbf{x}\cdot\mathbf{v}_s \equiv \kappa = n_1\kappa_0\,, \quad (31.7)$$

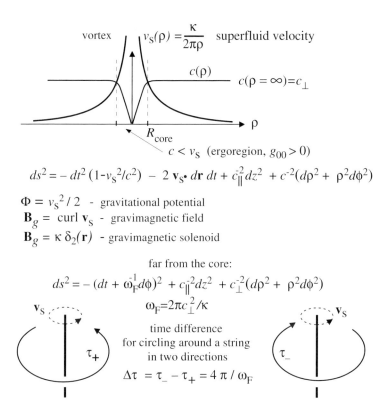

FIG. 31.2. Quantized vortex as spinning string – the cosmic string with the angular momentum consentrated in the core. *Top*: The effective metric produced by superflow circulating around a vortex is similar to the metric produced by a spinning string. *Bottom*: As in the case of a spinning string, there is a constant time difference for a quasiparticle circling with the 'speed of light' around a vortex in clockwise and counterclockwise directions. The string serves as a gravimagnetic solenoid. Quasiparticles experience the gravitational Aharonov–Bohm effect leading to an extra force acting on a vortex from the system of quasiparticles – the so-called Iordanskii force.

where the circulation quantum is $\kappa_0 = \pi\hbar/m$ in superfluid ^{3}He and $\kappa_0 = 2\pi\hbar/m$ in superfluid ^{4}He. Let us recall that in ^{3}He-A, n_1 can be half-integer: $n_1 = 1/2$ for the half-quantum vortex – the Alice string – in Sec. 15.3.1.

31.2.2 Lense–Thirring effect

The circulating motion of the superfluid vacuum around a vortex leads to the local Doppler shift of the quasiparticle energy which in the low-energy limit modifies the contravariant effective metric according to eqn (9.11). The corresponding covariant metric $g_{\mu\nu}$ in eqn (9.13) describing the effective spacetime viewed by quasiparticles leads to the following interval in the frame where the vortex is at rest (the texture-comoving frame):

$$ds^2 = -dt^2 + \frac{\rho^2}{c^2(\rho)}\left(d\phi - \frac{\kappa}{2\pi\rho^2}dt\right)^2 + \frac{d\rho^2}{c^2(\rho)} + \frac{dz^2}{c_\parallel^2}. \tag{31.8}$$

Far from the core, where $c(\rho) = c_\perp$, the same metric is applicable to phonons propagating around the vortex in superfluid ^{4}He after the isotropic 'speed of light' is introduced, $c_\perp = c_\parallel = c$. The vierbein in eqn (31.6) and thus the metric in eqn (31.8) have singularities at $\rho = 0$, where the vierbein is degenerate and the effective speed of light in the direction transverse to the vortex axis vanishes, $c(\rho \to 0) \to 0$. The vortex axis is thus the line of the infinite red shift.

The second term in the metric (31.8) indicates that the local superfluid-comoving frame is moving with time. In general relativity the term proportional to $(d\phi - \omega_{\text{LT}}(\rho)dt)^2$ also describes frame dragging, which occurs around a gravitating object rotating about its axis, such as a rotating planet, a star or a black hole. The phenomenon of frame dragging by a rotating body is known as the Lense–Thirring (1918) effect, and the function $\omega_{\text{LT}}(\rho)$ is called the Lense–Thirring angular velocity. It is the velocity with which the observer, who has zero angular momentum, orbits around a rotating body. In our case of a vortex, from eqn (31.8) it follows that the Lense–Thirring angular velocity is

$$\omega_{\text{LT}}(\rho) = \frac{\kappa}{2\pi\rho^2}. \tag{31.9}$$

31.2.3 Spinning string

Far from the vortex the quadratic terms $v_s^2/c_\perp^2$ can be neglected, and one obtains the following asymptotic interval:

$$ds^2 = -\left(dt + \frac{d\phi}{\omega_F}\right)^2 + \frac{1}{c_\perp^2}(d\rho^2 + \rho^2 d\phi^2) + \frac{1}{c_\parallel^2}dz^2 , \quad \omega_F = \frac{2\pi c_\perp^2}{\kappa}. \tag{31.10}$$

The connection between the time and the azimuthal angle ϕ in the interval suggests that there is another characteristic angular velocity ω_F, which is constant for the effective metric produced by a vortex. On the quantum level, this connection between the time and the angle implies that the physical quantities which depend on the quasiparticle energy should be periodic functions of the energy with the period equal to $\hbar\omega_F$. Thus if the interval (31.10) is valid everywhere one has a fundamental frequency ω_F.

For phonons around the vortex in superfluid ^{4}He this fundamental frequency is $\omega_F = mc^2/n_1\hbar$, while for the ^{3}He-A fermions $\omega_F = 2mc_\perp^2/n_1\hbar$. Note that in both cases such a microscopic quantity as the bare mass of ^{3}He and ^{4}He atoms explicitly enters the effective metric. Let us recall that the quantization of circulation of superfluid velocity around the vortex, as well as the topological stability of the vortex, are the properties of the 'trans-Planckian' physics. They do not follow from the effective RQFT arising in ^{3}He-A. The inner observer living in the relativistic 'sub-Planckian' world, if he (or she) finds the vortex loop somewhere, can extract the information on the mass of the atoms comprising the quantum vacuum.

In relativistic theories the metric with rotation similar to that in eqn (31.10) was obtained for the so-called spinning cosmic string in 3+1 spacetime, and also for the spinning particle in 2+1 gravity (see Mazur 1986, 1996; Staruszkiewicz 1963; Deser *et al.* 1984). The spinning cosmic string is the string which has its rotational angular momentum concentrated in the string core. The metric outside such a string is

$$ds^2 = -\left(dt + \frac{d\phi}{\omega_F}\right)^2 + \frac{1}{c^2}(dz^2 + d\rho^2 + \rho^2 d\phi^2) \,, \quad \omega_F = \frac{1}{4JGc} \,, \qquad (31.11)$$

where G is the gravitational constant, and J is the angular momentum per unit length of the string. Thus vortices in superfluids simulate the spinning cosmic strings (Davis and Shellard 1989) with the following correspondence between the circulation κ around a vortex and the angular momentum density J of a cosmic string:

$$\kappa = \frac{8\pi}{\sqrt{-g}} JG \,. \qquad (31.12)$$

31.2.4 *Asymmetry in propagation of light*

The effect peculiar to the spinning string, which was modeled in condensed matter, is the gravitational Aharonov–Bohm effect (Mazur 1986). Outside the string the metric, which enters the interval ds, is locally flat. But there is a time difference for particles propagating around the spinning string in opposite directions (Fig. 31.2 *bottom*). Let us consider the classical propagation of light along the circumference of radius R assuming that there is a confinement potential – a mirror – which keeps the light within the circumference. Then the trajectories of phonons (null geodesics) at $\rho = R$ and $z = 0$ are described by the equation $ds^2 = 0$. At large distances from the core, $R \gg c/\omega_F$, these trajectories correspond to the cw and ccw rotations with angular velocities

$$\dot{\phi}_\pm = \frac{1}{\pm \frac{R}{c} - \frac{1}{\omega_F}} \qquad (31.13)$$

The difference in the periods $T_\pm = 2\pi/\dot{\phi}_\pm$ for cw and ccw motion of the photon (as well as the phonon or Bogoliubov quasiparticle) is thus related to the fundamental frequency (Harari and Polychronakos 1988):

$$2\tau = \frac{4\pi}{\omega_F} \,. \qquad (31.14)$$

The apparent 'speed of light' measured by an inner observer is also different for 'light' propagating in opposite directions: $c_\pm \approx c(1 \pm c/\omega_F R)$.

This asymmetry between quasiparticles moving on different sides of a vortex is the origin of the Iordanskii force acting on a vortex from the heat bath of quasiparticles ('matter'), which we discuss in Section 31.3.

31.2.5 Vortex as gravimagnetic flux tube

Using the definition of the gravimagnetic field in eqn (31.1) and expressing it through the effective metric in superfluid ^{3}He-A one obtains

$$\mathbf{B}_g = \nabla \times \mathbf{G} = -\frac{1}{c_\perp^2}\nabla \times \mathbf{v}_\text{s} = -\hat{\mathbf{z}}\frac{\kappa}{c_\perp^2}\delta_2(\rho) = -\hat{\mathbf{z}}\frac{2\pi}{\omega_\text{F}}\delta_2(\rho) \ . \quad (31.15)$$

The gravimagnetic field is concentrated in the core of a vortex or a spinning string, which plays the role of a gravimagnetic flux tube.

It is also instructive to derive the correspondence between the magnetic flux tube and the gravimagnetic flux tube, since it allows us to consider the analog of the Aharonov–Bohm effect and finally to find the Iordanskii force acting on the vortex. Let us consider the simplest case of isotropic superfluid ^{4}He, where quasiparticles are phonons with energy $\tilde{E} = cp + \mathbf{p}\cdot\mathbf{v}_\text{s}(\mathbf{r})$ in the texture-comoving frame, or $(\tilde{E} - \mathbf{p}\cdot\mathbf{v}_\text{s}(\mathbf{r}))^2 = c^2 p^2$. Taking into account that for the stationary metric of the texture-comoving frame the energy $\tilde{E}$ is a conserved quantum number and substituting $\mathbf{p} = -i\hbar\nabla$, one obtains the wave equation for the sound waves propagating outside the vortex core

$$\frac{1}{c^2}\left(\tilde{E} + i\hbar\mathbf{v}_\text{s}\cdot\nabla\right)^2 \alpha + \hbar^2\nabla^2\alpha = 0 \ . \quad (31.16)$$

In the asymptotic region far from the vortex core, the quadratic terms $\mathbf{v}_\text{s}^2/c^2$ can be neglected and this equation can be rewritten as (Sonin 1997)

$$\tilde{E}^2\alpha - c^2\left(-i\hbar\nabla + \frac{\tilde{E}}{c^2}\mathbf{v}_\text{s}(\mathbf{r})\right)^2 \alpha = 0 \ . \quad (31.17)$$

This equation maps the problem under discussion to the Aharonov–Bohm (1959) problem for the magnetic flux tube with the vector potential $\mathbf{A} = -\mathbf{v}_\text{s}/c^2$ and the energy $\tilde{E}$ playing the part of the electric charge of the particle, $e \equiv \tilde{E}$ (Mazur 1987; Jensen and Kučera 1993; Gal'tsov and Letelier 1993). The effective magnetic field $\mathbf{B} = \nabla \times \mathbf{A} = -\hat{\mathbf{z}}(\kappa/c^2)\delta_2(\rho) = -\hat{\mathbf{z}}(2\pi/\omega_\text{F})\delta_2(\rho)$ concentrated in the vortex core reproduces eqn (31.15) for the gravimagnetic field.

31.3 Gravitational Aharonov–Bohm effect and Iordanskii force on a vortex

As we discussed in Secs 18.3 and 18.4, in superfluids, with their two-fluid hydrodynamics for the superfluid vacuum and quasiparticles forming 'matter', there are three different topological contributions to the force acting on a quantized vortex in Fig. 18.3. The more familiar Magnus force arises when the vortex moves with respect to the superfluid vacuum. For the relativistic cosmic string this force is absent since the corresponding superfluid density of the quantum physical vacuum is zero. However, the analog of this force appears if the cosmic string moves in the uniform background charge density (Davis and Shellard 1989; Lee 1994).

The other two forces of topological origin also have analogs for the cosmic strings: one of them comes from the analog of the axial anomaly discussed in Sec 18.3 and Chapter 25; and another one – the Iordanskii (1964, 1966) force – comes from the analog of the gravitational Aharonov–Bohm effect experienced by the spinning cosmic string discussed by Mazur (1986, 1987, 1996). The connection between the Iordanskii force and the conventional Aharonov–Bohm effect was developed by Sonin (1975, 1997) and Shelankov (1998, 2002), while the connection with the gravitational AB effect on a spinning string was discussed by Volovik (1998a) and Stone (2000a). A peculiar spacetime metric around the spinning string leads to the asymmetry in the scattering of quasiparticles on a spinning string and finally to the Iordanskii lifting force acting on a vortex moving with respect to the heat bath of quasiparticles ('matter').

31.3.1 Symmetric scattering from the vortex

Let us start with phonons scattered by a vortex in ^{4}He. Following the same reasoning as for the conventional AB effect, one finds that the symmetric part of the scattering cross-section of a quasiparticle with energy $E(p) = cp$ on a vortex is (Sonin 1997)

$$\frac{d\sigma_\parallel}{d\theta} = \frac{\hbar c}{2\pi E(p)} \cot^2 \frac{\theta}{2} \sin^2 \frac{\pi E(p)}{\hbar \omega_F} \;, \quad E(p) = cp \;. \tag{31.18}$$

This equation satisfies the periodicity of the cross-section as a function of energy with the period determined by the fundamental frequency, $\Delta E = \hbar \omega_F$, as is required by the spinning string metric in eqn (31.11). In superfluids, the relevant quasiparticle energies are typically smaller than $\hbar \omega_F$, which is comparable to the Planck energy. For small $E \ll \omega_F$ the result in eqn (31.18) was obtained by Fetter (1964). The generalization of the Fetter result for quasiparticles with arbitrary spectrum $E(p)$ – rotons in ^{4}He and the Bogoliubov–Nambu fermions in superconductors – in the range $E(p) \ll \hbar \omega_F$ was suggested by Demircan et al. (1995). In our notation it is

$$\frac{d\sigma_\parallel}{d\theta} = \frac{\kappa^2 \hbar p}{8\pi v_G^2} \cot^2 \frac{\theta}{2} \;, \tag{31.19}$$

where $v_G = dE/dp$ is the group velocity of quasiparticles.

The equation for the particles scattered by a cosmic spinning string in relativistic theory was suggested by Mazur (1987, 1996). If the mass density of the spinning string is not taken into account the result is

$$\frac{d\sigma_\parallel}{d\theta} = \frac{\hbar c}{2\pi E \sin^2(\theta/2)} \sin^2 \frac{\pi E}{\hbar \omega_F} \;, \quad E = cp \;. \tag{31.20}$$

There is a deviation from the result in eqn (31.18) for the vortex, since the metrics of the vortex and spinning string coincide only asymptotically. However, eqn (31.20) preserves the most important properties of eqn (31.18): periodicity in E and the same singular behavior at small scattering angle θ.

31.3.2 Asymmetric scattering from the vortex

This singularity at $\theta = 0$ is an indication of the existence of the transverse (or asymmetric) cross-section (Shelankov 1998, 2002). For the conventional AB effect this asymmetric part leads to the Lorentz force, which acts on the magnetic flux tube in the presence of an electric current carried by excitations outside the tube. The Lorentz force is transverse to the current, and it corresponds to the Iordanskii transverse force, which acts on the vortex in the presence of the mass current carried by the normal component of the liquid. The asymmetric part in the scattering of the quasiparticles on the velocity field of the vortex has been calculated by Sonin (1997) for phonons and rotons in ^{4}He and by Cleary (1968) for the Bogoliubov–Nambu quasiparticles in conventional superconductors. In the case of 'relativistic' phonons the transverse cross-section is, according to Sonin (1997) and Shelankov (1998, 2002),

$$\sigma_\perp(p) = \frac{\hbar c}{E(p)} \sin \frac{2\pi E(p)}{\hbar \omega_F} \;, \quad E(p) = cp \;. \tag{31.21}$$

It is periodic in the phonon energy E again with the fundamental period ω_F.

31.3.3 Classical derivation of asymmetric cross-section

The periodicity of the cross-section as a function of E is the result of quantum interference. At low $E \ll \omega_F$, when $\sin x = x$ the Planck constant $\hbar$ completely drops out from eqn (31.21), which means that for the calculation in this limit we do not need quantum mechanics. In the low-energy limit the anomalous cross-section can be calculated within purely classical mechanics, using a simple classical theory of scattering (Sonin 1997). And it can be easily generalized for arbitrary energy spectrum $E(p)$, if $E(p) \ll \omega_F$.

Far from the vortex, where the circulating velocity is small, the trajectory of a quasiparticle is almost a straight line parallel, say, to the axis y, with the distance from the vortex line being the impact parameter $x = b$. The quasiparticle moves along this line with almost constant momentum $p_y \approx p$ and almost constant group velocity $v_G(p) = dE/dp$. The change in the transverse momentum during this motion is determined by the Hamiltonian equations $dp_x/dt = -\partial \tilde{E}/\partial x = -p_y \partial v_{sy}/\partial x$, or $dp_x/dy = -(p/v_G)\partial v_{sy}/\partial x$. The deflection angle $\Delta\theta = \Delta p_x/p$ is obtained by integration of $d\theta = dp_x/p$ along the trajectory

$$\Delta\theta(b,p) = \frac{1}{v_G(p)} \partial_x \int_{-\infty}^{+\infty} dy\, v_{sy} = \frac{\kappa}{2 v_G(p)} \partial_x \text{sign}(x) = \frac{\kappa}{v_G(p)} \delta(b) \;. \tag{31.22}$$

The deflection angle is zero except for the singularity at the origin. The gravimagnetic AB solenoid looks like a fast-rotating infinitely thin string: the quasiparticle does not feel the string except when it hits the string directly. Then the quasiparticle gets a kick and scatters in all directions with the preferable direction along $\kappa \hat{\mathbf{z}} \times \mathbf{p}$. The total momentum change after scattering is linearly proportional to the circulation κ and this is the origin of the Iordanskii force.

The transverse cross-section is obtained by integration of deflection angle $\Delta\theta(b,p)$ over the impact parameter b:

$$\sigma_\perp(p) = \int_{-\infty}^{+\infty} db\, \Delta\theta(b,p) = \frac{\kappa}{v_G(p)}\,. \tag{31.23}$$

For phonons with $v_G(p) = c$ this is the result of eqn (31.21) at $E \ll \omega_F$. In order to check that the singularity at $b=0$ was treated correctly, we can regularize this singularity by introducing a correction to the vortex velocity field, e.g. $\mathbf{v}_s = (\kappa/2\pi)\hat{\phi}\rho/(\rho^2 + R_{\text{core}}^2)$, where R_{core} is the core radius. Then

$$\Delta\theta(b,p) = \frac{\kappa}{2\pi v_G(p)} \partial_b \int_{-\infty}^{+\infty} dy \frac{b}{y^2 + b^2 + R_{\text{core}}^2} = \frac{\kappa}{2 v_G(p)} \partial_b \left(\frac{b}{\sqrt{b^2 + R_{\text{core}}^2}} \right). \tag{31.24}$$

Now the deflection angle is a smooth function of b, and integrating it over b one obtains the same equation (31.23) for the transverse cross-section, which does not depend of the profile of $\Delta\theta(b)$. This confirms the topological origin of the anomalous part of the cross-section.

31.3.4 Iordanskii force on spinning string

This asymmetric part of scattering, $\sigma_\perp(p)$, describes the momentum transfer in the transverse direction, $\delta\mathbf{P} = \sigma_\perp(p)\mathbf{p} \times \hat{\mathbf{z}}$, due to scattering by a vortex. Integration over the distribution $f(\mathbf{p})$ of quasiparticles gives rise to the momentum transfer per unit time, and thus to the force acting on a vortex (per unit length) from the system of quasiparticles:

$$\begin{aligned}\mathbf{F}_{\text{Iordanskii}} &= \int \frac{d^3p}{(2\pi\hbar)^3} \sigma_\perp(p) v_G(p) f(\mathbf{p})\mathbf{p} \times \hat{\mathbf{z}} \\ &= -\kappa\hat{\mathbf{z}} \times \int \frac{d^3p}{(2\pi\hbar)^3} f(\mathbf{p})\mathbf{p} = \kappa\mathbf{P} \times \hat{\mathbf{z}}\,.\end{aligned} \tag{31.25}$$

Being of topological origin this force does not depend on details: it is determined by the momentum density $\mathbf{P}$ carried by excitations (matter) and by the circulation κ around the vortex.

In the case of thermal distribution of quasiparticles, the net momentum density arises if there is a counterflow, the flow of the normal component with respect to the superfluid vacuum: $\mathbf{P} = mn_n(\mathbf{v}_n - \mathbf{v}_s)$. Substituting this into eqn (31.25) one obtains eqn (18.28) for the Iordanskii force, which we used to study the vortex dynamics in ^{3}He-A and ^{3}He-B. The same Iordanskii force must act on a spinning cosmic string when it moves with respect to the matter.

The Iordanskii force has been experimentally identified in rotating superfluid ^{3}He-B. According to the theory for the transport of vortices in ^{3}He-B (Sec. 18.4), at low T, where the spectral flow (and thus the effect of the axial anomaly) is completely suppressed, the Iordanskii force completely determines the so-called mutual friction parameter $d_\perp$ (see Sec. 25.2.4). The Iordanskii force dictates that it must be negative: $d_\perp \approx -n_n/n_s$. This is in accordance with the experimental data by Bevan *et al.* (1995) and Hook *et al.* (1996) in Fig. 25.1 which demonstrated that $d_\perp$ does become negative at low T. At higher T the spectral flow becomes dominating which leads to the sign reversal of $d_\perp$. The observed negative sign of $d_\perp$ at low T thus provides the experimental verification of the analog of the gravitational AB effect in ^{3}He-B.

31.4 Quantum friction in rotating vacuum

31.4.1 *Zel'dovich–Starobinsky effect*

Now we turn to the effective gravimagnetic field in superfluids caused by rotation with constant angular velocity. Let us consider a cylinder of radius R rotating with angular velocity Ω inside the vortex-free superfluid vacuum which is asymptotically at rest in the inertial frame (Fig. 31.3). We are interested in how the angular momentum of a rotating body transfers to the liquid at $T = 0$, when there is no viscous normal component, which usually causes the rotational friction.

When the body rotates in the superfluid vacuum it represents the time-dependent potential which disturbs the vacuum. In the case of a cylinder, this potential is caused by imperfections on the surface of the cylinder. If the surface is atomically smooth, the time-dependent field is provided by the atomic structure of the body, say by the crystal lattice. Due to the perturbations which are time dependent in the inertial frame of the superfluid, the quasiparticle energy is not the conserved quantity in this frame. As a result the rotating cylinder must radiate quasiparticles. The emission of quasiparticles, phonons and rotons in superfluid ^{4}He and pairs of Bogoliubov fermions in superfluid ^{3}He leads to the rotational friction experienced by the body even at $T = 0$. This corresponds to the quantum friction experienced by a body rotating in the quantum vacuum first discussed by Zel'dovich (1971*b*) and Starobinskii (1973).

Zel'dovich (1971*a*) was the first to predict that the rotating body (say, the dielectric cylinder) amplifies those electromagnetic modes which satisfy the condition

$$\omega - L_z \Omega < 0 \ . \qquad (31.26)$$

Here ω is the frequency of the mode, L_z is its azimuthal quantum number along the rotation axis, and Ω is the angular velocity of the rotating cylinder. This amplification of the incoming radiation is referred to as superradiance (see e.g. Bekenstein and Schiffer 1998). The wave extracts energy and angular momentum from the rotating body, slowing down the rotation.

The other aspect of this phenomenon is that due to quantum effects, the cylinder rotating in the quantum vacuum spontaneously emits the electromagnetic modes satisfying eqn (31.26) (Zel'dovich 1971*a*)). The same occurs for any body rotating in the quantum vacuum, including the rotating black hole as was discussed by Starobinskii (1973), if the above condition is satisfied. The dissipation of energy and angular momentum due to interaction with the quantum vacuum is an example of the quantum friction.

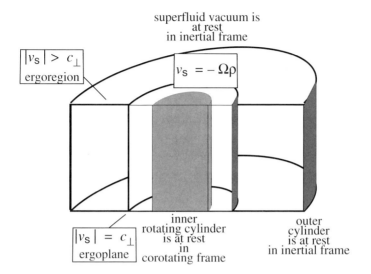

Effective metric in corotating frame:
$$ds^2 = -dt^2 + c_\perp^{-2}\rho^2(d\phi + \Omega dt)^2 + c_\perp^{-2}d\rho^2 + c_\parallel^{-2}dz^2$$

FIG. 31.3. Rotational quantum friction in superfluids as simulation of Zel'dovich–Starobinsky effect. The inner cylinder rotates in superfluid vacuum forming the preferred rotating reference frame. In this frame the effective metric has an ergoregion, where the negative energy levels are empty. The process of filling of these levels is similar to the radiation from the rotating black hole.

It is instructive to consider the quantum friction in the frame rotating with the inner cylinder.

31.4.2 *Effective metric for quasiparticles under rotation*

Let us first ignore the radiation from the rotating body and thus the slowing down of the rotating cylinder. Then in the rotating frame, the cylinder and the superfluid vacuum are stationary. If the velocity of rotation is not too high, so that the superfluid velocity $\mathbf{v}_s$ with respect to the surface of the cylinder does not exceed the Landau criterion, all the perturbations of the superfluid caused by the surface roughness of the cylinder are also stationary in the rotating frame. Hence in the rotating frame the quasiparticle energy is a good quantum number. Moreover, if the outer cylinder is far away, the inner cylinder represents the proper environment for the quantum liquid, and the rotating frame becomes the frame of the environment dictating the equilibrium conditions.

Far from the body the superfluid is at rest in the inertial frame. Thus in the rotating frame the superfluid velocity is $\mathbf{v}_s = -\mathbf{\Omega} \times \mathbf{r}$. This is valid far enough from the body, since close to the body there are space-dependent perturbations of the $\mathbf{v}_s$-field caused by the surface roughness. For the moment we ignore these perturbations.

Let us first discuss the superfluid ^{4}He whose quasiparticles – phonons – obey the effective acoustic metric. Substituting $\mathbf{v}_s = -\mathbf{\Omega} \times \mathbf{r}$ into eqn (5.2) one obtains that far from the cylinder the line element, which determines the propagation of phonons in the rotating frame of the environment, corresponds to the conventional metric (31.2) of flat space in the rotating frame:

$$ds^2 = -\left(1 - \frac{\Omega^2 \rho^2}{c^2}\right) dt^2 + 2\frac{\Omega}{c^2}\rho^2 d\phi dt + \frac{1}{c^2} d\mathbf{r}^2$$
$$= -dt^2 + \frac{1}{c^2}\left(\rho^2 (d\phi + \Omega dt)^2 + dz^2 + d\rho^2\right). \tag{31.27}$$

But now it is the acoustic metric where the role of speed of light c is played by the speed of sound. This metric contains the Lense–Thirring angular velocity $\omega_{\mathrm{LT}} = -\Omega$, which is independent of whether the metric is fundamental or acoustic.

In cylindrically symmetric systems the quasiparticle spectrum is described by two good quantum numbers: the angular momentum L_z and the linear momentum p_z. If the radial motion is treated in the quasiclassical approximation, the energy spectrum of phonons in the rotating frame becomes

$$\tilde{E} = E(p) + \mathbf{p}\cdot\mathbf{v}_s = c\sqrt{\frac{L_z^2}{\rho^2} + p_z^2 + p_\rho^2} - \Omega L_z, \tag{31.28}$$

which also follows from the acoustic metric (31.27).

The rotons in ^{4}He (Fig. 6.1) and Bogoliubov fermionic quasiparticles in ^{3}He-B are not described by the effective relativistic theory. For them the energy spectrum in the rotating frame is found directly from the Doppler shift, $\tilde{E} = E(p) + \mathbf{p}\cdot\mathbf{v}_s$, which gives

$$\tilde{E}(p) = \Delta + \frac{(p - p_0)^2}{2m_0} - \Omega L_z, \quad \text{roton}, \tag{31.29}$$

$$\tilde{E}(p) = \sqrt{\Delta_0^2 + v_F^2(p - p_F)^2} - \Omega L_z, \quad \text{Bogoliubov fermion}. \tag{31.30}$$

Here p_0 marks the roton minimum in superfluid ^{4}He; m_0 is the roton mass; Δ is the roton gap (Fig. 6.1); and $p^2 = p_z^2 + p_\rho^2 + L_z^2/\rho^2$.

31.4.3 Ergoregion in rotating superfluids

In the acoustic metric (31.27), the metric element g_{00} becomes zero at the radius $\rho_{e\text{ rel}} = c/\Omega$, and becomes positive when $\rho > \rho_{e\text{ rel}}$. On the other hand, at $\rho > \rho_{e\text{ rel}}$ the energy of phonons in eqn (31.28) becomes negative for those momenta for which $p_z^2 + p_\rho^2 < L_z^2(\rho_{e\text{ rel}}^{-2} - \rho^{-2})$. Thus $\rho_{e\text{ rel}} = c/\Omega$ marks the position of the ergosurface arising in the relativistic theory (Fig. 31.4).

In the non-relativistic case, the ergosurface is determined by the Landau velocity (26.1) and occurs at $\rho_e = v_{\text{Landau}}/\Omega$. This coincides with the relativistic

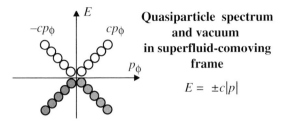
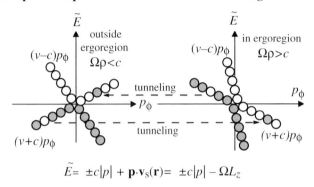

FIG. 31.4. Fermionic vacuum seen by an external observer in the frame corotating with the cylinder is different from that viewed by an inner observer in the superfluid-comoving frame, which in our case coincides with the laboratory frame. The states which are occupied in the vacuum viewed by an inner observer in the superfluid-comoving frame are shaded. In the ergoregion, where $v_s > c$, some states which are occupied in the superfluid-comoving frame, have positive energy in the corotating frame. The quasiparticles occupying these levels must be radiated away. At $T=0$ the radiation occurs via quantum tunneling from the surface of the inner cylinder to the ergoregion. The rate of tunneling reproduces the Zel'dovich–Starobinsky effect of radiation from the rotating black hole.

ergosurface only in case of the fully relativistic spectrum, where $v_{\rm Landau} = c$. In real superfluid ^{4}He, the Landau velocity is determined by the roton spectrum (31.29) in Fig. 6.1. If $m\Delta \ll p_0^2$, the Landau velocity for the creation of rotons $v_{\rm Landau} \approx \Delta/p_0$, and it is less than the Landau velocity $v_{\rm Landau} = c$ for creation of phonons. However, we shall see that though the relativistic phonon ergosurface is further away from the cylinder than the non-relativistic roton ergosurface, indexergosurface!relativistic it plays the main role in quantum friction.

The existence of the ergoregion, where quasiparticles have negative energy in the rotating frame, means that the original vacuum as seen by an inner observer in the superfluid-comoving frame no longer is the vacuum when viewed by the external observer corotating with the inner cylinder. In Fig. 31.4 *top* the quasiparticle spectrum is shown in the superfluid-comoving inertial frame, and in Fig. 31.4 *bottom* in the rotating frame. For simplicity we consider here fermionic quasiparticles with momentum along the azimuthal direction, $p_\phi = L_z/\rho$. One can see that beyond the ergosurface the states on the branch $\tilde{E} = \mathbf{p} \cdot \mathbf{v}_\mathrm{s} + cp$ have different signs of the energy in the two frames. The states, which are occupied in the vacuum viewed by an inner observer in the superfluid-comoving frame, have positive energy for the external observer in the rotating frame, while the negative energy states in the rotating frame are empty.

Thus from the point of view of the corotating external observer the original state is highly excited. For him (or her), it is necessary to fill all the negative energy states and radiate away all the positive energy states. The process of filling and emptying the energy levels is viewed by the inner superfluid-comoving observer as the radiation of quasiparticles and quasiholes by the rotating body.

Let us recall that in this consideration we assume that the ergosurface is close to the rotating cylinder and far from the outer cylinder, which is at rest in the inertial frame. That is why the influence of the rotating cylinder on the quasiparticle behavior is dominating, and thus the frame of the environment which dictates the equilibrium conditions is the rotating frame.

31.4.4 *Radiation to the ergoregion as a source of rotational quantum friction*

Let us consider how the radiation of quasiparticles occurs if the angular velocity of rotation Ω is small (Calogeracos and Volovik 1999a). If $\Omega R \ll v_\mathrm{Landau}$, one has $\rho_e \gg R$, i.e. the ergoregion is far from the surface of the cylinder. On the other hand, in the vicinity of the surface of the cylinder the superfluid velocity is much less than the Landau critical velocity, $v_\mathrm{s} = \Omega R \ll v_\mathrm{Landau}$, and thus quasiparticles can never be nucleated there. From the energy consideration they can be nucleated only in the ergoregion.

How do the quasiparticles in the ergoregion know that they can have negative energy and thus must be nucleated? Only by the interaction of the system with the environment which is the surface of the rotating cylinder: in the absence of the interaction any connection to the rotating reference frame is lost and there is no radiation.

One scenario of the radiation is demonstrated in Fig. 31.4 *bottom*. A quasiparticle which belongs to the original vacuum in the inertial frame, but have the positive energy in the ergoregion in the rotating frame moves to the empty state with the same energy outside the ergoregion. As a result the quasiparticle–quasihole pair is created in the original vacuum. In this process the azimuthal projection p_ϕ changes sign. This means that the angular momentum $L_z = \rho p_\phi$ must change sign, which is prohibited if the angular momentum is conserved. The non-conservation of the angular momentum occurs only due to interaction of a quasiparticle with the rough surface of the inner cylinder. The quasiparticle must first move to the classically forbidden region, reach the surface of the cylinder, interact with it, reversing the angular momentum, and then move to the final state in the ergoregion, again through the classically forbidden region. In this scenario a quasiparticle must tunnel twice through the classically forbidden region.

There is another process which is substantially more effective, since it is enough to tunnel only once and thus the tunneling exponent is two times smaller. Due to the surface roughness, there are regions near the surface of a rotating cylinder where there is a finite density of quasiparticle states with zero energy. These regions serve as a reservoir of quasiparticles with energy $\tilde{E} = 0$. These quasiparticles can tunnel to the ergosurface, where their energy $\tilde{E} = 0$, and $E = \hbar\omega = \hbar\Omega L_z$. In the quasiclassical approximation the tunneling probability is e^{-2S}, where

$$S = \text{Im} \int d\rho \, p_\rho(\tilde{E} = 0) . \tag{31.31}$$

For phonons with $p_z = 0$, according to eqn (31.28) one has

$$S = L \int_R^{\rho_{e\text{ rel}}} d\rho \sqrt{\frac{1}{\rho^2} - \frac{1}{\rho_{e\text{ rel}}^2}} \approx L_z \ln \frac{\rho_{e\text{ rel}}}{R} . \tag{31.32}$$

Thus all the phonons with $L_z > 0$ are radiated, but the radiation probability decreases at higher L_z. If the superfluid velocity at the surface of the inner cylinder is much less than c, i.e. $\Omega R \ll c$, the probability of radiation of phonons with the frequency $\omega = \Omega L$ becomes

$$w \propto e^{-2S} = \left(\frac{R}{\rho_{e\text{ rel}}}\right)^{2L_z} = \left(\frac{\Omega R}{c}\right)^{2L_z} = \left(\frac{\omega R}{cL}\right)^{2L_z} , \quad \Omega R \ll c . \tag{31.33}$$

Since each emitted phonon carries the angular momentum L_z, the cylinder rotating in the superfluid vacuum (at $T = 0$) is losing its angular momentum, which means quantum rotational friction. The quantum friction due to phonon emission becomes the dominating mechanism of dissipation at low enough temperature where the conventional viscous friction between the rotating cylinder and the normal component of the liquid is small.

For a cylinder rotating in the relativistic quantum vacuum the speed of sound c must be substituted by the speed of light, and eqn (31.33) appears to be proportional to the superradiant amplification of the electromagnetic waves by the rotating dielectric cylinder derived by Zel'dovich (1971a,b) (see also Bekenstein and Schiffer 1998).

31.4.5 Emission of rotons

Since the tunneling rate exponentially decreases with L_z, only the lowest L_z is important. In the case of rotons, whose spectrum is given by eqn (31.29), L_z is restricted by the Zel'dovich condition (31.26) which is satisfied only for $L_z > \Delta/\Omega$. Thus the momentum of the radiated roton is $L_{\min} = \Delta/\Omega \approx v_{\text{Landau}} p_0/\Omega$, where $v_{\text{Landau}} \approx \Delta/p_0$ is the Landau critical velocity for the emission of rotons (Fig. 6.1). The tunneling trajectory for the roton with $\tilde{E} = 0$ is determined by the equation $p = p_0$. For $p_z = 0$ it is $p_\rho(\rho) = i\sqrt{|p_0^2 - L_{\min}^2/\rho^2|}$, and for the tunneling exponent e^{-2S} one obtains

$$S = \operatorname{Im} \int d\rho\, p_\rho = L_{\min} \int_R^{\rho_e} dr \sqrt{\frac{1}{\rho^2} - \frac{1}{\rho_e^2}} \approx L_{\min} \ln \frac{\rho_e}{R}\ . \qquad (31.34)$$

Here the position of the non-relativistic ergosurface is $\rho_e = v_{\text{Landau}}/\Omega \approx L_{\min}/p_0$. The same is applicable to the emission of pairs of Bogoliubov fermions in ^{3}He-B in eqn (31.30).

Since the rotation velocity Ω is always much smaller than the gap Δ, the momentum $L_{\min}$ of the radiated roton is big. That is why the radiation of rotons is exponentially suppressed, compared to the emission of phonons. Thus the quantum rotational friction in superfluid ^{4}He is governed by the phonon emission, which simulates the quantum rotational friction in the relativistic quantum vacuum.

31.4.6 *Discussion*

Motion with constant angular velocity is another realization of the Unruh (1976) effect. In the Unruh effect a body moving in the quantum vacuum with constant rectilinear acceleration a radiates particles whose spectrum can be interpreted as thermal with temperature $T_U = \hbar a/2\pi c$ called the Unruh temperature. On the other hand the observer comoving with the body interprets the Minkowski vacuum as a canonical ensemble of states (see references in Audretsch and Müller 1994). The observed thermal spectrum is characterized by the Planck or the Fermi–Dirac distribution depending not only on the quantum statistics of the emitted particles, but also on the space dimension (Ooguri 1986). In condensed matter it is difficult to simulate the motion at constant proper acceleration (hyperbolic motion). On the other hand a body rotating in the superfluid vacuum simulates the uniform circular motion of a body with constant centripetal acceleration. Such motion in the quantum vacuum was also extensively discussed in the literature, see the latest references in Davies *et al.* (1996), Leinaas (1998) and Unruh (1998).

In our case the role of the Minkowski vacuum is played by the initial vacuum in the superfluid-comoving frame, while the detector corotating with the body is played by the collection of quasiparticles attached to the surface of the cylinder and rotating together with the cylinder. Their distribution is disturbed by the tunneling to the ergoregion, which is seen by the superfluid-comoving observer as radiation. Distinct from the linearly accelerated body, the radiation by a rotating body does not look thermal. Also, the rotating observer does not see the Minkowski vacuum as a thermal bath. This means that the matter of the body excited by interaction with the quantum fluctuations of the Minkowski vacuum cannot be described in terms of a canonical ensemble of states with intrinsic temperature depending on the angular velocity of rotation. Nevertheless, if one assumes that the detector matter is big enough and the interaction with the external world is weak, the thermalization processes inside the detector will lead to the well-defined temperature of the detector matter.

Instead of a body rotating in superfluid one can consider a cluster of vortices in Fig. 14.1. Even if the outer cylinder is not rotating, the cluster rotates as a solid body with respect to the superfluid vacuum. This situation is non-equilibrium, but, if the radiation is ignored, the detector matter within the cluster is in a local equilibrium characterized by the local temperature. In some limit case the temperature in the center of the cluster is expressed through the rotation velocity $T = (2/\pi)v_{\text{cluster}}p_F$, where $v_{\text{cluster}} = \Omega R$ is the velocity on the periphery of the cluster and R is the radius of the cluster (Volovik 1995b, 1995a).

The radiation of bosonic (quasi)particles and pairs of fermions by a rotating body produces quantum friction, which leads to deceleration of the cylinder or vortex cluster rotating in the superfluid vacuum, or of an object rotating in the quantum vacuum (a dielectric body discussed by Zel'dovich (1971a,b) or a black hole discussed ny Starobinskii (1973).

The quantum rotational friction experienced by a body rotating in the quantum vacuum occurs because in the presence of the ergoregion the vacuum in the rotating frame is not well-defined. This provides an example of when the vacuum state cannot be determined uniquely, since it is different for different observers. Whose vacuum is more relevant depends on the real physical interactions with the environment. If, for example, the inner cylinder is absolutely perfect so that the interaction between the cylinder and the superfluid vacuum vanishes, there will be no radiation, and the original 'Minkowski' vacuum, as viewed in the superfluid-comoving frame by an inner observer, will persist for ever despite the existence of the ergoregion in the rotating frame. The discussed radiation and the quantum rotational friction are absent also if a superfluid is confined within a rotating cylinder, whose radius $R < v_{\text{Landau}}/\Omega$: in this case there is simply no ergoregion in the superfluid, and the quantum friction occurs by the quantum tunneling of quasiparticles between the cylinders. The more violent process of the decay of the ergoregion based on the non-relativistic counterpart in Chapter 27 will be discussed in Sec. 32.3.

In this chapter we discussed the case when the superfluid vacuum flows along the ergosurface. That is why there is no event horizon. In the next chapter we consider the two competing vacua in the presence of the event horizon.

32

ANALOGS OF EVENT HORIZON

A black hole is the region from which the observer who is outside the hole cannot obtain any information. The event horizon represents the boundary of the black-hole region. In astrophysics, it is believed that black holes can be formed by gravitational collapse. There are several candidate black holes. These are compact stars in double-star systems whose estimated mass exceeds the maximum possible mass of a neutron star above which the collapse is inevitable. In addition there is growing observational evidence of existence of supermassive black holes in galactic bulges. At the moment, however, there is no definite proof that any of these candidates for black holes possesses an event horizon. That is why it is not excluded that gravitational collapse never produces the black hole.

According to Hawking (1974), if an isolated black hole is formed, it slowly radiates away its mass by emitting a thermal flux with the Hawking temperature $T_H = (\hbar c/2\pi) E_{g\{h\}}$, where $E_{g\{h\}}$ is the 'surface gravity' – the gravity at the place of a horizon.

Analogs of the black-hole horizon can be realized in such condensed matter where the effective metric arises for quasiparticles. The simplest way to do this is to exploit the liquids moving with velocities exceeding the local maximum attainable speed of quasiparticles. Then an inner observer, who uses only quasiparticles as a means of transferring the information, finds that some regions of space are not accessible for observation. For this observer, who lives in the quantum liquid, these regions are black holes. The first to propose such an artificial black hole was Unruh (1981, 1995). Unruh considered the propagation of acoustic waves in an effective metric produced by moving liquid: he called the region from which the sound cannot escape the 'dumb hole'.

There are several schemes of how the event horizon can be realized in quantum liquids. Let us start with the arrangement when the flow velocity is constant but the 'speed of light', i.e. the maximum attainable speed for quasiparticles, changes in space and in some region becomes less than the flow velocity of the liquid.

32.1 Event horizons in vierbein wall and Hawking radiation

32.1.1 *From infinite red shift to horizons*

First we return to the surface of the infinite red shift which is simulated by the vierbein wall discussed in Sec. 30.1, and consider how the effective metric is modified if there is a superflow across the vierbein wall with constant superfluid velocity $\mathbf{v}_s = v_s \hat{\mathbf{x}}$ (Fig. 32.1). Such a situation occurs if we move the vierbein wall with respect to the superfluid vacuum, or fix the wall and push the superfluid vacuum through it. These two arrangements are equivalent since we ignore the

EVENT HORIZONS IN VIERBEIN WALL AND HAWKING RADIATION

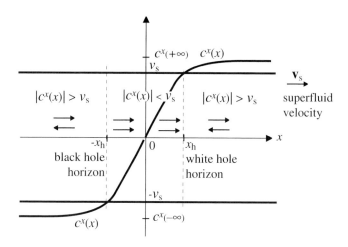

FIG. 32.1. When there is a superflow across the vierbein wall, it splits into a pair of horizons: black hole and white hole. Between the horizons the superfluid velocity exceeds the 'speed of light' of quasiparticles and g_{00} becomes positive. Since their speed c^x is smaller that the velocity of the superfluid vacuum, the 'relativistic' quasiparticles within the horizon can move only along the streamlines as shown by arrows. The horizon at $x = -x_h$ is the black-hole horizon, since no information can be extracted from the region behind this horizon, if the low-energy quasiparticles are used for communication.

normal component of the liquid and/or the walls of container to avoid dissipation due to conventional friction between the vierbein wall and the normal component.

Since the external environment is absent the proper environment frame is the texture-comoving frame (the frame of the vierbein wall). In this frame the metric is stationary, and the superflow across the wall leads to the Doppler shift of the quasiparticle spectrum. As a result the line element of the effective spacetime in eqn (30.10) becomes (Jacobson and Volovik 1998b, Jacobson and Koike 2002)

$$ds^2 = -dt^2 + \frac{1}{c^2(x)}(dx - v_s dt)^2 , \quad c(x) = c_\perp \tanh\frac{x}{d} . \tag{32.1}$$

For simplicity we consider the 1+1 dimension of spacetime ignoring the coordinates y and z along the vierbein wall.

Equation (32.1) can be rewritten in the form

$$ds^2 = -\left(1 - \frac{v_s^2}{c^2(x)}\right)\left(dt + \frac{v_s dx}{c^2(x) - v_s^2}\right)^2 + \frac{dx^2}{c^2(x) - v_s^2} , \tag{32.2}$$

which leads to a natural temptation to perform the coordinate transformation

$$\tilde{t} = t + \int^x \frac{v_s dx}{c^2(x) - v_s^2} , \tag{32.3}$$

to obtain the more familiar metric

$$ds^2 = -\left(1 - \frac{v_s^2}{c^2(x)}\right) d\tilde{t}^2 + \frac{dx^2}{c^2(x) - v_s^2} \,. \tag{32.4}$$

Equation (32.4) is similar to the radial part of the Schwarzschild metric for the black hole. The roles of the Newton gravitational potential $\Phi(x)$ and the gravitational field $E_g = -\partial_x \Phi(x)$ are played by

$$\Phi(x) = -\frac{1}{2} \frac{v_s^2}{c^2(x)} \,, \quad E_g = -\frac{v_s^2}{c^3(x)} \partial_x c \,. \tag{32.5}$$

The effective gravitational field attracts quasiparticles to singularity at $x = 0$.

The metric in eqn (32.4) shows that there are two horizons within the vierbein wall where the superfluid velocity equals the local speed of 'light': at the plane $x = x_h$ where $c(x_h) = v_s$ and at the plane $x = -x_h$ where $c(-x_h) = -v_s$. The gravitational field is regular at horizons: $E_g(\pm x_h) = -(\partial_x c)_{\pm x_h}/c(\pm x_h)$.

The original metric in eqn (32.1) is well determined everywhere except for the physical singularity at $x = 0$, where $c_x(x) = 0$ and the tetrad field is degenerate. On the contrary, the 'Schwarzschild' metric (32.4) obtained after coordinate transformation has coordinate singularities at both horizons. Thus this metric is not determined globally: it can describe the effective spacetime only in one of three region: (i) at $x < -x_h$; (ii) at $-x_h < x < x_h$; or (iii) at $x > x_h$. Thus the effective spacetime is reduced compared to the physical spacetime. The coordinate transformation (32.3) is also ill defined in the presence of horizons since the integral is not well determined.

This is another example demonstrating that general coordinate transformations are not so innocent, especially if they contain singularities, since they lead to the spacetime which is physically non-equivalent to the original one. The previous example of coordinate transformation which removed the half of the physical spacetime was discussed in Sec. 30.1. As distinct from fundamental general relativity, in the effective gravity the symmetry under general coordinate transformations is the symmetry emerging in the low-energy corner. It does not exist at the fundamental level. That is why the use of general coordinate transformations becomes tricky especially if it leads to the extension or contraction of the physical spacetime.

In our case we know the underlying physics which allows us to determine whether or not the given metric is physically correct. In the effective gravity in superfluids, the primary quantity is the energy spectrum of quasiparticles, which in the low-energy corner becomes 'relativistic' and acquires the Lorentzian form. This form determines the effective contravariant metric $g^{\mu\nu}$ as a function of the 'absolute spacetime' of the laboratory. In turn, the contravariant metric gives rise to the secondary object – the covariant metric $g_{\mu\nu}$ which describes the effective spacetime. This physical $g_{\mu\nu}$ is determined everywhere except the manifolds, where $g^{\mu\nu}$ is degenerate. All this means that the effective spacetimes are physical if they came from the physically reasonable quasiparticle spectrum.

From this point of view the metric (32.1) is physical, while the metric (32.4) obtained after unphysical singular transformation is not.

The spectrum of the ^{3}He-A fermionic quasiparticles is the Doppler-shifted equation (30.8):

$$\tilde{E}(p_x) = E(p_x) + p_x v_s \ , \ E(p_x) = \pm\sqrt{M^2 + c^2(x) p_x^2} \ . \tag{32.6}$$

It can be written in the relativistic form

$$g^{\mu\nu} p_\mu p_\nu + M^2 = 0 \ , \tag{32.7}$$

which gives rise to the physical effective metric in eqn (32.1).

Inspection of the energy spectrum (32.6) or of the metric (32.1) shows that the two horizons, at $x = x_h$ and at $x = -x_h$, are essentially different. Let us consider the motion of quasiparticles with $M = 0$ in the region between the horizons, at $-x_h < x < x_h$. The group velocity of a quasiparticle in the wall frame is $v_{Gx} = d\tilde{E}/dp_x = v_s \pm c$. It is positive for both directions of the quasiparticle momentum p_x (see Fig. 32.1, where the superfluid velocity v_s is chosen positive). All quasiparticles between the horizons move to the right, and thus cannot cross the plane $x = -x_h$ from the inside. This indicates that this plane is the black-hole horizon. The inner observer living in the region $x < -x_h$ cannot obtain any information from the region $x > -x_h$, if he (or she) uses the 'relativistic' quasiparticles for communication. On the other hand, all the quasiparticles will finally cross the plane $x = x_h$, which means that this plane is the white-hole horizon.

The appearance of pairs of white-hole/black-hole horizons is typical of superfluid systems. The region between the horizons belongs to the ergoregion, since for some quasiparticles the energy $\tilde{E}(p_x)$ in the frame of the environment is negative even for the positive square root in eqn (32.6). Note that this definition of the ergoregion differs from the standard definition in general relativity, where it is the region between the ergosurface and horizon. One must extend the ergoregion into the region behind the horizon too, if one wants to use this definition in case the general covariance is violated at high energy. This is simply because the true horizon disappears if the superluminal motion is allowed at high energy, while the ergosurface remains. Now the whole region behind the ergosurface is responsible for the phenomenon of superradiance and thus it is the ergoregion.

32.1.2 Vacuum in the presence of horizon

In the ergoregion between the horizons the notion of the vacuum state becomes subtle, since it depends on the reference frame. As we know, there are two important reference frames in superfluids: the superfluid-comoving frame where the quasiparticle energy $E(\mathbf{p})$ is velocity independent; and the frame of the environment where the quasiparticle energy is $\tilde{E}(\mathbf{p}) = E(\mathbf{p}) + \mathbf{p} \cdot \mathbf{v}_s$. Depending on the physical situation the frame of the environment can be the laboratory frame, the frame of the container, or the texture-comoving frame such as the frame of the moving vierbein wall discussed here. The vacua as viewed in the two frames do not coincide in the ergoregion. And also, as we know, there are two principally different observers: an inner observer who is made of quasiparticles and lives in the superfluid; and an external observer who lives in the Galilean world of the laboratory and does not obey the very restrictive effective metric experienced by the low-energy quasiparticles.

Let us consider the perfect but not very realistic case when the walls of the container are very far away and the texture $c(x)$ is very smooth. Let us start to move the vierbein wall with respect to the superfluid vacuum, assuming that nothing dramatic happens with the initial superfluid vacuum except that the two horizons are formed. This is actually a rather non-trivial assumption, since we shall see that in the presence of horizons the vacuum will inevitably start to reconstruct. Our assumption means that such reconstruction occurs slowly, so that there is some time after the motion started when we can ignore any modification of the original Minkowski vacuum in the superfluid-comoving frame in which the states with the $E(p) < 0$ are occupied, while the states with $E(p) > 0$ are empty.

How are the horizons and the initial vacuum viewed by different observers? The inner observer can be chosen as a massive object with the energy spectrum $\tilde{E}(\mathbf{p}) = \sqrt{c^2 p^2 + M^2} + \mathbf{p} \cdot \mathbf{v}_s$, who is made of low-energy 'relativistic' quasiparticles. In his (or her) measurements he (or she) uses rods and clocks made of the same quasiparticles. Further we assume that the rest energy M of the observer does not depend on position in the superfluid. Since the inner observer is very restrictive in his (or her) observations, we need two such observers: (i) the inner observer who is at rest in the texture-comoving frame; and (ii) the inner observer who is at rest in the superfluid-comoving frame.

The external observer lives in the Galilean world of the laboratory, and is made of atoms which are not necessarily the ^{3}He atoms comprising the vacuum of the quantum liquid. The observer uses conventional rods and clocks made of atoms. The external observer can make observations in any frame, but we prefer that this observer is at rest in the frame of the environment. We leave the more detailed discussion until Sec. 32.2.2.

(1) *Inner observer comoving with the vierbein wall.* The inner observer can be at rest with the vierbein wall in the regions outside the horizons only: in the region between the horizons all quasiparticles have positive velocity and thus the inner observer will be dragged away from this region by the flow of the superfluid vacuum. For the far-distant inner observer living at $x = -\infty$ in the texture-comoving frame, the spacetime is seen as given by the Schwarzschild metric in eqn (32.4). In the texture-comoving frame, the states with negative energy are modified by the velocity (gravity) field, $E(\mathbf{p}) \rightarrow \tilde{E}(\mathbf{p}) = E(\mathbf{p}) + \mathbf{p} \cdot \mathbf{v}_s$, but they do not leave the negative energy continuum. That is why the vacuum which this inner observer sees coincides with the original Minkowski vacuum.

For this observer, nothing happens with the vacuum if horizons appear, if we neglect the tiny quantum effect of Hawking radiation. As for the vacuum behind the black-hole horizon, the observer simply cannot obtain any information from that region since he (or she) uses rods and clocks made of quasiparticles which cannot propagate through the black-hole horizon.

(2) *Inner observer comoving with the superfluid vacuum.* If the observer's momentum is $p=0$, then from the Hamilton equations of motion

$$\frac{d\mathbf{p}}{dt} = -\frac{\partial \tilde{E}}{\partial \mathbf{r}} \quad , \quad \frac{d\mathbf{r}}{dt} = \frac{\partial \tilde{E}}{\partial \mathbf{p}} \, , \tag{32.8}$$

it follows that the free observer will be comoving with the superfluid vacuum, i.e. moving with velocity $\dot{\mathbf{r}} = \mathbf{v}_s$. If the inner observer starts to move at $x = -\infty$, he (or she) will be freely falling to the black hole, and will cross the horizon. However, the observer will not see any dramatic change when crossing. For the freely falling observer the original vacuum is locally Minkowskian. The states with $E(p) < 0$ in eqn (32.6) are occupied in this vacuum, while the states with $E(p) > 0$ are empty (see Fig. 32.4 for a somewhat different arrangement). When the observer crosses the horizon, this vacuum does not change for him (or her). Finally, the observer will see the same Minkowski vacuum even when crossing the physical singularity (of course, if he (or she) remains alive after that).

So, as distinct from the texture-comoving observer, the superfluid-comoving (freely falling) observer sees the Minkowski vacuum in the whole physical spacetime. However, the vacuum viewed in the superfluid-comoving frame can be determined only locally. During his (or her) motion, the inner observer detects the smooth monotonous increase of the gravitational field with time all the way from $-\infty$ to the phyisical singularity. In such a time-dependent environment, the quasiparticle energy $E(p)$ is not a conserved quantity. That is why the inner observer can expect that in the time-dependent field the vacuum will be disturbed, i.e. quasiparticles will be created from the vacuum. As we shall see below, this actually does occur if horizons are present.

(3) *External observer in the frame of the environment.* The external observer is the most knowledgeable person. For him (or her), all the physical space $-\infty < x < +\infty$ is accessible even if he (or she) is stationary in the texture-comoving frame of the vierbein wall. In this frame the energy $\tilde{E}(\mathbf{p})$ is conserved, and thus it is the proper environment frame in which one can try to construct the vacuum state in the whole spacetime. However, the initial superfluid-comoving Minkowski vacuum cannot serve for such purposes: in the ergoregion between the quasiparticle horizons, this initial vacuum is seen by the external observer as a highly excited system. Some energy states with the negative root in eqn (32.6), $E(\mathbf{p}) < 0$, which are occupied in the initial vacuum, have positive energy, $\tilde{E}(\mathbf{p}) > 0$, in the frame of the environment. The quasiparticles filling such states must be emitted to reach the equilibrium conditions in this frame. Thus, in the presence of horizons, the initial Minkowski vacuum must dissipate. The dissipation process at $T = 0$ means the quantum friction which decelerates the motion of the vierbein wall with respect to the superfluid. The inner observer who is stationary in the texture frame sees the decay of the Minkowski vacuum as Hawking radiation of quasiparticles from the horizon which will be discussed in Sec. 32.2.5.

Thus the situation is very similar to that in the presence of the astronomical black hole (if it exists), with one important exception. We know that in the case of quantum liquids there exist an external observer who is not restricted by the quasiparticle 'speed of light' and thus is able to observe all the physical space and all the energy scales including the trans-Planckian physics. What is the analog of an external observer in our real world? This is the observer who has access to the high-energy physics above the Planck energy scale. In principle, we can become such observers in the future. But this can also be an observer who does not belong to our vacuum. This is like a foreign body in liquid ^{3}He, which is not constructed from ^{3}He atoms, and thus does not belong to the vacuum of the quantum liquid.

32.1.3 *Dissipation due to horizon*

If the walls of the container are far away and the interaction with them can be neglected, the radiation of quasiparticles and quasiholes at $T = 0$ occurs only due to the presence of the spatial inhomogeneity of the metric (the texture). In the limit of zero gradients, the inner observer, who moves with the superfluid velocity, does not know at $T = 0$ whether the liquid is moving or not. The preferred reference frame of the environment – the texture-comoving frame – is lost, and there is no reason for emission and dissipation. The uniformly moving vacuum cannot dissipate. Thus the smooth dissipation process must be determined by the spatial derivatives of the hydrodynamic fields, or (in the relativistic domain) by the space-dependent effective metric (the gravitational field) which establish the preferred reference frame.

An example of the dissipation, caused by the gradients of the metric, is the Hawking (1974) radiation of (quasi)particles from the black-hole horizon, which will be discussed in Sec. 32.2.5. The Hawking radiation is characterized by the Hawking temperature proportional to the surface gravity E_{gh} (Sec. 32.2.5). In case of a moving vierbein wall where the 'gravitational field' is given by eqn (32.5), the role of E_{gh} is played by the gradient of the 'speed of light' at the horizon:

$$T_{\rm H} = \frac{\hbar c}{2\pi} E_{gh} = \frac{\hbar}{2\pi} \left(\frac{dc}{dx} \right)_{\rm h} . \tag{32.9}$$

If the profile of the 'speed of light' is given by eqn (32.2), the Hawking temperature depends on the velocity $v_{\rm s}$ of superflow across the vierbein wall in the following way:

$$T_{\rm H}(v_{\rm s}) = T_{\rm H}(v_{\rm s} = 0) \left(1 - \frac{v_{\rm s}^2}{c_\perp^2} \right) , \quad T_{\rm H}(v_{\rm s} = 0) = \frac{\hbar c_\perp}{2\pi d} . \tag{32.10}$$

The Hawking radiation leads to energy dissipation and thus to quantum friction, which decreases the velocity $v_{\rm s}$ of the domain wall with respect to the superfluid. Due to the deceleration of the wall motion, the Hawking temperature increases with time. The distance between horizons, $2x_{\rm h}$, decreases until the complete stop of the domain wall when the two horizons merge. In a given example the Hawking temperature approaches its asymptotic value $T_{\rm H}(v_{\rm s} = 0)$ in eqn (32.10); but when the horizons merge, the Hawking radiation disappears:

there is no more ergoregion and the stationary, topologically stable domain wall becomes non-dissipative. Nevertheless the asymptotic Hawking temperature is still important for the stationary domain wall, where it determines communication between sister Universes across the singularity – the surface of the infinite red shift – as was discussed in Sec. 30.1.4.

For the real domain walls in thin films, the above consideration is valid when the quasiparticle spectrum in the background of the vierbein wall can be considered as continuous. This takes place only if the horizons are far apart, i.e. the distance between them is much larger than the 'Planck length', whose role is played by the superfluid coherence length, $x_h \gg d \sim \xi$. This condition is satisfied when the relative velocity between the domain wall and the superfluid vacuum is close to $c_\perp$, so that $x_h \sim (d/2) \ln[1/(1 - v_s^2/c_\perp^2)] \gg d$. When velocity decreases, the dissipation must stop when the distance between the horizons becomes comparable to the 'Planck length'.

The Hawking flux of radiation could in principle be detected by quasiparticle detectors. However, even the most optimistic estimate for the possible domain walls in thin ^{3}He-A films place Hawking temperature below 1 μK. This is still too low for current experiments. That is why, at the moment, we can use the above example only in a gedanken experiment. Possibly Bose–Einstein condensates in laser-manipulated traps will be more favorable for observation of the Hawking effect. Also the Hawking radiation is only one of the possible mechanisms of the dissipation: the real scenario of relaxation of the horizons can depend on the details of the Planckian physics, i.e. on the back reaction of the superfluid vacuum to the filling of the negative energy states. The Hawking radiation can dominate only at the early stage of the decay of the local Minkowski vacuum. One of the possible mechanisms of relaxation is the amplification of the bosonic modes reflected from the horizons. According to Corley and Jacobson (1999) the mode bouncing between the black-hole and white-hole horizons is amplified after each reflection and is growing exponentially leading to the lasing effect. This can produce the fast instability of the original Minkowski vacuum in the presence of the two horizons. Another mechanism of dissipation is related to the ergoregion instability discussed in Chapter 27 for non-relativistic systems. The application to relativistic systems will be considered in Sec. 32.3.

32.1.4 Horizons in a tube and extremal black hole

Similar scenario of formation of the pair of black-hole and white-hole horizons can be realized using the superflow through the so-called Laval nozzle with converging and then diverging flow (Fig. 32.2 *left*). In the middle the velocity can exceed the 'speed of light', and the white and black horizons are formed.

Such geometry has been considered also for the Bose condensate in the laser traps of special form (Garay *et al.* 2000, 2001) and in ^{3}He-A (Volovik 1999a). However, in case of the Bose gases and Bose liquids, there is a severe restriction on the formation of the horizons imposed by the hydrodynamics of compressible liquid. It happens, for instance, that the spherically symmetric 'dumb hole' suggested by Unruh (1981, 1995) is inconsistent with hydrodynamic equations

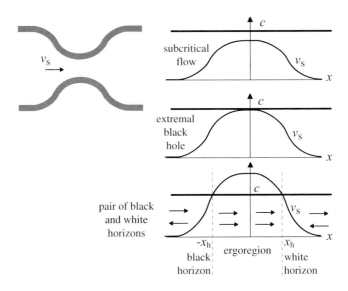

FIG. 32.2. Evolution of the effective spacetime when the superfluid velocity v_s through the orifice increases continuously. When v_s exceeds the 'speed of light' c the black-hole/white-hole pair appears (*right bottom*). Arrows show possible directions of the quasiparticle motion in the low-energy 'relativistic' limit. Between the horizons these quasiparticles can move only to the right and thus cannot escape from the region through the black horizon. *Right middle*: The intermediate state between subcritical and supercritical regimes when the velocity profile first touches the 'speed of light'. The effective metric for quasiparticles is equivalent to the metric in the vicinity of the horizon of an extremal black hole.

(see the review paper by Volovik 2001b). Probably the Laval nozzle is the only possible geometry for the physical realization of the dumb-hole horizon in Bose systems. However, even in this geometry there is a restriction: the hydrodynamic equations require that the acoustic horizon – the plane at which the velocity of the liquid crosses the speed of sound – must be just at the narrowest cross-section of the nozzle (Sakagami and Ohashi 2002). It is the same requirement as for the position of the shock wave in a pipe (note that the shock wave represents the real physical singularity at which the hydrodynamic variables – the metric field – exhibit discontinuity; here we consider only smooth configurations, and thus do not discuss the shock waves and their gravitational analogs).

In superfluid ^{3}He-A, the relevant speed of 'light' c is much smaller than the speed of sound in the underlying liquid, that is why the quasiparticle horizon does not coincide with the acoustic horizon and is not prohibited by the hydrodynamics. In the geometry of Laval the nozzle the 'speed of light' is constant, $c = c_\perp$, while the gravitational field is simulated by the space-dependent superfluid velocity $\mathbf{v}_s(\mathbf{r})$. If the cross-section of the chanel changes smoothly, the problem is reduced to the 1+1 spacetime dimension with the following interval in the reference frame of the tube which serves as the frame of the environment, where the metric is time independent:

$$ds^2 = -\left(1 - \frac{v_s^2(x)}{c^2}\right)dt^2 - 2\frac{v_s(x)}{c^2}dxdt + \frac{1}{c^2}dx^2 . \tag{32.11}$$

When the maximal velocity $v_{\max}$ of the superflow exceeds the 'speed of light' c the pair of horizons, black and white, are formed (Fig. 32.2 *right bottom*). This is similar to the case discussed in Sec. 32.1.

When $v_{\max} = c$ (Fig. 32.2 *right middle*) one obtains another remarkable metric. Close to the middle of the nozzle (at $x = 0$) the superfluid velocity can be expanded near the maximum, $v_s^2 \approx c^2(1 - x^2/x_0^2)$, and one obtains the following effective interval:

$$ds^2 = -\frac{x^2}{x_0^2}d\tilde{t}^2 + \frac{x_0^2}{x^2}dx^2 , \quad \tilde{t} = t - \frac{x_0^2}{cx} . \tag{32.12}$$

The plane $x = 0$, where the black and white horizons merge, marks the bridge between the two spaces, at $x < 0$ and at $x > 0$. It is the surface of the infinite red shift. In general relativity the metric in eqn (32.12) corresponds to the radial part of the metric of an extremal black hole in the vicinity of a horizon (bridge). The extremal black hole is the particular case of the electrically charged Reissner–Nordström black hole:

$$ds^2 = -\frac{(r - r_+)(r - r_-)}{r^2}dt^2 + \frac{r^2}{(r - r_+)(r - r_-)}dr^2 + r^2 d\Omega , \tag{32.13}$$

$$r_+ r_- = \frac{Q^2}{G^2} , \quad r_+ + r_- = \frac{2\mathcal{M}}{G} . \tag{32.14}$$

Here $\hbar = c = 1$; G is the Newton constant; $\mathcal{M}$ is the black-hole mass; Q is its electric charge in normalized units; and r_- and r_+ are inner and outer horizons. In the extreme limit when Q approaches $\mathcal{M}$, the two horizons merge and the extremal black hole is obtained:

$$ds^2 = -\left(1 - \frac{r_h}{r}\right)^2 dt^2 + \left(1 - \frac{r_h}{r}\right)^{-2} dr^2 + r^2 d\Omega . \tag{32.15}$$

Here $r_h = r_+ = r_- = G/\mathcal{M}$. Expanding $r = r_h + x$ in the vicinity of the horizon (bridge), one obtains the metric (32.12) with $x_0 = r_h$.

Since horizons merge and the ergoregion disappears, the vacuum in the presence of the extremal black hole becomes well-defined together with the global thermal equilibrium states. Though, according to Tolman's law in eqn (5.29), the effective temperature, $T_{\text{eff}} = Tx_0/|x|$, is strongly diverging at the bridge, this divergence is cured by the superluminal dispersion in eqn (26.4). The characteristic momenta $\mathbf{p}$ of quasiparticles in a global thermal equilibrium, whose energy in the frame of the environment $\tilde{E}(\mathbf{p}) \sim T$, are still by a factor $(T/E_{\text{Planck}})^{1/3}$ smaller than the Planck momentum. That is why in the expansion $E^2(p) = c^2 p^2 (1 + \gamma p^2 c^2 + \ldots)$ in eqn (26.4) only the first non-relativistic correction with coefficient $\gamma \sim 1/E_{\text{Planck}}^2$ is needed to determine the thermodynamics of the extremal black hole. This is an example where the non-relativistic high-energy physics is involved, but it is not necessary to know the exact 'atomic' structure of the Planck medium. For the 3D bridge corresponding to the extremal black hole, this first non-relativistic correction to the particle energy spectrum leads to the following thermal entropy related to the bridge (see the review paper by Volovik 2001b):

$$S_{\text{bridge}} \sim T^2 r_{\text{h}}^3 \frac{E_{\text{Planck}}}{\hbar^3 c^3}. \tag{32.16}$$

Thus the superluminal dispersion smooths the jump at $T = 0$ from the Hawking–Bekenstein entropy $\pi E_{\text{Planck}}^2 r_+^2/(\hbar c)^2$ of the horizon of the Reissner–Nordström black hole at $Q < M$ to the zero entropy when $Q > M$ and horizons no longer exist. The coefficient γ can be considered as the phenomenological parameter of the effective theory, but it can be different for different (quasi)particles.

32.2 Painlevé–Gullstrand metric in superfluids

The (3+1)-dimensional equivalent of the spherically symmetric black hole in flowing liquids – a 'dumb hole' suggested by Unruh (1981, 1995) – can be, at least in principle, realized in Fermi superfluids, where the maximum attainable velocity of quasiparticles is less than the speed of sound, and thus the hydrodynamics does not prevent the formation of a horizon.

32.2.1 *Radial flow with event horizon*

Let us consider the spherically symmetric radial flow of the superfluid vacuum, which is time independent in the frame of the environment (Fig. 32.3). For simplicity let us assume the isotropic 'speed of light' for quasiparticles. Then in the laboratory frame the dynamics of quasiparticles, propagating in this velocity field, is given by the line element provided by the effective metric in eqn (5.2):

$$ds^2 = -\left(1 - \frac{v_{\text{s}}^2(r)}{c^2}\right) dt^2 - 2\frac{v_{\text{s}}(r)}{c^2} dr\, dt + \frac{1}{c^2}(dr^2 + r^2 d\Omega^2). \tag{32.17}$$

If the 'superflow' is inward, and the velocity profile is $v_{\text{s}}(r) = -c(r_{\text{h}}/r)^{1/2}$, this equation corresponds to the line element for the black hole obtained by Painlevé (1921) and Gullstrand (1922). Among the other metrics used for the description of the black hole (including the Schwarzschild metric in eqn (32.13) with $Q = 0$), the Painlevé–Gullstrand metric has many advantages (Kraus and Wilczek 1994; Parikh and Wilczek 2000; Martel and Poisson 2001; Schützhold 2001). It is inspiring that such a metric naturally arises in the condensed matter analogs of gravity. This metric is stationary: it is invariant under translation of time $t \to t + t_0$. But it is not static: under the time reversal operation $t \to -t$ the non-diagonal element of the metric changes sign, $Tv_{\text{s}} = -v_{\text{s}}$. As distinct

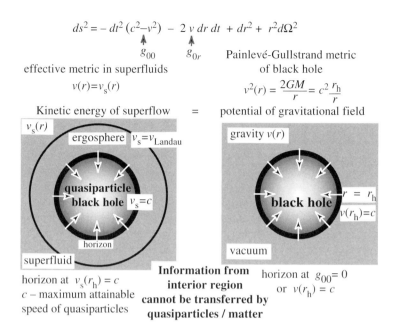

FIG. 32.3. Unruh analog of a black hole in superfluids. At the horizon, the superfluid velocity v_s reaches the maximum attainable speed of 'relativistic' quasiparticles. Such a relativistic horizon does not necessarily coincide with the non-relativistic ergosurface – the surface where v_s reaches the Landau velocity v_{Landau}. For example, in superfluid ^{4}He the Landau velocity for emission of rotons is smaller than the maximum attainable speed of quasiparticles in the low-energy corner. However, in a smooth field the roton emission is exponentially suppressed (see Sec. 31.4.5), and thus can be made negligibly small compared to the Hawking radiation of phonons from a horizon.

from the Schwarzschild metric this metric has no coordinate singularity at the horizon and thus it allows us to consider the motion of (quasi)particles through the horizon to the physical singularity at $r = 0$.

The velocity field $v_s(r) = -c(r_h/r)^{1/2}$ has a simple interpretation: it is the velocity of the observer who freely falls along the radius toward the center of the black hole with zero initial velocity at infinity. As can be found from the Hamiltonian dynamics in eqn (32.8), the motion of the observer obeys the Newton's laws all the way from infinity through the horizon to the singularity:

$$\frac{d^2 r}{dt^2} = -\frac{G\mathcal{M}}{r^2}, \qquad (32.18)$$

where $\mathcal{M}$ is the mass of the black hole. This gives the velocity of the observer for his (or her) radial motion inward

$$\frac{dr}{dt} = -\sqrt{\frac{2G\mathcal{M}}{r}} = -c\sqrt{\frac{r_\mathrm{h}}{r}} \equiv v_\mathrm{s}(r) \ . \tag{32.19}$$

The time coordinate t is the local proper time for the freely falling observer who drags the local coordinate frame with him (or her).

Thus in this simple case of a non-rotating electrically neutral gravitating object, the only result of gravity is the dragging effect according to Newtonian gravity. The same dragging effect is provided by the moving superfluid vacuum if it obeys Newtonian dynamics. This is one of several examples discussed by Trautman (1966) when general relativity is closely related to Newtonian dynamics (see also Czerniawski 2002).

In the analogy between gravity and moving superfluid one has the following correspondence. The freely falling observer corresponds to an inner observer comoving with the superfluid vacuum. The far-distant observer corresponds to an inner observer who is at rest in the frame of the environment; we shall see in Sec. 32.2.7 that for this inner observer the effective metric is the Schwarzschild metric. The Painlevé–Gullstrand metric corresponds to the effective metric for quasiparticles measured by the external observer who belongs to the world of the laboratory with its absolute space and absolute time; the Painlevé–Gullstrand metric in liquids is the function of the absolute coordinates.

Beyond the horizon, at $r < r_\mathrm{h}$, the velocity of the frame-dragging exceeds the speed of light. In relativistic language the radial coordinate r becomes time-like, because a quasiparticle beyond the horizon can move along the r coordinate only in one direction, toward the singularity, and the connection with the outside world is lost. This occurs in superfluids too, if the connection with the environment is realized by low-energy 'relativistic' quasiparticles only.

The outward superflow with the velocity field $v_\mathrm{s}(r) = +c(r_\mathrm{h}/r)^{1/2}$ reproduces the white hole. This velocity field equals the velocity of the observer freely escaping from the white hole with zero final velocity at infinity.

For the general spherically symmetric flow of superfluids with radial superfluid velocity $v_\mathrm{s}(r)$, the Schwarzschild radius r_h is determined as $v_\mathrm{s}(r_\mathrm{h}) = \pm c$. The 'surface gravity' at the Schwarzschild radius and the Hawking temperature of the black hole are, respectively,

$$E_{gh} = \frac{1}{2c^2}\left(\frac{dv_\mathrm{s}^2}{dr}\right)_{r_\mathrm{h}} \ , \quad T_\mathrm{H} = \frac{\hbar c}{2\pi} E_{gh} = \frac{\hbar}{2\pi}\left(\frac{dv_\mathrm{s}}{dr}\right)_{r_\mathrm{h}} \ . \tag{32.20}$$

32.2.2 Ingoing particles and initial vacuum

The energy of a quasiparticle in the laboratory frame (the frame of the environment) is expressed in terms of the invariant (velocity-independent) energy $E(p)$ in the superfluid-comoving frame as $\tilde{E}(\mathbf{p}) = E(p) + \mathbf{p}\cdot\mathbf{v}_\mathrm{s}$. Let us first consider a massless quasiparticle moving in the radial direction from infinity to the black-hole horizon and further to the singularity. This is called the ingoing particle. Its classical motion is characterized by the radial momentum $p_r < 0$. Since the

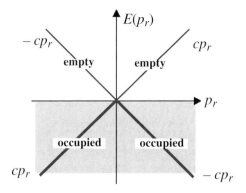

FIG. 32.4. The initial vacuum: the quasiparticle states with $E(\mathbf{p}) < 0$ are empty, while those with $E(\mathbf{p}) > 0$ are occupied. The energy spectrum of quasiparticles is shown as a function of the radial momentum p_r for zero transverse momentum $\mathbf{p}_\perp = 0$. The same vacuum is seen by the inner observer moving with the superfluid velocity $\mathbf{v}_s$ through the black-hole horizon toward the singularity. It corresponds to the local Minkowski vacuum in the freely falling frame, and it is the same outside and inside the horizon.

Painlevé–Gullstrand metric is stationary, the energy of a moving quasiparticle is conserved in the laboratory frame: $E(p_r) + p_r v_s(r) = \tilde{E} =$ constant. For the ingoing particle, its radial momentum $p_r < 0$ and thus its energy in the superfluid-comoving frame has the following coordinate dependence:

$$E(r) = -cp_r(r) = \frac{\tilde{E}}{1 - v_s(r)/c} = \frac{\tilde{E}}{1 + \sqrt{\frac{r_h}{r}}} \cdot \qquad (32.21)$$

It has no pathology when the quasiparticle crosses the black-hole horizon where $v_s(r_h) = -c$.

The same occurs for massive particles or for the massive inner observer. If the observer moves toward the black hole starting from rest at infinity, he (or she) moves with the superfluid velocity and smoothly crosses the horizon, as we discussed in Sec. 32.1.2. For this superfluid-comoving inner observer, the vacuum state is determined by the sign of the energy $E(p)$ in the superfluid-comoving frame. The vacuum seen by the observer corresponds to the state where the levels with $E(p) < 0$ are occupied. This is the initial state of the liquid outside and inside the horizon, and it corresponds to the vacuum in the freely falling frame (Fig. 32.4).

The evolution and reconstruction of the initial vacuum due to the presence of a horizon is one of the most important problems in the black-hole physics. We have already realized that in the presence of a horizon the initial state of

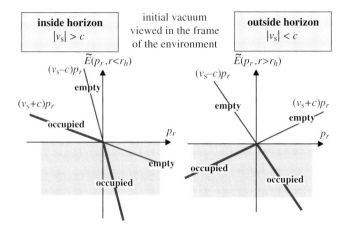

FIG. 32.5. The initial vacuum – the vacuum in the superfluid-comoving (freely falling) frame – as seen in the laboratory frame of the environment by an external observer who belongs to the trans-Planckian world. In the frame of the environment the quasiparticle energy is $\tilde{E}(\mathbf{p}) = E(\mathbf{p}) + \mathbf{p} \cdot \mathbf{v}_s(\mathbf{r})$. *Right*: Outside the horizon the initial vacuum coincides with the vacuum in the laboratory frame. *Left*: Beyond the horizon the initial vacuum represents a highly excited system: many states with $\tilde{E}(\mathbf{p}) < 0$ are empty, while many states with $\tilde{E}(\mathbf{p}) > 0$ are occupied. Here also the energy spectrum of quasiparticles is considered at zero transverse momentum $\mathbf{p}_\perp = 0$.

the liquid – the vacuum in the superfluid-comoving (freely falling) frame – represents a highly excited state in the frame of the environment (Fig. 32.5). For an external (trans-Planckian) observer of the laboratory world, the vacuum state is determined by the sign of the energy $\tilde{E}(p_r)$ in the environment frame, and he (or she) sees that in the initial state of the liquid the huge amount of levels with $\tilde{E}(p_r) < 0$ are not occupied in the region beyond the horizon. This is absolutely unbearable for him (or her), since any interaction with the environment must lead to the decay of this highly excited state. The problem is what is the most important process of the collapse of the superfluid-comoving vacuum.

32.2.3 Outgoing particle and gravitational red shift

Let us now consider the outgoing quasiparticle, a quasiparticle moving in the radial direction from the black-hole horizon to infinity. Its classical motion is again characterized by the radial momentum, but now with $p_r > 0$. From energy conservation in the laboratory frame, $E(p_r) + p_r v_s(r) = \tilde{E} = \text{constant}$, one obtains that the momentum p_r and the energy $E(p) = cp = cp_r$ of a massless particle in the superfluid-comoving (freely falling) frame obey

$$E(r) = cp_r(r) = \frac{\tilde{E}}{1 + v_s(r)/c} = \frac{\tilde{E}}{1 - \sqrt{\frac{r_h}{r}}}. \quad (32.22)$$

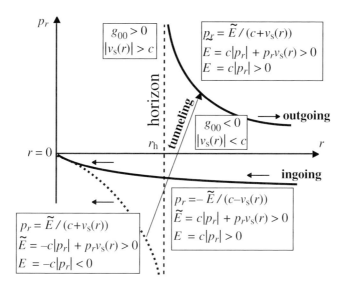

FIG. 32.6. Radial trajectories of massless 'relativistic' quasiparticles in artificial black hole ($v_s(r) < 0$). Only those trajectories are shown on which quasiparticles have positive energy in the laboratory frame, $\tilde{E} > 0$. Arrows show the direction of motion. Ingoing quasiparticles smoothly cross the horizon. On the other two trajectories quasiparticles have diverging energy in the vicinity of the horizon (infinite blue shift) which must be cut off by Planckian physics. Quasiparticles moving from the horizon to the singularity along the trajectory marked by the dotted line have different sign of energy in superfluid-comoving and laboratory frames. In the initial vacuum, the states on this branch are occupied, since in the superfluid-comoving (freely falling) frame the energy of these quasiparticles is negative, $E(p) < 0$. However, in the laboratory frame the energy of quasiparticles is positive, $\tilde{E} > 0$, and thus they must be emitted. The process of emission is the tunneling to the outgoing branch, which represents the Hawking radiation.

The energy $E(r)$ of a quasiparticle in the superfluid-comoving frame is very large close to the horizon, but becomes less and less when the quasiparticle moves away from the horizon (Fig. 32.6). This is the same phenomenon as the gravitational red shift in general relativity superimposed on the Doppler shift (see the book by Landau and Lifshitz 1975), since the emitter is freely falling with velocity $v = v_s(r)$. The frequency of the spectral line measured by the observer at infinity is

$$\tilde{\omega} = \omega \sqrt{-g_{00}} \frac{\sqrt{1 - \frac{v^2}{c^2}}}{1 - \frac{v}{c}} = \omega \left(1 - \sqrt{\frac{r_h}{r}}\right), \qquad (32.23)$$

where ω is the nominal frequency of this line.

32.2.4 Horizon as the window to Planckian physics

The horizon at $r = r_{\rm h}$ represents a surface of the infinite red shift: at this surface the energy of an outgoing quasiparticle in eqn (32.22) diverges. This means that if we observe particles coming to us from the very vicinity of the horizon, these outgoing particles originally had a huge energy approaching the Planck energy scale. Thus the event horizon can serve as a magnifying glass which allows us to see what happens at the Planck length scale. At some point the low-energy relativistic approximation inevitably becomes invalid and the deviations from the linear (relativistic) spectrum can be observed.

There is another class of trajectories – the dotted line in Fig. 32.6 – which probe the Planckian physics. These are the trajectories of quasiparticles propagating from the horizon to the singularity. The most important properties of quasiparticles on these trajectories is that their energies in the superfluid-comoving and laboratory frames, $E(p)$ and $\tilde{E}(\mathbf{p})$, have different sign. The negative energy states, $E < 0$, which are originally occupied in the vacuum of the superfluid-comoving frame, have the positive energy $\tilde{E} > 0$ in the laboratory frame of the environment, and thus quasiparticles occupying these states must be emitted. In the process of emission, the initial local Minkowski vacuum is decaying. In principle there can be more violent processes of the decay of the vacuum behind horizon, one of which is discussed in Sec. 32.3.

In any case the final destiny of the system in the presence of a horizon, i.e. the formation of a new vacuum state as viewed in the laboratory frame, depends on Planck-scale physics. Such a vacuum state does not exist in the relativistic domain: the number of negative energy levels to be filled in the new vacuum is infinite, and we need the Planck energy cut-off to restrict the number of negative levels. In the process of filling these states the horizon will or will not be destroyed depending on the Planckian physics. The situation is very similar to the fate of the supercritical superflow: when the flow velocity exceeds the Landau critical value $v_{\rm Landau}$ the superfluidity will (Fig. 26.1 *middle*) or will not (Fig. 26.1 *left*) be destroyed depending on the microscopic physics.

32.2.5 Hawking radiation

Irrespective of the final destiny of a horizon, the first stage of the dissipation of the initial vacuum can start with the Hawking-like radiation, provided some other, more violent, process of vacuum instability does not intervene. Let us consider the Hawking radiation using the semiclassical description, which is valid when the quasiparticle energy is much larger than the crossover temperature of order of the Hawking temperature $T_{\rm H}$. In this case the Hawking radiation can be described as the quantum tunneling between classical trajectories in Fig. 32.6.

We consider the positive energy states, $\tilde{E} > 0$, as viewed in the laboratory frame, where the velocity field is time independent. The tunneling exponent is determined by the usual quasiclassical action $S = {\rm Im} \int p_r(r) dr$. At low energy, when the non-relativistic corrections are neglected, the momentum as a function of r on the classical trajectory has a pole at the horizon $p(r) \approx \tilde{E}/v'_\text{s}(r-r_\text{h})$. Here v'_s is the derivative of the superfluid velocity at the horizon, which is equivalent to the surface gravity according to eqn (32.20). Shifting the contour of integration into the upper half-plane of the complex variable r, one obtains for the tunneling action

$$S = \text{Im} \int p_r(r) dr = \frac{\tilde{E}}{v'_\text{s}} \text{Im} \int \frac{dr}{r - r_\text{h}} = \frac{\pi \tilde{E}}{v'_\text{s}} \ . \qquad (32.24)$$

The probability of tunneling,

$$\exp\left(-\frac{2S}{\hbar}\right) = \exp\left(-\frac{2\pi \tilde{E}}{\hbar v'_\text{s}}\right) = \exp\left(-\frac{\tilde{E}}{T_\text{H}}\right) , \qquad (32.25)$$

reproduces the thermal radiation with the Hawking temperature in eqn (32.20).

In the process of quantum tunneling from the occupied to empty levels pairs of fermions are created: (quasi)particles outside the horizon and (quasi)holes beyond the horizon.

32.2.6 *Preferred reference frames: frame for Planckian physics and absolute spacetime*

The energy $E(p)$ of the outgoing quasiparticle in the superfluid-comoving (freely falling) frame in eqn (32.22) diverges at the horizon, where the Planckian physics intervenes. The first manifestation of the Planckian physics is the deviation from the linear relativistic law $E(p) = cp$. The quantum liquids give us examples where the dispersion of the quasiparticle spectrum $E(p)$ becomes 'superluminal' at high energy, i.e. the velocity dE/dp exceeds the maximum attainable speed c of the low-energy quasiparticles (see Sec. 26.1.5). If the same happens in the quantum vacuum of RQFT there are many important physical consequences.

This can be illustrated by the following example. At the level of the effective relativistic theory, the space-like coordinate r becomes time-like beyond the horizon, since in this region all (quasi)particles can move only in one direction of r – toward the singularity. At the fundamental level, because of the superluminal dispersion, (quasi)particles can go back and forth even beyond the horizon. Thus the superluminal dispersion restores the space-like nature of the r coordinate even beyond the horizon.

With our present knowledge there is an essential gap between the quantum vacuum in quantum liquids and the quantum vacuum of our Universe. In superfluids, as we know, there are two preferred reference frames. One of them is an 'absolute' spacetime of the laboratory frame (the frame of the environment). In the effective gravity in quantum liquids, the effective metric appears as a function of the coordinates $(\mathbf{r}, t)$ of this absolute spacetime. Another one is the frame locally comoving with the superfluid vacuum. In this frame the local metric is a Minkowski one. It is in this frame that the velocity-independent energy $E(p)$ of quasiparticles is introduced. When $E(p)$ is large enough the low-energy 'relativistic' spectrum acquires the non-linear 'non-relativistic' correction which contains the Planck energy scale (see eqn (26.4)). This means that it is the local superfluid-comoving frame where the Planck energy cut-off is introduced.

As for the quantum vacuum of our Universe, we are still at such a low energy that we cannot say whether any of the two preferred reference frames exist. For us it is an open question in whose reference frame the Planck energy scale must be introduced, i.e. what the analog is of the superfluid-comoving frame. Also we cannot say whether there is an absolute spacetime which is the analog of the laboratory frame. The magnifying glass of the event horizon could serve as a possible source for spotting these two reference frames, if they exist.

The existence of the absolute spacetime in quantum liquids allows us to resolve between different metrics that are equivalent for inner observers. Let us suppose that we managed to construct the 'perfect' quantum liquid – the liquid with a favorable hierarchy of Planck scales. In such a liquid the integrals over fermions are restricted within the region where fermions are still relativistic, and thus the obtained effective action for the effective gravity obeys general covariance, i.e. it is the Einstein action. Being covariant the Einstein action does not depend on the choice of the reference frame, and thus the Einstein equations can be solved in any coordinate system. However, in the presence of the horizon or ergoregion some of the solutions are not determined in the whole spacetime of the liquid. In these cases discrimination between different solutions arises and one must choose between them. In the quantum liquid the choice is natural: we know that at high energy the general covariance disappears and for a correct description we must use absolute coordinates in the preferred frame of the laboratory. In general relativity, the ambiguity in the presence of a horizon requires proper choice of the solution, which is not known within the effective theory and is not known in general since we do not know the fundamental 'microscopic' background. One can thus only guess what is the proper solution of the Einstein equations.

It is rather clear that Schwarzschild solution cannot be the proper choice, since the whole spacetime is not covered by Schwarzschild coordinates: the region beyond the horizon is missing (see Sec. 32.2.7). Let us accept the quantum liquid scenario as a working hypothesis, which we shall use for description of the black-hole interior in Sec. 32.4. We assume that it is the Painlevé–Gullstrand metric with the frame dragging inward which is the proper stationary solution, with $(\mathbf{r}, t)$ considered as absolute coordinates, while the local frame of the freely falling observer serves as an analog of the superfluid-comoving frame in which the Planck energy physics is introduced. We also assume that the Planckian physics is superluminal. Particles arriving to us from the black hole could come not only from the very vicinity of the horizon, but also from beyond the horizon, since this is allowed by the superluminal dispersion, and even from the vicinity of the singularity at $r = 0$. In this case the magnifying glass will allow us to probe the trans-Planckian physics in the vicinity of the singularity. Let us stress, however, that this is only a guess: being well within the effective theory we cannot discriminate between different choices of the preferred frames.

32.2.7 Schwarzschild metric in effective gravity

The Schwarzschild line element

$$ds^2 = -\left(1 - \frac{v_s^2(r)}{c^2}\right)d\tilde{t}^2 + \frac{1}{c^2 - v_s^2(r)}dr^2 + \frac{1}{c^2}r^2 d\Omega^2 \tag{32.26}$$

can be obtained from the Painlevé–Gullstrand line element in eqn (32.17) by the coordinate transformation

$$\tilde{t} = t + \int^r dr \frac{v_s}{c^2 - v_s^2}. \tag{32.27}$$

This coordinate transformation is forbidden at the fundamental level, but is valid for the low-energy quasiparticles obeying general covariance, which live outside the horizon. Thus in quantum liquids the Schwarzschild metric has some, though very limited, physical sense: it is the metric viewed by the far-distant inner observer.

Let us consider such an inner observer who is at rest at some point $R \gg r_h$ far from the artificial black hole where the superfluid velocity is almost zero. It appears that the observer's time is given just by $\tilde{t}$ in eqn (32.27), and thus the metric which he (or she) observes is the Schwarzschild line element in eqn (32.26). How does the observer find this? He (or she) sends the pulse of quasiparticles, which plays the role of light, and looks for the reflection signal from the body which is at point r. If the observer sends the signal at the moment t_1, it arrives at point r at $t = t_1 + \int_r^R dr/|v_-|$ of the absolute (laboratory) time, where v_+ and v_- are absolute (laboratory) velocities of radially propagating quasiparticles, moving outward and inward respectively:

$$v_\pm = \frac{dr}{dt} = \frac{d\tilde{E}}{dp_r} = \pm c + v_s. \tag{32.28}$$

From the point of view of the inner observer the speed of light c is an invariant quantity and does not depend on direction of propagation. Thus for the inner observer the moment of arrival of pulse to r is not t but $\tilde{t} = (t_1 + t_2)/2$, where t_2 is the time when the pulse reflected from r returns to the observer at R. Since $t_2 - t_1 = \int_r^R dr/|v_-| + \int_r^R dr/|v_+|$, one obtains that the time measured by the inner observer is

$$\tilde{t}(r,t) = \frac{t_1 + t_2}{2} = t + \frac{1}{2}\left(\int_r^R \frac{dr}{v_+} + \int_r^R \frac{dr}{v_-}\right)$$

$$= t + \left(\frac{2r_h}{v_s(r)} + \frac{r_h}{c}\ln\frac{c - v_s(r)}{c + v_s(r)}\right) - \left(\frac{2r_h}{v_s(R)} + \frac{r_h}{c}\ln\frac{c - v_s(R)}{c + v_s(R)}\right). \tag{32.29}$$

This is just eqn (32.27).

The Schwarzschild metric is not applicable if one tries to extend the consideration to the space beyond the horizon and to physics at higher energy,

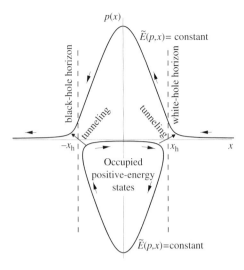

FIG. 32.7. Trajectories of massless 'relativistic' quasiparticles in the presence of pair of white and black horizons in Laval-nozzle-type geometry in Fig. 32.2 *right bottom* according to Fischer and Volovik (2001). Only those trajectories are shown on which quasiparticles have positive energy in the laboratory frame, $\tilde{E} > 0$. As distinct from Fig. 32.6 the non-linear superluminal dispersion of the quasiparticle spectrum is included. The trajectories in Fig. 32.6 which terminated at nonsingular points in space are now closed or smoothly cross the horizons. As a result the spacetime is 'geodesically complete'. The Hawking process of emission by tunneling is also shown.

while the Painlevé–Gullstrand metric is suitable for that. One may argue that the Painlevé–Gullstrand metric is also restricted since it is not geodesically complete. However, this is not necessary for the effective theory. The effective spacetime is not geodesically complete in the presence of the horizon, simply because it exists only in the low-energy 'relativistic' corner. The trajectories in Fig. 32.6 do not terminate at the horizon: they escape to a non-relativistic domain when their energy increases beyond the relativistic, linear approximation regime and the trajectory continues (Jacobson and Volovik 1998*a*,*b*; Jacobson and Koike 2002). Such example, when the superluminal dispersion makes the trajectories 'geodesically complete', is shown in Fig. 32.7 for the black-hole/white-hole pair arising in the Laval nozzle geometry in Fig. 32.2. The trajectories are either closed or continue across the horizons.

Another example of the incomplete spacetime in effective gravity is provided by the vierbein domain walls – the walls with the degenerate metric – discussed in Sec. 30.1. For the inner observer who lives in one of the domains, the effective spacetime is flat and complete. But this is only half of the real (absolute) spacetime: the other domains, which do really exist in the absolute spacetime, remain unknown to the inner observer.

FIG. 32.8. Whirlpool simulating the rotating black hole. The radial velocity of the flow is directed toward the center of the black hole.

These examples also show the importance of the superluminal dispersion at high energy. The high energy dispersion of the relativistic particles was exploited in the black-hole physics by Jacobson (1996, 1999), Corley and Jacobson (1996, 1999) and Corley (1998).

32.2.8 Discrete symmetries of black hole

Superfluids can also simulate the rotating black hole. An example is shown in Fig. 32.8. The types of the condensed matter black holes, ergoregions and surfaces of the infinite red shift can now be classified in terms of the symmetry of the superfluid velocity field $\mathbf{v}_s$. There are three important elements of discrete symmetries which form the group $Z_2 \times Z_2$. One of them is time reversal symmetry T; another one is the π rotation U_2 around the axis perpendicular to the axis of rotational symmetry; and the third one is the combined symmetry operation TU_2. Let us enumerate possible types of spacetime configurations in terms of these symmetries.

(1) T-, U_2- and TU_2-symmetric. This is the most symmetric configuration: all the elements of the $Z_2 \times Z_2$ group are preserved. This occurs if there is no superfluid velocity field. The surfaces of the infinite red shift with zero value of the non-diagonal metric elements $g_{i0} = 0$, corresponding to $\mathbf{v}_s = 0$, has been discussed in condensed matter by Chapline et al. (2001, 2002).

(2) U_2-symmetric. Analogs of the non-rotating black holes obtained by the dragging effect of the radial flow of the liquid violate the time reversal symmetry T, since $T\mathbf{v}_s = -\mathbf{v}_s$. The U_2 symmetry is preserved, simply because of the full spherical symmetry of these black holes. The symmetry operations T and TU_2 transform the non-rotating black hole into the non-rotating white hole.

(3) TU_2-symmetric. This describes the dragging effect, produced by a vortex, by a rotating cylinder, or by a rotating vortex cluster discussed in the previous chapter, giving rise to the ergosurface without horizon. The azimuthal flow of the liquid violates both T and U_2 symmetries: these symmetry operations reverse the sense of rotation. But the combined symmetry TU_2 is preserved.

(4) Finally the whirlpool in Fig. 32.8 simulating the rotating black hole violates both T and TU_2. For example, the symmetry operation T transforms a rotating black hole into a white hole rotating with the same angular velocity but in the opposite direction. The operation TU_2 transforms a rotating black hole into a white hole rotating in the same direction.

In all these cases the space parity P of the velocity field is preserved: $P\mathbf{v}_s(\mathbf{r}) = -\mathbf{v}_s(-\mathbf{r}) = \mathbf{v}_s(\mathbf{r})$.

Can this classification be applied to astronomical black holes? If so, then which sign of the non-diagonal element g_{0i} in the Painlevé–Gullstrand metric corresponds to the physical situation: black hole or white hole? One can argue (Czerniawski 2002) that, if physical gravitation corresponds to the dragging inward, then only one of the two metrics must be chosen, namely that with the minus sign for the dragging velocity. However, how can one exclude the possibility that the hole which one observes in the vacuum is the white one, if one does not know the prehistory of the formation of the hole and cannot detect the accretion of the interstellar medium, since the vacuum state is assumed to be outside the horizon.

Why can we resolve the direction of the radial flow in the liquid, yet are not able do this for the gravitational black and white holes? The difference occurs only because of the different nature of the observers. For the quantum liquid gravity we are the external observers who do not belong to the vacuum of the quantum liquid, but belong to the 'trans-Planckian' world. Our maximum attainable speed is much larger than the internal 'speed of light' in the vacuum of the quantum liquid, and we do not obey the relativistic laws of the low-energy physics of this vacuum. This is the reason why we can resolve the direction of the flow. An inner observer, who lives in the quantum liquid and uses the low-energy quasiparticles for exchange of information, is not able to determine the direction of the flow of the liquid. Moreover, this observer cannot resolve between the black hole with broken T symmetry discussed here and the T-symmetric hole discussed by Chapline *et al.* (2001, 2002) and Mazur and Mottola (2001), since for the inner observer both are described by the Schwarzschild metric.

This can be resolved only when the trans-Planckian physics is introduced. The physics beyond the horizon as observed by the external trans-Planckian observer essentially depends on T symmetry. The most prominent physical consequence of broken time reversal symmetry is, as we shall see in Sec. 32.4, that if the horizon survives after the vacuum reconstruction the vacuum inside the horizon acquires a Fermi surface and thus undergoes the quantum phase transition to a different universality class of fermionic vacua (Huhtala and Volovik 2001).

32.3 Horizon and singularity on AB-brane

The experimental realization of two superfluid liquids sliding along each other by Blaauwgeers *et al.* (2002) discussed in Sec. 17.3 and Chapter 27 opened another route for construction of an analog of the black-hole event horizon and also of the singularity in the effective Lorentzian metric. Now it is the effective metric experienced by the collective modes living on the AB-interface – the AB-brane – separating two different superfluid vacua, ^{3}He-A and ^{3}He-B. Schützhold and Unruh (2002) suggested use of the capillary–gravity waves (ripplons) on the surface of a liquid flowing in a shallow basin. In the long-wavelength limit the energy spectrum of the surface modes becomes 'relativistic', which allows us to describe the propagating modes in terms of the effective Lorentzian metric. This idea can be modified to the case of ripplons propagating along the interface between two superfluids – the AB-brane.

There are many advantages when one uses the superfluid liquids instead of conventional ones: (i) The superfluids can slide along each other without any friction until the critical velocity is reached, and thus all the problems related to viscosity disappear. (ii) The superfluids, as we know, represent the quantum vacua similar to that in RQFT. That is why the quantum effects related to the vacuum in the presence of an exotic metric can be simulated. (iii) The interface between two different superfluid vacua is analogous to the brane in modern RQFT (see Part V), and one can study the brane physics, in particular the interaction between brane matter and matter living in the higher-dimensional space outside the brane. Here, in an example of the AB-brane, we show that the interaction of the brane with the bulk environment can lead to vacuum instability of the brane vacuum in the presence of the ergoregion beyond the horizon. (iv) By reducing the temperature one can make the time of development of the instability long enough to experimentally probe the singularity within the black hole (the so-called physical singularity).

32.3.1 *Effective metric for modes living on the AB-brane*

Let us consider surface waves – ripplons – propagating along the AB-brane in the slab geometry shown in Fig. 32.9. Two superfluids, ^{3}He-A and ^{3}He-B, separated by the AB-brane are moving along the brane with velocities $\mathbf{v}_1 \equiv \mathbf{v}_{sB}$ and $\mathbf{v}_2 \equiv \mathbf{v}_{sA}$ in the container frame (the frame of the environment). The normal components of the liquids – the systems of quasiparticles on both sides of the interface, which play the role of the bulk matter outside the brane – are at rest with respect to the container walls in a global thermodynamic equilibrium, $\mathbf{v}_n = 0$. The dispersion relation for ripplons which represent the matter living in the 2D brane world can be obtained by modification of eqn (27.8) to the slab geometry:

$$M_1(k)(\omega - \mathbf{k} \cdot \mathbf{v}_1)^2 + M_2(k)(\omega - \mathbf{k} \cdot \mathbf{v}_2)^2 = F + k^2\sigma - i\Gamma\omega . \qquad (32.30)$$

Here, as before, σ and F are the surface tension of the AB-brane and the force stabilizing the position of the brane respectively. The quantities $M_1(k)$ and $M_2(k)$ are those k-dependent masses of the liquids which are forced into motion by the oscillating brane:

$$M_1(k) = \frac{\rho_1}{k \tanh kh_1} , \quad M_2(k) = \frac{\rho_2}{k \tanh kh_2} , \qquad (32.31)$$

where h_1 and h_2 are the thicknesses of the layers of two superfluids; ρ_1 and ρ_2 are mass densities of the liquids (we assume that the temperature is low enough

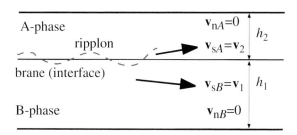

FIG. 32.9. The brane – the interface between two moving superfluids, ^{3}He-A and ^{3}He-B. $\mathbf{v}_{sA}$ and $\mathbf{v}_{sB}$ are the superfluid velocities of two liquids sliding along the brane, while the normal components of the liquids – the analogs of matter living outside the brane – are at rest in the frame of the container, $\mathbf{v}_{nA} = \mathbf{v}_{nB} = 0$. The dashed line demonstrates the propagating surface wave (ripplon) which represents matter living on the brane.

so that the normal fraction of each of the two superfluid liquids is small, $\rho_{s1} \approx \rho_1$ and $\rho_{s2} \approx \rho_2$).

Finally Γ is the frame-fixing parameter – the coefficient in front of the friction force in eqns (27.6) and (29.18) experienced by the AB-brane when it moves with respect to the 3D environment along the normal $\hat{\mathbf{z}}$ to the brane, $\mathbf{F}_{\text{friction}} = -\Gamma(\mathbf{v}_{AB} - \mathbf{v}_n)$ (in the frame of the container $\mathbf{v}_n = 0$). The friction term in eqn (32.30) containing the parameter Γ is the only term which couples the 2D brane with the 3D environment consisting of liquids in the bulk and the container walls. If $\Gamma = 0$, the connection with the 3D environment is lost, and the brane subsystem becomes Galilean invariant. The Γ term violates Galilean invariance of the 2D world of the AB-brane due to the interaction with the higher-dimensional environment.

In experiments conducted by Blaauwgeers *et al.* (2002) one has $kh_1 \gg 1$ and $kh_2 \gg 1$, and eqn (27.8) is restored. In the opposite limit of a thin slab, where $kh_1 \ll 1$ and $kh_2 \ll 1$, one obtains

$$\alpha_1 (\omega - \mathbf{k} \cdot \mathbf{v}_1)^2 + \alpha_2 (\omega - \mathbf{k} \cdot \mathbf{v}_2)^2 = c^2 k^2 \left(1 + \frac{k^2}{k_0^2}\right) - 2i\tilde{\Gamma}(k)\omega , \qquad (32.32)$$

where

$$\alpha_1 = \frac{h_2 \rho_1}{h_2 \rho_1 + h_1 \rho_2} , \quad \alpha_2 = \frac{h_1 \rho_2}{h_2 \rho_1 + h_1 \rho_2} , \quad \alpha_1 + \alpha_2 = 0 , \qquad (32.33)$$

$$k_0^2 = \frac{F}{\sigma} , \quad c^2 = \frac{F h_1 h_2}{h_2 \rho_1 + h_1 \rho_2} , \quad \tilde{\Gamma}(k) = \frac{\Gamma}{2} k^2 \frac{h_1 h_2}{h_2 \rho_1 + h_1 \rho_2} . \qquad (32.34)$$

For $k \ll k_0$ and $\omega \ll ck_0$ the dominating, quadratic in $\mathbf{k}$ and ω, terms in eqn (32.32) can be rewritten in the Lorentzian form

$$g^{\mu\nu} k_\mu k_\nu = 2i\omega \tilde{\Gamma}(k) - c^2 k^4 / k_0^2 , \qquad (32.35)$$

$$k_\mu = (-\omega, k_x, k_y) \ , \ k = \sqrt{k_x^2 + k_y^2} \ . \tag{32.36}$$

The right-hand side of eqn (32.35) contains the remaining small terms violating the effective Lorentz invariance – the attenuation of ripplons due to the interaction with the higher-dimensional environment and their non-linear dispersion. The quantities k_0 and ck_0 play the roles of the Planck momentum and Planck energy for ripplons: they determine the scales where the Lorentz symmetry of ripplons is violated. Both terms on the rhs of eqn (32.35) come from the physics which is 'trans-Planckian' for the ripplons. The Planck energy scale of the 2D physics in the brane is typically much smaller than the Planck energy scales in eqn (7.31) for the bulk 3D superfluids outside the brane. The frame-fixing parameter Γ is determined by the physics of 3D quasiparticles scattering on the brane as discussed in Sec. 29.3.1. We shall consider velocities $\mathbf{v}_1$ and $\mathbf{v}_2$, which are comparable to the 'speed of light' c on the brane. But they are small compared to the maximum attainable speeds $c_\parallel$ and $c_\perp$ in the 3D world outside the brane, i.e. $c_{D=2} \ll c_{D=3}$.

At sufficiently small k both non-Lorentzian terms – attenuation and non-linear dispersion on the rhs of eqn (32.35) – can be ignored, and the dynamics of ripplons living on the AB-brane is described by the following effective contravariant metric in the frame of the environment $g^{\mu\nu}$:

$$g^{00} = -1 \ , \ g^{0i} = -\alpha_1 v_1^i - \alpha_2 v_2^i \ , \ g^{ij} = c^2 \delta^{ij} - \alpha_1 v_1^i v_1^j - \alpha_2 v_2^i v_2^j \ . \tag{32.37}$$

Introducing two vectors, $\mathbf{U}$ and $\mathbf{W}$, which describe the relative motion of two superfluids and the mean velocity of the two superfluids with respect to the container, respectively,

$$\mathbf{U} = \sqrt{\alpha_1 \alpha_2}(\mathbf{v}_1 - \mathbf{v}_2) \ , \ \mathbf{W} = \alpha_1 \mathbf{v}_1 + \alpha_2 \mathbf{v}_2 \ , \tag{32.38}$$

one obtains the following expression for the time independent effective contravariant metric for ripplons:

$$g^{00} = -1 \ , \ g^{0i} = -W^i \ , \ g^{ij} = c^2 \delta^{ij} - W^i W^j - U^i U^j \ . \tag{32.39}$$

The corresponding effective covariant metric of the (2+1)-dimensional spacetime is

$$g_{ij} = \frac{1}{c^2}\left(\delta^{ij} + \frac{U^i U^j}{c^2 - U^2}\right) \ , \ g_{00} = -1 + g_{ij} W^i W^j \ , \ g_{0i} = -g_{ij} W^j \ . \tag{32.40}$$

32.3.2 Horizon and singularity

The criterion of the original Kelvin–Helmholtz instability, which takes place in the absence of the environment, i.e. at $\Gamma = 0$, is modified in the slab geometry. If $kh_1 \ll 1$ and $kh_2 \ll 1$, then instead of eqn (27.1) or the third of eqns (27.13) one obtains that the classical KH instability takes place when $U = c$, i.e. when the relative velocity of motion of the two liquids reaches the critical value

$$|\mathbf{v}_1 - \mathbf{v}_2| = v_{\text{KH}} = \frac{c}{\sqrt{\alpha_1 \alpha_2}} \ . \tag{32.41}$$

From eqn (32.40) it follows that at this velocity the determinant of the metric tensor

$$g = -\frac{1}{c^2 (c^2 - U^2)} \tag{32.42}$$

has a physical singularity: it crosses the infinite value and changes sign.

However, before U reaches c, the other important thresholds can be crossed where analogs of the ergosurface and horizon in general relativity appear. Let us consider the simplest situation when velocities $\mathbf{U}$ and $\mathbf{W}$ are parallel to each other (i.e. $\mathbf{v}_1 \parallel \mathbf{v}_2$); these velocities are radial and depend only on the radial coordinate r along the flow (Fig. 32.10 *left*). Then the interval of the effective 2+1 spacetime in which ripplons move along the geodesic curves is

$$ds^2 = \frac{-(c^2 - W^2(r) - U^2(r))dt^2 - 2W(r)dtdr + dr^2}{c^2 - U^2(r)} + r^2 d\phi^2 \ , \tag{32.43}$$

or

$$ds^2 = -d\tilde{t}^2 \frac{c^2 - W^2(r) - U^2(r)}{c^2 - U^2(r)} + \frac{dr^2}{c^2 - W^2(r) - U^2(r)} + r^2 d\phi^2 \ , \tag{32.44}$$

$$d\tilde{t} = dt + \frac{W(r)dr}{c^2 - W^2(r) - U^2(r)} \ . \tag{32.45}$$

The circle $r = r_{\text{h}}$, where $g_{00} = 0$, i.e. where $W^2(r_{\text{h}}) + U^2(r_{\text{h}}) = c^2$, marks the 'co-ordinate singularity' which is the black-hole horizon if the velocity W is inward (see Fig. 32.10 *left*). In such radial-flow geometry the horizon also represents the ergosurface (ergoline in 2D space) which is determined as the surface bounding the region where the ripplon states can have negative energy. As before, we call the whole region behind the ergosurface the ergoregion. This definition differs from that accepted in general relativity, but we must extend the notion of the ergoregion to the case when the Lorentz invariance and general covariance are violated, and the absolute reference frame appears (see Sec. 26.1.5). At the ergosurface, the Landau critical velocity for generation of ripplons is reached. And also, as follows from eqn (32.35) (see also Sec. 27.1.2), the ergoregion coincides with the region where the brane fluctuations become unstable, since not only the real but also the imaginary part of the ripplon spectrum crosses zero at the ergosurface. It becomes positive in the ergoregion (Fig. 32.10 *right*).

32.3.3 Brane instability beyond the horizon

The positive value of the imaginary part of the ripplon spectrum beyond the horizon means that the brane vacuum becomes unstable in the presence of the

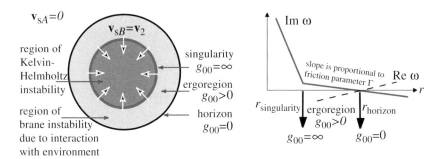

FIG. 32.10. Artificial black hole in AB-brane. *Left*: Horizon and singularity in the effective metric for ripplons on the brane (AB-interface). We assume that the A-phase is at rest, while the B-phase is radially moving to the center as shown by the arrows. *Right*: Real and imaginary parts of the ripplon spectrum cross zero values at the horizon. In the ergoregion (the region beyond the horizon) the attenuation of ripplons transforms to the amplification leading to the instability of the brane world. The time of development of this instability is long at low T, where Γ is small. On the contrary, in the region beyond the singularity, the Kelvin–Helmholtz instability develops rapidly and $r_{\text{singularity}} \to 0$.

ergoregion. The slope of Im $\omega(r)$ in Fig. 32.10 *right* is proportional to Γ, which means that instability develops only at non-zero frame-fixing parameter Γ, i.e. due to the interaction of the 2D ripplons with the 3D quasiparticles living in bulk superfluids on both sides of the brane. Outside the horizon, the interaction of the brane with the bulk environment leads to the attenuation of the propagating ripplons: the imaginary part of the ripplon spectrum is negative there. In the ergoregion, the imaginary part of the spectrum of ripplons becomes positive, i.e. the attenuation transforms to amplification of surface waves with negative Im ω. The similar mechanism of the instability of the rotating black hole has been discussed by Press (1998).

Since the instability of the interface with respect to exponentially growing surface fluctuations develops in the presence of the shear flow, this instability results in the formation of vortices observed by Blaauwgeers *et al.* (2002) (see Chapter 27). In these experiments, however, the conditions $kh_1 \ll 1$ and $kh_2 \ll 1$ were not satisfied, i.e. the relativistic description was not applicable. Also, in the rotating cryostat the superfluids flow in the azimuthal direction instead of the radial one. That is why there was no horizon in the experiment. However, the notion of the non-relativistic ergosurface and of the ergoregion beyond this ergosurface, where the ripplon energy becomes negative in the container frame, is applicable. According to Fig. 27.3 the observed threshold velocity for the instability which leads to vortex formation exactly corresponds to the appearance of the ergosurface (ergoline) in the reference frame of the bulk environment. The same physics of the brane vacuum instability in the ergoregion will hold for shallow superfluids in the presence of a 'relativistic' horizon.

There are two ingredients which cause vacuum instability within the ergoregion: (i) the existence of the absolute reference frame of the environment outside the brane; and (ii) the interaction of the brane with this environment ($\Gamma \neq 0$) which violates Galilean invariance (and Lorentz invariance in the 'relativistic' case) within the brane. They lead to attenuation of the ripplon in the region outside the ergosurface (horizon). Beyond the ergosurface (horizon) this attenuation transforms to amplification which destabilizes the vacuum there. This mechanism may have an important implication for the astronomical black hole. If there is any intrinsic attenuation of, say, photons (either due to superluminal dispersion, or due to the interaction with the higher-dimensional environment), this may lead to catastrophical decay of the black hole due to instability beyond the horizon, which we discuss in Sec. 32.4.1.

Let us estimate the time of development of such instability, first in the artificial black hole within the AB-brane and then in the astronomical black hole. According to Sec. 29.3.3 the frame-fixing parameter in the friction force experienced by the AB-brane due to Andreev scattering of quasiparticles living in the bulk superfluid on the A-phase side of the brane is $\Gamma \sim T^3 m^*/\hbar^3 c_\perp c_\parallel$ at $T \ll T_c$. Let us recall that $c_\perp$ and $c_\parallel$ are the 'speeds of light' for 3D quasiparticles living in anisotropic ^{3}He-A, and these speeds are much larger than the typical 'speed of light' c of quasiparticles (ripplons) living on the 2D brane; the superfluid transition temperature $T_c \sim \Delta_0$ also marks the 3D Planck energy scale. Assuming the most pessimistic scenario in which the instability is caused mainly by the exponential growth of ripplons with the 2D 'Planck' wave number k_0, one obtains the following estimate for the time of development of the instability in the ergoregion far enough from the horizon: $\tau \sim 1/\tilde{\Gamma}(k_0) \sim 10(T_c/T)^3$ s. Thus at low T the state with the horizon can live for a long time (minutes or even hours), and this lifetime of the horizon can be made even longer if the threshold is only slightly exceeded.

This gives a unique possibility to study the horizon and the region beyond the horizon. The physical singularity, where the determinant of the metric is singular, can also be easily constructed and investigated.

At lower temperature $T < m^* c_\perp^2$ the temperature dependence of Γ changes (see Sec. 29.3.2), $\Gamma \sim T^4/\hbar^3 c_\perp^3 c_\parallel$, and at very low T it becomes temperature independent, $\Gamma \sim \hbar k^4$, which corresponds to the dynamical Casimir force acting on the 2D brane (mirror) moving in the 3D vacuum (Davis and Fulling 1976). Such intrinsic attenuation of ripplons, transforming to the amplification in the ergoregion, can lead to instability of the brane vacuum beyond the horizon even at $T = 0$.

32.4 From 'acoustic' black hole to 'real' black hole

32.4.1 *Black-hole instability beyond the horizon*

Now let us suppose that the same situation takes place in our world, i.e. our world is the 3D brane in, say, 4D space. The modes of our 3D brane world (photons, or gravitons, or fermionic particles) may have finite lifetimes due to the interaction with the 4D environment in the bulk. In this case this will lead to the instability of the vacuum beyond the horizon of the astronomical black holes. This instability can be considered using eqn (32.35) which incorporates both terms violating the Lorentz invariance of our world at high energy. Following the analogy, we can write the intrinsic width of the particle spectrum due to the interaction with the bulk environment as a power law $\tilde{\Gamma}(k) \sim \mu(ck/\mu)^n$, where the parameter μ contains the Planck energy scale in 4D space, and thus is well above the Planck energy scale E_{Planck} of our 3D brane world, $\mu \gg E_{\text{Planck}}$. The exponent $n = 6$ if the analogy is exact. In principle, the role of the 4D environment can be played by the trans-Planckian physics, which prvides the absolute reference frame. In this case the broadening of the particle spectrum characterized by $\tilde{\Gamma}(k)$ can be caused by the superluminal upturn of the spectrum which leads to the decay of particles. The non-linear dispersion of the particle spectrum was exploited both in the black-hole physics and cosmology (Jacobson 1996, 2000; Corley and Jacobson 1996, 1999; Corley 1998; Starobinsky 2001; Niemeyer and Parentani 2001).

We shall use the Painlevé–Gullstrand metric in eqn (32.17), which, together with the superluminal dispersion of the particle spectrum, allows us to consider the region beyond the horizon:

$$g^{00} = -1 \ , \ g^{0i} = -v_s^i \ , \ g^{ij} = c^2 \delta^{ij} - v_s^i v_s^j \ , \ \mathbf{v_s} = -\hat{\mathbf{r}}\sqrt{\frac{2G\mathcal{M}}{r}} \ . \tag{32.46}$$

Here G is the Newton constant and $\mathcal{M}$ is the mass of the black hole. This metric coincides with the 3D generalization of the metric of ripplons on the AB-brane in eqn (32.43) in the case when $\mathbf{v}_1 = \mathbf{v}_2$. Equation (32.43) gives the following energy spectrum for particles living in the brane:

$$(\tilde{E} - \mathbf{p} \cdot \mathbf{v_s})^2 = E^2(p) - 2i\tilde{\Gamma}(p)\tilde{E} \ , \tag{32.47}$$

$$E^2(p) = M^2 + c^2 p^2 + c^2 \frac{p^4}{p_0^2} \ , \tag{32.48}$$

or

$$\tilde{E}(\mathbf{p}) = \mathbf{p} \cdot \mathbf{v_s} - i\tilde{\Gamma}(p) \pm \sqrt{E^2(p) - \tilde{\Gamma}^2(p) - 2i\tilde{\Gamma}(p)\mathbf{p} \cdot \mathbf{v_s}} \ . \tag{32.49}$$

For massless particles and small $\tilde{\Gamma}(p) \ll cp$, the imaginary part of the energy spectrum is

$$\text{Im } \tilde{E}(\mathbf{p}) = -i\tilde{\Gamma}(p)\left(1 \pm \frac{\mathbf{p} \cdot \mathbf{v_s}}{E(p, M=0)}\right) \ . \tag{32.50}$$

Beyond the horizon, where $W > c$, the imaginary part becomes positive for $|\mathbf{p} \cdot \mathbf{v_s}| > E(p, M=0)$ (or $p^2 < p_0^2/(v_s^2/c^2 - 1)$), i.e. attenuation transforms to amplification of waves with these $\mathbf{p}$. This demonstrates the instability of the vacuum with respect to exponentially growing electromagnetic or other fluctuations in the ergoregion. Such an instability is absent when the frame-fixing parameter

$\tilde{\Gamma} = 0$, i.e. if there is no interaction with the trans-Planckian or extra-dimensional environment.

The time of the development of instability within the conventional black hole is determined by the region far from the horizon, where the relevant $p \sim p_0 \sim p_{\text{Planck}}$. Thus $\tau \sim 1/\tilde{\Gamma}(p_0) \sim \mu^{n-1}/E_{\text{Planck}}^n$. If μ is of the same order as the brane Planck scale, the time of development of instability is the Planck time. That is why the astronomical black hole can exist only if $\mu \gg E_{\text{Planck}}$, which takes place when the 3D and 2D Planck scales are essentially different, as happens in the case of the AB-brane. Let us compare this ergoregion instability with the decay of the black hole due to Hawking radiation. The latter corresponds to $n = 4$ and $\mu = \mathcal{M}$, where $\mathcal{M}$ is the black-hole mass. This always leads to the astronomical time for the decay of the astronomical black hole.

32.4.2 Modified Dirac equation for fermions

In the presence of the horizon the Minkowski vacuum decays via Hawking radiation or by more violent processes such as the ergoregion instability discussed in Sec. 32.4.1. Due to the vacuum decay inside the horizon, the horizon can be destroyed. However, in principle it is possible that the reconstruction of the vacuum will preserve the horizon. This is possible for fermionic vacua, when the new vacuum can be formed after fermions finally fill all the negative energy levels in the ergoregionas. This happens in superfluids where the new stable vacuum can be reached when the Landau criterion is exceeded and the ergoregion is formed (Fig. 26.1 left).

What kind of vacuum is formed beyond the horizon in this hypothetical case of a stable black hole? In Sec. 26.1.3 we have seen that in the region where $v_{\text{Landau}} < v_s < v_c$ (i.e. where the superfluid velocity exceeds the Landau criterion but the superfluidity is not destroyed) the new superfluid vacuum acquires a Fermi surface (see Fig. 26.1 left for ^{3}He-B and Fig. 26.1 right for ^{3}He-A). This suggests that beyond the horizon the vacuum state (if it exists) has fermion zero modes which form a Fermi surface. This would occur only for horizons with broken time reversal symmetry, as in the Painlevé–Gullstrand metric which can be described in terms of the flow velocity.

Let us consider how the fermion zero modes of co-dimension 1 can appear in the interior of a black hole. We need to invoke Planck-scale physics which is not known. But probably the final result – the existence of the Fermi surface beyond the horizon – is universal because of the universality of zeros of co-dimension 1. So let us use such an extension of the Standard Model and gravity to high energy as suggested by the quantum Fermi liquids, and exploit as a guide the analog of the horizon and ergoregion in these liquids. This means that we make the following crucial assumptions.

(1) We assume that the Painlevé–Gullstrand metric with the frame dragging inward is the proper stationary solution of the Einstein equations, whose spacetime coordinates are absolute even on the Planck scale. It is in this absolute frame of the trans-Planckian environment that the true vacuum must be found. In quantum liquids this corresponds to the abslolute spacetime of the laboratory

environment, in which the velocity field (the gravitational field) and the horizon are stationary.

(2) The local frame of the observer who is freely falling toward the singularity will be considered as the counterpart of the superfluid-comoving frame. In this preferred frame the non-relativistic corrections due to Planck energy physics are introduced which violate the Lorentz invariance. In the freely falling frame these corrections do not depend on the gravitational field. The freely falling frame for the Planckian physics was also suggested by Corley and Jacobson (1996).

(3) The energy spectrum of particles in this freely falling frame is given by eqn (32.48).

(4) In the absolute frame of the trans-Planckian environment their energy spectrum is

$$\tilde{E} = \mathbf{p} \cdot \mathbf{v}_s \pm c\sqrt{M^2 + c^2 p^2 + c^2 \frac{p^4}{p_0^2}} \,, \tag{32.51}$$

where $\mathbf{v}_s = -\hat{\mathbf{r}} c \sqrt{r_h/r}$ is the frame-dragging field. In the limit of the effective theory $p \ll p_0$ this corresponds to the spectrum of relativistic particles, $g^{\mu\nu} p_\mu p_\nu + M^2 = 0$, in the black-hole background described by the Painlevé–Gullstrand metric in eqn (32.46).

The Painlevé–Gullstrand metric describes the spacetime in both exterior and interior regions. This spacetime, though not static, $g_{\mu\nu}(t) \neq g_{\mu\nu}(-t)$, is stationary in the absolute frame, $\partial_t g_{\mu\nu} = 0$. That is why the energy in the interior region is well determined for any value of the momentum $\mathbf{p}$. This allows us to determine the ground state (vacuum) of the Standard Model in the interior region of the black hole. This is the analog of the Boulware vacuum in general relativity (Boulware 1975), the divergence of the energy density of the Boulware vacuum at the horizon (and beyond) is cured by the superluminal dispersion.

(5) Since we do not know the Planckian physics, the back reaction of the gravitational field to the establishment of a new vacuum is not known. That is why we simply fix the gravitational background, assuming that $\mathbf{v}_s = -\hat{\mathbf{r}} c \sqrt{r_h/r}$ always holds. Since the Fermi surface is a topologically stable object, modification of the $\mathbf{v}_s(\mathbf{r})$-field due to the back reaction should not destroy the Fermi surface.

(6) We consider here only the fermionic vacuum of the Standard Model. We assume that, as in fermionic quantum liquids, the bosonic fields in the Standard Model are the collective modes of the fermionic vacuum. The Dirac equation for fermions in the gravitational background of the Painlevé–Gullstrand metric can be written using the tetrad formalism (see Doran 2000). The violation of the Lorentz invariance at high energy can be introduced by adding the non-linear γ_5 term, which gives the superluminal dispersion. As a result one obtains the following modified Dirac equation (Huhtala and Volovik 2001):

$$i\partial_t \Psi = -ic\alpha^i \partial_i \Psi + M\gamma_0 \Psi + H_P \Psi + H_G \Psi \,. \tag{32.52}$$

Here H_P and H_G are Hamiltonians coming from the Planckian physics and from the gravitational field correspondingly:

$$H_P = -\frac{\hbar^2 c}{p_0}\gamma_5 \partial_i^2 \ , \ H_G = i\hbar c \sqrt{\frac{r_{\rm h}}{r}} \left(\frac{3}{4r} + \partial_r\right) \ . \qquad (32.53)$$

The Dirac matrices used are

$$\alpha^i = \begin{pmatrix} 0 & \sigma^i \\ \sigma^i & 0 \end{pmatrix} \quad \gamma^0 = \begin{pmatrix} 1 & 0 \\ 0 & -1 \end{pmatrix} \quad \gamma_5 = \begin{pmatrix} 0 & -i \\ i & 0 \end{pmatrix} \ . \qquad (32.54)$$

The equation (32.52) is the starting point for calculation of the fermion zero modes and new vacuum within the black hole. Here we shall use the fact that the main contribution to the new vacuum comes from the negative energy fermions, whose wavelength is on the order of Planck length (Huhtala and Volovik 2001). For such fermions the semiclassical approximation works well and we can use the classical energy spectrum in eqn (32.51). Also the masses M of fermions can be neglected at such scales, and we return to the chiral fermions of the Standard Model, but with non-linear dispersion.

32.4.3 *Fermi surface for Standard Model fermions inside horizon*

Because of the possibility of superluminal propagation the surface $r = r_{\rm h}$ is not the true horizon: particles can penetrate this surface from both sides. However, this surface has a physical meaning: it marks the boundary of the ergoregion. At $r < r_{\rm h}$, some particles with positive square root in eqn (32.51) have negative energy. As a result, at $r < r_{\rm h}$ the Fermi surface appears – the surface in the 3D momentum space, where the energy of particles is zero, $\tilde{E}(\mathbf{p}) = 0$. After the vacuum reconstruction, when all the negative energy states are occupied, the horizon becomes the physical surface separating quantum vacuua with different topology of the fermionic spectrum. The coordinate singularity of the metric becomes the real physical singularity.

For the spectrum in eqn (32.51) with $M = 0$ the Fermi surface is given by an equation which expresses the radial momentum p_r in terms of the transverse momentum $p_\perp$:

$$p_r^2(p_\perp) = \frac{1}{2}p_0^2\left(\frac{v_{\rm s}^2}{c^2} - 1\right) - p_\perp^2 \pm \sqrt{\frac{1}{4}p_0^4\left(\frac{v_{\rm s}^2}{c^2} - 1\right)^2 - p_0^2 p_\perp^2 \frac{v_{\rm s}^2}{c^2}} \ . \qquad (32.55)$$

This surface of co-dimension 1 exists at each point $\mathbf{r}$ within the ergoregion, where $v_{\rm s}^2 > c^2$. It exists only in a restricted range of the transverse momenta, with the restriction provided by the cut-off parameter p_0:

$$p_\perp < \frac{1}{2}p_0 \left|\frac{v_{\rm s}}{c} - \frac{c}{v_{\rm s}}\right| \ . \qquad (32.56)$$

This means that the Fermi surface formed by the fermion zero modes beyond the horizon is a closed surface in the 3D momentum $\mathbf{p}$-space (see Fig. 32.11 *right*).

The Fermi surface of fermion zero modes provides the finite density of fermionic states (DOS) at $\tilde{E} = 0$:

$$N(\tilde{E} = 0) = N_F \sum_{\mathbf{p},r} \delta(\tilde{E}(\mathbf{p})) \qquad (32.57)$$

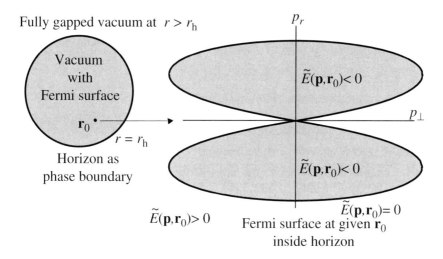

FIG. 32.11. Formation of the event horizon can be accompanied by the topological phase transition in momentum space: the Fermi surface for the Standard Model fermions can be formed beyond the horizon. The Fermi surface $\tilde{E}(\mathbf{p}, \mathbf{r}_0) = 0$ is shown for a given point $\mathbf{r}_0$ inside the black hole: $0 < |\mathbf{r}_0| < r_\mathrm{h}$ (*right*). The Fermi surface is absent for $r > r_\mathrm{h}$, where all the fermions of the Standard Model are massive. Thus the event horizon serves as the interface between the two vacua (*left*). In condensed matter, the quantum phase transition at $T = 0$ with formation or topological reconstruction of the Fermi surface is called the Lifshitz transition.

$$= \frac{4\pi N_F}{(2\pi\hbar)^3} \int_0^{r_\mathrm{h}} r^2 dr \int d^3p\, \delta\left(p_r v_\mathrm{s} + c\sqrt{p^2 + \frac{p^4}{p_0^2}}\right) \quad (32.58)$$

$$= \frac{N_F}{\pi\hbar^3} \int_0^{r_\mathrm{h}} r^2 dr \int_0^{p_\perp^2(r)} \frac{d(p_\perp^2)}{|v_G|} . \quad (32.59)$$

Here $N_F = 16 N_g$ is the number of massless chiral fermionic species in the Standard Model with N_g generations; v_G is the radial component of the group velocity of particles at the Fermi surface:

$$v_G(\tilde{E} = 0) = \frac{dE}{dp_r} = \mp c\sqrt{\left(\frac{v_\mathrm{s}}{c} - \frac{c}{v_\mathrm{s}}\right)^2 - 4\frac{p_\perp^2}{p_0^2}} . \quad (32.60)$$

Integration over $p_\perp^2$ in eqn (32.59) gives the following density of states of fermion zero modes

$$N(\dot{E} = 0) = \frac{N_F p_0^2}{\pi\hbar^3 c} \int_0^{r_\mathrm{h}} r^2 dr \left|\frac{v_\mathrm{s}}{c} - \frac{c}{v_\mathrm{s}}\right| = \frac{4 N_F}{35\pi\hbar^3 c} p_0^2 r_\mathrm{h}^3 . \quad (32.61)$$

The main contribution to the DOS and thus to the thermodynamics comes from the energies comparable to the cut-off energy $E_0 = cp_0$, which is much larger than fermion masses. That is why all the masses of fermions were neglected.

32.4.4 Thermodynamics of 'black-hole matter'

Fermion zero modes with their finite DOS $N(\tilde{E} = 0)$ determine the thermodynamics of the black-hole matter at $T \neq 0$. The thermal energy $\mathcal{E}(T)$ carried by the Standard Model fermions in the interior of the black hole at non-zero temperature is

$$\mathcal{E}(T) = N(0) \int d\tilde{E}\, \tilde{E} f(\tilde{E}/T) = \frac{\pi^2}{6} N(0) T^2 = \frac{16\pi}{105\hbar^3} N_F G^3 E_0^2 \mathcal{M}^3 T^2 , \quad (32.62)$$

where $f(x) = 1/(e^x + 1)$ is the Fermi distribution function. The entropy of the black-hole matter $\mathcal{S}(T) = -d\mathcal{F}/dT$ (where $\mathcal{F}(T) = \mathcal{E}(T) - T\mathcal{S}(T)$ is the free energy) is

$$\mathcal{S}(T) = \frac{\pi^2}{3} N(0) T = \frac{32\pi}{105\hbar^3} N_F G^3 E_0^2 \mathcal{M}^3 T , \quad (32.63)$$

where $\mathcal{M} = r_\mathrm{h}/2Gc$ is the black-hole mass; and G is the gravitational constant.

Let us compare the entropy of fermion zero modes to the phenomenological Bekenstein–Hawking (BH) entropy of the black hole. In the phenomenological approach the black hole is in the thermodynamic equilibrium with the environment if the Hawking temperature T_H coincides with the temperature T of the environment. This equilibrium is, however, unstable since it corresponds to the saddle-point thermodynamic state – the sphaleron – with a negative specific heat. The BH entropy of such a black-hole sphaleron is

$$\mathcal{S}_\mathrm{BH} = \frac{4\pi G}{\hbar} \mathcal{M}^2 , \quad T_\mathrm{H} = \frac{\hbar}{8\pi G \mathcal{M}} . \quad (32.64)$$

The black-hole sphaleron is similar to the vortex-ring sphaleron in Fig. 26.5 with $R \to r_\mathrm{h}$, $mw/\ln(R/R_\mathrm{core}) \to 2\pi T/c$ and $\tilde{E}(R) = E(R) - \mathbf{w} \cdot \mathbf{p}(R) \to \mathcal{M}(r_\mathrm{h}) - T\mathcal{S}_\mathrm{BH}(r_\mathrm{h})$ (here $\mathbf{p}$ is the linear momentum of a vortex ring and $\mathbf{w}$ is the counterflow velocity). The energy and entropy of the sphaleron determine the rate of black hole creation by thermal activation: $e^{-\tilde{E}_\mathrm{sph}/T} \to e^{\mathcal{S}_\mathrm{BH}} e^{-\mathcal{M}/T} = e^{-\mathcal{S}_\mathrm{BH}}$ (see Hawking et al. 1995). An analogy between the vortex-loop and black-hole thermodynamics was discussed by Copeland and Lahiri (1995), Volovik (1995b) and Kopnin and Volovik (1998a); in both cases the entropy is proportional to the area.

Equations for the microscopic and phenomenological entropy, (32.63) and (32.64) respectively, are consistent with each other if

$$\frac{1}{105\pi} N_F E_0^2 = \frac{\hbar}{G} \equiv E_\mathrm{Planck}^2 . \quad (32.65)$$

In eqn (32.65) the cut-off energy E_0 and the number of chiral fermionic species N_F are combined to form the gravitational constant G. The same occurs in

Sakharov's (1967a) effective gravity, where all the fermionic species must add to produce the inverse effective gravitational constant: $G^{-1} \sim N_F E_0^2/\hbar$ (see also eqn (10.52) for the Newton constant in the effective gravity of ^{3}He-A, where the role of the cut-off parameter E_0 is played by Δ_0).

Equation (32.65) reflects the deep phenomenon of the insensitivity of the effective phenomenological theory to the details of the microscopic physics. In a given microscopic theory, the black-hole entropy comes from the fermionic microstates, and thus it must be proportional to the number of fermionic species N_F in the Standard Model. On the phenomenological level, eqns (32.64) are completely determined within the effective theory, and thus cannot explicitly contain such microscopic parameters as E_0 and N_F. This apparent contradiction was resolved by Jacobson (1994) who noticed that the number of fermions N_F and the cut-off parameter E_0 enter the microscopic entropy in the same combination as in the effective gravitational constant G. As a result the black-hole entropy is completely determined by the parameter G of the effective theory.

32.4.5 Gravitational bag

Let us show that in a given microscopic model the micro- and macro- entropies coincide, i.e. that the relation (32.65) between the Planck scale parameter E_0 and the Newton gravitational constant does hold. We apply the condensed matter conjecture that the equilibrium vacuum is not gravitating. If so, then the whole gravitational mass $\mathcal{M}$ of the black hole is provided by the 'matter' above the vacuum, i.e. by thermal fermionic microstates in the black-hole interior. Thus $\mathcal{M}$ is determined by the thermal energy $\mathcal{E}$ of fermion zero modes and by their pressure p:

$$\mathcal{M} = \mathcal{E} + 3pV , \qquad (32.66)$$

where $V = (4\pi/3)r_h^3$ is the volume within the horizon. Since the equation of state of thermal fermions forming the Fermi surface is

$$\mathcal{E} = \frac{1}{2}ST = pV , \qquad (32.67)$$

one obtains

$$\mathcal{M} = 4\mathcal{E} . \qquad (32.68)$$

The black-hole matter can be in equilibrium with the external environment if its temperature coincides with the temperature of the Hawking radiation: in this case absorption and emission compensate each other. Substituting the Hawking temperature $T = \frac{\hbar}{8\pi G \mathcal{M}}$ into eqn (32.68) and using eqn (32.62) for the thermal energy, one obtains the relation (32.65) between N_F, E_0 and G. This automatically leads to the Bekenstein–Hawking entropy (32.64) of the black hole if it is in equilibrium with the environment.

In this consideration the black hole represents some kind of self-consistent gravitational bag in Fig. 32.11 *left*. In the exterior region, one has either the Standard Model vacuum of the Fermi point universality class, or the fully gapped Dirac vacuum if the temperature is below the electroweak transition. The interior of the bag consists of the new vacuum belonging to the universality class of Fermi surface formed by fermion zero modes. The bag is self-sustaining in the sense that the vacuum in the interior is adjusted to the gravitational field determined by the Painlevé–Gullstrand metric, while thermal excitations inside the bag serve as a source of this gravitational field. The bag boundary – the black-hole horizon – represents the interface between two vacua of different universality classes.

Of course, there are many weak points in this description of the gravitational bag. First of all, the reconstruction of the vacuum within the black hole involves the Planck energy in the whole volume of the black hole. That is why it is very unlikely that such a bag can arise as a result of the gravitational collapse of ordinary matter. However, the rearrangement of the vacuum can be triggered by instabilities near the physical singularity, and then it will propagate to the whole interior of the black hole if the latter survives after such a violent process. Second, it is not clear whether the traditional description of the black hole is applicable in the presence of the new vacuum. One may expect that it is applicable for the gravitational field outside the horizon, while the principle of the non-gravitating vacuum can provide the connection between the low-energy degrees of freedom of the new vacuum inside the hole and the gravitational field outside. Also we did not consider the change of the Painlevé–Gullstrand metric due to the spatial distribution of thermal fermions inside the horizon. However, if the gravitational bag is stable this will not change the universality class of the vacuum: it will contain the Fermi surface, and thus the T^2 dependence of the thermal energy of fermion zero modes will persist, and fermionic microstates will be responsible for the black-hole entropy.

The main reason for the formation of the Fermi surface beyond the horizon is the violated time reversal symmetry of the Painlevé–Gullstrand metric. In a similar way the violated time reversal symmetry leads to the formation of the Fermi surface in the flowing superfluids when the Landau velocity is exceeded (see Sec. 26.1.3). In principle, the large density of states on the Fermi surface may generate further symmetry breaking in the black-hole interior, as it occurs in condensed matter. In this case the universality class of the vacuum can change spontaneously.

33

CONCLUSION

According to the modern view the elementary particles (electrons, neutrinos, quarks, etc.) are excitations of some more fundamental medium called the quantum vacuum. This is the new ether of the 21st century. The electromagnetic and gravitational fields, as well as the fields transferring the weak and the strong interactions, all represent different types of collective motion of the quantum vacuum.

Among the existing condensed matter systems, the particular quantum liquid – superfluid ^{3}He-A – most closely resembles the quantum vacuum of the Standard Model. This is the collection of ^{3}He atoms condensed into the liquid state like water. But as distinct from water, the behavior of this liquid is determined by the quantum mechanical zero-point motion of atoms. Due to the large amplitude of this motion, the liquid does not solidify even at zero temperature. At zero temperature this liquid represents a coherent vacuum system similar to a Bose condensate. This is the analog of our quantum vacuum. When the temperature is non-zero this 'vacuum' is excited. Excitations of the quantum liquid consist of fermionic quasiparticles and bosonic collective modes. Although quasiparticles are not the real atoms of the liquid, they behave as real particles and serve as the 'elementary particles' of this liquid. Interacting fermionic and bosonic quantum fields constitute the quantum field theory in the quantum liquid. As distinct from the fundamental quantum field theories with their fundamental elementary particles, there are no 'bare' quasiparticles in quantum liquids. One cannot say that the bare quasiparticle interacting with the quantum vacuum becomes dressed: quasiparticles are excitations of the quantum vacuum of a quantum liquid and thus do not exist without the vacuum.

The most important property of this particular quantum liquid – superfluid ^{3}He-A – is that its quasiparticles are very similar to the chiral elementary particles of the Standard Model (electrons and neutrino), while its collective modes are very similar to gravitational, electromagnetic and $SU(2)$ gauge fields, and the quanta of these collective modes are analogs of gravitons, photons and weak bosons. In the low-energy corner the Lorentz invariance emerges and the quantum field theory becomes a relativistic quantum field theory. The reason for this similarity between the two systems is a common momentum space topology.

This momentum space topology (Chapter 8) is instrumental for classifying of universality classes of fermionic vacua in terms of their fermionic and bosonic zero modes. It provides the topological protection for the low-energy properties of systems of a given class: the character of the fermionic spectrum, collective modes and leading symmetries. What unites superfluid ^{3}He-A and the Standard

Model into one universality class of fermionic vacua is the existence of topologically stable fermion zero modes in the vicinity of Fermi points – zeros of co-dimension 3. As a result the quantum liquids belonging to this universality class reproduce many fragments of the Standard Model and gravity. Chirality, relativistic spin, Weyl fermions, gauge fields and gravity emerge in the vicinity of a Fermi point together with the physical laws and corresponding symmetries which include Lorentz symmetry and local $SU(N)$ symmetry. Such emergent behavior supports the 'anti-grand-unification' idea that the Standard Model and even GUT are effective theories gradually emerging in the infrared limit together with the corresponding symmetries.

This similarity based on common momentum space topology allows us to provide analogies between many phenomena in quantum liquids and in the quantum vacuum of the Standard Model. These phenomena have the same physics but in many cases are expressed in different languages and can be visualized in terms of different observables. However, in the low-energy corner they are described by the same equations if written in a covariant and gauge invariant form.

The main advantage of quantum liquids is that these systems are complete being described by, say, the BCS theory. This theory incorporates not only the 'relativistic' infrared regime, but also several successive scales of the short-distance physics, which correspond to different ultraviolet 'trans-Planckian' ranges of high energy. Since in the microscopic theory there is no need for a cut-off imposed by hand, there is no need for the ultraviolet renormalization and all subtle issues related to the ultraviolet cut-off in quantum field theory can be resolved on physical grounds (as was demonstrated in the example of effects related to axial anomaly on ^{3}He-A, quantum liquids can also help to resolve some subtle issues of infrared problems).

Comparison to quantum liquids can show the proper physical way from the low-energy effective theory toward a more fundamental one. This can help us to distinguish between different schemes which cannot be resolved within the effective theory. For example, using only the arguments based on the common momentum space topology one finds that some of the unification schemes of the strong and electroweak interactions are more preferable than others: in particular, the Pati–Salam group G(224) = $SU(4)_C \times SU(2)_L \times SU(2)_R$ has many advantages from the condensed matter point of view compared to the more traditional $SO(10)$ unification (Sec. 12.2). This means that the condensed matter analogy of emergent RQFT provides us with some kind of selection rule. By comparing a given theory in particle physics to the picture of how the discussed phenomenon could occur in condensed matter, one is able to judge whether this theory is consistent with the condensed matter point of view or not.

Our ultimate goal is to reveal the still unknown structure of the ether (the quantum vacuum) using our experience with quantum liquids. Unfortunately, liquid ^{3}He-A cannot serve as a perfect model for the quantum vacuum: though it belongs to the right universality class of co-dimension 3 and thus reproduces many fragments of the Standard Model vacuum, the full pattern is missing. The main disadvantage of ^{3}He-A is that the effective gravity there is far from being fully covariant: the effect of the non-covariant 'trans-Planckian' physics shows up even in the low-energy corner, and as a result the gravitational field does not obey the Einstein equations.

The realization of a quantum liquid with the completely covariant effective theory at low energy requires some effort. We need such a 'perfect' quantum liquid, where in the low-energy corner the symmetries become 'exact' to a very high precision, as we observe today in our Universe. The natural question is: are there any guiding principles to obtain the perfect quantum vacuum? The experience with ^{3}He-A, especially with its hierarchy of 'Planck' energy scales there, shows possible routes. In ^{3}He-A, the Lorentz invariance experienced by fermionic quasiparticles is violated at the lowest 'Planck' scale, $E_{\text{Lorentz}} = E_{\text{Planck 1}}$. The largest Planck scale $E_{\text{cutoff}} = E_{\text{Planck 3}}$, which is separated from the lowest one by six orders of magnitude, provides the natural cut-off for the divergent integrals over the fermionic field in the effective action for the gauge and gravitational fields. Since in ^{3}He-A $E_{\text{Lorentz}} \ll E_{\text{cutoff}}$, for most terms in the effective action for bosonic fields the main contribution comes from the energy region where the fermions are 'non-relativistic'; that is why these terms are non-relativistic and non-symmetric. This is very crucial for the effective gravity in ^{3}He-A, which is mostly non-Einsteinean.

The situation is somewhat better for the gauge fields because of the logarithmic divergence of the effective action, and one finds that in the leading logarithmic approximation the action for the effective $U(1)$ gauge field is gauge invariant and obeys general covariance. However, logarithmic accuracy is not enough to explain the really huge accuracy of the symmetries in the low-energy corner of our vacuum.

In principle, it is clear how to properly 'correct' the ^{3}He-A to make it perfect: one must somehow interchange the Planck scales so that $E_{\text{Lorentz}} \gg E_{\text{cutoff}}$. In this case the main terms in the bosonic action will have all the symmetries of our world. According to Bjorken, 'the emergence can only work if there is an extremely small expansion parameter in the game'. This small parameter can be the ratio of Planck scales: $E_{\text{cutoff}}/E_{\text{Lorentz}}$. The precision of symmetries will be determined by some power of $E_{\text{cutoff}}/E_{\text{Lorentz}}$. Since the theory is effective, the non-covariant terms will always remain there. But these remaining terms become vanishingly small in the low-energy limit, either due to the small parameter $E_{\text{cutoff}}/E_{\text{Lorentz}}$, or since they contain the Planck energy cut-off in denominator. An example of such a 'non-renormalizable' term in ^{3}He-A, which is the remnant of the 'trans-Planckian' physics, is presented in Sec. 19.3.2. It corresponds to the mass term for the $U(1)_Y$ gauge field of the hyperphoton which violates gauge invariance, but contains the Planck energy cut-off in the denominator.

The quantum liquids with Fermi points show one of possibly many routes from the low-energy 'relativistic' to the high-energy 'trans-Planckian' physics. Of course, one might expect many different routes to high energy, since the systems of the same universality class are similar only in the vicinity of the fixed point: they can and will diverge far from each other at higher energies. Nevertheless, the first non-renormalizable corrections could also be universal, and in many cases only this first non-renormalizable or the non-relativistic correction completely describes the physics of the system if it cannot be determined within the purely relativistic domain. An example is provided by the entropy of the extremal black hole discussed in Sec. 32.1.4.

In many condensed matter systems (even of different universality classes and even in bosonic quantum liquids), the propagating bosonic or fermionic modes obey an effective Lorentzian metric. That is why, if one ignores the dynamics of the gravity and considers only its effect on the 'matter', one finds that the analog of the gravitational field acting on the matter can be easily constructed. The gravitational field (i) can be simulated by flowing liquids: normal fluids, superfluids, and Bose–Einstein condensates; (ii) it can also be reproduced by elastic strains, dislocations and disclinations in solids (see the review paper by Dzyaloshinskii and Volovick (1980), the recent paper by Schmidt and Kohler (2001) and references therein); (iii) in Fermi systems with Fermi points the same effective gravitational field interacts with both fermions and bosons; (iv) an effective 2+1 metric field arises on surfaces, interfaces and membranes, for example, on a curved surface of a superfluid liquid (Andreev and Kompaneetz, 1972), this metric field plays the role of gravity for quasiparticles living within the surface; (v) an effective 2+1 metric field for ripplons on the interface between two sliding superfluids (Sec. 32.3), etc. Though the full dynamical realization of gravity can take place only in fermionic condensed matter of Fermi point universality class and with a proper hierarchy of Planck scales, gravity with non-Einsteinean dynamics can serve to simulate many different phenomena related to the marriage of gravity and quantum theory.

The analog of gravity in superfluids shows a possible way of how to solve the cosmological constant problem without having to invoke supersymmetry or any fine tuning (Sec. 29.4). The naive calculations of the energy density of the superfluid ground state in the framework of the effective theory suggests that it is on the order of the zero-point energy of the bosonic field in Bose superfluids and on the order of the energy of the Dirac vacuum of fermionic quasiparticles in Fermi superfluids. If one translates this into the language of RQFT, one obtains the energy density on the order of E^4_{Planck}, with the sign being determined by the fermionic and bosonic content. The standard field theoretical estimate gives the same huge value for the energy of the RQFT vacuum and thus for the cosmological constant, which is in severe contradiction with real life.

However, an exact treatment of the trans-Planckian physics in quantum liquids gives an exact vanishing of the appropriate vacuum energy density of the liquid. This follows solely from a stability analysis of the ground state of the isolated liquid combined with a thermodynamic identity. This analysis is beyond the effective theory: the low-energy inner observer who lives in the quantum liquid is not able to obtain this result.

At the same time the vacuum of the liquid, which looks empty for an inner observer, is not 'empty' at all: it is densely populated with the underlying atoms, which, however, do not contribute to the cosmological constant. The underlying liquid starts to contribute when it is excited. This is manifested, for example, in

Casimir forces that arise from the distortion of the superfluid vacuum. Applying this to RQFT one may conclude that the equilibrium quantum vacuum does not gravitate. Such principle of the non-gravitating vacuum cannot be derived or justified within the effective theory. On the other hand, in the underlying microscopic physics this result of the complete nullification does not depend on the microscopic details. It occurs for any relation between the Planck scales, E_{Lorentz} and E_{cutoff}, and thus is applicable to the case when the gravity is obeying the Einstein equations. The universality of this result suggests that it can be applicable to the fermionic vacuum of the Standard Model too.

But what happens if a phase transition occurs in which the symmetry of the vacuum is broken, as is supposed to happen in the early Universe when, say, the electroweak symmetry was broken? In the effective theory, such a transition must be accompanied by a change of the vacuum energy, which means that the vacuum has a huge energy either above or below the transition. However, in the exact microscopic theory of the liquid, the phase transition does not disturb the zero value of the vacuum energy. After the liquid relaxes to a new equilibrium vacuum state, its vacuum energy will be zero again. The energy change is completely compensated by the change of the chemical potential of the underlying atoms of the liquid which comprise the vacuum state; such quantities as atoms of the vacuum and the related chemical potential are not known by an inner observer who uses the effective theory.

The analogy with quantum liquids also shows that in the perturbed vacuum the energy density must be on the order of the energy of the perturbations. In particular, if the perturbations are caused by quasiparticles created at non-zero temperature T, which comprise the matter, the vacuum energy density is of the order of the energy density of matter. This is in agreement with modern astrophysical observations. Thus the cosmological constant is not a constant at all but is the dynamical quantity which is either continuously or in a stepwise manner adjusted to perturbations.

Another phenomenon where the marriage of gravity and quantum theory is important is the black hole. Having many systems for simulating of gravity, we can expect in the near future that the analogs of an event horizon could be constructed in the laboratory. In ^{3}He-A the 'maximum attainable velocity' of quasiparticles, which plays the role of the speed of light, is rather small: about 3 cm s^{-1}. Quasiparticles cannot escape from the region of liquid which moves faster than they can propagate. Such regions serve as black holes. Thus, in principle, the analog of an event horizon can be reproduced, and the Hawking radiation of quasiparticles from the horizon can be measured. The AB-brane in superfluid ^{3}He (the interface between the two sliding superfluids in Sec. 32.3) and the Bose condensates in laser-manipulated traps will probably be the first condensed matter systems where the event horizon will be realized.

The condensed matter analogs of horizons may exhibit Hawking radiation, but in addition other, unexpected, effects related to the quantum vacuum could arise, such as instabilities experienced by the vacuum in the presence of horizons and ergoregions. The non-relativistic version of the ergoregion instability was experimentally investigated using the AB-interface (Chapter 27). With the AB-brane one can also simulate the interaction of particles living on the brane with those living in the bulk, i.e. in the higher-dimensional space outside the brane. This interaction leads to the decay of the brane vacuum in the region beyond the horizon. This mechanism can be crucial for astronomical black holes, if this analogy is applicable and we live in the brane world. If the matter fields in the brane are properly coupled to, say, gravitons in the bulk, this may lead to fast collapse of the black hole.

Due to instabilities the vacuum is reconstructed. This leads either to destruction of the horizon, or to the horizon as an interface between vacua of different universality classes (Sec. 32.4). Since the short-distance physics is explicitly known in condensed matter this helps to clarify all the problems related to the vacuum in the presence of a horizon, or in the background of other exotic effective metrics, such as surfaces of infinite red shift, surfaces or lines where the metric is degenerate, etc.

In the presence of such exotic metrics the definition of the vacuum state becomes subtle from the point of view of the effective theory. It depends on the reference frame thus disturbing the equivalence between the frames obtained by a coordinate transformation. It is impossible to resolve which reference frame is more physical from within the effective theory. But looking at the theory from the outside, from the Planck energy scale, where the symmetries of the effective theory are violated, one can resolve between the frames. The inner low-energy observer living in a quantum liquid has the same restrictions imposed on him (or her) by the relativistic physics as we have in our vacuum. For an external observer outside the liquid, these restrictions disappear together with Lorentz invariance and general covariance, and he (or she) can judge what is the true vacuum. The vacuum in a non-trivial background is the proper area where the Planckian physics emerges, and we need the experience from the quantum liquids where this Planckian physics is known in order to understand what the meaning of different 'equivalence' schemes is. The frames which are equivalent from the point of view of an inner observer are not equivalent in the Planckian physics.

At the moment only one of the exotic metrics has been experimentally simulated in superfluid ^{3}He. This is the metric induced by a spinning cosmic string, which is reproduced in the background of a quantized vortex. The spinning string represents the Aharonov–Bohm geometry in which the flux of the gravimagnetic field is concentrated in the core of the string (vortex). In the presence of the gravimagnetic flux, particles experience the analog of the Aharonov–Bohm effect, called the gravitational Aharonov–Bohm effect. The analog of this effect has been experimentally confirmed in superfluids by measurement of the Iordanskii force acting on quantized vortices (Sec. 31.3).

As for the other (non-gravitational) analogies, the most interesting ones are also related to the interplay between the vacuum and matter. They can be fully investigated in quantum liquids, because of the absence of the cut-off problem. These are the anomalies which are at the origin of the exchange of the fermionic charges between the vacuum and matter. Such anomalies are the attributes of

Fermi systems in the universality class of Fermi points: the Standard Model and ^{3}He-A. The spectral flow of the fermionic levels, which carries the fermionic charge from the vacuum to matter, occurs just through the Fermi point. That is why the anomaly can be calculated in the vicinity of the Fermi point, where the equation for the anomaly in quantum liquids become fully gauge invariant and fully obeys general covariance with absolute precision. It is the same relativistic equation found by Adler (1969) and Bell and Jackiw (1969) in RQFT (see Chapter 18). ^{3}He-A provided the first experimental confirmation of the Adler–Bell–Jackiw equation to a precision of a few percent. The anomalous nucleation of the baryonic or leptonic charge from the vacuum described by Adler–Bell–Jackiw equation is at the basis of modern theories of baryogenesis in the early Universe.

The modified equation is obtained for the nucleation of fermionic charge by the moving string – the quantized vortex. The transfer of the fermionic charge from the 'vacuum' to 'matter' is mediated by the spectral flow of fermion zero modes of co-dimension 1 living on vortices – analogs of matter living in the brane world. This condensed matter illustration of the cancellation of anomalies in 1+1 and 3+1 systems (the Callan–Harvey effect) has also been experimentally verified in quantum liquids – now in ^{3}He-B (Chapter 25). The relativistic counterpart of this phenomenon is baryogenesis by cosmic strings, which thus has been experimentally probed. In general, physics of fermion zero modes of the homogeneous or inhomogeneous quantum vacuum governs practically all phenomena in the low-temperature limit, and the most important of them are fermion zero modes with co-dimension 1 and co-dimension 3 which are topologically protected.

The other effect related to the axial anomaly – the helical instability of the superfluid/normal counterflow in ^{3}He-A – is described by the same physics and by the same equations as the formation of the (hyper)magnetic field due to the helical instability experienced by the vacuum in the presence of the heat bath of the right-handed electrons (Sec. 19.3). That is why its observation in ^{3}He-A provided experimental support to the Joyce–Shaposhnikov scenario on the genesis of the primordial magnetic field in the early Universe. In the future the macroscopic parity-violating effects suggested by Vilenkin (Sec. 20.1) must be simulated in ^{3}He-A. In both systems, ^{3}He-A and the Standard Model, they are described by the same mixed axial–gravitational Chern–Simons action.

We have found that practically all the physics of the Standard Model and gravity emerge in the vicinity of the Fermi point, including even the quantum field theory. We started with a system of many atoms which obeys the conventional quantum mechanics and is described by the many-body Schrödinger wave function. The creation and annihilation of atoms is strongly forbidden at low energy scale, and thus there is no quantum field theory for the original atoms. The quantum field theory emerges for excitations – fermionic and bosonic quasiparticles – which can be nucleated from the vacuum. In the systems with Fermi points, this emergent quantum field theory becomes relativistic.

The scheme of the emergent phenomena discussed in this book is not complete: quantum mechanics is still fundamental. It is the only ingredient which does not emerge in condensed matter. Quantum mechanics is already there from the very beginning governing the dynamics of the original bare atoms. Planck's constant $\hbar$ is the only constant which is fundamental: it is the same for the high-energy atomic physics and for the low-energy RQFT. However, in exploring the quantum liquids with Fermi points, we are probably on the right track toward understanding the properties of the quantum vacuum and the origin of quantum mechanics.

REFERENCES

Abrikosov A. A. (1957). 'On the magnetic properties of superconductors of the second group', Sov. Phys. JETP **5**, 1174.

Abrikosov A. A. (1998). 'Quantum magnetoresistance', Phys. Rev. **B 58**, 2788–2794.

Abrikosov A. A. and Beneslavskii S. D. (1971). 'Possible existence of substances intermediate between metals and dielectrics', Sov. Phys. JETP **32**, 699–708.

Abrikosov A. A., Gorkov L. P. and Dzyaloshinskii I. E. (1965). *Quantum field theoretical methods in statistical physics*, Pergamon, Oxford.

Achúcarro A., Davis A. C., Pickles M. and Urrestilla J. (2001). 'Vortices in theories with flat directions', hep-th/0109097.

Achúcarro A. and Vachaspati T. (2000). 'Semilocal and electroweak strings', Phys. Rep. **327**, 347–426.

Adler S. (1969). 'Axial-vector vertex in spinor electrodynamics', Phys. Rev. **177**, 2426–2438.

Adler S. L. (1999). 'Fermion-sector frustrated $SU(4)$ as a preonic precursor of the Standard Model', Int. J. Mod. Phys. **A 14**, 1911–1934.

Aharonov Y. and Bohm D. (1959). 'Significance of electromagnetic potentials in the quantum theory', Phys. Rev. **115**, 485–491.

Akama K. (1983). 'Pregeometry', in: Lecture Notes in Physics, **176**, *Gauge Theory and Gravitation*, ed. N. Nakanishi and H. Nariai, Springer-Verlag, Berlin, pp. 267–271;K. Akama, 'An early proposal of "brane world"', hep-th/0001113.

Akama K., Chikashige Y., Matsuki T. and Terazawa H. (1978). 'Gravity and electromagnetism as collective phenomena: a derivation of Einstein's general relativity', Prog. Theor. Phys **60**, 868.

Alama S., Berlinsky A. J., Bronsard L. and Giorgi T. (1999). 'Vortices with antiferromagnetic cores in the $SO(5)$ model of high-temperature superconductivity', Phys. Rev. **B 60**, 6901–6906.

Alford M. G., Benson K., Coleman S., March-Russell J. and Wilzcek F. (1990). 'Interactions and excitations of non-Abelian vortices', Phys. Rev. Lett. **64**, 1632–1635.

Alford M., Rajagopal K. and Wilczek F. (1998). 'QCD at finite baryon density: nucleon droplets and color superconductivity', Phys. Lett. **B 422**, 247–256.

Amelino-Camelia G., Bjorken J. D. Larsson and S. E. (1997). 'Pion production from baked-Alaska disoriented chiral condensate', Phys. Rev. **D 56**, 6942–6956.

Anderson P. W. (1963). 'Plasmons, gauge invariance, and mass'. Phys. Rev. **130**, 439–442.

Anderson P. W. (1984). *Basic notions of condensed matter physics*, Benjamin/Cummings, New York.

Anderson P. W. and Toulouse G. (1977). 'Phase slippage without vortex cores: vortex textures in superfluid ^{3}He', Phys. Rev. Lett. **38**, 508–511.

Andreev A. F. (1964). 'The thermal conductivity of the intermediate state in superconductors', JETP **19**, 1228–1231.

Andreev A. F. (1998). 'Superfluidity, superconductivity, and magnetism in mesoscopics', Usp. Fiz. Nauk **168**, 655–664.

Andreev A. F. and Kompaneetz D. A. (1972). 'Surface phenomena in a superfluid liquid', JETP **34**, 1316–1323.

Aranson I. S., Kopnin N.B. and Vinokur V.M. (1999). 'Nucleation of vortices by rapid thermal quench', Phys. Rev. Lett. **83**, 2600–2603.

Aranson I. S., Kopnin N. B. and Vinokur V. M. (2001). 'Dynamics of vortex nucleation by rapid thermal quench', Phys. Rev. **B 63**, 184501.

Aranson I. S., Kopnin N .B. and Vinokur V. M. (2002). 'Nucleation of vortices in superfluid ^{3}He-B by rapid thermal quench', in: *Vortices in unconventional superconductors and superfluids*, eds. R. P. Huebener, N. Schopohl and G. E. Volovik, Springer Series in Solid-State Science **132**, Springer-Verlag, Berlin, pp. 49–64.

Arovas D. P., Berlinsky A. J., Kallin C. and Zhang S.-C. (1997). 'Superconducting vortex with antiferromagnetic core', Phys. Rev. Lett. **79**, 2871–2874.

Atiyah M. F., Patodi V. F. and Singer I. M. (1976). 'Spectral asymmetry and Riemann geometry 1.', Math. Proc. Cambridge Philos. Soc. **78**, 405.

Atiyah M. F., Patodi V. F. and Singer I. M. (1980). 'Spectral asymmetry and Riemann geometry 3.', Math. Proc. Cambridge Philos. Soc. **79**, 71.

Atiyah M. F. and Singer I. M. (1968). 'The index of elliptic operators. 1', Ann. Math. **87**, 484–530.

Atiyah M. F. and Singer I. M. (1971). 'The index of elliptic operators. 4', Ann. Math. **93**, 119–138.

Audretsch J. and Müller R. (1994). 'Spontaneous excitation of an accelerated atom: the contributions of vacuum fluctuations and radiation reaction', Phys. Rev. **A 50**, 1755–1763.

d'Auria R. and Regge T. (1982). 'Gravity theories with asymptotically flat instantons', Nucl. Phys. **B 195**, 308.

Avenel O., Ihas G. G. and Varoquaux E. (1993). 'The nucleation of vortices in superfluid ^{4}He: answers and questions', J. Low Temp. Phys. **93**, 1031–1057.

Avenel O., Hakonen P. and Varoquaux E. (1997). 'Detection of the rotation of the earth with a superfluid gyrometer', Phys. Rev. Lett. **78**, 3602–3605.

Avenel O., Hakonen P. and Varoquaux E. (1998). 'Superfluid gyrometers: present state and future prospects', J. Low Temp. Phys. **110**, 709–718.

Avenel O. and Varoquaux E. (1996). 'Detection of the Earth rotation with a superfluid double-hole resonator', Czech. J. Phys. **46-S6**, 3319–3320.

Avron J. E., Seiler R. and Simon B. (1983). 'Homotopy and quantization in condensed matter physics', Phys. Rev. Lett. **51**, 51–53.

Axenides M., Perivolaropoulos L. and Tomaras T. (1998). 'Core phase structure of cosmic strings and monopoles', Phys. Rev. **D 58**, 103512.

Balachandran A. P. and Digal S. (2002). 'Topological string defect formation during the chiral phase transition', Int. J. Mod. Phys. **A 17**, 1149–1158.

Balinskii A. A., Volovik G. E. and Kats E. I. (1984). 'Disclination symmetry in uniaxial and biaxial nematic liquid crystals', Sov. Phys. JETP **60**, 748–753.

Banados M., Teitelboim C. and Zanelli J. (1992). 'Black hole in three-dimensional spacetime', Phys. Rev. Lett., **69**, 1849–1852.

Barriola M. (1995). 'Electroweak strings produce baryons', Phys. Rev. **D 51**, 300–304.

Bartkowiak M., Daley S. W. J., Fisher S.N., Guénault A. M., Plenderleith G. N., Haley R. P., Pickett G. R. and Skyba P. (1999). 'Thermodynamics of the A-B phase transition and the geometry of the A-phase gap nodes', Phys. Rev. Lett. **83**, 3462–3465.

Bäuerle C., Bunkov Yu. M., Fisher S. N., Godfrin H. and Pickett G. R. (1996). 'Laboratory simulation of cosmic string formation in the early Universe using superfluid He-3', Nature **382**, 332–334.

Baym G. and Chandler E. (1983). 'The hydrodynamics of rotating superfluids. I. Zero-temperature, nondissipative theory', J. Low Temp. Phys. **50**, 57–87.

Bekenstein J. D. and Schiffer M. (1998). 'The many faces of superradiance', Phys. Rev. **D 58**, 064014.

Belavin A. A., Polyakov A. M., Schwarz A. S. and Tyupkin Yu. S. (1975). 'Pseudoparticle solutions of the Yang-Mills equations', Phys. Lett. **B 59**, 85–87.

Bell J. S. and Jackiw R. (1969). 'A PCAC puzzle: $\pi_0 \to \gamma\gamma$ in the σ model', Nuovo Cim. **A 60**, 47–61.

Ben-Menahem S. and Cooper A. R. (1992). 'Baryogenesis from unstable domain walls' Nucl. Phys. **B 388**, 409–434.

Bengtsson I. (1991). 'Degenerate metrics and an empty black hole', Classical Quantum Gravity **8**, 1847–1858.

Bengtsson I. and Jacobson T. (1998). 'Degenerate metric phase boundaries', Classical Quantum Gravity **14** 3109–3121 (1997); Erratum, **15**, 3941–3942.

Berry M.V. (1984). 'Quantal phase factors accompanying adiabatic changes', Proc. R. Soc. **A 392**, 45–57.

Bevan T. D. C., Manninen A. J., Cook J. B., Alles H., Hook J. R. and Hall H. E. (1997a). 'Vortex mutual friction in superfluid ^{3}He vortices', J. Low Temp. Phys. **109**, 423–459.

Bevan T. D. C., Manninen A. J., Cook J. B., Armstrong A. J., Hook J. R., and Hall H. E. (1995). 'Vortex mutual friction in rotating superfluid ^{3}He-B', Phys. Rev. Lett., **74**, 750–753.

Bevan T. D. C., Manninen A. J., Cook J. B., Hook J. R., Hall H. E., Vachaspati T. and Volovik G. E. (1997b). 'Momentogenesis by ^{3}He vortices: an experimental analog of primordial baryogenesis', Nature, **386**, 689–692.

Birkhoff G. (1962). 'Helmholtz and Taylor instability', in: *Hydrodynamic Instability*, Proceedings of Symposia in Applied Mathematics, Vol. XIII, ed G.

Birkhoff, R. Bellman and C.C. Lin, American Mathematical Society 1962, pp. 55–76.

Bjorken J. D. (1997). 'Disoriented chiral condensate: theory and phenomenology', Acta Phys. Pol. **B 28**, 2773–2791.

Bjorken J. D. (2001*a*). 'Standard model parameters and the cosmological constant', Phys. Rev. **D 64**, 085008.

Bjorken J. D. (2001*b*). 'Emergent gauge bosons', hep-th/0111196.

Blaauwgeers R., Eltsov V. B., Eska G., Finne A. P., Haley R. P., Krusius M., Ruohio J. J., Skrbek L. and Volovik G. E. (2002). 'Shear flow and Kelvin–Helmholtz instability in superfluids', Phys. Rev. Lett. **89**, 155301.

Blaauwgeers R., Eltsov V. B., Götz H., Krusius M., Ruohio J. J., Schanen R. and Volovik G. E. (2000). 'Double-quantum vortex in superfluid ^{3}He-A', Nature **404**, 471–473.

Blagoev K. B. and Bedell K. S. (1997). 'Luttinger theorem in one dimensional metals', Phys. Rev. Lett. **79**, 1106–1109.

Blaha, S. (1976). 'Quantization rules for point singularities in superfluid ^{3}He and liquid crystals', Phys. Rev. Lett. **36**, 874–876.

Blatter G., Feigel'man M. V., Geshkenbein V. B., Larkin A. I. and van Otterlo A. (1996). 'Electrostatics of vortices in type-II superconductors', Phys. Rev. Lett. **77**, 566–569.

Blatter G., Feigel'man M. V., Geshkenbein V. B., Larkin A. I. and Vinokur V. M. (1994). 'Vortices in high-temperature superconductors', Rev. Mod. Phys. **66**, 1125–1388.

Bogoliubov N. N. (1947). J. Phys. (USSR) **11**, 23.

Boulware D. G. (1975). 'Quantum field theory in Schwarzschild and Rindler spaces', Phys. Rev. **D 11**, 1404–1423.

Bradley D. I., Bunkov Yu. M., Cousins D. J., Enrico M. P., Fisher S. N., Follows M. R., Guénault A. M., Hayes W. M., Pickett G. R. and Sloan T. (1995). 'Potential dark matter detector? The detection of low energy neutrons by superfluid ^{3}He', Phys. Rev. Lett. **75**, 1887–1890.

Bravyi S. and Kitaev A. (2000). 'Fermionic quantum computation', quant-ph/0003137.

Brout R. (2001). 'Who is the Inflaton?', gr-qc/0103097.

Bulgac A. and Magierski P. (2001). 'Quantum corrections to the ground state energy of inhomogeneous neutron matter', Nucl. Phys. **A 683**, 695–712.

Bunkov Yu. M. (2000). '"Aurore de Venise" – cosmological scenario for A-B phase transition in superfluid ^{3}He', in: *Topological Defects and the Non-Equilibrium Dynamics of Symmetry Breaking Phase Tranitions*, eds. Yu. M. Bunkov and H. Godfrin, Kluwer, Dodrecht, pp. 121–138.

Bunkov Yu. M. and Timofeevskaya O. D. (1998). 'Cosmological scenario for A-B phase transition in superfluid ^{3}He', Phys. Rev. Lett. **80**, 4927–4930.

Burlachkov L. I. and Kopnin N.B. (1987). +Magnetic properties of triplet superconductors in the non-unitary state'. Sov. Phys. JETP **65**, 630.

Caldwell R. R., Dave R. and Steinhardt P. J. (1998). 'Cosmological imprint of an energy component with general equation of state', Phys. Rev. Lett. **80**,

1582–1585.

Callan C. G., Dashen R. and Gross D. J. (1977). 'A mechanism for quark confinement', Phys. Lett. **B 66**, 375.

Callan C. G., Jr. and Harvey J. A. (1985). 'Anomalies and fermion zero modes on strings and domain walls', Nucl. Phys. **B 250**, 427.

Calogeracos A., Dombey N. and Imagawa K. (1996). 'Spontaneous fermion production by a supercritical potential well', Phys. Atom. Nucl. **59**, 1275–1289.

Calogeracos A. and Volovik G. E. (1999a). 'Rotational quantum friction in superfluids: radiation from object rotating in superfluid vacuum', JETP Lett. **69**, 281–287.

Calogeracos A. and Volovik G. E. (1999b). 'Critical velocity in ^{3}He-B vibrating wire experiments as analog of vacuum instability in a slowly oscillating electric field', JETP **88**, 40–45.

Carney J. P., Guénault A. M., Pickett G. R. and Spencer G. F. (1989). 'Extreme nonlinear damping by the quasiparticle gas in superfluid ^{3}He-B in the low-temperature limit', Phys. Rev. Lett., **62**, 3042–3045.

Caroli C., de Gennes P. G. and Matricon J. (1964). 'Bound fermion states on a vortex line in a type II superconductor', Phys. Lett. **9**, 307–309.

Carter B. and Davis A. C. (2000). 'Chiral vortons and cosmological constraints on particle physics', Phys. Rev. **D 61**, 123501.

Casimir H. G. B. (1948). 'On the attraction between two perfectly conducting plates ...', Kon. Ned. Akad. Wetensch. Proc. **51**, 793.

Castelijns C. A. M., Coates K. F., Guénault A. M., Mussett S. G. and Pickett G. R. (1986). 'Landau critical velocity for a macroscopic object moving in superfluid ^{3}He-B: evidence for gap suppression at a moving surface', Phys. Rev. Lett. **56**, 69–72.

Cerdonio M. and Vitale S. (1984). 'Superfluid ^{4}He analog of the rf superconducting quantum interference device and the detection of inertial and gravitational fields', Phys. Rev. **B 29**, 481–483.

Chadha S. and Nielsen H. B. (1983). 'Lorentz invariance as a low-energy phenomenon', Nucl. Phys. **B 217**, 125–144.

Chapline G. (1999). 'The vacuum energy in a condensate model for spacetime', Mod. Phys. Lett. **A 14**, 2169–2178.

Chapline G., Hohlfeld E., Laughlin R. B. and Santiago D. I. (2001). 'Quantum phase transitions and the breakdown of classical general relativity', Philos. Mag. **B 81**, 235–254.

Chapline G., Laughlin R. B. and Santiago D. I. (2002). 'Emergent relativity and the physics of black hole horizons', in: *Artificial Black Holes*, eds. M. Novello, M. Visser and G. Volovik, World Scientific, Singapore, pp. 179–198.

Chechetkin V.R. (1976). 'Types of vortex solutions in superfluid ^{3}He', Sov. Phys. JETP, **44**, 766–772.

Chen C.-W. (1977). *Magnetism and Metallurgy of Soft Magnetic Materials*, North-Holland, Amsterdam.

Cleary R.M. (1968). 'Scattering of single-particle excitations by a vortex in a clean type-II superconductor', Phys. Rev. **175**, 587-596.

Cognola G., Elizalde E. and Kirsten K. (2001). 'Casimir energies for spherically symmetric cavities', J. Phys. **A 34**,7311–7327.

Coleman S. and Glashow S. L. (1997). 'Cosmic ray and neutrino tests of special relativity', Phys. Lett. **B 405**, 249–252.

Combescot R. and Dombre T. (1986). 'Twisting in superfluid ^{3}He-A and consequences for hydrodynamics at $T=0$', Phys. Rev. **B 33**, 79–90.

Copeland E. J. and Lahiri A. (1995). 'How is a closed string loop like a black hole?', Classical Quantum Gravity **12**, L113–117.

Corley S. (1998). 'Computing the spectrum of black hole radiation in the presence of high frequency dispersion: an analytical approach', Phys. Rev. **D 57**, 6280–6291.

Corley S. and Jacobson T. (1996). 'Hawking spectrum and high frequency dispersion', Phys. Rev. **D 54**, 1568–1586.

Corley S. and Jacobson T. (1999). 'Black hole lasers', Phys. Rev. **D 59**, 124011.

Cornwall, J. M. (1999). 'Center vortices, nexuses, and the Georgi-Glashow model', Phys. Rev. **D 59**, 125015.

Cross M. C. (1975). 'A generalized Ginzburg–Landau approach to the superfluidity of ^{3}He', J. Low Temp. Phys. **21**, 525–534.

Czerniawski J. (2002). 'What is wrong with Schwarzschilds coordinates?' gr-qc/0201037.

Davis A. C. and Martin A. P. (1994). 'Global strings and the Aharonov–Bohm effect', Nucl. Phys. **B 419**, 341–351.

Davies P. C. W., Dray T. and Manogue C. A. (1996). 'Detecting the rotating quantum vacuum', Phys. Rev. **D 53**, 4382–4387

Davis P. C. W. and Fulling S. A. (1976). 'Radiation from a moving mirror in two-dimensional space-time conformal anomaly', Proc. R. Soc. **A 348**, 393–414.

Davis R. L. (1992). 'Quantum nucleation of vorticity', Physica **B 178**, 76–82.

Davis R. L. and Shellard E. P. S. (1989). 'Global string lifetime: never say forever!', Phys. Rev. Lett. **63**, 2021–2024.

Davis S. C., Davis A. C. and Perkins W. B. (1997). 'Cosmic string zero modes and multiple phase transitions', Phys. Lett. **B 408**, 81–90.

Demircan E., Ao P. and Niu Q. (1995). 'Interactions of collective excitations with vortices in superfluid systems', Phys. Rev. **B 52**, 476–482.

Deser S., Jackiw R. and 't Hooft G. (1984). 'Three-dimenional Einstein gravity: dynamics of flat space', Ann. Phys. **152**, 220–235.

Dirac P. A. M. (1931). 'Quantized singularities in the electromagnetic field', Proc. R. Soc. **133**, 60–72.

Dirac, P. A. M. (1937). 'The cosmological constants', Nature **139** 323.

Dirac, P. A. M. (1938). 'New basis for cosmology', Proc. R. Soc. **A 165**, 199–208.

Dobbs E. R. (2000). *Helium Three*, Oxford University Press.

Doran C. (2000). 'New form of the Kerr solution', Phys. Rev. **D 61**, 067503.

Duan J. M. (1994). 'Mass of a vortex line in superfluid ^{4}He: effects of gauge-symmetry breaking, Phys. Rev. Lett. **B 49**, 12381–12383.

Duan J. M. and Leggett A. J. (1992). 'Inertial mass of a moving singularity in a Fermi superfluid', Phys. Rev. Lett. **68**, 1216–1219.

Duff M. J., Okun L. B. and Veneziano G. (2002). 'Trialogue on the number of fundamental constants', J. High Energy Phys. **0203**, 023.

Dvali G., Liu H. and Vachaspati T. (1998). 'Sweeping away the monopole problem', Phys. Rev. Lett. **80** 2281–2284.

Dziarmaga J. (2002). 'Low-temperature effective electromagnetism in superfluid ^{3}He-A', JETP Lett. **75**, 273–277.

Dziarmaga J., Laguna P. and Zurek W. H. (1999). 'Symmetry breaking with a slant: topological defects after an inhomogeneous quench', Phys. Rev. Lett. **82**, 4749–4752.

Dzyaloshinskii I. E. and Volovick G. E. (1980). 'Poisson brackets in condensed matter', Ann. Phys. **125**, 67–97.

Einstein A. (1917). 'Kosmologische Betrachtungen zur allgemeinen Relativitätstheorie', Sitzungsber. Konigl. Preuss. Akad. Wiss., **1**, 142–152.

Einstein A. (1924). Sitzber. Preuss. Akad. Wiss. 261; *ibid.* 3 (1925).

Eltsov V. B., Kibble T. W. B., Krusius M., Ruutu V. M. H. and Volovik G. E. (2000). 'Composite defect extends analogy between cosmology and ^{3}He', Phys. Rev. Lett. **85**, 4739–4742.

Esposito G., Kamenshchik A. Yu. and Kirsten K. (1999). 'On the zero point energy of a conducting shell', Int. J. Mod. Phys. **A 14**, 281–300.

Esposito G., Kamenshchik A. Yu. and Kirsten K. (2000). 'Casimir energy in the axial gauge', Phys. Rev. **D 62**, 085027.

Falomir H., Kirsten K. and Rebora K. (2001). 'Divergencies in the Casimir energy for a medium with realistic ultraviolet behavior', hep-th/0103050.

Fateev V. A., Frolov I. V. and Shwarts A. S. (1979). 'Quantum fluctuations of instantons in the nonlinear σ model', Nucl. Phys. **B 154**, 1–20.

Fetter A. L. (1964). 'Scattering of sound by a classical vortex', Phys. Rev. **136**, A1488–A1493.

Feynman R. P. (1955). 'Application of quantum mechanics to liquid helium', in: *Progress in Low Temperature Physics* 1, ed. C. J. Gorter, North-Holland, Amsterdam, pp. 17–53.

Fischer U. R. and Volovik G. E. (2001). 'Thermal quasi-equilibrium states across Landau horizons in the effective gravity of superfluids', Int. J. Mod. Phys. **D 10**, 57–88.

Fisher S. N., Hale A. J., Guénault A. M. and Pickett G. R. (2001). 'Generation and detection of quantum turbulence in superfluid ^{3}He-B', Phys. Rev. Lett. **66**, 244–247.

Foot R. and Silagadze Z. K. (2001). 'Do mirror planets exist in our solar system?', Acta Phys. Pol. **B 32**, 2271–2278.

Foot R., Lew H. and Volkas R.R. (1991). 'Models of extended Pati–Salam gauge symmetry', Phys. Rev. **D 44**, 859–864.

Forste S. (2002). 'Strings, branes and extra dimensions', Fortsch. Phys. **50**, 221–403.
Froggatt C. D. and Nielsen H. B. (1991). *Origin of Symmetry*, World Scientific, Singapore.
Froggatt C. D. and Nielsen H. B. (1999). 'Why do we have parity violation?', in: Proceedings *What comes beyond the Standard Model?*, eds. N. M. Borstnik, H. B. Nielsen and C. Froggatt, Ljubljana, Slovenia, DMFA, hep-ph/9906466.
Frolov V. and Fursaev D. (1998). 'Thermal fields, entropy, and black holes', Classical Quantum Gravity **15**, 2041–2074.
Furusaki A., Matsumoto M. and Sigrist M., (2001). 'Spontaneous Hall effect in a chiral p-wave superconductor', Phys. Rev. **B 64**, 054514.
Gago A. M., Nunokawa H. and Zukanovich Funchal R. (2001). 'Violation of equivalence principle and solar neutrinos', Nucl. Phys. Proc. Suppl. **100**, 68–70.
Gal'tsov D. V. and Letelier P. S. (1993). 'Spinning strings and cosmic dislocations', Phys. Rev. **D 47**, 4273–4276.
Garay L. J., Anglin J. R., Cirac J. I. and Zoller P. (2000). 'Sonic analog of gravitational black holes in Bose–Einstein condensates', Phys. Rev. Lett. **85**, 4643–4647.
Garay L. J., Anglin J. R., Cirac J. I. and Zoller P. (2001). 'Sonic black holes in dilute Bose–Einstein condensates', Phys. Rev. **A 63**, 023611.
Garriga J. and Vachaspati T. (1995). 'Zero modes on linked strings', Nucl. Phys. **B 438**, 161–181.
Gershtein S. S. and Zel'dovich Ya. B. (1969). ZhETF, **57**, 674 [JETP, **30**, 358 (1970)].
Geshkenbein V., Larkin A. and Barone A. (1987). 'Vortices with half magnetic flux quanta in "heavy-fermion" superconductors', Phys. Rev. **B 36**, 235–238.
Giovannini M. and Shaposhnikov E. M. (1998). 'Primordial hypermagnetic fields and triangle anomaly', Phys.Rev. **D 57** 2186–2206.
Girvin S. M. (2000). 'Spins and isospin: exotic order in quantum Hall ferromagnets', Phys. Today **53**, 39–45.
Gisiger T. and Paranjape M. B. (1998). 'Recent mathematical developments in the Skyrme model', Phys. Rep. **306**, 109–211.
Glashow S. L., Halprin A., Krastev P. I., Leung C. N. and Pantaleone J. (1997). 'Remarks on neutrino tests of special relativity', Phys. Rev. **D 56**, 2433–2434.
Gorter C. J. and Mellink J. H. (1949). 'On the irreversible processes in liquid helium II', Physica **15**, 285–304.
Goryo J. and Ishikawa K. (1999). 'Observation of induced Chern–Simons term in P- and T-violating superconductors', Phys. Lett **A 260**, 294–299.
Grinevich P. G. and Volovik G. E. (1988). 'Topology of gap nodes in superfluid ^{3}He', J. Low Temp. Phys. **72**, 371–380.
Gullstrand A. (1922). 'Allgemeine Lösung des statischen Einkörperproblems in der Einsteinschen Gravitationstheorie', Ark. Mat. Astron. Fys. **16**, 1–15.

Hagen C. R. (2000). 'Casimir energy for spherical boundaries', Phys. Rev. **D 61**, 065005.

Hagen C. R. (2001). 'Cutoff dependence and Lorentz invariance of the Casimir effect', quant-ph/0102135.

Hakonen P. J., Ikkala O. T., Islander S. T., Lounasmaa O. V. and Volovik G. E. (1983a). 'NMR experiments on rotating superfluid ^{3}He-A and ^{3}He-B and their theoretical interpretation,' J. Low Temp. Phys. **53**, 425–476.

Hakonen P. J., Krusius M., Salomaa M. M., Simola J. T., Bunkov Yu. M., Mineev V. P. and Volovik G. E. (1983b). 'Magnetic vortices in rotating superfluid ^{3}He-B', Phys. Rev. Lett. **51**, 1362–1365.

Halperin B. I. (1982). 'Quantized Hall conductance, current-carrying edge states, and the existence of extended states in a two-dimensional disordered potential', Phys. Rev. **B 25**, 2185–2190.

Hanson A. J. and Regge T. (1978). 'Torsion and quantum gravity', in: *Proceedings of the Integrative Conference on Group Theory and Mathematical Physics*, University of Texas at Austin.

Harari D. and Polychronakos A. P. (1988). 'Gravitational time delay due to a spinning string', Phys. Rev. **D 38**, 3320–3322.

Harari H. and Seiberg N. (1981). 'Generation labels in composite models for quarks and leptons', Phys. Lett. **B 102**, 263–266.

Hawking S. W. (1974). 'Black hole explosions?', Nature **248**, 30–31.

Hawking S. W., Horowitz G. T. and Ross S. F. (1995). 'Entropy, area, and black hole pairs', Phys. Rev. **D 51**, 4302–4314.

Heeger A. J., Kivelson S., Wu W. P. and Schrieffer J. R. (1988). 'Solitons in conducting polymers', Rev. Mod. Phys. **60**, 781–850.

Heinilä M. and Volovik G. E. (1995). 'Bifurcations in the growth process of a vortex sheet in rotating superfluid', Physica, **B 210**, 300–310.

von Helmholtz H. L. F. (1868). 'Über discontinuierliche Flüssigkeitsbewegungen', Monatsber. königl. Akad. Wiss. Berlin, **4**, 215–228.

Hindmarsh M. B. and Kibble T. W. B. (1995). 'Cosmic strings', Rep. Prog. Phys. **58**, 477–562.

Ho T.-L. (1998). 'Spinor Bose condensates in optical traps', Phys. Rev. Lett. **81**, 742–745.

Ho T.-L., Fulco J. R., Schrieffer J. R. and Wilczek F. (1984). 'Solitons in superfluid ^{3}He-A: bound states on domain walls', Phys. Rev. Lett. **52**, 1524–1527.

't Hooft G. (1974). 'Magnetic monopoles in unified gauge theories', Nucl. Phys. **79**, 276–284.

't Hooft G. (1999). 'Quantum gravity as a dissipative deterministic system', Classical Quantum Gravity **16**, 3263–3279.

Hoogenboom B. W., Kugler M., Revaz B., Maggio-Aprile I., Fischer O. and Renner Ch. (2000). 'Shape and motion of vortex cores in $Bi_2Sr_2CaCu_2O_{8+\delta}$', Phys. Rev. **B 62**, 9179–9185.

Hook J. R., Manninen A. J., Cook J.B. and Hall H. E. (1996). 'Vortex mutual friction in rotating superfluid ^{3}He', Czech. J. Phys. **46** Suppl. S6, 2930–

2936.

Horowitz G. T. (1991). 'Topology change in classical and quantum gravity', Classical Quantum Gravity **8**, 587–602.

Hoyle C. D., Schmidt U., Heckel B. R., Adelberger E. G., Gundlach J. H., Kapner D. J. and Swanson H. E. (2001). 'Submillimeter test of the gravitational inverse-square law: a search for "large" extra dimensions', Phys. Rev. Lett. **86**, 1418–1421.

Hu B. L. (1996). 'General relativity as geometro-hydrodynamics', Expanded version of an invited talk at 2nd International Sakharov Conference on Physics, Moscow, 20–23 May 1996, gr-qc/9607070.

Huhtala P. and Volovik G. E. (2001). 'Fermionic microstates within Painlevé–Gullstrand black hole', JETP **94**, 853–861.

Ikkala O. T., Volovik G. E., Hakonen P. J., Bunkov Yu. M., Islander S. T. and Kharadze G. A. (1982). 'NMR on rotating superfluid ^{3}He-B', JETP Lett. **35**, 416–419.

Iliopoulus J., Nanopoulus D. V. and Tomaras T. N. (1980). 'Infrared stability or anti-grandunification', Phys. Lett. **B 55**, 141–144.

Iordanskii S. V. (1964). 'On the mutual friction between the normal and superfluid components in a rotating Bose gas', Ann. Phys., **29**, 335–349.

Iordanskii S. V. (1965). 'Vortex ring formation in a superfluid', Sov. Phys. JETP **21**, 467–471.

Iordanskii S. V. (1966). 'Mutual friction force in a rotating Bose gas', Sov. Phys. JETP **22**, 160–167.

Iordanskii S. V. and Finkelstein A. M. (1972). 'Effect of quantum fluctuations on the lifetimes of metastable states in solids', JETP **35**, 215–221.

Ishida K., Mukuda H., Kitaoka Y., Asayama K., Mao Z. Q., Mori Y. and Maeno Y. (1998). 'Spin-triplet superconductivity in Sr_2RuO_4 identified by ^{17}O Knight shift', Nature **396**, 658–660.

Ishikawa K. and Matsuyama T. (1986). 'Magnetic field induced multi component QED in three-dimensions and quantum Hall effect', Z. Phys. C **33**, 41–45.

Ishikawa K. and Matsuyama T. (1987). 'A microscopic theory of the quantum Hall effect', Nucl. Phys. **B 280**, 523–548.

Isoshima T., Nakahara M., Ohmi T. and Machida K. (2000). 'Creation of a persistent current and vortex in a Bose-Einstein condensate of alkali-metal atoms', Phys. Rev. A **61**, 63610.

Ivanov D. A. (2001). 'Non-Abelian statistics of half-quantum vortices in p-wave superconductors', Phys. Rev. Lett. **86**, 268–271.

Ivanov D. A. (2002). 'Random-matrix ensembles in p-wave vortices', in: *Vortices in unconventional superconductors and superfluids*, eds. R. P. Huebener, N. Schopohl and G. E. Volovik, Springer Series in Solid-State Science **132**, Springer-Verlag, Berlin, pp. 253–265.

Jackiw R. (2000). 'When radiative corrections are finite but undetermined', Int. J. Mod. Phys. **B 14**, 2011–2022.

Jackiw R. and Rebbi C. (1976a). 'Solitons with fermion number 1/2', Phys. Rev. **D 13**, 3398–3409.

Jackiw R. and Rebbi C. (1976b). 'Spin from isospin in a gauge theory', Phys. Rev. Lett. **36**, 1116–1119.

Jackiw R. and Rebbi C. (1984). *Solitons and Particles*, World Scientific, Singapore.

Jacobson T. (1994). 'Black hole entropy and induced gravity', gr-qc/9404039.

Jacobson T. (1996). 'On the origin of the outgoing black hole modes', Phys. Rev. **D 53**, 7082–7088.

Jacobson T. A. (1999). 'Trans-Planckian redshifts and the substance of the spacetime river', Prog. Theor. Phys. Suppl. **136**, 1–17.

Jacobson T. and Koike T. (2002). 'Black hole and baby universe in a thin film of ^{3}He-A', in: *Artificial Black Holes*, eds. M. Novello, M. Visser and G. Volovik, World Scientific, Singapore, pp. 87–108.

Jacobson T. A. and Volovik G. E. (1998a). 'Event horizons and ergoregions in ^{3}He', Phys. Rev. **D 58**, 064021.

Jacobson T. A. and Volovik G. E. (1998b). 'Effective spacetime and Hawking radiation from a moving domain wall in a thin film of ^{3}He-A', JETP Lett. **68**, 874–880.

Jegerlehner F. (1998). 'The "ether-world" and elementary particles', in: *Theory of Elementary Particles*: Proceedings of the 31st International Symposium, Ahrenshoop, September 2–6, 1997, Buckow, Germany, eds. H. Dorn, D. Lüst, G. Weigt, Wiley-VCH, Berlin, p. 386, hep-th/9803021.

Jensen B. and Kučera J. (1993). 'On a gravitational Aharonov–Bohm effect', J. Math. Phys. **34**, 4975–4985.

Joyce M. and Shaposhnikov M. (1997). 'Primordial magnetic fields, right electrons, and the abelian anomaly', Phys. Rev. Lett., **79**, 1193–1196.

Kagan M. Yu. (1986). 'Hydrodynamic equations for the interface boundary between quantum liquid and quantum crystal ($T = 0$)', JETP **63**, 288.

Kaluza T. (1921). 'On the problem of unity in physics' Sitzungsber. Preuss. Akad. Wiss. K**1**, 966–972.

Kamerlingh Onnes H. (1911a). Leiden Commun. **108**; Proc. R. Acad. Amsterdam **11** 168.

Kamerlingh Onnes H. (1911b). Proc. R. Acad. Amsterdam **13** 1903.

Kang G. (1997). 'Quantum aspects of ergoregion instability', Phys. Rev. **D 55**, 7563–7573.

Kapitza P. L. (1938). 'Viscosity of liquid helium below the λ-point', Nature **141**, 74.

Kapitza P. L. (1941). ZhETF **11**, 1; *ibid.* **11**, 58.

Kardar M. and Golestanian R. (1999). 'The "friction" of vacuum, and other fluctuation-induced forces', Rev. Mod. Phys. **71**, 1233–1245.

Lord Kelvin (Sir William Thomson) (1910). *Mathematical and physical papers*, Vol. **4**, *Hydrodynamics and General Dynamics*, Cambridge University Press.

Khalatnikov I. M. (1965). *An Introduction to the Theory of Superfluidity*, Benjamin, New York.

Khazan M. V. (1985). 'Analog of the Aharonov–Bohm effect in superfluid ^{3}He-A', JETP Lett. **41**, 486–488.

Khodel V. A. and Shaginyan V. R. (1990). 'Superfluidity in system with fermion condensate', JETP Lett. **51**, 553–555.

Khomskii D. I. and Freimuth A. (1995). 'Charged vortices in high temperature superconductors', Phys. Rev. Lett. **75**, 1384–1387.

Kibble T. W. B. (1976). 'Topology of cosmic domains and strings', J. Phys. **A 9**, 1387–1398.

Kibble T. W. B. (2000). 'Classification of topological defects and their relevance to cosmology and elsewhere', in: *Topological Defects and the Non-Equilibrium Dynamics of Symmetry Breaking Phase Tranitions*, eds. Yu. M. Bunkov and H. Godfrin, Kluwer, Dodrecht, pp. 7–31.

Kibble T. W. B. and Volovik G. E. (1997). 'On phase ordering behind the propagating front of a second-order transition', JETP Lett. **65**, 102–107.

Kirtley J. R., Tsuei C. C. , Rupp M., Sun Z., Lock See Yu-Jahnes, Gupta A., Ketchen M. B., Moler K. A. and Bhushan M. (1996). 'Direct imaging of integer and half-integer Josephson vortices in high-T_c grain boundaries', Phys. Rev. Lett., **76**, 1336–1339.

Kita T. (1998). 'Angular momentum of anisotropic superfluids at finite temperatures', J. Phys. Soc. Jpn. **67**, 216.

Klein O. (1926). 'Quantum theory and five-dimensional theory of relativity', Z. Phys. **37**, 895–906 [Surveys High Energ. Phys. **5**, 241–244 (1986)].

Kléman M. (1983). *Points, Lines and Walls in Liquid Crystals, Magnetic Systems and Various Ordered Media*, Wiley, New York.

Kleman M. and Lavrentovich O. D. (2003). *Soft Matter Physics: An Introduction*, Springer-Verlag, New York.

Kohmoto M. (1985). 'Topological invariant and the quantization of the Hall conductance', Ann. Phys. **160**, 343–354.

Kondo Y., Korhonen J. S., Krusius M., Dmitriev V. V., Mukharskiy Yu. M., Sonin E. B. and Volovik G. E. (1991). 'Observation of the nonaxisymmetric vortex in ^{3}He-B', Phys. Rev. Lett. **67**, 81–84.

Kondo Y., Korhonen J. S., Krusius M., Dmitriev V. V., Thuneberg E. V. and Volovik G. E. (1992). 'Combined spin–mass vortices with soliton tail in superfluid ^{3}He-B', Phys. Rev. Lett. **68**, 3331–3334.

Kopnin N. B. (1978). 'Frequency-dependent dissipation in the mixed state of pure type-II superconductors at low temperatures', JETP Lett., **27**, 390.

Kopnin N. B. (1987). 'Movement of the interface between the A and B phases in superfluid helium-3: linear theory', JETP **65**, 1187–1192.

Kopnin N. B. (1993). 'Mutual friction in superfluid ^{3}He. II. Continuous vortices in ^{3}He-A at low temperatures', Phys. Rev. **B 47**, 14354–14363.

Kopnin N. B. (1995). 'Theory of mutual friction in superfluid ^{3}He at low temperatures', Physica **B 210**, 267.

Kopnin N. B. (2001). *Theory of Nonequilibrium Superconductivity*, Clarendon Press, Oxford.
Kopnin N. B. and Kravtsov V. E. (1976). 'Conductivity and Hall effect of pure type-II superconductors at low temperatures', JETP Lett. **23**, 578–581.
Kopnin N. B. and Salomaa M. M. (1991). 'Mutual friction in superfluid ^{3}He: effects of bound states in the vortex core', Phys. Rev. **B 44**, 9667–9677.
Kopnin N. B. and Thuneberg E. V. (1999). 'Time-dependent Ginzburg-Landau analysis of inhomogeneous normal-superfluid transitions', Phys. Rev. Lett. **83**, 116–119.
Kopnin N. B. and Vinokur V. M. (1998). 'Dynamic vortex mass in clean Fermi superfluids and superconductors', Phys. Rev. Lett. **81**, 3952–3955.
Kopnin N. B. and Volovik G. E. (1997). 'Flux-flow in d-wave superconductors: low temperature universality and scaling', Phys. Rev. Lett. **79**, 1377–1380.
Kopnin N. B. and Volovik G. E. (1998a). 'Rotating vortex core: an instrument for detecting the core excitations', Phys. Rev. **B 57**, 8526–8531.
Kopnin N. B. and Volovik G. E. (1998b). 'Critical velocity and event horizon in pair-correlated systems with "relativistic" fermionic quasiparticles', JETP Lett. **67**, 140–145.
Korshunov S. E. (1991). 'Instability of superfluid helium free surface in the presence of heat flow', Europhys. Lett. **16**, 673–675.
Korshunov S. E. (2002). 'Analog of Kelvin–Helmholtz instability on a free surface of a superfluid liquid', JETP Lett. **75**, 423–425.
Kostelecky V. A. and Mewes M. (2001). 'Cosmological constraints on Lorentz violation in electrodynamics', Phys. Rev. Lett. **87**, 251304.
Kosterlitz J. M. and Thouless D. J. (1978). 'Two-dimensional physics', in: Progress in Low Temperature Physics, ed. D. F. Brewer, North-Holland, Amsterdam, **7**b, pp. 371–433.
Kraus P. and Wilczek F. (1994). 'Some applications of a simple stationary element for the Schwarzschild geometry', Mod. Phys. Lett. **A 9**, 3713–3719.
Krusius M., Thuneberg E. V. and Parts Ü. (1994). 'A–B phase transition in rotating superfluid ^{3}He', Physica **B 197**, 376–389.
Kuznetsov E. A. and Lushnikov P. M. (1995). 'Non-linear theory of excitation of waves by Kelvin–Helmholtz instability', ZhETF **108**, 614–630.
Lambert C. J. (1990). 'On the approach to criticality of a vibrating. macroscopic object in superfluid ^{3}He-B', Physica **B 165-166**, 653–654.
Lambrecht A. and Reynaud S. (2002). 'Recent experiments on the Casimir effect: description and analysis', Sem. Poincaré **1**, 79–92.
Landau L. D. (1941). 'Theory of superfluidity of helium II', J. Phys. USSR **5**, 71.
Landau L. D. (1956). 'Theory of Fermi liquid', Sov. Phys. JETP **3**, 920.
Landau L. D. and Lifshitz E. M. (1955). 'On rotation of liquid helium', Dokl. Akad. Nauk, **100**, 669–672.
Landau L. D. and Lifshitz E. M. (1975). *Classical Fields*, Pergamon Press, Oxford.

Langacker P. and Pi S.-Y. (1980). 'Magnetic monopoles in grand unified theories', Phys. Rev. Lett. **45**, 1–4.

Laughlin R. B. and Pines D. (2000). 'The Theory of Everything', Proc. Natl Acad. Sci. USA **97**, 28–31.

Law C. K. (1994). 'Resonance response of the quantum vacuum to an oscillating boundary', Phys. Rev. Lett. **73**, 1931–1934.

Lee K.-M. (1994). 'Vortex dynamics in self-dual Maxwell-Higgs systems with a uniform background electric charge density', Phys. Rev. **D 49**, 4265–4276.

Lee T. D. and Yang C. N. (1957). 'Many-body problem in quantum mechanics and quantum statistical mechanics, Phys. Rev. **105**, 1119–1120.

Leggett A. J. (1977a). 'Macroscopic parity non-conservation due to neutral current?', Phys. Rev. Lett. **39**, 587–590.

Leggett A. J. (1977b). 'Superfluid ^{3}He-A is a liquid ferromagnet', Nature **270**, 585–586.

Leggett A. J. (1992). 'The ^{3}He A-B interface', J. Low Temp. Phys. **87**, 571–593.

Leggett A. J. (1998). 'How can we use low-temperature systems to shed light on questions of more general interest', J. Low Temp. Phys. **110**, 718–728.

Leggett A. J. (2002). 'High-energy low-temperature physics: production of phase transitions and topological defects by energetic particles in superfluid ^{3}He', J. Low Temp. Phys. **126**, 775–804.

Leggett A. J. and Takagi S. (1978). 'Orientational dynamics of superfluid ^{3}He: a "Two-Fluid" model. II. Orbital dynamics', Ann. Phys. **110**, 353–406.

Leggett A. J. and Yip S. (1990). 'Nucleation and growth of ^{3}He-B in the supercooled A-phase', in: **Helium Three**, eds. W. P. Halperin and L. P. Pitaevskii, Elsevier Science, pp. 523–607.

Leinaas J. M. (1998). 'Accelerated electrons and the Unruh effect', Talk given at 15th Advanced ICFA Beam Dynamics Workshop on Quantum Aspects of Beam Physics, Monterey, CA, 4–9 Jan. 1998, hep-th/9804179.

Lense J. and Thirring H. (1918). 'Über den Einfluss der Eigenrotation der Zentralkörper auf die Bewegung der Planeten und Monde nach der Einsteinschen Gravitationstheorie', Phys. Z. **19**, 156–163.

Leonhardt U. and Volovik G. E. (2000). 'How to create Alice string (half-quantum vortex) in a vector Bose–Einstein condensate', JETP Lett. **72**, 46–48.

Liberati S., Sonego S. and Visser M. (2000). 'Unexpectedly large surface gravities for acoustic horizons?', Classical Quantum Gravity **17**, 2903–2923.

Liberati S., Sonego S. and Visser M. (2002). 'Faster-than-c signals, special relativity, and causality', Ann. Phys. **298**, 167–185.

Lifshitz E. M. and Kagan Yu. (1972). 'Quantum kinetics of phase transitions at temperatures close to absolute zero', JETP **35**, 206–214.

Lin C. C. and Benney D. J. (1962). 'On the instability of shear flows', in: *Hydrodynamic Instability*, Proceedings of Symposia in Applied Mathematics, Vol. XIII, eds. G. Birkhoff, R. Bellman and C. C. Lin, American Mathematical Society, pp. 1–24.

Linde A. (1990). *Particle Physics and Inflationary Cosmology*, Harwood Academic, Chur.

Lo H.-K. and Preskill H. (1993). 'Non-Abelian vortices and non-Abelian statistics', Phys. Rev. **D 48**, 4821–4834.

London F. (1938). 'The λ-phenomenon of liquid helium and the Bose–Einstein degeneracy', Nature **141**, 643–644.

London H. (1946). In: *Report of International Conference on Fundamental Particles and Low Temperatures*, Vol. II, Physical Society, London, p. 48.

London H. and London F. (1935). Proc. R. Soc. **A 149**, 71; Physica **2**, 341.

Lubensky T. C., Pettey D., Currier N. and Stark H. (1997). 'Topological defects and interactions in nematic emulsions', Phys. Rev. **E 57**, 610–625.

Luck'yanchuk I. A. and Zhitomirsky M. E. (1995). 'Magnetic properties of unconventional superconductors', Superconductivity Rev. **1**, 207–255.

Luttinger J. M. (1960). 'Fermi surface and some simple equilibrium properties of a system of interacting fermions', Phys. Rev. **119**, 1153–1163.

Lynden-Bell D. and Nouri-Zonoz M. (1998). 'Classical monopoles: Newton, NUT space, gravomagnetic lensing, and atomic spectra', Rev. Mod. Phys. **70**, 427–446.

Madison, K. W., Chevy F., Bretin V. and Dalibard J. (2001). 'Stationary states of a rotating Bose–Einstein condensate: routes to vortex nucleation', Phys. Rev. Lett. **86,** 4443–4446.

Maia Neto P. A. and Reynaud S. (1993). 'Dissipative force on a sphere moving in vacuum', Phys. Rev. **A 47**, 1639–1646.

Majumdar D., Raychaudhuri A. and Sil A. (2001). 'Solar neutrino results and violation of the equivalence principle: an analysis of the existing data and predictions for SNO', Phys. Rev. **D 63**, 073014.

Makhlin Yu. G. and Misirpashaev T. Sh. (1995). 'Topology of vortex-soliton intersection – invariants and torus homotopy', JETP Lett. **61**, 49–55.

Makhlin Yu. G. and Volovik G. E. (1995). One-dimensional Fermi liquid and symmetry breaking in the vortex core, JETP Lett. **62**, 737–744.

Malozemoff A. and Slonczewski J. C. (1979). *Magnetc Domain Walls in Bubble Materials*, Academic Press, New York.

March-Russel J., Preskill J. and Wilczek, F. (1992). 'Internal frame dragging and a global analog of the Aharonov–Bohm effect', Phys. Rev. Lett. **68**, 2567–2571.

Marchetti P. A., Su Zhao-Bin and Lu Yu (1996). 'Dimensional reduction of $U(1) \times SU(2)$ Chern–Simons bosonization: application to the $t-J$ model', Nucl. Phys. **B 482**, 731–757.

Martel K. and Poisson E. (2001). 'Regular coordinate systems for Schwarzschild and other spherical spacetimes', Am. J. Phys. **69**, 476–480.

Marzlin K .P., Zhang W. and Sanders B. C. (2000). 'Creation of skyrmions in a spinor Bose–Einstein condensate', Phys. Rev. **A 62**, 013602.

Matsuda Y. and Kumagai K. (2002). 'Charged vortices in high-T_c superconductors', in: *Vortices in unconventional superconductors and superfluids*, eds. R. P. Huebener, N. Schopohl, and G. E. Volovik, Springer Series in

Solid-State Science **132**, Springer-Verlag, Berlin, pp. 283–299.
Matsumoto M. and Sigrist M. (1999). 'Quasiparticle states near the surface and the domain wall in a $p_x \pm i p_y$-wave superconductor', J. Phys. Soc. Jpn. **68**, 994.
Matsuyama T. (1987). 'Quantization of conductivity induced by topological structure of energy–momentum space in generalized QED_3', Prog. Theor. Phys. **77**, 711–730.
Matthews M. R., Anderson B. P., Haljan P. C., Hall D. S., Wieman C. E. and Cornell E. A. (1999). Phys. Rev. Lett. **83**, 2498–2501.
Mazur P. O. (1986). 'Spinning cosmic strings and quantization of energy', Phys. Rev. Lett. **57**, 929–932.
Mazur P. O. (1987). 'Mazur replies to Comment on "Spinning cosmic strings and quantization of energy"', Phys. Rev. Lett. **59**, 2380.
Mazur P. O. (1996). Reply to Comment on "Spinning Cosmic Strings and Quantization of Energy", hep-th/9611206.
Mazur P. O. and Mottola E. (2001). 'Gravitational condensate stars: an alternative to black holes', gr-qc/0109035.
Mermin N. D. (1977). 'Surface singularities and superflow in ^{3}He-A', in: *Quantum Fluids and Solids*, eds. S. B. Trickey, E. D. Adams and J. W. Dufty, Plenum, New York, pp. 3–22.
Mermin N. D. (1979). 'The topological theory of defects in ordered media', Rev. Mod. Phys. **51**, 591–648.
Mermin N. D. and Ho T.-L. (1976). 'Circulation and angular momentum in the A phase of superfluid ^{3}He', Phys. Rev. Lett. **36**, 594–597.
Michel L. (1980). 'Symmetry defects and broken symmetry. Configurations. Hidden symmetry', Rev. Mod. Phys. **52**, 617–651.
Milton K. A. (2000). 'Dimensional and dynamical aspects of the Casimir effect: understanding the reality and significance of vacuum energy', hep-th/0009173.
Mineev V. P. (1998). *Topologically Stable Defects and Solitons in Ordered media*, ed. I. M. Khalatnikov, Harwood Academic, Chur.
Mineev V. P. and Volovik G. E. (1978). 'Planar and linear solitons in superfluid ^3He,' Phys. Rev. B **18**, 3197–3203.
Misirpashaev T. Sh. (1991). 'The topological classification of defects at a phase interface', JETP **72**, 973–982.
Misirpashaev T. Sh. and Volovik G. E. (1995). 'Fermion zero modes in symmetric vortices in superfluid ^{3}He', Physica B **210**, 338–346.
Mohazzab M. (2000). 'Second sound horizon', J. Low Temp. Phys. **121**, 659–664.
Mortensen N. A., Ronnow H. M., Bruus H. and Hedegard P. (2000). 'Magnetic neutron scattering resonance of high-T_c superconductors in external magnetic fields: an $SO(5)$ study', Phys. Rev. B **62**, 8703–8706.
Mukharsky Yu., Avenel O. and Varoquaux E. (2000). 'Rotation measurements with a superfluid ^{3}He gyrometer', Physica B **280**, 287–288.

Murakami S., Nagaosa N. and Sigrist M. (1999). 'An $SO(5)$ model of p-wave superconductivity and ferromagnetism', Phys. Rev. Lett. **82**, 2939–2942.

Muzikar P. and Rainer D. (1983). 'Nonanalytic supercurrents in ^{3}He-A', Phys. Rev. **B 27**, 4243–4250.

Naculich S. G. (1995). 'Fermions destabilize electroweak strings', Phys. Rev. Lett. **75**, 998–1001.

Nagai K. (1984). 'Nonanalytic properties and normal current of superfluid ^{3}He-A at $T = 0$ K', J. Low Temp. Phys. **55**, 233–246.

Nambu Y. (1977). 'String-like configurations in the Weinberg-Salam theory', Nucl. Phys. **B 130**, 505.

Nambu Y. (1985). 'Fermion-boson relations in the BCS-type theories', Physica **D 15**, 147–151.

Nambu Y. and Jona-Lasinio G. (1961). 'Dynamical model of elementary particles based on an analogy with superconductivity. I.', Phys. Rev. **122**, 345–358; 'Dynamical model of elementary particles based on an analogy with superconductivity. II.', Phys. Rev. **124**, 246–254.

Nielsen H. B. and Ninomiya M. (1981). 'Absence of neutrinos on a lattice. I - Proof by homotopy theory', Nucl. Phys. **B 185**, 20 [Erratum, Nucl. Phys. **B 195**, 541 (1982)]; 'Absence of neutrinos on a lattice. II - Intuitive homotopy proof', Nucl. Phys. **B 193**, 173.

Nielsen H. B. and Ninomiya M. (1983). 'The Adler–Bell–Jackiw anomaly and Weyl fermions in a crystal', Phys. Lett. **130 B**, 389–396.

Nielsen H. B. and Olesen P. (1973). 'Vortex line models for dual strings', Nucl. Phys. **B 61**, 45–61.

Niemeyer J. C. and Parentani R. (2001). 'Trans-Planckian dispersion and scale-invariance of inflationary perturbations', Phys. Rev. **D 64**, 101301.

Novikov S. P. (1982). 'The Hamiltonian formalism and a multivalued analog of Morse theory' (in Russian), Usp. Mat. Nauk **37**, 3–49.

Okun L. B. (2002). 'Cube or hypercube of natural units', hep-ph/0112339, to be published in *Multiple Facets of Quantization and Supersymmetry*, Michael Marinov Memorial Volume, eds. M. Olshanetsky and A. Vainshtein, World Scientific, Singapore.

Onsager L. (1949). (discussion on paper by C. J. Gorter), Suppl. Nuovo Cimento **6**. 249–250; reprinted in the book by Thouless (1998), pp. 177–178.

Onsager L. (unpublished); see also F. London, *Superfluids*, Vol. II, Wiley, New York, p. 151.

Ooguri H. (1986). 'Spectrum of Hawking radiation and the Huygens principle', Phys. Rev. **D 33**, 3573–3580.

Osheroff, D. D., Richardson R. C. and Lee D. M. (1972). 'Evidence for a new phase of solid ^{3}He', Phys. Rev. Lett. **28**, 885–888.

Oshikawa M. (2000). 'Topological approach to Luttinger's theorem and the Fermi surface of a Kondo lattice', Phys. Rev. Lett. **84**, 3370–3373.

Packard R. E. (1998). 'The role of the Josephson-Anderson equation in superfluid helium', Rev. Mod. Phys. **70**, 641–651.

Packard R. E. and Vitale S. (1992). 'Principles of superfluid-helium gyroscopes', Phys. Rev. **B 46**, 3540–3549.

Padmanabhan T. (1998). 'Conceptual issues in combining general relativity and quantum theory', hep-th/9812018.

Painlevé P. (1921). 'La mécanique classique et la théorie de la relativité', C. R. Hebd. Acad. Sci. (Paris) **173**, 677–680.

Palmeri J. (1990). 'Superfluid kinetic equation approach to the dynamics of the ^{3}He A-B phase boundary', Phys. Rev. **B 42**, 4010–4035.

Parikh M. K. and Wilczek F. (2000). 'Hawking radiation as tunneling', Phys. Rev. Lett. **85**, 5042–5045.

Parts Ü., Karimäki J. M., Koivuniemi J. H., Krusius M., Ruutu V. M. H., Thuneberg E. V. and Volovik G. E. (1995a). 'Phase diagram of vortices in superfluid ^{3}He-A', Phys. Rev. Lett. **75**, 3320–3323.

Parts Ü., Krusius M., Koivuniemi J.H., Ruutu V. M. H., Thuneberg E. V. and Volovik G. E. (1994a). 'Bragg reflection from the vortex-sheet planes in rotating ^{3}He-A', JETP Lett. **59**, 851–856.

Parts Ü., Ruutu V. M. H., Koivuniemi J .H., Bunkov Yu. N., Dmitriev V. V., Fogelström M., Huenber M., Kondo Y., Kopnin N. B., Korhonen J .S., Krusius M., Lounasmaa O. V., Soininen P. I. and Volovik G. E. (1995b). 'Single-vortex nucleation in rotating superfluid ^{3}He-B', Europhys. Lett. **31**, 449–454.

Parts Ü., Thuneberg E. V., Volovik G .E., Koivuniemi J. H., Ruutu V. M. H., Heinilä M., Karimäki J. M. and Krusius M. (1994b). 'Vortex sheet in rotating superfluid ^{3}He-A', Phys. Rev. Lett. **72**, 3839–3842.

Pati J. C. (2000). 'Discovery of proton decay: a must for theory, a challenge for experiment', hep-ph/0005095.

Pati J. C. (2002). 'Confronting the conventional ideas of grand unification with fermion masses, neutrino oscillations and proton decay', hep-ph/02042405.

Pati J. C. and Salam A. (1973). 'Is baryon number conserved?', Phys. Rev. Lett. **31**, 661–664.

Pati J. C. and Salam A. (1974). 'Lepton number as the fourth color', Phys. Rev. **D 10**, 275–289.

Peccei R.D. (1999). 'Discrete and global symmetries in particle physics', in: Proceedings *Broken Symmetries*, eds. L. Mathelitsch and W. Plessas, Springer-Verlag, Berlin (Springer Lecture Notes in Physics, Vol. 521).

Pekola J. P. , Torizuka K., Manninen A. J., Kyynäräinen J. M. and Volovik G. E. (1990) 'Observation of a topological transition in the ^{3}He-A vortices', Phys. Rev. Lett. **65**, 3293–3296.

Perez-Victoria M. (2001). 'Physical (ir)relevance of ambiguities to Lorentz and CPT violation in QED', J. High Energy Phys. **0104**, 032.

Perlmutter S. *et al.* Supernova Cosmology Project (1999). 'Measurements of Ω and Λ from 42 high redshift supernovae', Astrophys. J. **517**, 565–586.

Peshkov V. P. (1944). Dokl. Akad. Nauk. **45**, 365 [J. Phys. (Moscow) **8**, 381 (1944)].

Pogosian L. and Vachaspati T. (2000). 'Interaction of magnetic monopoles and domain walls', Phys. Rev. **D 62**, 105005.

Polyakov A. M. (1974). 'Particle spectrum in quantum field theory', JETP Lett. **20**, 194–195.

Polyakov A. M. (1981) 'Quantum geometry of bosonic strings', Phys. Lett. **103B**, 207–210.

Polyakov A.M. (1987). *Gauge Fields and Strings*, Contemporary Concepts in Physics **3**, Harwood Academic, Chur.

Preskill J. (1992). 'Semilocal defects', Phys. Rev. **D 46**, 4218–4231.

Press W. H. (1998). 'Table-top model for black hole electromagnetic instabilities,' in *Black Holes and High Energy Astrophysics*, Proceedings of the XLIX Yamada Conference in Kyoto, Japan, April, 1998, Universal Academy Press, Tokyo; http://www.lanl.gov/dldstp/yconfpaper.pdf.

Rajantie A. (2002). 'Formation of topological defects in gauge field theories', Int. J. Mod. Phys. **A 17**, 1–43.

Randall L. and Sundrum R. (1999). 'Large mass hierarchy from a small extra dimension', Phys. Rev. Lett. **83**, 3370–3373.

Rasetti M. and Regge T. (1975). Physica **A 80**, 217.

Ravndal F. (2000). 'Problems with the Casimir vacuum energy', hep-ph/0009208.

Lord Rayleigh (J. W. Strutt) (1899). *Scientific papers*, Vol. **1**, Cambridge University Press.

Read N. and Green D. (2000). 'Paired states of fermions in two dimensions with breaking of parity and time-reversal symmetries, and the fractional quantum Hall effect', Phys. Rev. **B 61**, 10267–10297.

Revaz B., Genoud J.-Y., Junod A., Neumaier K., Erb A. and Walker E. (1998). 'd-wave scaling relations in the mixed-state specific heat of $YBa_2Cu_3O_7$', Phys. Rev. Lett. **80**, 3364–2267.

Rice M. (1998). 'Superconductivity: an analog of superfluid ^{3}He', Nature **396**, 627–629.

Riess A. G. *et al.* (2000). 'Tests of the accelerating universe with near-infrared observations of a high-redshift type Ia supernova', Astrophys. J. **536**, 62–67.

Rovelli C. (2000). 'Notes for a brief history of quantum gravity', gr-qc/0006061.

Rubakov V. A. and Shaposhnikov M. E. (1983). 'Do we live inside a domain wall?' Phys. Lett. **B 125**, 136–138.

Ruutu V. M. H., Eltsov V. B., Gill A. J., Kibble T. W. B., Krusius M., Makhlin Yu. G., Plaçais B., Volovik G. E. and Wen Xu (1996a). 'Vortex formation in neutron-irradiated superfluid ^{3}He as an analog of cosmological defect formation', Nature **382**, 334–336.

Ruutu V. M. H., Eltsov V. B., Krusius M., Makhlin Yu. G., Plaçais B. and Volovik G. E. (1998). 'Vortex nucleation in quench-cooled superfluid phase transition', Phys. Rev. Lett. **80**, 1465–1468.

Ruutu V. M. H., Kopu J., Krusius M., Parts Ü., Plaçais B., Thuneberg E. V. and Xu W. (1996b). 'Critical velocity of continuous vortex formation in rotating ^{3}He-A', Czech. J. Phys. **46** Suppl. S1, 7–8.

Ruutu V. M. H., Kopu J., Krusius M., Parts Ü., Plaçais B., Thuneberg E. V. and Xu W. (1997). 'Critical velocity of vortex nucleation in rotating superfluid ^{3}He-A', Phys. Rev. Lett. **79**, 5058–5061.

Ruutu V. M. H., Parts Ü., Koivuniemi J. H., Krusius M., Thuneberg E. V. and Volovik G. E. (1994). 'The intersection of a vortex line with a transverse soliton plane in rotating ^{3}He-A: π_3 topology', JETP Lett. **60**, 671–678.

Ryder L. H. and Mashhoon B. (2001). 'Spin and rotation in general relativity', gr-qc/0102101.

Sagnac M. G. (1914). 'Effet tourbillonnaire optique. La circulation de l'e luminaux dans un interferographe tournant', J. Phys. Theor. Appl. **4**, 177–195.

Sakagami Masa-aki and Ohashi A. (2002). 'Hawking radiation in laboratories', Progr. Theor. Phys. **107**, 1267–1272.

Sakharov A. D. (1967*a*). 'Vacuum quantum fluctuations in curved space and the theory of gravitation', Dokl. Akad. Nauk **177**, 70–71 [Sov. Phys. Dokl. **12**, 1040-41 (1968)]; reprinted in Gen. Relative. Gravity **32**, 365–367 (2000).

Sakharov A. D. (1967*b*). 'Violation of CP invariance, C asymmetry, and baryon asymmetry of the Universe', JETP Lett. **8**, 24–27.

Salomaa M. M. and Volovik G. E. (1983). 'Vortices with ferromagnetic superfluid core in ^{3}He-B', Phys. Rev. Lett. **51**, 2040–2043.

Salomaa M. M. and Volovik G. E. (1986). 'Vortices with spontaneously broken axisymmetry in ^{3}He-B', Phys. Rev. Lett. **56**, 363–366.

Salomaa M. M. and Volovik G. E. (1987). 'Quantized vortices in superfluid ^{3}He', Rev. Mod. Phys. **59**, 533–613.

Salomaa M. M. and Volovik G. E. (1988). 'Cosmic-like domain walls in superfluid ^{3}He-B: instantons and diabolical points in $(\mathbf{k}, \mathbf{r})$ space', Phys. Rev. **B 37**, 9298–9311.

Salomaa M. M. and Volovik G. E. (1989). 'Half-solitons in superfluid ^{3}He-A: novel $\pi/2$-quanta of phase slippage', J. Low Temp. Phys. **74**, 319–346.

Schmidt J. and Kohler C. (2001). 'Torsion degrees of freedom in the Regge calculus as dislocations on the simplicial lattice', Gen. Relativ. Gravity **33**, 1799–1808.

Schopohl N. and Volovik G. E. (1992). 'Schwinger pair production in the orbital dynamics of ^{3}He-B', Ann. Phys. **215**, 372–385.

Schulz H. J., Cuniberti G. and Pieri P. (2000). 'Fermi liquids and Luttinger liquids', in *Field Theories for Low-Dimensional Condensed Matter Systems*, eds G. Morandi *et al.*, Springer Solid State Sciences Series, Springer-Verlag, Berlin.

Schützhold R. (2001). 'On the Hawking effect', Phys. Rev. **D 64**, 024029.

Schützhold R. and Unruh W. G. (2002). 'Gravity wave analogs of black holes', Phys. Rev. **D 66**, 044019.

Schwab K., Bruckner N. and Packard R. E. (1997). 'Detection of the Earth's rotation using superfluid phase coherence', Nature **386**, 585–587.

Schwarz A. S. (1982). 'Field theories with no local conservation of the electric charge', Nucl. Phys. **B 208**, 141–158.

Schwarz A. S. and Tyupkin Yu. S. (1982). 'Grand Unification and mirror particles', Nucl. Phys. **209**, 427–432.

Senthil T., Marston J. B. and Fisher M. P. A. (1999). 'Spin quantum Hall effect in unconventional superconductors', Phys. Rev. **B 60**, 4245–4254.

Shafi Q. and Tavartkiladze Z. (2001). '$SU(4)_c \times SU(2)_L \times SU(2)_R$ model from 5D SUSY $SU(4)_c \times SU(4)_{L+R}$', hep-ph/0108247.

Shelankov A. L. (1998). 'Magnetic force exerted by the Aharonov–Bohm line', Europhys. Lett. **43**, 623–628.

Shelankov A. L. (2002). 'The Lorentz force exerted by the Aharonov–Bohm line', in: *Vortices in unconventional superconductors and superfluids*, eds. R. P. Huebener, N. Schopohl and G. E. Volovik, Springer Series in Solid-State Science **132**, Springer-Verlag, Berlin, pp. 147–166.

Sigrist M. and Agterberg D.F. (1999). 'The role of domain walls on the vortex creep dynamics in unconventional superconductors', Prog. Theor. Phys. **102**, 965–981.

Sigrist M., Bailey D. B. and Laughlin R. B. (1995). 'Fractional vortices as evidence of time-reversal rymmetry breaking in high-temperature superconductors', Phys. Rev. Lett. **74**, 3249–3252.

Sigrist M., Rice T. M. and Ueda K. (1989). 'Low-field magnetic response of complex superconductors', Phys. Rev. Lett. **63**, 1727–1730.

Silagadze Z. K. (2001). 'TEV scale gravity, mirror universe, and ... dinosaurs', Acta Phys. Pol. **B 32**, 99–128.

Sinha S. and Castin Y. (2001). 'Dynamic instability of a rotating Bose–Einstein condensate', Phys. Rev. Lett. **87**, 190402.

Skyrme T. H. R. (1961). 'A non-linear field theory', Proc. R. Soc. **A 260**, 127–138.

Sokolov D. D. and Starobinsky A. A. (1977). 'On the structure of curvature tensor on conical singularities', Dokl. Akad. Nauk **234**, 1043–1046 [Sov. Phys. - Dokl. **22**, 312 (1977)].

Sonin E. B. (1973). 'Critical velocities at very low temperatures, and the vortices in a quantum Bose fluid', JETP **37**, 494–500.

Sonin E. B. (1975). 'Friction between the normal component and vortices in rotating superfluid helium', JETP **42**, 469–475.

Sonin E. B. (1997). 'Magnus force in superfluids and superconductors', Phys. Rev. **B 55**, 485–501.

Sonin E. B., Geshkenbein V. H., van Otterlo A. and Blatter G. (1998). 'Vortex motion in charged and neutral superfluids: a hydrodynamic approach', Phys. Rev. **B 57**, 575–581.

Sonoda H. (2000). 'Chiral QED out of matter', hep-th/0005188; 'QED out of matter', Nucl. Phys. **B 585**, 725–740.

Starkman G.D. and Vachaspati T. (1996). 'Galactic cosmic strings as sources of primary antiprotons', Phys. Rev. **D 53**, 6711–6714.

Starobinskii A. A. (1973). 'Amplification of waves during reflection from a rotating "black hole"', JETP **37**, 28–32.

REFERENCES

Starobinsky A. A. (1999). Plenary talk at Cosmion-99, Moscow, 17–24 October 1999.

Starobinsky A. A. (2001). 'Robustness of the inflationary perturbation spectrum to trans-Planckian physics', JETP Lett. **73**, 371–374.

Staruszkievicz A. (1963). 'Gravitation theory in three-dimensional space', Acta Phys. Pol. **24**, 735–740.

Steel J. V. and Negele J. W. (2000). 'Meron pairs and fermion zero modes', Phys. Rev. Lett. **85**, 4207–4210.

Stone M. (1990). 'Schur functions, chiral bosons, and the quantum-Hall-effect edge states', Phys. Rev. **B 42**, 8399–8404.

Stone M. (1996). 'Spectral flow, Magnus force, and mutual friction via the geometric optics limit of Andreev reflection', Phys. Rev. **B 54**, 13222–13229.

Stone M. (2000a). 'Iordanskii force and the gravitational Aharonov–Bohm effect for a moving vortex', Phys. Rev. **B 61**, 11780–11786.

Stone M. (2000b). 'Acoustic energy and momentum in a moving medium', Phys. Rev. **B 62**, 1341–1350.

Stone M. (2002). 'Phonons and forces: momentum versus pseudomomentum in moving fluids', in: *Artificial Black Holes*, eds. M. Novello, M. Visser and G. Volovik, World Scientific, pp. 335–364.

Su W. P., Schrieffer J. R. and Heeger A. J. (1979). 'Solitons in polyacetylene', Phys. Rev. Lett. **42**, 1698–1701.

Tait T. M. P (1999). 'Signals for the electroweak symmetry breaking associated with the top quark', Ph.D. Thesis, hep-ph/9907462.

Terazawa H. (1999). 'High energy physics in the 21st century', in: *Proceedings of 22nd International Workshop on the Fundamental Problems of High Energy Physics and Field Theory*, KEK Preprint 99-46, July 1999.

Terazawa H., Chikashige Y., Akama K. and Matsuki T. (1977). 'Simple relation between the fine-structure and gravitational constants', Phys. Rev. **D 15**, 1181–1183.

Thouless D. J. (1998). *Topological Quantum Numbers in Nonrelativistic Physics*, World Scientific, Singapore.

Thuneberg E. V. (1986). 'Identification of vortices in superfluid ^{3}He-B', Phys. Rev. Lett. **56**, 359–362.

Thuneberg E. V. (1987a). 'Spin-current vortices in superfluid ^{3}He-B' Europhys. Lett. **3**, 711–715.

Thuneberg E. V. (1987b). 'Ginzburg-Landau theory of vortices in superfluid ^{3}He-B', Phys. Rev. B **36**, 3583–3597.

Tisza L. (1938). 'Transport phenomena in helium II', Nature **141**, 913.

Tisza L. (1940). J. Physique et Rad. **1** 165; *ibid.* 350; 'The theory of liquid helium', Phys. Rev. **72**, 838–854 (1947).

Tolman R. C. (1934). *Relativity, Thermodynamics and Cosmology*, Clarendon Press, Oxford.

Tornkvist O. (2000). 'Cosmic magnetic fields from particle physics', astro-ph/0004098.

Toulouse G. (1979). 'Gauge concepts in condensed matter physics', in: *Recent Developments In Gauge Theories*, 331–362.
Trautman A. (1966). 'Comparison of Newtonian and relativistic theories of spacetime', in: *Perspectives in Geometry and Relativity*, Indiana University Press.
Trebin H.-R. and Kutka R. (1995). 'Relations between defects in the bulk and on the surface of an ordered medium: a topological investigation', J. Phys. A: Math. Gen. **28**, 2005–2014.
Trodden M. (1999). 'Electroweak baryogenesis', Rev. Mod. Phys. **71**, 1463–1500.
Troshin S. M. and Tyurin N. E. (1997). 'Hyperon polarization in the constituent quark model', Phys. Rev. **D 55**, 1265–1272.
Turner M. S. and White M. (1997). 'CDM models with a smooth component', Phys. Rev. **D 56**, R4439–R4443.
Turok N. (1992). 'Electroweak Baryogenesis', in: *Perspectives on Higgs physics*, ed. G. L. Kane, World Scientific, Singapore, pp. 300–329.
Unruh W. G. (1976). 'Notes on black hole evaporation', Phys. Rev. **D 14**, 870–892.
Unruh W. G. (1981). 'Experimental black-hole evaporation?', Phys. Rev. Lett. **46**, 1351–1354.
Unruh W. G. (1995). 'Sonic analog of black holes and the effects of high frequencies on black hole evaporation', Phys. Rev. **D 51**, 2827–2838.
Unruh W. G. (1998). 'Acceleration radiation for orbiting electrons', Phys. Rep. **307**, 163–171.
Uwaha M. and Nozieres P. (1986). 'Flow-induced instabilities at the superfluid-solid interface', J. Physique **47**, 263–271.
Uzan J.-P. (2002). 'The fundamental constants and their variation: observational status and theoretical motivations', hep-ph/0205340.
Vachaspati T. and Achúcarro A. (1991). 'Semilocal cosmic strings', Phys. Rev. **D 44**, 3067-3071.
Vachaspati T. and Field G. B. (1994). 'Electroweak string configurations with baryon number', Phys. Rev. Lett. **73**, 373–376; **74**, 1258(E) (1995).
van Otterlo A., Feigel'man M. V., Geshkenbein V. B. and Blatter G. (1995). 'Vortex dynamics and the Hall anomaly: A microscopic analysis', Phys. Rev. Lett. **75**, 3736–3739.
Varoquaux E., Avenel O., Ihas G. and Salmelin R. (1992). 'Phase slippage in superfluid ^{3}He-B', Physica **B 178**, 309–317.
Vilenkin A. (1979). 'Macroscopic parity-violating effects: neutrino fluxes from rotating black holes and in rotating thermal radiation', Phys. Rev. **D 20**, 1807–1812.
Vilenkin A. (1980a). 'Quantum field theory at finite temperature in a rotating system' **D 21**, 2260–2269.
Vilenkin A. (1980b). 'Equilibrium parity-violating current in a magnetic field', Phys. Rev. **D 22**, 3080–3084.

Vilenkin A. and Leahy D. A. (1982). 'Parity non-conservation and the origin of cosmic magnetic fields', Astrophys. J. **254**, 77–81.

Vilenkin A. and Shellard E. P. S. (1994). *Cosmic strings and other topological defects*, Cambridge University Press.

Visser M. (1998). 'Acoustic black holes: horizons, ergospheres, and Hawking radiation', Classical Quantum Gravity **15**, 1767–1791.

Vollhardt D., Maki K. and Schopohl N. (1980). 'Anisotropic gap distortion due to superflow and the depairing critical current in superfluid ^{3}He-B', J. Low Temp. Phys. **39**, 79–92.

Vollhardt D. and Wölfle P. (1990). *The superfluid phases of helium 3*, Taylor and Francis, London.

Volovik G. E. (1972). 'Quantum mechanical creation of vortices in superfluids', JETP Lett. **15**, 81–83.

Volovik G. E. (1978). 'Topological singularities on the surface of an ordered system', JETP Lett. **28**, 59–62.

Volovik G. E. (1986a). 'Chiral anomaly and the law of conservation of momentum in ^{3}He-A', JETP Lett. **43**, 55–554.

Volovik G. E. (1986b). 'Wess-Zumino action for the orbital dynamics of ^{3}He-A, JETP Lett. **44**, 185–189.

Volovik G. E. (1987). 'Linear momentum in ferromagnets', J. Phys. **C 20**, L83–L87.

Volovik G. E. (1988). 'Analog of quantum Hall effect in superfluid ^{3}He film, JETP **67**, 1804–1811.

Volovik G. E. (1990a). 'Symmetry in superfluid ^{3}He', in: **Helium Three**, eds. W. P. Halperin and L. P. Pitaevskii, Elsevier Science, pp. 27–134.

Volovik G. E. (1990b). 'Half-quantum vortices in the B phase of superfluid ^{3}He', JETP Lett. **52**, 358–363.

Volovik G. E. (1990c). 'Defects at interface between A and B phases of superfluid ^{3}He', JETP Lett. **51**, 449–452.

Volovik G. E. (1991). 'A new class of normal Fermi liquids', JETP Lett. **53**, 222–225.

Volovik G. E. (1992a). *Exotic properties of superfluid ^{3}He*, World Scientific, Singapore.

Volovik G. E. (1992b). 'Hydrodynamic action for orbital and superfluid dynamics of ^{3}He-A at $T = 0$', JETP **75**, 990–997.

Volovik G. E. (1993a). 'Vortex motion in fermi superfluids and Callan–Harvey effect', JETP Lett. **57**, 244–248.

Volovik G.E. (1993b). 'Action for anomaly in fermi superfluids: quantized vortices and gap nodes', JETP **77**, 435–441.

Volovik G.E. (1993c). 'Superconductivity with lines of gap nodes: density of states in the vortex', JETP Lett. **58**, 469–473.

Volovik G.E. (1995a). 'Three non-dissipative forces on a moving vortex line in superfluids and superconductors', JETP Lett. **62**, 65–71.

Volovik G.E. (1995b). 'Is there analogy between quantized vortex and black hole?', gr-qc/9510001.

Volovik G. E. (1996). 'Cosmology, particle physics and ^{3}He', Czech. J. Phys. **46** S6, 3048–3055.

Volovik G. E. (1997a). 'On edge states in superconductor with time inversion symmetry breaking', JETP Lett. **66**, 522–527.

Volovik G. E. (1997b). 'Comment on vortex mass and quantum tunneling of vortices', JETP Lett. **65**, 217–223.

Volovik G. E. (1998a). 'Vortex vs spinning string: Iordanskii force and gravitational Aharonov–Bohm effect', JETP Lett. **67**, 881–887.

Volovik G. E. (1998b). 'Axial anomaly in ^{3}He-A: simulation of baryogenesis and generation of primordial magnetic field in Manchester and Helsinki', Physica B **255**, 86–107.

Volovik G. E. (1999a). 'Simulation of Painlevé–Gullstrand black hole in thin ^{3}He-A film', JETP Lett. **69**, 705–713.

Volovik G. E. (1999b). 'Fermion zero modes on vortices in chiral superconductors', JETP Lett. **70**, 609–614.

Volovik G. E. (1999c). 'Vierbein walls in condensed matter', JETP Lett. **70**, 711–716.

Volovik G. E. (2000a). 'Momentum space topology of Standard Model', J. Low Temp. Phys. **119**, 241–247.

Volovik G. E. (2000b). 'Monopoles and fractional vortices in chiral superconductors', Proc. Natl Acad. Sci. USA **97**, 2431–2436.

Volovik G. E. (2001a). 'Reentrant violation of special relativity in the low-energy corner', JETP Lett. **73**, 162–165.

Volovik G. E. (2001b). 'Superfluid analogies of cosmological phenomena', Phys. Rep. **351**, 195–348.

Volovik G. E. and Gor'kov L. P. (1984). 'Unconventional superconductivity in UBe$_{13}$', JETP Lett. **39**, 674–677.

Volovik G. E. and Gor'kov L. P. (1985). 'Superconductivity classes in the heavy fermion systems', JETP **61**, 843–854.

Volovik G. E. and Kharadze G.A. (1990). 'Decay of the nonsingular vortex in superfluid ^{3}He-A near the transition to the A_1-phase', JETP Lett. **52**, 419–424.

Volovik G. E. and Khazan M. V. (1982). 'Dynamics of the A-phase of ^{3}He at low pressure', JETP **55**, 867–871.

Volovik G. E. and Konyshev V. A. (1988). 'Properties of the superfluid systems with multiple zeros in fermion spectrum', JETP Lett. **47**, 250–254.

Volovik G. E. and Mineev V. P. (1976a). 'Vortices with free ends in superfluid ^{3}He-A', JETP Lett. **23**, 593–596.

Volovik G. E. and Mineev V. P. (1976b). 'Linear and point singularities in superfluid ^{3}He', JETP Lett. **24**, 561–563.

Volovik G. E. and Mineev V. P. (1977). 'Investigation of singularities in superfluid ^{3}He and liquid crystals by homotopic topology methods', JETP **45** 1186–1196.

Volovik G. E. and Mineev V. P. (1981). '^{3}He-A vs Bose liquid: orbital angular momentum and orbital dynamics', JETP **54**, 524–530.

Volovik G. E. and Mineev V. P. (1982). 'Current in superfluid Fermi liquids and the vortex core structure', JETP **56**, 579–586.

Volovik G. E. and Salomaa M. M. (1985). 'Spontaneous breaking of axial symmetry in v-vortices in superfluid ^{3}He-B,' JETP Lett. **42**, 521–524.

Volovik G. E. and Vachaspati T. (1996). 'Aspects of ^{3}He and the standard electroweak model', Int. J. Mod. Phys. **B 10**, 471–521.

Volovik G. E. and Vilenkin A. (2000). 'Macroscopic parity-violating effects and ^{3}He-A', Phys. Rev. **D 62**, 025014.

Volovik G. E. and Yakovenko V. M. (1989). 'Fractional charge, spin and statistics of solitons in superfluid ^{3}He film', J. Phys.: Condens. Matter **1**, 5263–5274.

Volovik G. E. and Yakovenko V. M. (1997). 'Comment on "Hopf term for a two-dimensional electron gas"', Phys. Rev. Lett. **79** 3791–3791.

Von Neumann J. and Wigner E. P. (1929). Phys. Z. **30**, 467.

Weinberg S. (1983). 'Overview of theoretical prospects for understanding the values of fundamental constants', in: *The Constants of Physics*, W.H. McCrea and M.J. Rees editors, Phil. Trans. R. Soc. London **A 310**, 249.

Weinberg S. (1989). 'The cosmological constant problem', Rev. Mod. Phys. **61**, 1–23.

Weinberg S. (1995). *The Quantum Theory of Fields*, Cambridge University Press.

Weinberg S. (1999). 'What is quantum field theory, and what did we think it is?' in: *Conceptual Foundations of Quantum Field Theory* ed. T. Y. Cao, Cambridge University Press, pp. 241–251, hep-th/9702027.

Wen X. G. (1990a). 'Electrodynamical properties of gapless edge excitations in the fractional quantum Hall states', Phys. Rev. Lett. **64**, 2206–2209.

Wen X. G. (1990b). 'Metallic non-Fermi-liquid fixed point in two and higher dimensions', Phys. Rev. **B 42**, 6623–6630.

Wen X. G. (2000). 'Continuous topological phase transitions between clean quantum Hall states', Phys. Rev. Lett. **84**, 3950–3953.

Wexler C. and Thouless D. J. (1996). 'Effective vortex dynamics in superfluid systems', cond-mat/9612059.

Wilczek F. (1998). 'From Notes to Chords in QCD', Nucl. Phys. **A 642**, 1–13.

Wilczek F. and Zee A. (1983). 'Linking numbers, spin and statistics of solitons', Phys. Rev. Lett. **51**, 2250–2252.

Wilkinson D. and Goldhaber A. S. (1977). 'Spherically symmetric monopoles', Phys. Rev. **D 16**, 1221–1231.

Witten E. (1985). 'Superconducting strings', Nucl. Phys. **B249**, 557–592.

Woo C. W. (1976). 'Microscopic calculations for condensed phases of helium', in *The Physics of Liquid and Solid Helium*, Part I, eds. K. H. Bennemann and J. B. Ketterson, Wiley, New York.

Yakovenko V. M. (1989). 'Spin, statistics and charge of solitons in (2+1)-dimensional theories', Fizika (Zagreb) **21**, suppl. 3, 231, cond-mat/9703195.

Yakovenko V. M. (1993). 'Metals in a high magnetic field: a universality class of marginal Fermi liquid', Phys. Rev. **B 47**, 8851–8857.

Ying S. (1998). 'The quantum aspects of relativistic fermion systems with particle condensation', Ann. Phys. **266**, 295–350.

Yip S. and Leggett A. J. (1986). 'Dynamics of the ^{3}He A-B phase boundary', Phys. Rev. Lett., **57**, 345–348.

Zel'dovich Ya. B. (1967). 'Interpretation of electrodynamics as a consequence of quantum theory', JETP Lett. **6**, 345–347.

Zel'dovich Ya. B. (1971*a*). 'Amplification of cylindrical electromagnetic waves from a rotating body', JETP **35**, 1085–1087.

Zel'dovich Ya. B. (1971*b*). 'Generation of waves by a rotating body', JETP Lett. **14**, 180–181.

Zhang J., Childress S., Ubchaber A. and Shelley M. (2000). 'Flexible filaments in a flowing soap film as a model of one-dimensional flags in a two-dimensional wind', Nature **408**, 835–839.

Zhang S. C. and Hu J. (2001). 'A four-dimensional generalization of quantum Hall effect', Science, **94**, 823–828.

Zhang X., Huang T. and Brandenberger R. H. (1998). 'Pion and η' strings', Phys. Rev. **D 58**, 027702.

Zhitomirsky M. E. (1995). 'Dissociation of flux line in unconventional superconductor', J. Phys. Soc. Jpn. **64**, 913–921.

Zurek W. H. (1985). 'Cosmological experiments in superfluid helium?', Nature **317**, 505.

INDEX

action
  absence of action in hydrodynamic systems, 52–54
  area law for strings, 334
  effective, 26, 36
    for superfluid vacuum, 36, 37
  Einstein–Hilbert, 8, 11, 12, 15, 42, 442
  multi-valued action, 55, 335
  non-local, 54
  Novikov–Wess–Zumino, 55
    for Fermi points, 246
    for ferromagnets, 55
  Pauli term, 270
  topological term, 6
    vortex dynamics, 246, 334, 336, 337
  volume law
    for Fermi points, 246
    for vortices, 246, 334, 336, 337
adiabatic invariant, 337
  Bohr–Sommerfeld quantization, 297, 298
Andreev reflection, 293, 342, 375, 377, 378, 452
anti-grand-unification, 3–7, 184, 462
asymptotic freedom, 115
axiplanar phase, 78, 82, 86, 99, 114, 116, 117, 303

Baked Alaska mechanism
  in B-phase, 357
  in high-energy physics, 357
baryogenesis, 1, 6, 212, 241, 263, 312, 467
Berry phase, 138, 140, 141, 297
Bianchi identities, 12, 387
black hole
  candidates, 424
  entropy, 434, 458, 459
    dependence on number of fields, 459
  extremal, 401, 431, 433, 434
    entropy of, 434
  Painleve–Gullstrand, 434, 436, 437, 442–444, 446, 453–455, 460
  Reissner–Nordstrom, 433
  Schwarzschild, 385, 426, 434
  sphaleron, 458
  supermassive, 424
Bloch line, 194, 207
Bloch wall, 207
boojum, 218, 220, 222, 224, 226
  at interface, 219, 222, 225, 226, 228, 229
  classification, 160, 219–221, 225
  in liquid crystal, 219
  in rotating container, 221, 228
  linear boojum, 220
Bose–Einstein condensate, 3, 21, 59, 349, 376, 431, 461, 464, 465
  2-component, 183, 201
  of Cooper pairs, 102
  spinor, 201
  with F=1, 194, 204
brane, 6, 139, 275, 282, 286, 339, 447–454, 465–467
bridge, 433, 434

charge
  baryonic, 122, 124, 125, 164, 191, 240, 252
    non-conservation, 52, 124, 148, 191, 239–241, 250, 252, 312
  chiral, 165
  electric, 20, 111, 114, 146–149, 152, 177, 180, 187, 188, 237–239, 251, 285, 326, 433
    analog of, 69, 105, 106, 281, 283, 412
    fractional, 300
    of skyrmion, 271
    of string, 374
    of vortex, 374
    running, 111
  fermionic, 122, 124, 136, 253, 264, 270, 281
    fractional, 90
    in A-phase, 124–127, 241, 244, 254, 259
    instability of, 257
    non-conservation, 124, 235, 236, 238, 239, 250, 252, 255, 259, 285, 312, 313, 466, 467
    of texture, 6, 242, 244, 248, 252–254, 259, 266, 271, 276, 307, 338, 375
chemical potential
  effective, 123–125
  of particles, 18, 123, 125
  of quasiparticles, 123, 124
Chern–Simons term, 122, 252, 254, 255, 257, 265–267, 269–272
  axial–gravitational, 260, 261, 263, 264, 467
Cheshire charge, 191

498 INDEX

chiral anomaly, 6, 52, 123
chirality, 124
   emergence of, 5, 106, 109
   generalization of, 106
   protected by symmetry, 106
Clebsch variables, 54
clocks and rods
   of external observer, 40
   of inner observer, 40
collapse
   gravitational, 424, 460
   vacuum, 438
collective mode
   capillary-gravity wave, 339, 341, 344, 447
   electromagnetic waves, 113
   of vacuum, 5
   orbital waves, 112, 113
   spin wave, 57, 162, 164, 202, 204, 210, 370, 395, 401
composite
   defect, 94, 167, 168, 170, 189, 212, 215, 217, 218, 226
   fermion, 94, 149, 150
condensate
   Bose, 21
   fermionic, 104, 180
confinement
   color, 204
   of half-quantum vortices, 180, 189
   in momentum space, 104, 180
   of monopole and nexus, 217
   of monopoles in superconductor, 217
   topological, 170, 189, 217
conical space, 404
Cooper pair, 67, 102
core
   of hedgehog, 174
      formation of Dirac mass in core, 175
      hard core, 175
      instability, 173
      soft core, 176
   of monopole
      instability, 174
      symmetry, 172
   of string
      instability, 180
      superconductivity, 180
   of vortex
      antiferromagnetism, 179
      broken parity, 179
      broken rotational symmetry, 179
      ferromagnetism, 179
      Goldstone boson, 180
      gyromagnetism, 179
      hard core, 165, 186, 201, 216
      soft core, 165, 186, 199–202, 205, 213, 245, 250, 303
      twisted, 180
coset space, 160
cosmological constant, 13, 31
   before and after phase transition, 30
   estimate using RQFT, 8, 13, 14, 56
   experiment, 14
   from microphysics, 8
   in quantum liquids, 15, 26, 37
   nullification, 29–31, 37
   problem, 1, 5, 14, 16, 30
   second problem, 49
counterflow
   energy, 126
   normal–superfluid, 59
   velocity
      as effective chemical potential, 123, 125, 263
      definition, 45
      double role, 123, 125
creation and annihilation operators, 18
critical velocity for vortex nucleation
   A-phase, 332
   B-phase, 332
   by propagating front, 359, 362
   by surface instability, 339, 347
   helium-4, 332
   under radiation, 351, 362
cut-off
   GUT, 111
   infrared, 57, 111, 112, 167, 306, 329, 462
   Planck, 13, 15, 19, 25, 73, 74, 110, 111, 114, 133, 256, 257, 370, 381, 394, 395, 440, 442, 456, 458, 459, 463
   supersymmetry, 14
   ultraviolet, 13, 15, 56, 57, 72, 111, 130, 134, 168, 287, 462, 463

dark matter, 14, 382
degenerate
   Fermi point, 106, 108, 114, 115, 142, 144, 145, 153, 156, 183
   fermion zero modes, 300
   metric, 70, 397, 398, 404, 408, 410, 426, 444
   order parameter, 69
   states of hedgehog, 173
   states of vortex core, 179
   vacuum, 21, 78
dimensional reduction, 135, 136, 238, 266, 271, 276, 283–286, 294
disclination, 172, 173, 185–187, 189, 194, 225, 397, 404, 405, 464

Doppler shift, 33, 39, 43, 106, 307, 315, 321, 326–328, 378, 379, 409, 418, 425, 427, 439
double counting of degrees of freedom, 31, 71, 77, 111, 313
dumb hole, 424, 431, 434
dumbbell, 217

edge states, 90, 136, 276, 277
effect
  Aharonov–Bohm, 190, 412, 413, 466
    gravitational, 243, 248, 318, 406, 411, 413, 416, 466
  Callan–Harvey, 312, 314, 317
  Casimir, 12, 57, 369, 372, 391, 393, 395
    dynamic, 375
    for vortices, 168
    mesoscopic, 18, 371, 390, 393, 394, 396
  Corley–Jacobson, 431
  fountain, 59
  Hall, 269
    anomalous, 269, 286
    fractional quantum, 272
    quantum, 53, 135, 204, 286
  Josephson, 407
  Joyce–Shaposhnikov, 255
  Lense–Thirring, 410
  Meissner, 217
  Sagnac, 406–408
  Unruh, 379, 422
  Zel'dovich–Starobinsky, 343, 416, 421, 423
  zero-charge, 58, 111, 115
electromagnetic field
  emergence of, 69
energy–momentum tensor
  gas of relativistic quasiparticles, 44
  ill defined, 55
  in ferromagnets, 55
  in local equilibrium, 46
  of gravity field, 52
  of matter fields, 12, 13
    in quantum liquids, 41
  of vacuum, 13
  of vacuum field, 52
  pseudotensor, 52
equal spin pairing (ESP), 67, 76, 82
equation
  Adler–Bell–Jackiw, 124, 238, 239, 244, 248, 250, 253, 312, 314, 315, 317, 467
    for higher-dimensional anomaly, 285
  Boltzmann, 315, 316
  Caroli–de Gennes–Matricon, 288, 290, 294, 296

  continuity, 37, 44, 53, 54, 310, 347
  Dirac, 151, 454
  Einstein, 4, 5, 12, 16, 41, 42, 44, 45, 50, 75, 109, 286, 382, 383, 386, 387, 397, 404, 442, 454–456, 463, 465
  Euler, 37, 53, 54
  Euler–Lagrange, 173, 330
  force balance, 249, 306, 317
  Ginzburg–Landau, 178, 330
    time-dependent, 358–360, 364
  Gor'kov, 314
  Hamilton, 54, 414, 429
  hydrodynamic, 53, 244, 306, 340, 432
  kinetic
    for fermion zero modes, 315
  Landau–Khalatnikov, 5
  Landau–Lifshitz, 55
    for vortex sheet, 210
  London, 37, 118
  Maxwell, 214
  of state
    for cold matter, 387
    for network of domain walls, 384
    for radiation, 383, 387
    for space curvature, 386, 387
    for stiff matter, 385
    for surface tension, 384
    for texture, 385
    for thermal fermions, 459
    for vacuum, 27, 385, 387, 391
    sine-Gordon, 385
  Sagnac, 408
  Schroedinger, 298
  wave, 412
  Weyl, 107, 109
  Yang–Mills–Higgs, 186
ergoregion, 16, 49, 264, 340, 341, 343, 375, 420, 423, 427–429, 431, 433, 442, 445, 456
  definition, 343, 427, 450
  ergoregion and Landau criterion, 324, 325, 344, 345, 454
  ergoregion instability, 49, 343–348, 423, 431, 451–453, 465
  in relativistic theory, 49
  in rotating frame, 419, 420, 423
  tunneling to ergoregion, 422
ergosurface
  non-relativistic, 324, 325, 418, 419, 422, 427, 450, 451
  relativistic, 49, 324, 325, 418, 427, 445, 450
expansion of the Universe
  radiation-dominated, 133
experiment

'cosmological', 202, 250, 259
AB-brane, 221, 226, 228, 229, 340, 347–349, 372, 376, 448
AB-interface, 446
Callan–Harvey effect, 317
chiral anomaly, 202, 249
composite defect, 170
ergoregion instability, 347
gravitational A–B effect, 318
helical instability, 202, 259
interface instability, 221, 340, 345, 347, 349, 446, 448, 451
Iordanskii force, 318
Kamiokande, 146
Kapitza, 59, 61
Kibble mechanism, 80, 351, 356, 357, 362
Michelson–Morley, 40, 327
vortex
  -skyrmion, 201–203, 206, 213, 228, 245, 248, 249, 259, 332
  asymmetric core, 171, 179, 180, 298
  core transition, 171, 213
  double-skyrmion vortex, 206
  ferromagnetic core, 179
  sheet, 194, 207–210
  spin–mass, 80, 170, 226, 357
  topological transition, 206

Fermi liquid
  Landau theory, 17, 19, 61, 92, 315
  with Fermi surface, 19, 61, 92
Fermi point, 5, 72
  as a generic degeneracy point, 101
  as hedgehog in momentum space, 95
  collective modes, 3–5, 19, 100, 101, 109, 110, 113, 116, 117, 131, 286, 461
  degenerate, 106, 108, 114, 115, 142, 144, 145, 153, 156, 183
  discrete symmetry, 5, 115
  effective theory, 5
  elementary, 99, 109, 114
  emergence of chirality, 106
  emergence of effective metric, 70
  emergence of electromagnetic field, 69
  emergence of Yang–Mills field, 115, 116, 183
  emergent RQFT, 99–101, 109
  fermion doubling, 102
  in A-phase, 5, 82, 94, 99, 105, 111, 123, 124
  in A1-phase, 81, 83
  in axiplanar phase, 5, 83, 99
  in combined space, 292, 293, 301, 303
  in model BCS system, 69, 102
  in planar phase, 5, 84, 99, 175
  in semiconductors, 101, 113
  in Standard Model, 5, 85, 94, 116
  isospin, 108
  marginal, 84, 85, 99, 149, 175
  protected by symmetry, 85
  protected by topology, 86
  spin–statistics connection, 108
  topological charge, 97, 99, 109, 114
  topological invariant, 96, 97, 103, 106
  topological stability, 70, 97, 98, 100, 101
  universality class, 3, 5, 8, 19, 62, 70, 99, 101, 111, 130, 145, 286, 459, 462, 467
  zero of co-dimension 3, 103, 238, 247, 284, 286, 293, 301
Fermi surface, 5, 66, 67, 69, 83, 86, 90, 91, 93–95, 127, 128, 135, 197, 235, 236, 238, 276, 279, 285, 294, 303, 391, 455
  collective modes, 91
  density of states, 128, 307, 322, 324, 456, 457
  in angular momentum space, 288, 314
  in mixed state of d-wave superconductor, 128
  in non-Landau liquids, 93, 94
  in supercritical flow, 128, 322, 323, 454
  in vortex core, 301, 303, 308, 314
  inside horizon, 65, 86, 446, 454, 456–460
  Luttinger theorem, 92, 93
  multi-dimensional, 92, 302
  protected by topology, 61, 86
  topological invariant, 91, 93, 103
  topological stability, 66, 87–91, 101
  universality class, 5, 61, 65, 91, 276, 460
  zero of co-dimension 1, 103, 135, 238, 276, 285, 290, 308
fermion zero modes, 275
  anomalous, 290, 300
  as normal component, 307, 310
  co-dimension, 103, 108
  CPT-symmetry, 297, 298
  current carrying, 282
  density of states, 310, 456–458, 460
  effective Hamiltonian, 297, 298
  effective theory, 4, 89, 90, 295, 297, 307, 312–315
  entropy, 338, 458, 460
  in asymmetric vortex, 298, 299
  in black hole, 454, 456, 458–460
  in domain wall, 275–278, 281–283
  in magnetic field, 238, 284, 285
  in quantum Hall effect, 277
  in smooth core, 308, 310

INDEX 501

in vortex core, 187, 275, 281, 283, 288, 290–292, 294, 296, 297, 300, 307, 308, 312, 314, 315, 338, 467
index theorem, 276, 277, 280, 281, 286, 289, 298, 315
kinetic equation, 315
multi-dimensional, 103
of co-dimension 0, 104, 290, 300
of co-dimension 1, 103, 135, 238, 276, 285, 300, 308, 454, 467
of co-dimension 2, 103
of co-dimension 3, 103, 238, 284, 286, 292, 294, 300, 301, 303, 467
of co-dimension 5, 103, 284
of quantum vacuum, 4, 5, 61, 85, 87, 89, 90, 93, 95, 96, 101, 103, 108, 110, 133
on brane, 275, 283, 286
on half-quantum vortex, 299
on monopole, 252
on strings, 180, 188, 290, 313
on surface, 136
on topological defects, 4, 6, 90
protected by symmetry, 85, 89
protected by topology, 61, 85, 87, 89, 103
pseudo-zero mode, 288, 289, 295, 312
spectral flow, 238, 285, 312, 313, 315, 316, 338, 467
topological charge, 89, 276, 277
vortex mass, 306–308, 311
flat brane, 275
flat directions, 82, 116
flat membrane
 chiral particle as, 102
flat spacetime, 376, 400, 401, 403, 404, 411, 418, 444
flat Universe, 386, 387
flatness of Universe, 1, 387, 388
fluid helicity, 54
flux
 flux quantum, 187, 191–193, 214, 217
 flux tube, 214, 406, 408, 412, 414
 fractional flux, 191–194, 211, 217
 gravimagnetic flux, 408, 412, 466
 magnetic flux, 141, 180, 187, 191, 192, 194, 214, 217, 238, 406, 412, 414
force
 Casimir, 167, 168, 452, 465
  between hedgehogs, 168
  between spin and mass vortices, 168, 170, 189
 Coriolis, 406

friction force on AB-interface, 341, 342, 377, 379, 380, 448, 452
friction force on vortex, 245, 248, 249, 316, 317
fundamental, 15
gravitational, 52
Iordanskii, 243, 248, 249, 317, 318, 411–415, 466
Kopnin, 55, 235, 246–248, 314, 317
Lorentz, 414
Magnus, 168, 171, 243, 247–249, 317, 334, 347, 412
mutual friction, 242
nuclear, 163
on mirror moving in vacuum, 375, 379
spectral-flow, 55, 235, 245–249, 313, 314, 316, 317
four-vector
 cut-off momentum, 57
 group velocity of quasiparticles, 43
 quasiparticle momentum, 43
 temperature, 47, 48
 velocity of quasiparticle matter, 46
fractional
 charge, 300
  topological, 140, 142, 189, 191
 defect, 94, 185, 189, 194, 217
  in momentum space, 94, 104, 189
 entropy, 299
 fermion number, 393
 flux, 191, 193, 194
 quantum Hall effect, 272
 spin, 300
 statistics, 300
frame dragging, 33, 324, 410, 436, 442, 445, 454, 455
friction
 quantum, 429, 430
fundamental constants
 for inner observer, 40
 hierarchy of, 20
 in BCS model, 73, 75, 119, 129, 130
 in Bose gas, 25
 in effective theory, 6, 11, 19, 20, 23, 37, 75, 91, 334, 370, 371
 in Theory of Everything, 20, 73
 of nature, 40, 132, 133
fundamental frequency, 410, 411

gauge field
 from discrete symmetry, 116
Gershtein–Zel'dovich mechanism, 326, 327
grand unification, 1, 3, 111, 115, 145, 147, 148, 184, 212, 214, 228, 241, 462

gravimagnetic field, 226, 261, 406, 408, 412, 416, 466
gravineutrality, 387
gravitational bag, 459
gravitational red shift, 438, 439

Hamiltonian
  Bogoliubov–Nambu, 77, 78, 82, 84, 94, 136–138, 143, 144, 153, 154, 283, 291, 295, 298, 398
  Dirac, 79, 137, 143, 151
  effective
    for fermion zero modes, 297, 298
  Weyl, 94, 108, 141
harmonic oscillators, 22
Hawking radiation, 441
  as quantum friction, 398, 430
  as quantum tunneling, 440, 441
hedgehog, 3, 165, 168, 172–176, 184, 185, 189, 205, 213–218, 220, 222, 229
  hard core, 174
  in momentum space, 95, 97, 137, 140, 142, 292, 293
  soft core, 174, 176
hierarchy of Planck scales, 7, 25, 73, 129, 134, 287, 442, 463, 464
homotopy group, 89
  fundamental, 89
    in momentum space, 65, 89
    in real space, 89, 159, 185, 186, 198, 224
  linear defects, 165
  relative, 90, 135, 160, 170, 195–198, 200, 211, 214, 218–220, 223
  second, 90
    in momentum space, 137
    in real space, 135, 160, 173, 174, 176, 184, 185, 213, 214
  third, 90
    in combined space, 301
    in momentum space, 65, 98
    in real space, 135, 160
  zeroth, 197
horizon, 16, 47, 65, 87, 197, 213, 324, 355, 375, 401, 423, 424, 426–428, 431, 434–437, 440–446, 450, 453, 454, 456, 459, 460, 465
  acoustic, 432
  as interface, 466
  at brane, 447, 450–453, 465
  black-hole, 1, 6, 324, 340, 378, 403, 424, 427, 428, 430, 431, 433, 436–438, 440, 446, 450, 460
  coordinate singularity at, 401, 426, 435, 450
  cosmological, 384, 385
  dumb-hole, 431
  entropy of, 434, 458, 459
  for quasiparticles, 5, 47, 49, 50, 324, 325, 424, 427–432, 434, 439, 441
  inner, 433
  outer, 433
  vacuum beyond horizon, 454, 458, 460, 466
  white-hole, 427, 431, 433, 436, 446
hydrodynamics
  quantum, 56, 370, 395
  two-fluid, 17, 32, 42, 59, 61

impact parameter, 292, 295
index theorem, 276, 277, 279–282, 289, 313, 315
  Atiyah-Singer, 277
inflation, 30
inflaton, 30
instanton, 160
  collective coordinates description, 333, 336
  in RQFT, 333
  in superfluids, 333, 335
  quantum field description, 333
interface
  AB-interface, 220–224, 226, 228, 229, 347, 372, 375–377, 379, 447–449, 452, 454, 465, 466
  between sliding liquids, 339
  between sliding superfluids, 340, 342, 345–347, 446, 464, 465
  between two vacua, 221, 223, 224, 231, 275–277, 280–283, 286, 339, 341, 372–374, 376, 397, 447, 460, 466
  solid–liquid, 349
  superfluid–normal, 349, 358, 359, 362
  superfluid–superfluid, 362, 447
  symmetry classification, 222
  topological defects at, 221–224, 228, 229
  vacuum manifold of, 222–224
invariance
  Galilean
    broken, 35, 161
    modified, 33, 34
  gauge, 2, 3, 5, 8
    violated, 2
  Lorentz, 2, 5, 11, 39, 67, 80, 99, 106, 109, 116, 130, 144, 156
    violation, 2, 6, 117, 153, 154, 156, 261, 287, 391, 449, 450, 452, 453, 455, 461, 463, 466

INDEX 503

translational, 35
  for particles, 35
  for quasiparticles, 35

Killing vector
  definition, 48
  time-like, 48

Landau criterion, 60, 62, 128, 322–326, 328, 329, 331, 344, 350, 351, 364, 417, 418, 420, 421, 440, 450, 454, 460
  for single superfluid, 321, 344
  general, 321, 345
Laval nozzle, 431, 444
law
  conservation law, 32
    covariant, 12, 15, 41, 42, 45, 51
    energy, 51, 52
    mass, 230, 231
    momentum, 51, 52
    particle number, 18, 21, 37, 44, 390
    quasiparticle number, 23, 123
    spin, 230, 231
  Newton, 236, 238, 275, 435
  physical laws, 1, 2, 389, 462
    emergence of, 5
    symmetry of, 160, 198
  Tolman, 48, 49, 433
Leggett angle, 169
length
  coherence length (definition), 119
  dipole, 145, 169, 176, 195, 202, 214, 385
  London penetration length, 187
  Planck, 7, 370, 381, 431, 440, 456
  screening length
    of hypermagnetic field, 187
    of magnetic field, 187
Lense–Thirring angular velocity, 410, 418
line of infinite redshift, 410
Lorentz–Fitzgerald contraction, 40

marginal vacuum
  Dirac vacuum in 2+1, 142
  planar state, 84, 85, 99, 149, 175
  polar state, 85
  Standard Model, 85, 149
Maslov index, 298
mass
  associated, 309–311
  bare mass, 17, 33, 36, 129–131, 261, 305, 310, 370, 410
  black hole, 433, 435, 453, 454, 458
  Dirac mass, 84, 111, 143–146, 149, 175, 258
  effective, 91, 119, 129, 142, 306

formation of, 85, 116, 117, 149, 152, 169
hydrodynamic, 306, 309
inertial, 305, 310, 311
Kopnin mass, 307–311
mass density, 53, 247, 334, 387, 404, 413
mass protection, 85, 149, 151, 152
mass tensor, 305
mass–energy relation, 305, 306, 311
  of gauge boson, 116, 187
  of hyperphoton, 145, 256, 258, 259, 463
  quark mass, 163, 164
  roton mass, 418
  vortex mass, 306, 307, 309–311
    relativistic, 306, 307
Mermin–Ho relation, 107, 121, 185, 199, 200, 205, 216, 217, 254, 304
  generalized, 304
meron, 204, 207, 211
metric
  acoustic, 4, 15, 19, 38, 39, 42, 305, 306, 381, 418
  determinant, 25, 73, 112, 381, 397, 450, 452
  effective, 5, 12, 15, 16, 19, 32, 33, 36, 38, 39, 42, 46, 73, 75, 106, 107, 112, 124, 159, 261, 305, 381, 389, 397, 404, 409, 410, 412, 418, 424, 427, 428, 430, 434, 436, 441, 447, 466
  for ripplons, 447, 449, 450, 453, 464
  Painleve–Gullstrand, 434, 436, 437, 442–444, 446, 453–455, 460
  Reissner–Nordstrom, 433
  Robertson–Walker, 385, 386, 388
  Schwarzschild, 426, 428, 434–436, 442, 443, 446
metric singularity
  coordinate singularity, 400, 401, 426, 435, 450, 456
  physical singularity, 426, 429, 432, 435, 447, 450, 452, 456, 460
minigap, 288, 295, 296, 299, 300, 315, 317
mixed state, 128
monopole
  't Hooft–Polyakov, 172, 173, 184, 214
  Dirac monopole, 187, 214, 215, 217, 226
    in momentum space, 140, 141
  electroweak, 186, 217
  gravimagnetic, 226
  hypermagnetic, 217
  in superconductor, 217, 218
  magnetic, 109, 141, 173, 174, 205, 214, 215, 217, 228, 328
  monopole erasure, 228, 229
  monopole terminating string, 185, 186, 212, 215, 217

monopoles in GUT, 214, 217, 218, 228
overabundance of monopoles, 212, 228

Newton constant, 11–13, 404, 411, 433, 453, 458
  in Dirac cosmology, 133
  in induced gravity, 15, 110, 133, 459
    dependence on number of fermion zero modes, 15, 111, 133, 458, 459
    in A-phase, 111, 132, 133, 459
    temperature dependence, 133
    time dependence, 133
  in units of energy, 12, 110
nexus, 215, 217, 218, 226
nodal line
  in polar phase, 85
  instability, 70, 85, 86, 103
  protected by symmetry, 85
  zero of co-dimension 2, 103, 138
normal component, 42
  at T=0, 126
  density (definition), 45
  tensor, 45

obsever
  external, 429, 430, 446
    definition, 39, 428
  freely falling, 429, 436, 455
  inner, 325, 424, 428, 436, 446, 464, 466
    definition, 39
order parameter
  2-component Bose condensate, 201, 202
  A-phase vacuum, 68
  Alice string, 189, 299
  B-phase vacuum, 78, 162
  chiral quark condensate, 163
  d-wave superconductor, 193
  p-wave spinless superfluid, 67
  p-wave superconductor, 191
  p-wave superfluid, 161
  planar phase, 68
  polar phase, 68
  scalar, 21, 67, 177
  spinor, 183, 201, 202
  Standard Model, 183, 201, 202
  vector, 67, 202

pair breaking, 326
pair creation, 18
  by Hawking radiation, 441
  by rotating cylinder, 416, 420, 422, 423
  in alternating electric field, 326, 327
  in B-phase, 326, 327
  in strong electric field, 325, 326
  of black holes, 458
  Schwinger, 326
partial pressure, 29
  of curvature, 386
  of matter, 383
  of quintessence, 384
  of texture, 384, 385
  of vacuum, 13, 383–385
Pati–Salam model, 1, 147, 462
phase ordering, 351, 362
phase transition
  A–B, 30
  electroweak, 459
  in early Universe, 2–4
  Lifshitz, 324, 458
  quantum, 36, 90, 137, 138, 140–142, 269, 322, 324, 354, 446, 456, 457
  superfluid, 3, 53
Poisson brackets
  for ferromagnets, 54
  for hydrodynamics of normal liquid, 54
  scheme, 53
principle
  equivalence principle, 11, 15, 16, 156, 389
  of homogeneity of Universe, 388
  of non-gravitating vacuum, 30, 56, 460, 465
  of vacuum stability, 30, 31
pseudomomentum, 35
pseudorotations, 22

quantization
  of gravity, 6, 56
quantum phase transition, 36, 90, 102, 104, 137, 138, 140–142, 269, 322, 324, 354, 446, 456, 457
quantum statistics, 49, 61, 67, 74, 380, 422
  fractional, 300
  non-Abelian, 299
  spin–statistics connection, 67, 108
quasiparticle
  antibaryon, 191, 240
  antilepton, 240
  antiquark, 163, 240
  as flat membrane, 102
  baryon, 14, 52, 80, 135, 146, 148, 164, 191, 239–241, 250, 253, 259, 382, 467
  Bogoliubov fermion, 418
  C-boson, 149
  chiral, 282

INDEX

Dirac fermion, 86, 327
electron, 4, 20, 36, 58, 65, 66, 76, 77
  left, 80, 81
  right, 80, 81
electronic neutrino, 156
eta meson, 164
exotic fermion, 155
gauge boson, 116, 187, 461
Goldstone boson, 113, 116
graviton, 7, 8, 38, 42, 56, 80, 382, 453, 461, 466
holon, 93, 94, 149, 150, 189
hyperphoton, 145, 256, 258, 259, 463
in domain wall, 276, 281, 282
lepton, 90, 94, 145, 146, 148, 188, 239–241, 250, 252, 253, 259, 260, 467
Majorana fermion, 71, 77, 299
muon, 139
muonic neutrino, 156
neutrino, 146, 156, 184, 239, 461
neutron, 65, 163, 252, 351, 356–358, 362, 424
phonon, 4, 7, 15, 19, 23, 24, 27, 31, 32, 37–39, 42, 46, 56, 57, 60, 61, 370, 371, 380, 381, 395, 410–416, 418, 419, 421, 422
photon, 110, 113, 240, 407, 411, 452, 461
pion, 163, 164
proton, 148, 163, 252, 356
pseudo-Goldstone bosons, 163, 164
quark, 80, 90, 94, 125, 146, 148, 163, 164, 187, 239, 240, 253, 389, 461
  down, 148, 163, 164, 240
  strange, 164
  top, 152
  up, 148, 163, 164, 240
ripplon, 341–343, 447, 449–453, 464
  critical, 343, 345, 347
roton, 32, 38, 60, 61, 325, 413, 414, 416, 418, 419, 421, 422
slave boson, 149
spinon, 93, 94, 149, 150, 189
tau lepton, 139
w-fermion, 149
Weyl fermion, 8, 86, 87, 108, 109, 133, 140, 312, 462

radiation
  Hawking, 398, 424, 428–431, 440, 454, 459, 465
  neutron, 351, 356, 357
  phonon, 421

quasiparticle, 327, 335, 379, 420, 423, 430
roton, 422
to ergoregion, 420
reference frame
  absolute, 426, 436, 441, 442, 453–455
  competing, 335, 427
  container, 248, 261, 326, 329, 330, 345, 346
  environment, 33, 48, 49, 106, 121, 126, 127, 321, 322, 324, 329, 336, 341, 343–346, 349, 356, 417, 425, 427, 429, 430, 433, 434, 436, 441, 447, 451, 454
  frame-fixing parameter, 342, 344, 346, 349, 448, 451, 454
  freely falling, 436, 437, 442, 455
  heat bath, 246, 261, 313, 330, 331, 377
  inertial, 228, 356, 408, 420
  laboratory, 48, 438
    quasiparticle spectrum in, 33, 34, 321
  lost, 324, 343, 420, 448
  preferred, 57, 128, 161, 261, 321, 324, 336, 341, 430, 442, 455
  rotating, 261, 263, 362, 406, 407, 417, 418, 420
  superfluid-comoving, 34, 45, 306, 324, 344, 345, 410, 427, 436, 441, 442, 455
    definition, 33
    quasiparticle spectrum in, 33
    vacuum in, 428, 429, 437
  texture-comoving, 48, 245, 307, 310, 315, 324, 377, 410, 412, 425, 427, 428
  vacuum in, 422, 428, 429, 440
Ricci scalar curvature, 11
running coupling
  in A-phase, 58, 111, 112, 256
  in QED, 111
  in Standard Model, 147, 205

shock wave, 432
skyrmion, 135, 201, 203, 204, 214, 245, 246, 267, 269
  fermionic charge of, 266, 271
  in liquid crystal, 160
  in momentum space, 135, 137
  point-like, 160, 202, 204, 252, 266, 267
  proton as, 252
  quantum statistics of, 267–269
  spin of, 268
  vortex-skyrmion, 159, 199–207, 213, 214, 219, 221, 228, 229, 245–248, 303, 308, 309, 312, 314, 317, 331, 332

506   INDEX

soliton, 3, 90, 135, 139, 160, 163, 169–171,
        176, 195, 196, 204, 208, 210,
        211, 306, 384, 385
  kink on, 206–208, 211
  non-topological, 180
  terminated by string, 163, 169, 189,
        194, 195, 197, 206, 212
  topological stability, 169, 170, 196–199
  twist soliton, 384
spacetime
  absolute, 39, 426, 441, 442, 444, 454
  continuum, 4
  empty, 34
  foam, 1
  Galilean, 4, 39
  geodesically complete, 444
  geodesically incomplete, 444
  Minkowski, 14, 406
    effective, 12
  physical, 400, 426, 429
spectral asymmetry, 282
  index, 278, 279, 281
spectral flow, 80, 92, 93, 238, 240,
        243–246, 248, 250, 263, 285,
        312–317, 338, 415, 467
speed of light
  anisotropy, 12, 40, 305
  as fundamental constant, 11, 19, 25, 40
  as material parameter, 11, 19, 24
  coordinate, 40
  for external observer, 40, 305
  for inner observer, 40, 305
  invariance, 40
sphaleron
  black hole, 458
  definition, 330
  in RQFT, 241, 248
  in toroidal channel, 331
  vortex ring, 330, 331, 335, 458
spin
  emergence of, 5, 67, 108
  interchange of spin and isospin, 109
  relativistic, 67
  spin–orbit coupling, 19, 144, 145, 153,
        154, 161–164, 169, 174, 176,
        189, 195, 196, 198, 203, 205,
        212, 214, 220, 224, 258, 299,
        384
  spin–statistics connection, 67, 108, 266,
        267, 269, 276, 300
  spinless fermions, 67, 68
string
  Alice string, 82, 94, 174, 184, 185,
        189–191, 195, 197, 198, 229,
        231, 299, 409
  Dirac string, 184, 185, 214, 216, 217

  in momentum space, 141
  electroweak string, 186, 187, 217
  fundamental string, 8, 15, 334
  Nielsen–Olesen string, 186, 187, 275
  pion string, 163, 167, 169
  semilocal string, 205
  spinning string, 16, 411–413, 415, 466
  string crossing wall, 224–226, 228
  stringy texture, 312
  superconducting, 180
  terminated by monopole, 185, 186, 212,
        215, 217
  wall bounded by, 163, 169, 170, 189,
        194, 195, 197, 206, 212
  with supercurrent, 180
  Witten, 179, 180
superflow
  irrotational, 60
  supercritical, 321, 322, 440
superfluid velocity
  as electromagnetic field, 106
  as metric field, 38, 42
  as torsion field, 108
  Galilean transformation, 33, 107
  in A-phase, 107
  in helium-4, 32, 36
  irrotational, 36
superfluid density
  definition, 46
  tensor, 46
superfluidity/superconductivity
  anisotropic, 61
  chiral, 72, 163, 164, 188, 191, 193, 194,
        205, 211, 216–218, 265, 291,
        300
  color, 80, 164, 188
  in quantum crystal, 59
  in quark matter, 80
  of baryon charge, 164, 188
  of electric charge, 180, 187, 188
  of mass, 165, 188
  of spin, 188
  spin-triplet, 76–78, 101
superradiance, 416, 421, 427
surface of infinite redshift, 6, 277, 397,
        399–401, 424, 431, 433, 440,
        445, 466
symmetry
  approximate, 144, 163, 164, 212
  breaking, 3
  combined, 81, 82, 85, 108, 154, 177, 445
    parity as combined symmetry, 81, 82,
        85, 115, 153, 183
  electroweak, 2, 163
  general covariance, 3, 8, 11, 57, 116

INDEX 507

definition, 11
emerging, 52
for matter fields, 12, 39, 109, 110
protected, 134
violation, 114
GUT, 2, 3, 147, 148
hidden, 116, 117, 162
isotopic spin symmetry, 163
Lorentz
  as combined symmetry, 108
  effective, 11, 106, 108
  reentrant violation, 134, 145, 153–156
  violation, 80, 106, 153, 287, 391, 449, 450, 453, 455
of black hole, 445
of defects, 171
  AB-interface, 222
  broken, 171, 173, 178–180, 298
  Goldstone boson from symmetry breaking, 171
  hedgehog, 172
  monopole, 172
  vortex, 171, 176, 177, 179, 180
of disordered state, 160
of normal liquid, 1, 2, 161
of nuclear forces, 163
of physical laws, 78, 160
of Standard Model, 1, 145–148
residual, 160
  A-phase, 81
  A1-phase, 84
  B-phase, 161
  definition, 78, 160
  planar phase, 85
  polar phase, 85
  role of discrete symmetry, 5, 82, 85, 115, 116, 143–145, 150, 151, 153, 155, 156, 177, 300
scale invariance, 57
supersymmetry, 13, 14, 30, 116, 117, 281, 296, 464
time reversal, 135–138, 177, 192–194, 196, 270, 282, 291, 434, 445, 446, 454, 460

temperature
  covariant relativistic, 48, 49
  Debye, 26, 370, 381, 384
  effective, 46–49, 124, 433
  Hawking, 402, 424, 430, 431, 436, 440, 441, 458, 459
  local, 45
  measured by external observer, 48
  measured by inner observer, 48, 379
  Tolman, 48, 379
  Unruh, 379, 422

tetrad gravity, 99, 101, 108, 109, 115, 117, 286, 398, 426, 455
theory
  Bardeen–Cooper–Schrieffer (BCS), 7, 20, 66, 67, 314, 462
  Bogoliubov, 20, 21
  effective, 2, 4–8, 23, 56, 295, 462
    insensitivity to microphysics, 4–7, 37, 314, 387, 459
    paradoxes of, 52, 53, 235, 260, 262
  Einstein, 16, 32
  elasticity, 17
  Kaluza–Klein, 6, 275
  Landau of Fermi liquid, 17, 19, 61, 315
  Leggett, 17
  London, 17
  non-renormalizable, 56
  Sakharov baryonic asymmetry, 240
  Sakharov induced gravity, 15, 57, 110, 132, 459
  Theory of Everything, 53
    in Bose gas, 21
    in Fermi gas, 391
    in quantum liquids, 17, 18, 20, 36, 53, 161, 270, 369, 381
thermal equilibrium, 49, 240, 260, 346
  global, 48, 49, 316, 379, 433, 434
    conditions, 48, 315, 330
    definition, 48
    in relativistic system, 48
    violation in ergoregion, 49
  local, 45–47, 378
  unstable, 330
time dilation, 40
transformation
  Bogoliubov, 22, 68, 69
  Galilean
    for particles, 33, 34
    for quasiparticles, 33–35
  Lorentz, 22

universality
  classes, 5, 7
    of fermionic vacua, 5, 65, 81, 86, 87, 103, 372, 460, 461, 464, 466
  of gravity, 11
Universe
  accelerated expansion, 14
  closed, 386, 387
  flat, 1, 386, 387
  open, 386
  static, 386

vacuum
  axial, 72, 82, 84, 102, 107
  Boulware, 455
  chiral, 6
  decay, 429, 431, 438, 440, 466
  degenerate, 72, 78, 120, 159, 160, 162, 182
  Dirac, 71, 73, 79, 80, 140, 142, 381, 389, 459, 464
    marginal, 142
  energy, 1, 8, 12–14, 18, 23–26, 29–31, 36, 37, 49, 56, 67, 69–76, 111, 116, 120, 195, 369, 371–376, 380–394, 464, 465
  equilibrium, 8, 29, 56, 57, 70, 76, 122, 372, 459, 465
  foreign object, 305, 396, 430
  in condensed matter, 3
  inert, 130–132
  initial, 236, 331, 422, 428, 429, 436, 437, 440
  manifold, 159, 160
    A-phase, 184
    B-phase, 162
    definition, 160
  of quantum liquid, 396, 428, 430, 446, 464
    as empty space, 3
  of Standard Model, 5, 6, 19, 30, 65, 145, 149, 251, 455, 461, 462, 465
  polarization, 111, 115
  reconstruction, 67, 428, 437, 440, 454, 460
  superfluid, 4, 305
vacuum instability
  beyond horizon, 431, 440, 447, 452–454, 460, 466
  helical, 6, 202, 251, 255, 257–259
  hydrodynamic instability, 331
  in ergoregion, 343–348, 431, 450–454, 465
  Kelvin–Helmholtz, 228, 339–342, 449
vierbein, 99, 101, 109, 115, 117
  degenerate, 397, 398, 410, 444
  linear defect, 397, 404, 408, 410
  point defect, 397
  wall, 397, 398, 400, 401, 403, 404, 424, 426–431, 444
vortex
  Abrikosov vortex, 187, 191, 192, 217, 218, 275, 374
  Alice string, 194
  Anderson–Toulouse–Chechetkin vortex, 199, 201

continuous vortex, 36, 199–202, 205–208, 213, 214, 216, 219, 221, 226, 228, 246, 304, 308, 312, 314, 317, 331, 332
dipole-locked, 202, 206
dipole-unlocked, 206
double-skyrmion vortex, 206, 213
doubly quantized, 170, 202, 204, 217, 228, 245
fractional, 185, 189, 191, 194, 211, 217
half-quantum vortex, 82, 94, 174, 185, 189, 190, 194, 195, 198, 206, 212, 220, 229, 299, 409
mass vortex, 165, 167, 168, 170, 178, 185, 186, 189, 225, 230, 357
Mermin–Ho vortex, 204, 207, 226, 228
multiply quantized, 348
point vortex, 221, 291
  in momentum space, 93
ring, 60, 329–331, 336
  black hole analogy, 458
  collective coordinates, 333, 336, 338
  critical, 330
  energy, 328, 329
  energy spectrum, 328
  Landau criterion, 329
  macroscopic quantum tunneling, 330, 331, 333, 336, 338
  momentum, 328, 336
  nucleation, 330, 338
  sphaleron, 330, 331, 458
  thermal activation, 330, 331
spin vortex, 166–170, 189, 195, 197, 199, 206, 224, 230, 357
spin–mass vortex, 80, 166, 168, 170, 189, 197, 206, 212, 226, 357
vortex cluster, 165, 167, 170, 171, 179, 352, 354, 355, 423, 445
vortex loop, 199, 306, 330, 337, 356
  bosonic zero mode, 336
  destroying soliton, 199
  in momentum space, 89, 93
vortex pair, 165, 170, 351
vortex tangle, 243, 352, 362
vortex-skyrmion, 159, 199–207, 213, 214, 219, 221, 228, 229, 245–248, 303, 308, 309, 312, 314, 317, 331, 332
vortex sheet
  at AB-interface, 228, 340, 348
  in 3+1 spacetime, 89
  in A-phase, 194, 206–210
  in momentum space, 104

in superconductors, 194, 211
  unstable, 209, 339, 340, 363
vorton, 180

W-parity, 292, 296, 297, 299, 300
  definition, 292

zero-point
  energy, 12, 13, 24–27, 29–31, 71, 73, 371, 380, 464
  fluctuations, 30, 392
  motion, 24, 27, 29, 58, 461
  oscillations, 20, 27, 57